Biology and Management of Problematic Crop Weed Species

Edited by

Bhagirath Singh Chauhan
The Centre for Crop Science, Queensland Alliance for Agriculture and Food Innovation (QAAFI) and School of Agriculture and Food Sciences (SAFS), The University of Queensland, Gatton, QLD, Australia

ELSEVIER

ACADEMIC PRESS
An imprint of Elsevier

Academic Press is an imprint of Elsevier
125 London Wall, London EC2Y 5AS, United Kingdom
525 B Street, Suite 1650, San Diego, CA 92101, United States
50 Hampshire Street, 5th Floor, Cambridge, MA 02139, United States
The Boulevard, Langford Lane, Kidlington, Oxford OX5 1GB, United Kingdom

Notices
Knowledge and best practice in this field are constantly changing. As new research and experience broaden our understanding, changes in research methods, professional practices, or medical treatment may become necessary.

Practitioners and researchers must always rely on their own experience and knowledge in evaluating and using any information, methods, compounds, or experiments described herein. In using such information or methods they should be mindful of their own safety and the safety of others, including parties for whom they have a professional responsibility.

To the fullest extent of the law, neither the Publisher nor the authors, contributors, or editors, assume any liability for any injury and/or damage to persons or property as a matter of products liability, negligence or otherwise, or from any use or operation of any methods, products, instructions, or ideas contained in the material herein.

British Library Cataloguing-in-Publication Data
A catalogue record for this book is available from the British Library

Library of Congress Cataloging-in-Publication Data
A catalog record for this book is available from the Library of Congress

ISBN: 978-0-12-822917-0

For Information on all Academic Press publications
visit our website at https://www.elsevier.com/books-and-journals

Publisher: Charlotte Cockle
Acquisitions Editor: Nancy Maragioglio
Editorial Project Manager: Mariana L. Kuhl
Production Project Manager: Maria Bernard
Cover Designer: Christian J. Bilbow

Typeset by MPS Limited, Chennai, India

Biology and Management of Problematic Crop Weed Species

Contents

List of contributors

Taseer Ahmad Department of Plant Production and Technologies, Faculty of Agricultural Sciences and Technologies, Omer Halisdemir University, Nigde, Turkey

Sharif Ahmed International Rice Research Institute (IRRI) Bangladesh Office, Dhaka, Bangladesh

Daizy Rani Batish Department of Botany, Panjab University, Chandigarh, India

Hugh J. Beckie School of Agriculture and Environment, The University of Western Australia, Perth, WA, Australia

Catherine P.D. Borger Department of Primary Industries and Regional Development, Northam, WA, Australia

Anthony Brusa Department of Agronomy and Plant Genetics, University of Minnesota, St. Paul, MN, United States

Ankur Chaudhary Chaudhary Charan Singh Haryana Agricultural University Regional Research Station, Uchani, India

Bhagirath Singh Chauhan The Centre for Crop Science, Queensland Alliance for Agriculture and Food Innovation (QAAFI) and School of Agriculture and Food Sciences (SAFS), The University of Queensland, Gatton, QLD, Australia

Tse-Seng Chuah Faculty of Plantation and Agrotechnology, University of Teknologi Mara, Arau, Malaysia

Diego J. Contreras Department of Crop and Soil Sciences, North Carolina State University, Raleigh, NC, United States

Sahil Dahiya Queensland Alliance for Agriculture and Food Innovation (QAAFI), The University of Queensland, Gatton, QLD, Australia

Laura Davies ADAS, Battlegate Road, Boxworth, Cambridgeshire, United Kingdom

Het Samir Desai The University of Queensland, Gatton, QLD, Australia

Masilamany Dilipkumar Rice Research Center, Malaysian Agricultural Research and Development Institute (MARDI), MARDI Seberang Perai, Kepala Batas, Malaysia

Matthew T. Elmore Rutgers University, New Brunswick, NJ, United States

Seyed Vahid Eslami Associate Professor, Department of Agronomy, University of Birjand, Birjand, Iran

Wesley J. Everman Department of Crop and Soil Sciences, North Carolina State University, Raleigh, NC, United States

Singarayer Florentine Centre for Environmental Management, School of Health and Life Sciences, Federation University Australia, Ballarat, VIC, Australia

Travis W. Gannon North Carolina State University, Raleigh, NC, United States

Charles M. Geddes Agriculture and Agri-Food Canada Research Centre, Lethbridge, AB, Canada

Prabhu Govindasamy Division of Crop Production, Indian Grassland and Fodder Research Institute, Jhansi, India

Gerald M. Henry University of Georgia, Athens, GA, United States

Talia Humphries Centre for Environmental Management, School of Health and Life Sciences, Federation University Australia, Ballarat, VIC, Australia

Khawar Jabran Department of Plant Production and Technologies, Faculty of Agricultural Sciences and Technologies, Omer Halisdemir University, Nigde, Turkey

Marie Jasieniuk Department of Plant Sciences, University of California, Davis, Davis, CA, United States

Amit J. Jhala Department of Agronomy and Horticulture, University of Nebraska-Lincoln, Lincoln, NE, United States

Eric A.L. Jones Department of Crop and Soil Sciences, North Carolina State University, Raleigh, NC, United States

Amarpreet Kaur Department of Botany, Panjab University, Chandigarh, India

Shalinder Kaur Department of Botany, Panjab University, Chandigarh, India

Simerjeet Kaur Punjab Agricultural University, Ludhiana, India

Gurpreet Singh Khalsa Queensland Alliance for Agriculture and Food Innovation (QAAFI), The University of Queensland, Gatton, QLD, Australia

Ravinder Kumar Kohli Department of Botany, Panjab University, Chandigarh, India

Virender Kumar International Rice Research Institute, Los Baños, Philippines

Stephen Richard Moss Stephen Moss Consulting, Harpenden, United Kingdom

Adusumilli Narayana Rao Consultant Scientist (Weed Science) and General Secretary, Asian-Pacific Weed Science Society (APWSS), Hyderabad, India

George Newcombe University of Idaho, College of Natural Resources, Mowcow, Idaho

Kenneth M. Olsen Department of Biology, Washington University in St Louis, St. Louis, MO, United States

Sheng Qiang Weed Research Laboratory, Nanjing Agricultural University, Nanjing, China

Aritz Royo-Esnal Department d'Hortofructicultura, Botànica i Jardineria, Agrotecnio, Universitat de Lleida, Lleida, Spain

Debalin Sarangi Department of Agronomy and Plant Genetics, University of Minnesota, St. Paul, MN, United States

Harminder Pal Singh Department of Environment Studies, Panjab University, Chandigarh, India

Beng-Kah Song School of Science, Monash University Malaysia, Bandar Sunway, Malaysia

Caleb C. Squires University of Sydney, School of Life and Environmental Science, Sydney Institute of Agriculture, Camden, NSW, Australia

Alireza Taab Department of Agronomy and Plant Breeding, Faculty of Agriculture, Ilam University, Ilam, Iran

Breanne D. Tidemann Agriculture and Agri-Food Canada Research Centre, Lacombe, AB, Canada

Joel Torra Department d'Hortofructicultura, Botànica i Jardineria, Agrotecnio, Universitat de Lleida, Lleida, Spain

Michael J. Walsh University of Sydney, School of Life and Environmental Science, Sydney Institute of Agriculture, Camden, NSW, Australia

Sarah Ward Associate Professor, Department of Soil and Crop Sciences, Colorado State University, Fort Collins, CO, United States

Chapter 1

Alopecurus myosuroides

Taseer Ahmad[1], Khawar Jabran[1] and Stephen Richard Moss[2]
[1]Department of Plant Production and Technologies, Faculty of Agricultural Sciences and Technologies, Omer Halisdemir University, Nigde, Turkey, [2]Stephen Moss Consulting, Harpenden, United Kingdom

Introduction

Alopecurus myosuroides Huds. (black-grass, slender foxtail) is an annual grass weed which belongs to the family Poaceae and mainly grows in lowland areas (WFO, 2020). It is a competitive weed, has the potential to produce a large number of seeds, grows rapidly, and can cause considerable yield loss (Blair et al., 1999; Marechal et al., 2012). Due to its competitive nature, it can reduce yields of winter wheat by up to 45% (Vizantinopoulos and Katranis, 1998). Furthermore, this weed is able to grow at a low temperatures, for example, 2°C or 3°C (Colbach et al., 2002). Colbach et al. (2002) also reported that the minimum temperature required for germination of *A. myosuroides* is 0°C, suggesting that this weed is well adapted to relatively cold conditions. The infestations of this weed on the farm tend to be increased by adopting agronomic practices such as no-tillage, reduce tillage, continuously growing winter cereals in the same field (Colbach and Durr, 2003), and conserving stubbles in the field.

Alopecurus myosuroides is one of the most important arable weeds and grows in more than 60 countries worldwide, but is mostly a major problem in the countries of Western Europe, including Germany, Belgium, France, and England (Holm et al., 1997). It is mainly an autumn-germinating grass weed and, while mainly found in winter crops, can also be a problem in spring crops. Recently, it has become an emerging issue in Northern and Eastern Europe too, and resistance has been reported in Czech Republic in 2008, Poland in 2010, and Sweden in 2011. Although it is often considered a relatively new weed, it was recognized as problematic in wheat fields over 175 years ago (Sinclair, 1838). However, it has become much more widespread during the last 50 years.

The life cycle of *A. myosuroides* comprises four main stages, that is, soil seedbank, emergence of seedlings, mature plants, and new seed production.

Biodegradation and Biodeterioration at the Nanoscale. DOI: https://doi.org/10.1016/B978-0-12-822917-0.00003-3

Plants can emerge throughout the year, but the majority seedlings (typically 80%) emerge in autumn, with a smaller proportion emerging in spring. It can compete with crops at all growth stages and often completes its life span before the crop is harvested. It can produce large numbers of seeds ($>50{,}000\ m/m^2$) which are typically shed between June and August in winter cereals (Moss, 1980). Most seeds of *A. myosuroides* remain dormant for a relatively short period after shedding, although this is affected by environmental conditions during seed maturation (Swain et al., 2006). Shed seeds have less dormancy if they mature under hot/dry conditions as compared with cool/damp conditions. Innate dormancy is of relatively short duration, and given favorable conditions, many freshly shed seeds are capable of germinating in the autumn, typically from September to November (Moss, 1980). In the soil, its seeds are relatively short-lived, when compared to some other weeds, with an average seed decline of about 70% per year. But as the seedbank may contain 50,000 seeds/m^2 in severe infestations, an appreciable number will still be viable even after 4 years or more. It used to be thought that *A. myosuroides* required chilling temperatures (vernalization) for flowering, but exposure to cool temperatures seems to be adequate and chilling does not seem to be essential (Chauvel, 1991). Some seeds of *A. myosuroides* can also germinate in early season spring crops, including peas (*Pisum sativum* L.) and barley (*Hordeum vulgare* L.), but fewer seedlings tend to emerge in late sown spring crops, like sunflower (*Helianthus annuus* L.) or maize (*Zea mays* L.).

Failure to control this weed may lead to rapidly growing populations that produce dense infestations within a field. Mostly the weed grows in heavy clay or silt loam soils that have a higher water retention capacity, which shows that it prefers damp, but not waterlogged, conditions. Although this weed infests several crops crops, it has a large association with cereals and autumn-sown crops (Himme and Bulcke, 1975). Early emerging plants are the most competitive, at least in cereals, with later emerging plants being less successful at competing with well-established crops. Herbicide applications have often been considered to be the best option for controlling *A. myosuroides* at mature stages, but herbicides tend to be less effective against larger weeds, especially if resistant. The over-reliance on herbicides has resulted in widespread resistance, at least in some countries, hence there is a need for more integrated weed management (IWM) programs.

Weed ecology and biology

Systematics and morphology

Alopecurus myosuroides is a eukaryotic plant which belongs to the largest kingdom Plantae, phylum name is spermatophyte, subphylum angiospermae, class monocotyledonae, and order cyperales. It belongs to the family

Poaceae, which is one of the largest plant families, containing 759 genera and 11,554 species (The Plant List, 2013).

The genus of *Alopecurus* contains 47 species, including *myosuroides*, which are found in nontropical regions to alpine tropics. *Alopecurus myosuroides* contains two subspecies that are distinguished by awns (Doğan, 1999). The subspecies *myosuroides* contains awns that have a length of 8–12 mm, whereas the subspecies *tonus* is either awnless or has awns that are shorter in length.

Mature plants of *A. myosuroides* are tufted and slender shaped, attaining a height of 10–80 cm (Naylor, 1972a). Leaves are linear to lanceolate shaped that have been recorded to be more than 15 cm in length and 2–8 mm wide, with a smooth green leaf sheath and a purple base. A ligule is also present and can be 2–5 mm long (Partzsch et al., 2006) and is obtuse and membranous. Culms are smooth, rarely solitary, and often geniculate and purplish at base.

The panicles it produces are narrow, pointed, and spike-like in shape, containing typically about 100 spikelets. These each contain a single caryopsis; therefore spikelets are commonly referred to as "seeds." The color of the spikelets is brown or sometimes purplish tinged. The size of the spikelet is 4–7 mm and mostly flattened and are generally produced to have one spikelet per branch (i.e., single flowered). Glumes are also present, and the outermost glume has an oblong to lanceolate shape, often three nerved, connate for length, and has short cilia that are present on the nerves. Lemma is also lanceolate in shape, hairy, and contains four nerves. At the base of the lemma, awns are present, which are geniculate and double in length of the lemma. Lodicules and palea are absent in the flower. Stamens are three in number, and the size of anthers is 3–5 mm. Caryopsis is ovoid in shape, yellowish in color, and 2–3 mm in length.

A shallow root system is present in plants of this species, which covers large surface areas. As early seedling growth is started, the secondary roots emerge from nodes of the cotyledon and soon grow longer than the primary seminal roots (Hoshikawa, 1969). Due to the shallow root system, plants can easily tolerate the waterlogged conditions that prevail mostly in the autumn season (Naylor, 1972a). The shallow rooting structure shows that it is more susceptible to drought conditions than wheat (Stratonovitch et al., 2012).

Habitat and distribution

Alopecurus myosuroides is mostly found in heavy and loamy soil. It is an arable land weed mostly causing problems in cereal fields (Long, 1938; Morse and Palmer, 1925). Wet conditions are favored for growth, but not prolonged waterlogging. It is also called a lowland weed species because it is not recorded at higher altitudes, for example, above 1000 feet (Salisbury, 1962). Metcalfe et al. (2017) conducted a field experiment to check the distribution of *A. myosuroides* in winter wheat and concluded that seedlings of *A. myosuroides* are often patchy and the location of these patches mostly depends on the environmental conditions.

Alopecurus myosuroides is considered a problem mostly in different parts of Europe, especially in west European countries like Germany, England, France, and some areas of Belgium and the Netherlands (Holm et al., 1997; Fig. 1.1a,b). Recently, it has been increasing in the Scandinavian countries such as Sweden, Norway, and Denmark, primarily due to changes in cropping systems which favor its growth. In these European countries, *A. myosuroides* is found abundantly in winter wheat and winter oilseed rape crop and especially in early sown varieties of winter cereals (Melander, 1995; Moss, 1987a; Lutman et al., 2013).

According to CFIA (2014), it is a native species in Europe but also widely distributed in Central/South Asia and has reached Australia, South America, and North America by introduction. The main reason for this weed has established itself in these areas is due to the presence of more water-retentive soils than lighter, sandier soils.

Alopecurus myosuroides is categorized as one of the three most prevailing weeds found in France from the year 1960, due to the rapid demographic expansion during the postwar period (Guyot, 1970). It has also been found in different regions of Turkey and is now a troublesome weed there (Wille et al., 1998). Kadioğlu et al. (1998) found that *Avena* spp. and *A. myosuroides* are the most noxious weeds of different cereal growing regions of Turkey and the reason for their abundance is poor control. *Alopecurus myosuroides* has become a significant weed in the central and northern regions of Greece and is mostly found in areas with heavy soils with poor drainage capacity and intensive cultivation of winter cereals.

(A) *Alopecurus myosuroides* seedlings infesting field beans

(B) *Alopecurus myosuroides* seedlings infesting oilseed rape

FIGURE 1.1 Pictorial view of *Alopecurus myosuroides* seedlings in different crop plants. *Picture credit: Stephen Richard Moss.*

Its occurrence as a weed is not only strongly linked to cropping systems, but also to environmental factors such as temperature, photoperiod, rainfall, pH, soil type, and availability of moisture (Radosevich et al., 2007). The occurrence of drought conditions makes *A. myosuroides* less competitive as compared with moist conditions.

Germination, reproduction, and seed dispersal mechanism

Germination

Alopecurus myosuroides seedlings can only emerge from seeds within the top 5 cm of the soil which explains why it is favored by minimum tillage strategies. When sufficient soil moisture is available, many seeds will germinate and about 80% of seedling emergence occurs in the autumn, that is, from September to November; therefore cereals grown during that season tend to be more vulnerable to the weed. Seeds of *A. myosuroides* mature over a long period during the summer and have a relatively short period of innate dormancy; therefore they require a few months to lose their dormancy and have the capacity to germinate in autumn. The seeds that remain ungerminated go into a period of secondary dormancy, and some may then germinate in the spring, but a small proportion can survive and remain viable in the soil for several years (Barralis, 1970).

Different environmental conditions, such as availability of light and the occurrence of a short dry period, help the seeds break dormancy. The germination process of *A. myosuroides* is known to be stimulated by light (Colbach et al., 2002). Light increases the germination of seeds, especially freshly produced seeds, as well as those buried in soil for several months. Soil temperature and available moisture contents have a vital role in early growth stages of *A. myosuroides* (Marechal et al., 2012).

Cook and Hurry (2013) observed from their study that 36%–97% germination of *A. myosuroides* occurred during September as compared to 21%–91% in October and only 1%–7% in November. They also observed that seedling growth from seeds sown in September and October was much faster than from seeds sown later, and they started producing tillers within 30 days of sowing. In November-sown seeds, plants produced leaves only and did not start tillering until spring.

Spring-sown crops had lower populations of *A. myosuroides* than autumn-sown crops because more weed plants emerged before crop planting in spring, and these can be destroyed by presowing cultivations. The proportion of seedlings emerging in spring-sown crops is much lower than in autumn-sown crops, sometimes as low as 2%. Little germination occurs in the summer probably due to different factors including the presence of dormancy of newly shed seeds, higher atmospheric temperature, lack of sufficient moisture, and the presence of a crop cover.

Reproduction

Alopecurus myosuroides is a wind-pollinated and short-day species. It is an allogamous species, and flowers produced by it are protogynous (Johnsson, 1944). Flowering occurs in the early summer season, whereas high temperatures and prolonged wet conditions may adversely affect flowing and pollination. Pollen grains produced in the flowers are spherical in shape and have a diameter of 27–38 μm and contain a single, operculate-shaped pore. Anthesis starts in spikelets near the tip of the inflorescence and proceeds downwards (Fig. 1.2b,c). Time for completing the anthesis process for all the flowers may vary.

Usually one, or occasionally two, caryopses are produced within a spikelet but poor pollination conditions can result in many (50% or more)

(A) *Alopecurus myosuroides* inflorescence

(B) *Alopecurus myosuroides* inflorescence with anthers emerging from the tip and stigmas visible below.

(C) *Alopecurus myosuroides* seeds

FIGURE 1.2 Pictorial view of *Alopecurus myosuroides* seed and inflorescence. *Picture credit: Bayer and Stephen Richard Moss.*

"empty" spikelets, containing no caryopses. Spikelets that are produced early tend to have a higher number of seeds. A head may contain 20–300 spikelets, but typically contain 100 on average. Head sizes also vary on the same plant.

Pollen dispersal in more than 70% of weed plants was found within a range of 1 m, although the fertilization process may occur at a very large distance (e.g., 60 m from donor species if there is no physical barrier present to pollen propagation process) (Chauvel, 1991). Pollen is likely to travel much further from large weed infestations.

Seed production is also much higher from autumn-emerging plants as compared to spring-emerging plants. The reason for this is that plants that emerge in spring produce fewer tillers per plant, and the number of spikelets per head is also lower, which ultimately leads to a decrease in the total number of seeds. The reproductive rate and competitive ability of *A. myosuroides* are much higher as compared with many other weeds in cereal crops (Marechal et al., 2012). Due to this characteristic, *A. myosuroides* can produce a large number of seeds, and if not effectively controlled, populations can increase 10-fold or more per annum, leading to dense infestations.

Seed dispersal mechanism

The unit of dispersal is the spikelet, which consists of a caryopsis along with the lemma and glumes. Shedding of seeds from the plants mostly takes place during July and August. Seeds are shed singly after the completion of the ripening process.

The maximum number of seeds is dispersed by barochory (i.e., dispersed by using gravity) but the increase in wind speed increases the rate and distance of dispersal (Colbach and Sache, 2001). The mean distance covered by *A. myosuroides* seeds is 51 cm (Marechal et al., 2012). Seed can also cover large distances by small animals, including mice and vole, and this type of dispersion mechanism is called epizoochory. This process of dispersal is much more effective when the plant height is similar to the animal height to which it comes into contact with.

Plant height is considered a vital component in the dispersal of seed. The taller the plant, the greater the distance seeds will be dispersed. Light seeds are carried to great distances, due to low-gravity interaction with the seed surface; ultimately, horizontal movement of the seed is increased (Colbach et al., 2001). A leptokurtic distribution of seeds occurs from the point of seed production, but it was noted that in case of seed dispersal through harvesting tools, the distance covered is more than the distance achieved by the natural mechanism of seed dispersal (Colbach et al., 2001). In practice, long-distance movement of seeds is most likely to occur by contaminated crop seed, straw, manure, and harvest and cultivation machinery. The cultivation of soil also causes the vertical movement of *A. myosuroides* seeds in the soil profile (Moss, 1990). When *A. myosuroides* grows in a crop such as winter wheat, it matures earlier than the main crop and most seeds are shed prior to crop harvest. With earlier harvested crops, such as winter barley, an appreciable amount of seeds will remain unshed at harvest, which may contaminate the grain samples.

The complete life cycle of *A. myosuroides* is shown in Fig. 1.3.

Seed biology

The true seed of *A. myosuroides* is narrowly elliptical and laterally compressed in shape and varies from 2 to 3 mm long and 1.0–1.5 mm wide (Fig. 1.2c). Seed color or spikelet is often greenish brown to brown. The length of the embryo is one-fifth to one-third of caryopsis and is located along the dorsal edge of the caryopsis. An oval-shaped hilum is also present, which allows the entrance of water into the embryo before active germination. The style bases are generally fused together or sometimes remain visible on the apex of caryopsis.

Infestations in cultivated plants and yield losses

Alopecurus myosuroides is most commonly found as a weed of autumn-sown crops in Europe. It produces a large number of seeds in favorable conditions, but seed production depends greatly on the competitiveness of the crop. For

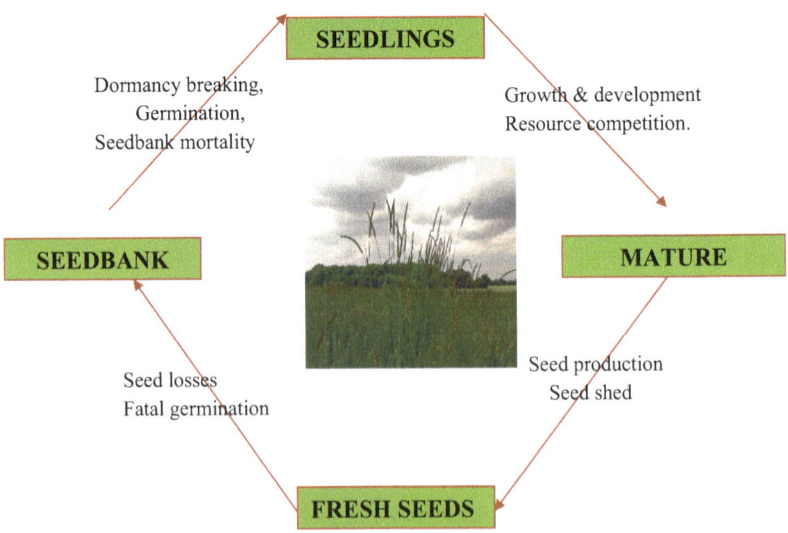

FIGURE 1.3 Complete life cycle of *Alopecurus myosuroides. Adopted from Metcalfe, H., 2017. Modelling the spatial variation in Alopecurus myosuroides for precision weed management (Doctoral dissertation). University of Reading, Berkshire, England, p. 12.*

example, barley grown in narrow (e.g., 12 cm) rows tends to be very competitive, whereas crops grown in wide (e.g., 50 cm) rows are much less competitive, at least at early growth stages. The critical period for crop/weed competition depends on when crops are sown and the growing conditions present at the time. In autumn-sown wheat in the United Kingdom, competition affecting yield was only significant once weed and crops started to grow more actively in the following spring, over 4 months after sowing (Moss, 1987a). Intensive cultivation prior to sowing spring cereals can also encourage germination and cause serious infestations of this weed. Even in competitive winter cereals, *A. myosuroides* can produce 200–2000 seeds on a single plant, and in high infestations, *A. myosuroides* can produce more than 50,000 seeds/m^2 area (Moss, 1985).

High infestations of *A. myosuroides* can cause a serious reduction in the yield and quality traits of the main crop. The economic threshold of *A. myosuroides* is 30–50 plants/m^2 in England (Cousens et al., 1985) and 20–42 plants/m^2 in Italy (Zanin et al., 1993). However, farmers need to reduce populations to much lower densities, ideally to less than 5 plants/m^2, due to the high potential rate of population increase. In winter cereals, 100–500 plants/m^2 can cause 20%–50% yield loss (Blair et al., 1999). Cousens (1985) reported that wheat yield may be reduced by 45% due to competition with *A. myosuroides*. It is also a troublesome weed in different regions of Turkey (Letouzé and Gasquez, 2000) as an increase in *A. myosuroides* density leads to a decrease wheat yield (Mennan and Işik, 2004). Mennan et al. (2003) concluded from their study that the economic threshold level for *A. myosuroides* is 16–33 plants/m^2. Wheat yield was also reduced by 0.4–0.8 t/ha when populations of *A. myosuroides* reaches 12–25 plants/m^2. Much higher wheat losses (e.g., 2 t/ha) were noted when plant density reached over 100 plants m/m^2.

Invasive properties and allelopathic potential

Invasive properties

Alopecurus myosuroides is a weed which is mostly found in crop fields and vineyards, and exceptional cases have been noted in different grasslands. The weed has also been seen along roadsides and on the gravelly soil. The invasiveness of *A. myosuroides* was firstly recognized by Brenchley (1920), who observed a huge amount of *A. myosuroides* seeds in peas and beans (21%), wheat (9%), oat (4%), barley (2%), and root crops (2%).

Due to the presence of the following characteristics, *A. myosuroides* is considered a major weed (CABI, 2020): This weed is fast growing in nature and has high genetic variability; adaptability and reproductive potential is also much higher; it has high-competitive ability and some seeds remain viable for several years.

Allelopathic potential

Allelopathy is the capability of a plant to affect its surrounding plants by releasing active biomolecules called allelochemicals. These allelochemicals positively and negatively influence the growth of other plants. *Alopecurus myosuroides* is generally considered one of the most competitive weeds, but it may have other characteristics including allelopathic behavior. The combination of both these characteristics has a greater role in influencing growth of the main crop. Research shows that a strong negative effect occurs on wheat yield, due to increment in competition between *A. myosuroides* and wheat plants, which results in a retardation in root growth through allelochemical produced by *A. myosuroides*. Bertholdsson (2012) concluded from his experiment that growing wheat cultivars that have high allelopathic potential could play an important role in IWM of *A. myosuroides*.

Herbicide resistance cases and mechanism

Alopecurus myosuroides is among the 15 most important herbicide-resistant weeds found worldwide (Heap, 2020). In Europe, it is currently the most important herbicide-resistant weed (Lutman et al., 2013), and resistance evolution tends to be much faster when compared with the time required to discover and approve a new herbicide. Since the late 1980s, no major new modes of action in herbicides have been identified (Beckie and Tardif, 2012) and none are expected in the next decade (Lamberth et al., 2013). Repeated use of herbicides to control different weeds has resulted in the widespread evolution of herbicide resistance.

Alopecurus myosuroides has evolved resistance against seven different sites of action, including acetyle coenzyme A carboxylase inhibitors (ACCase inhibitor), mitosis inhibitors, fatty acid inhibitors, photosystem II inhibitors, acetolactate synthase inhibitors (ALS inhibitor), and lipid biosynthesis inhibitors (Heap, 2020; Marshall and Moss, 2008). In *A. myosuroides*, ACCase and ALS inhibiting herbicides are the most important classes and 50% of control strategies rely on these modes of actions. Populations of *A. myosuroides* have evolved both target-site resistance (TSR) and nontarget site resistance (NTSR) mechanisms. The first case of resistance against fenoxaprop-P-ethyl was observed in a greenhouse experiment after repeated herbicide application on a susceptible population of *A. myosuroides* (Chauvel and Gasquez, 1994). Detailed studies have been conducted in countries like the United Kingdom and France on resistant populations of *A. myosuroides*. The studies found that 56% populations of *A. myosuroides* in France had TSR (Délye et al., 2007), whereas in the United Kingdom and Germany, 60% populations had NTSR (Drobny et al., 2006) (Fig. 1.4a,b).

Herbicide resistance in *A. myosuroides* has been confirmed in many countries in Europe, especially in the United Kingdom, Germany, England,

(A) (B)

FIGURE 1.4 Pictorial view of herbicide-resistant *Alopecurus myosuroides* infesting wheat fields. *Picture credit: Stephen Richard Moss.*

and France. In the United Kingdom, the first herbicide-resistant case of *A. myosuroides* was detected in 1982 (Heap, 2014). A survey conducted between 2009 and 2011 on randomly selected fields showed that 47% of populations had resistance to mesosulfuron plus iodosulfuron, 100% to clodinafop, 58% to cycloxydim, 79% to pendimethalin, and 47% to chlorotoluron (Moss et al., 2014).

There are three types of herbicide resistance mechanisms that are mostly observed in *A. myosuroides* (i.e., NTSR, ACCase TSR, and ALS TSR). The first NTSR enhanced metabolism resistance, which mostly occurs due to detoxification of the herbicide, and this is the most common resistance mechanism observed in *A. myosuroides*. Many different herbicides are affected by this mechanism to varying degrees, regardless of their mode of action, and in some cases, this results in complete failure to control weeds (i.e., loss of herbicide function). The second type of resistance mechanism is the ACCase TSR, which blocks the activity site specific to herbicides such as clodinafop, fluazifop-P, fenoxaprop, quizalofop, propaquizalofop, cycloxydim, tepraloxydim, and pinoxaden. The third type of resistance mechanism is ALS TSR, and it blocks the activity site specific to sulfonylurea and related herbicides, including flupyrsulfuron, pyroxsulam, and mesosulfuron + iodosulfuron. It is less common than ACCase TSR, but has been increasing rapidly recently (Moss, 2013).

Weed management practices

Cultural control

Cultural weed control varies with the farming systems, along with the specific crop type. In cultural control, a diverse range of different farming practices is used as the major tool to suppress weed growth and to increase crop growth. Some of these techniques include crop rotation, cultivations, delaying autumn sowing, use of competitive cultivars, increase in seed rate, and a focus on spring cropping and fallowing.

Plowing buries freshly shed seeds in the field to varying depths depending on soil conditions, plow design, and operation. But with *A. myosuroides*, most seeds will be buried to a depth of more than 5 cm from where seedlings are unlikely to emerge. Conversely, due to the relatively short persistence of *A myosuroides* seeds in the soil, fewer old seeds are likely to be returned to the soil surface during plowing. No-tillage systems mostly favor the seeds remaining close to the soil surface where they can germinate, and seedlings can readily emerge, causing potentially rapid population increases.

Delayed autumn sowing also plays an important role in controlling high infestations of *A. myosuroides*, as more of the weed plants emerge before sowing and these can easily be controlled by nonselective herbicides (e.g., glyphosate) or cultivations presowing. Growing competitive crops also influences the weed growth to varying degrees. For example, barley is more competitive than wheat so it can suppress more weeds. In the case of *A. myosuroides*, competitive crops are helpful in suppressing its growth (Cussans, 2009; Gehring et al., 2012). High seed rates of crops, especially winter cereals, also help in weed suppression. In the case of winter wheat, sowing rate and choice of appropriate cultivar can significantly affect the number of panicles produced by *A. myosuroides* (Blake, 2006).

In the past, the burning of crop residues considerably reduced the abundance of newly shed seeds, but burning had several adverse environmental impacts so it is no longer practiced in many countries (Moss, 1979, Froud-Williams, 1987). Wilson et al. (1989) found that burning crop residues killed more weed seeds than baling or any other straw disposal method. It has been noted that the disposal of straw by spreading and burning can reduce seeds in the soil seedbank by 50%, but it can also stimulate germination of some of the remaining seeds (Moss, 1987b). Baling of straw after seed shedding increased the population ninefold, whereas just a threefold increase was noted following burning. Grazing or mowing also prevents *A. myosuroides* seedlings from establishing and producing heads and therefore reduces the amount of seeds that return to the seedbank (Table 1.1).

Mechanical control

Alopecurus myosuroides is a weed that mainly infests autumn-sown crops and usually completes its life cycle prior to wheat harvest; therefore most weed seeds cannot be removed using harvest weed seed control techniques, which are more appropriate for some other weeds (Walsh et al., 2013). This makes the weed problematic. The use of the moldboard plow for seedbed preparation is effective for controlling *A. myosuroides* and other annual

TABLE 1.1 Cultural weed control.

Control method	Control percentage (%)	General comments
Plowing	69	By adopting rotational plowing, control efficiency can be increased.
Delay autumn sowing	31	Late sowing is considered effective but sometimes it can be risky.
Competitive crops	22	Marginal effects occur due to it.
Increase seed rate	26	Suppresses the weeds but lodging issue may arise.
Spring cropping	88	Very effective but limited number of herbicides are available.
Fallowing	70–80 per year	Absence of new seeding critical.

Source: Lutman, P.J.W., Moss, S.R., Cook, S., Welham, S.J., 2013. A review of the effects of crop agronomy on the management of Alopecurus myosuroides. Weed Res., 53, 299–313.

weeds, but it has its limitations. After crop harvest, shallow plowing and surface tillage will encourage nondormant *A. myosuroides* seeds to germinate freely which can then be destroyed by plowing or nonselective herbicides prior to crop sowing. Seedlings of *A. myosuroides* are very susceptible to mechanical injury and can easily be killed by plowing or cultivation (Cussans et al., 1996). In wheat, seedlings can also be killed by using harrows in dry conditions, but this is likely to cause crop damage. In wet conditions, compaction and trampling the soil near the young seedlings also inhibit their growth (Naylor, 1972b).

It has been observed that directly drilled wheat tends to have more serious infestations of *A. myosuroides* as compared to plowed land, if seed return is not prevented adequately (Cussans et al., 1979). In areas where seed shedding is prevented, lower weed densities may occur on direct drilled or minimally cultivated land because old seeds are not returned to the soil surface by cultivations. This highlights the importance of preventing seed return of *A. myosuroides*.

Chemical control

Use of herbicides is the main option for controlling *A. myosuroides* and many farmers adopt herbicides as the sole method for controlling the weed. It has been estimated that to prevent increasing weed populations, 88%

control of seed return is required in direct drilled crop whereas only 50% control is needed with effective annual plowing (Moss and Cussans, 1985). Both preemergence and postemergence herbicides are available for effective control of *A. myosuroides*. Preemergence herbicides, including pendimethalin, triallate, flufenacet, and prosulfocarb, are used to achieve control of *A. myosuroides* in cereals (Moss, 2013). The efficacy of preemergence herbicides is generally not seriously affected by resistance (Moss and Hull, 2009). As preemergence herbicides are directly applied to the soil soon after sowing, their efficacy depends upon the properties of the soil in the field where they are applied. For example, organic matter content in the soil influences the efficacy of applied herbicide, for example, soil that has more organic matter shows more adsorption and these soils may require a higher dose of herbicide (Nordmeyer, 2015).

Postemergence herbicides include clodinafop, chlorotoluron, fenoxaprop-p-ethyl, pinoxaden, pyroxysulam, mesosulfuron plus iodosulfuron and flupysulfuron; these are mostly used in cereal crops to control *A. myosuroides* (Lainsbury, 2012). More than 95% control of susceptible plants may be achieved by using these herbicides in cereals. In the case of noncereal crops, a broad list of herbicides is available that includes propaquizalofop, quizalofop, fluazifop-P, cycloxydim, clethodim, carbetamide, and propyzamide (Moss, 2013). In the past, isoproturon was used widely in cereals postemergence for controlling *A. myosuroides*. It is mostly absorbed by the roots or sometimes by the foliage, and it inhibits the photosynthesis process by affecting photosystem II (Orson, 1991). Clodinafop and fenoxaprop are applied as foliar sprays and they mostly affect the meristematic tissue. Unfortunately, *A. myosuroides* has evolved variable degrees of resistance to all the above-mentioned herbicides, with the exception of propyzamide, especially in the United Kingdom (Moss et al., 2007).

Integrated weed management

Sustainable long-term weed control cannot be achieved solely with herbicides or nonchemical strategies. Combined use of many different weed control measures, which are each only partially successful if applied solely, can improve the overall control of *A. myosuroides*. IWM is the most promising approach that keeps the weed population low (Gehring et al., 2012). Now, IWM is being promoted globally due to its several benefits, including reducing farmers' dependence on herbicides. However, the adoption of IWM strategies by farmers has generally been poor globally (Shaner and Beckie, 2014). This is not so much due to a lack of knowledge, but rather a lack of action, and the real challenge is how best to increase farmer uptake of IWM (Moss, 2019). This IWM system reduces the use of herbicides and increases the integration of cultural practices that leads to a faster decrease in *A. myosuroides* infestations.

According to Moss and Lutman (2013), key factors involved in controlling *A. myosuroides* through IWM are:

1. Keep the *A. myosuroides* populations as low as possible.
2. Use herbicides with a low risk of herbicide resistance evolution.
3. Use a herbicide mixture, it will improve the overall control and reduce the resistance risk.
4. Use a range of different nonchemical weed control methods to reduce dependence on herbicides.
5. Always pay attention to recommended rates and spray timings.
6. Adopt a proper crop rotation and add noncereal crops to the cropping system which helps in controlling *A. myosuroides* population through use of alternative herbicides.

Conclusion and future prospects

Alopecurus myosuroides is an annual grass weed that mostly infests winter cereals. Being a competitive weed, it competes for nutrients especially nitrogen and can cause large crop yield losses. Populations of this weed can increase rapidly. The biggest challenge is to control *A. myosuroides* in reduced cultivation systems, which have environmental benefits, while actively promoting IWM techniques and reducing the threat posed by herbicide resistance.

As *A. myosuroides* is mostly associated with specific soil types and typically found on poorly drained and heavy textured soils, changes in cropping patterns and reductions in control due to herbicide resistance may allow it to spread to lighter soils and cause problems there. Long-term and planned strategies are required which include effective IWM programs for the prevention and management of resistant *A. myosuroides*. The design of such IWM programs is complex, and there is a need for an in-depth understanding of weed biology, ecology, and dynamics in the varying agronomic systems that occur in different countries.

References

Barralis, G., 1970. Biology of *Alopecurus agrestis* L. 1. Primary dormancy and germination capacity. Rev. Generale de. Botanique 77, 429–443.

Beckie, H.J., Tardif, F.J., 2012. Herbicide cross resistance in weeds. Crop. Prot. 35, 15–28.

Bertholdsson, N., 2012. Allelopathy-A tool to improve the weed competitive ability of wheat with herbicide-resistant black-grass (*Alopecurus myosuroides* Huds.). Agronomy 2, 284–294.

Blair, A.M., Cussans, J.W., Lutman, P.J.W., 1999. A biological framework for developing a weed management support system for weed control in winter wheat: Weed competition and time of weed control. In Proceedings of the 1999 Brighton Conference—Weeds, Brighton, UK, 15–18 November 1999, pp. 753–760.

Blake, A., 2006. Weeds are a consideration. Farmers Wkly. 145 (14), 57.

Brenchley, W.E., 1920. On the relations between growth and the environmental conditions of temperature and bright sunshine. Ann. Appl. Biol. 6 (4), 211–244.

CABI, 2020. Alopecurus myosuroides. Invasive Species Compendium. CAB International, Wallingford, UK. <www.cabi.org/isc>, Accessed on 05 August 2020.

CFIA, 2014. Invasive plants—fact sheets. Canadian Food Inspection Agency, Ottawa, ON., <http://www.inspection.gc.ca/eng/1331614724083/1331614823132>. Accessed on 20 July 2020.

Chauvel, B., 1991. Polymorphism and selection for resistance to ureas substituted in Alopecurus myosuroides Huds (PhD thesis). University of Paris XI—Orsay, Orsay, France.

Chauvel, B., Gasquez, J., 1994. Relationships between genetic polymorphism and herbicide resistance within *Alopecurus myosuroides* Huds. Heredity 72 (4), 336–344.

Colbach, N., Durr, C., 2003. Effects of seed production and storage conditions on blackgrass (*Alopecurus myosuroides*) germination and shoot elongation. Weed Sci. 51, 708–717.

Colbach, N., Sache, I., 2001. Black-grass (*Alopecurus myosuroides* Huds.) seed dispersal from a single plant and its consequences on weed infestation. Ecol. Model. 139, 201–219.

Colbach, N., Chauvel, B., Durr, C., Richard, G., 2002. Effect of environmental conditions on *Alopecurus myosuroides* germination. I. Effect of temperature and light. Weed Res. 42, 210–221.

Cook, S., Hurry, S., 2013. Emergence pattern of black-grass (*Alopecurus myosuroides*). Conference: Crop Production in Southern Britain, Precision Decisions for Profitable Cropping. Volume: Aspects of Applied Biology, p. 127.

Cousens, R., 1985. A simple model relation yield loss to weed density. Ann. Appl. Biol. 107, 239–252.

Cousens, R., Wilson, B.J., Cussans, G.W., 1985. To spray or not to spray: the theory behind the practice. Proc. Br. Crop. Prot. Conf., Weeds 2, 671–678.

Cussans, G.W., 2009. The relative influence of agronomic factors on weed growth and crop yield. Asp. Appl. Biol. 91, 6. Crop. Prot. South. Br.

Cussans, G.W., Moss, S.R., Pollard, F., Wilson, B.J., 1979. Studies of the effects of tillage on annual weed populations. Proceedings of the EWRS Symposium, The influence of different factors on the development and control of weeds, pp. 115–122.

Cussans, G.W., Raudonius, S., Brain, P., Cumberworth, S., 1996. Effects of depth of seed burial and soil aggregate size on seedling emergence of *Alopecurus myosuroides*, *Galium aparine*, *Stellaria media* and wheat. Weed Res. 36, 133–141.

Délye, C., Menchari, Y., Guillemin, J.P., Matéjicek, A., Michel, S., Camilleri, C., et al., 2007. Status of black grass (*Alopecurus myosuroides*) resistance to acetyl-coenzyme A carboxylase inhibitors in France. Weed Res. 47 (2), 95–105.

Doğan, M., 1999. A concise taxonomic revision of the genus *Alopecurus* L. (Gramineae). Turk. J. Bot. 23 (4), 245–262.

Drobny, H., Salas, M., Claude, J.P., 2006. Management of metabolic resistant black-grass (*Alopecurus myosuroides* Huds.) populations in Germany - challenges and opportunities. J. Plant. Dis. Prot. 20, 65–72.

Froud-Williams, R.J., 1987. Survival and fate of weed seed population—interaction with cultural practice. Proceedings British Crop Protection Conference—Weeds, Brighton, UK, pp. 707–718.

Gehring, K., Balgheim, R., Meinlschmidt, E., Schleich-Saidfar, C., 2012. Prinzipien einer Anti-Resistenzstrategie bei der Bekämpfung von *Alopecurus myosuroides* und *Apera spica-venti* aus Sicht des Pflanzenschutzdienstes. In: Proceedings of 25th German Conference on Weed Biology and Weed Control, Braunschweig, Germany, Julius-Kühn-Archiv, 434, pp. 89–101.

Guyot, L., 1970. Current state of the weed problem in some regions of France. III Colloquium on the biology of weeds: Appendices. National School of Grignon, France. p. 31.

Heap, I., 2014. Herbicide resistant weeds. Integrated Pest Management. Springer, Dordrecht, pp. 281–301.

Heap I. 2020. The international survey of herbicide resistant weeds [online]. <http://www.weedscience.com>. Accessed on 26 August 2020.

Himme, M.V., Bulcke, R., 1975. The distribution, spread and importance of *Alopecurus myosuroides* Huds. in Europe. In Symposium on Status, Biology and Control of Grassweeds in Europe, organised by EWRS and COLUMA, Paris, 2, pp. 23–54.

Holm, L., Doll, J., Holm, E., Pancho, J.V., Herberger, J.P., 1997. World Weeds: Natural Histories and Distribution. John Wiley & Sons.

Hoshikawa, K., 1969. Underground organs of the seedlings and the systematics of Gramineae. Bot. Gaz. 130 (3), 192–203.

Johnsson, H., 1944. Meiotic aberrations and sterility in *Alopecurus myosuroides* Huds. Hereditas 30 (4), 469–566.

Kadioğlu, İ., Üremiş, İ., Uluğ, E., Boz, Ö., Uygur, F.N., 1998. Researches on the economic thresholds of wild oat (*Avena sterilis* L.) in wheat fields in Çukurova Region of Turkey. Turkish Weed Sci. 1 (2), 18–24.

Lainsbury, M.A., 2012. The UK Pesticide Guide 2012. CAB International/British Crop Production Council (BCPC), Wallingford, Oxfordshire, UK.

Lamberth, C., Jeanmart, S., Luksch, T., Plant, A., 2013. Current challenges and trends in the discovery of agrochemicals. Science 341 (6147), 742–746.

Letouzé, A., Gasquez, J., 2000. A pollen test to detect ACCase target-site resistance within *Alopecurus myosuroides* populations. Weed Res. 40 (2), 151–162.

Long, H.C., 1938. 2nd edn. Weeds of Grassland, 41. Weeds of Grassland.

Lutman, P.J.W., Moss, S.R., Cook, S., Welham, S.J., 2013. A review of the effects of crop agronomy on the management of *Alopecurus myosuroides*. Weed Res. 53, 299–313.

Marechal, P.Y., Henriet, F., Vancutsem, F., Bodson, B., 2012. Ecological review of black-grass (*Alopecurus myosuroides* Huds.) propagation abilities in relationship with herbicide resistance. Biotechnologie, Agronomie, Société et. Environ. 16 (1).

Marshall, R., Moss, S.R., 2008. Characterisation and molecular basis of ALS inhibitor resistance in the grass weed *Alopecurus myosuroides*. Weed Res. 48, 439–447.

Melander, B., 1995. Impact of drilling date on *Apera spica-venti* L. and *Alopecurus myosuroides* Huds, in winter cereals. Weed Res. 35, 157–166.

Mennan, H., Işik, D., 2004. The competitive ability of *Avena* spp. and *Alopecurus myosuroides* Huds. influenced by different wheat (*Triticum aestivum*) cultivars. Turk. J. Agric. For. 28 (4), 245–251.

Mennan, H., Bozoğlu, M., Işik, D., 2003. Economic thresholds of *Avena* spp., and *Alopecurus myosuroides* in winter wheat fields. Pak. J. Bot. 35 (2), 147–154.

Metcalfe, H., 2017. Modelling the spatial variation in Alopecurus myosuroides for precision weed management (Doctoral dissertation). University of Reading, Berkshire, England. p. 12.

Metcalfe, H., Milne, A.E., Murdoch, A.J., Storkey, J., 2017. Does variable soil pH have an effect on the within-field distribution of *A. myosuroides*. Asp. Appl. Biol. 134, 145–150.

Morse, R., Palmer, R., 1925. British weeds: their identification and control. British weeds: their identification and control.

Moss, S., 2019. Integrated weed management (IWM): why are farmers reluctant to adopt non-chemical alternatives to herbicides? Pest. Manage. Sci. 75, 1205–1211.

Moss, S.R., 1979. The influence of tillage and method of straw disposal on the survival and growth of black-grass, *Alopecurus myosuroides*, and its control by chlortoluron and isoproturon. Ann. Appl. Biol. 91, 91–100.
Moss, S.R., 1980. Some effects of burning cereal straw on seed viability, seedling establishment and control of *Alopecurus myosuroides* Huds. Weed Res. 20 (5), 271–276.
Moss, S.R., 1985. The survival of *Alopecurus myosuroides* Huds. seeds in soil. Weed Res. 25 (3), 201–211.
Moss, S.R., 1987a. Competition between *Alopecurus myosuroides* and winter wheat. In Proceedings of the British Crop Protection Conference-Weeds, Brighton, UK, 16–19.
Moss, S.R., 1987b. Influence of tillage, straw disposal system and seed return on the population dynamics of *Alopecurus myosuroides* Huds. in winter wheat. Weed Res. 27, 313–320.
Moss, S.R., 1990. Herbicide cross resistance in slender foxtail (*Alopecurus myosuroides*). Weed Sci. 38, 492–496.
Moss, S.R., 2013. Black-grass (*Alopecurus myosuroides*) everything you really wanted to know about black-grass but didn't know who to ask. A Rothamsted technical publication, Revised 2013, http://www.rothamsted.ac.uk.
Moss, S.R., Cussans, G.W., 1985. Variability in the susceptibility of *Alopecurus myosuroides* (black-grass) to chorotoluron and isoproturon. Asp. Appl. Biol. 9, 91–98.
Moss, S.R., Hull, R., 2009. The value of pre-emergence herbicides for combating herbicide-resistant *Alopecurus myosuroides* (black-grass). Asp. Appl. Biol. 91, 79–86.
Moss, S.R., Lutman, D.P., 2013. Black-grass: the potential of non-chemical control. A Rothamsted technical publication. Rothamsted Research, Harpenden, Herts, AL5 2JQ, UK. http://www.rothamsted.ac.uk
Moss, S.R., Perryman, s, Tatnell, L.V., 2007. Managing herbicide-resistant black-grass (*Alopecurus myosuroides*): theory and practice. Weed Technol. 21, 300–309.
Moss, S.R., Hull, R.I., Marshall, R., Perryman, S.A.M., 2014. Changes in the incidence of herbicide resistant Alopecurus myosuroides (black-grass) in England, based on sampling the same random fields on two occasions. Asp. Appl. Biol. 127, 39–48. Crop Production in Southern Britain, Precision decisions for profitable cropping.
Naylor, R.E.L., 1972a. Aspects of the population dynamics of the weed *Alopecurus myosuroides* Huds. in winter cereal crops. J. Appl. Ecol. 9, 127–139.
Naylor, R.E.L, 1972b. Alopecurus Myosuroides Huds. (A. Agrestis L.). J. Ecol. 611–622.
Nordmeyer, H., 2015. Herbicide application in precision farming based on soil organic matter. Am. J. Exp. Agric. 8 (3), 144–151.
Orson, J.H., 1991. Variability in the Control of Black-Grass with Herbicides in Winter Cereals—Harvest Years 1987–1989. HGCA Project Report No. 27. Home Grown Cereals Authority, London.
Partzsch, M., Cremer, J., Zimmermann, G., Goltz, H., 2006. Field and Garden Weeds. Agrimedia Gmbh, Bergen, Germany.
Radosevich, S.R., Holt, J.S., Ghersa, C.M., 2007. Ecology of Weeds and Invasive Plants: Relationship to Agriculture and Natural Resource Management. John Wiley & Sons.
Salisbury, E.J., 1962. The biology of garden weeds. Part. 2. J. R. Hortic. Soc. 87 (10), 458–470.
Shaner, D.L., Beckie, H.J., 2014. The future for weed control and technology. Pest. Manag. Sci. 70, 1329–1339.
Sinclair, C., 1838. Hill and Valley: Or, Hours in England and Wales. Robert Carter.
Stratonovitch, P., Storkey, J., Semenov, M.A., 2012. A process-based approach to modelling impacts of climate change on the damage niche of an agricultural weed. Glob. Change Biol. 18 (6), 2071–2080.

Swain, A.J., Hughes, Z.S., Cook, S.K., Moss, S.R., 2006. Quantifying the dormancy of *Alopecurus myosuroides* seeds produced by plants exposed to different soil moisture and temperature regimes. Weed Res. 46, 470–479.

The Plant List, 2013. Version 1.1. Published on the Internet; <http://www.theplantlist.org/>. Accessed on 23 July 2020.

Vizantinopoulos, S., Katranis, N., 1998. Management of blackgrass (*Alopecurus myosuroides*) in winter wheat in Greece. Weed Technol. 12 (3), 484–490.

Walsh, M., Newman, P., Powles, S., 2013. Targeting weed seeds in-crop: a new weed control paradigm for global agriculture. Weed Technol. 27 (3), 431–436.

WFO, 2020. *Alopecurus myosuroides* Huds. Published on the internet; <http://www.worldfloraonline.org/taxon/wfo-0000845620>. Accessed on 20 July 2020.

Wille, M.J., Thill, D.C., Price, W.J., 1998. Avena spp. (*Avena fatua*) seed production in spring barley (*Hordeum vulgare*) is affected by the interaction of *Avena* spp. density and herbicide rate. Weed Sci. 46, 336–343.

Wilson, B.J., Moss, S.R., Wright, K.J., 1989. Long term studies of weed populations in winter wheat as affected by straw disposal, tillage and herbicide use. Proceedings of the Brighton Crop Protection Conference—Weeds, Brighton, UK, 131–136.

Zanin, G., Berti, A., Taniolo, L., 1993. Estimation of economic thresholds for weed control in winter wheat. Weed Res. 33, 459–467.

Chapter 2

Amaranthus spp.

Debalin Sarangi[1], Amit J. Jhala[2], Prabhu Govindasamy[3] and Anthony Brusa[1]

[1]Department of Agronomy and Plant Genetics, University of Minnesota, St. Paul, MN, United States, [2]Department of Agronomy and Horticulture, University of Nebraska-Lincoln, Lincoln, NE, United States, [3]Division of Crop Production, Indian Grassland and Fodder Research Institute, Jhansi, India

Introduction

Genus *Amaranthus* is a genetically diverse group of herbaceous C4 plants that are found in warm to temperate regions around the world. *Amaranthus* spp. are widely found in North America and Europe, but have also been introduced in South America, Africa, China, and Australia (Assad et al., 2017). The number of species in *Amaranthus* genus is disputed and the estimates range from 60 (National Research Council, 1984) to 75 species (Ward et al., 2013). The *Amaranthus* genus has also been described as having untapped potential as a food resource; three grain amaranth species (*A. hypochondriacus* L., *A. cruentus* L., and *A. caudatus* L.) are grown as food crops in mountainous regions of South America (Pederson et al., 1990). When roasted and milled, the seeds of *Amaranthus* spp. are edible and rich in protein, whereas the leaves are high in Ca, Fe, and vitamin A (Becker et al., 1981). Despite these benefits, *Amaranthus* genus is known for having some of the most common and troublesome weeds in the world, including *A. palmeri* S. Wats (Palmer amaranth), *A. tuberculatus* (Moq.) Sauer (waterhemp), and *A. retroflexus* L. (redroot pigweed).

The *Amaranthus* genus contains both monoecious and dioecious weed species. The genus is composed of three subgenera, with 10 dioecious species in the subgenus Acnida (Steckel, 2007), whereas monoecious species fall into subgenera *Albersia* and *Amaranthus* (Mosyakin and Robertson, 2008). Notable monoecious species include *A. powellii* S. Wats (Powell's amaranth), *A. retroflexus*, and *A. hybridus* L. (smooth pigweed) (Assad et al., 2017). Dioecious amaranths include *A. palmeri* and *A. tuberculatus*. The dioecious biology of these two species

Biology and Management of Problematic Crop Weed Species. DOI: https://doi.org/10.1016/B978-0-12-822917-0.00010-0

combined with wind pollination allows them to maintain high levels of genetic variation, which plays a key role in the adaptation to diverse environmental conditions.

Identification of *Amaranthus* spp.

The species within the *Amaranthus* genus share several morphological and developmental characteristics. Leaf shapes tend to fall into one of the three categories: rounded (e.g., *A. blitoides* S. Wats, and *A. alba* L.), lanceolate (e.g., *A. tuberculatus* and *A. arenicola* I.M. Johnst), and ovate (e.g., *A. palmeri, A. spinosus* L., and *A. retroflexus*). The length of the petiole can be used to differentiate *A. palmeri* from other species, as the length of the petiole on *A. palmeri* is longer than the length of the leaf blade (Fig. 2.1). Additional characteristics for species identification include the presence or absence of pubescence on stem and leaves, with dioecious species such as *A. palmeri* being glabrous to sparsely hairy. Watermarks are present on some species (e.g., *A. palmeri*) but should not be used for the identification as they are not present all the time (Fig. 2.1).

Identifying characteristics of some key *Amaranths* (see Fig. 2.1):

Amaranthus hybridus: monoecious; glabrous or sparsely pubescent stem; ovate-shaped alternate leaves, wavy leaf margin; terminal panicles of numerous slender, cylindrical spikes, and slightly prickly.

Amaranthus palmeri: dioecious; glabrous stem and leaves; ovate-shaped alternate leaves, mature leaves have a petiole longer than the leaf blade; female bracts long, thick, and stiff.

Amaranthus retroflexus: monoecious; pubescent stem and leaves; alternate leaves, ovate, veins with trichomes on underside; large red taproot; spikes in thick inflorescence.

Amaranthus spinosus: monoecious; stems with two very sharp and stiff spines at most nodes; ovate leaves.

Amaranthus tuberculatus: dioecious; glabrous stem and leaves; glossy lanceolate alternate leaves.

FIGURE 2.1 Plants of weedy *Amaranthus* spp. (A) *A. palmeri*, (B) *A. tuberculatus*, (C) *A. retroflexus*.

Herbicide resistance mechanisms

Herbicide resistance in the *Amaranthus* genus is common, with the rapid spreading of herbicide-resistant *A. palmeri* and *A. tuberculatus* populations in the last 20 years (Heap, 2020). These herbicide-resistant traits include resistance to multiple herbicide sites of action caused by a variety of mechanisms including point mutation and metabolic resistance. The acetolactate synthase (ALS) inhibitor resistance is widespread and is generally conferred by target-site mutations in *A. palmeri* (Kohrt and Sprague, 2017; Küpper et al., 2017), *A. tuberculatus* (Chahal et al., 2015; Schultz et al., 2015), and *A. retroflexus* (Nandula et al., 2020). Resistance to protoporphyrinogen oxidase (PPO) inhibitors is also caused by target-site changes, conferred either by amino acid deletion (Lillie et al., 2019; Nie et al., 2019; Salas et al., 2016; Sarangi et al., 2019) or substitution (Nie et al., 2019; Salas-Perez et al., 2017; Varanasi et al., 2018).

Some of the recent reports suggested that cytochrome P450- or GST-driven metabolic resistance was found in *Amaranthus*. This includes resistance to 4-hydroxyphenylpyruvate dioxygenase (HPPD) inhibitors (Kaundun et al., 2017; Oliveira et al., 2018a,b) and Photosystem-II inhibitors (Huffman et al., 2015; Peterson et al., 2017; Vennapusa et al., 2018). Resistance to synthetic auxinic herbicides (such as dicamba and 2,4-D) is also metabolically mediated (Figueiredo et al., 2018; Shyam et al., 2019), although an *A. palmeri* population that was resistant to 2,4-D did not show resistance to Dicamba (Kumar et al., 2019).

Glyphosate, which inhibits the activity of 5-enolpyruvylshikimate-3-phosphate synthase (EPSPS) through competitive inhibition, is the most widely used herbicide worldwide. The resistance mechanism of *A. palmeri* to glyphosate is notable, as this weed has developed resistance to glyphosate through the duplication of the *EPSPS* gene and overexpression of EPSPS enzyme (Gaines et al., 2010). *Amaranthus tuberculatus* displays both *EPSPS* gene duplication (Lorentz et al., 2014) and target-site Pro-106-Ser substitution (Bell et al., 2013; Molin et al., 2016; Nandula et al., 2013). Reduced glyphosate absorption and translocation have also been reported to impart nontarget site resistance to glyphosate in *A. palmeri* (Dominguez-Valenzuela et al., 2017; Nandula et al., 2012).

Genetics, interspecific hybridization, and gene flow

The chromosome count of *Amaranthus* genus varies by species, where $2n$ = 32 or 34 (Murray, 1940). Estimates of genome size within the *Amaranthus* genus previously ranged from 930 Mb in *A. palmeri* to 1.39 Gb in *A. tuberculatus* (Rayburn et al., 2005), but more recent studies estimated the genome to be around half this size. *Amaranthus hybridus* has a genome size of 504 Mb (Montgomery et al., 2020), whereas *Amaranthus hypochondriacus* L.

(Prince-of-Wales Feather) has a genome size of 466 Mb with an estimated 48% of the genome being repeated elements (Clouse et al., 2016). The two important agronomic weed species fall on opposite ends of the spectrum when it comes to genome size: *A. palmeri* has one of the smallest genomes in this genus, estimated at only 422 Mb, whereas *A. tuberculatus* is the largest, at 675 Mb (Kreiner et al., 2019; Montgomery et al., 2020).

Hybridization and gene flow between *Amaranthus* species have been observed, both in greenhouse and field studies. *Amaranthus tuberculatus* is reported to form hybrids with *A. palmeri* (Steinau et al., 2003; Trucco et al., 2007; Wetzel et al., 1999), *A. hybridus* (Tranel et al., 2002; Trucco et al., 2005a,b), *A. powellii*, *A. retroflexus*, and *A. spinosus* (Murray, 1940). *Amaranthus palmeri and A. spinosus* are close relatives, sharing many genetic similarities including chromosome count ($2n = 34$) and genome size (Rayburn et al., 2005; Wassom and Tranel, 2005), so it is unsurprising that they can also form hybrids (Gaines et al., 2012; Molin et al., 2016). The interspecific hybridization success rate of *Amaranthus* spp. is low and varies by species, with *A. tuberculatus* × *A. hybridus* rates estimated under field conditions from 0.4% to 2.3% (Trucco et al., 2005b) whereas *A. palmeri* × *A. spinosus* rates were $< 1.4\%$ (Gaines et al., 2012).

A major concern is that their ability to form hybrids offers the potential for herbicide resistance traits to be transferred between species through introgression (Jhala et al., 2021). Observed examples of this include a Trp574Leu substitution in *A. palmeri* that confers resistance to ALS-inhibitors being transferred to an *A. spinosus x A. palmeri* hybrid (Molin et al., 2016). In a greenhouse study, Franssen et al. (2001b) reported that gene flow occurred between *A. tuberculatus* and *A. palmeri* at a low frequency and that transferred the ALS resistance traits. Oliveira et al. (2018b) conducted field experiments in Nebraska, United States, to determine interspecific and intraspecific transfer of metabolism-based HPPD resistance in dioecious *Amaranthus* species. In this study, gene flow was detected using a resistant *A. tuberculatus* population with enhanced mesotrione (a HPPD herbicide) detoxification via cytochrome P450 enzymes and the results showed that 0.1% hybridization between *A. tuberculatus* and *A. palmeri* occurred under field conditions.

Description of major weedy *Amaranthus* spp.

Amaranthus palmeri

Native to the southwestern United States, *A. palmeri* has been ranked as the most troublesome weed in agronomic cropping systems in the United States (Ward et al., 2013). In the last few decades, human activities such as agricultural development and seed and equipment transportation have led to the spread of *A. palmeri* to the northern United States (Culpepper et al., 2006). Since the first report of *A. palmeri* in Virginia in 1915, it has become one of the most problematic weeds in agronomic cropping systems in the United States (Chahal et al., 2017).

Growth and seed production

Amaranthus palmeri has a high photosynthetic rate leading to a high growth rate and considerable biomass accumulation compared with other *Amaranthus* species (Ehleringer, 1983; Jha and Norsworthy, 2009). *Amaranthus palmeri* leaves are oriented in a perpendicular fashion to incoming sunlight (Fig. 2.1), allowing the plant to absorb maximum light to fix the carbon at a faster rate (Ehleringer and Forseth, 1980). *Amaranthus palmeri* is a prolific seed producer (Keeley et al., 1987) and has been found to survive and produce a high number of seeds even under moisture-stressed situation (Chahal et al., 2018). Keeley et al. (1987) reported that a female *A. palmeri* plant can produce 200,000–600,000 seeds. Schwartz et al. (2016) reported that Palmer amaranth growing in soybean [*Glycine max* (L.) Merr.] field can produce as high as 800,000 seeds per female plant. Seed production of *A. palmeri* grown without crop competition can exceed one million seeds per plant (Keeley et al., 1987). *Amaranthus palmeri* seeds are usually dispersed by gravity forces; however, dispersal via irrigation, plowing, mowing, harvesting, birds, and mammals has been documented in multiple studies (Costea et al., 2004, 2005).

Competitiveness and yield loss

Continuous emergence pattern of *A. palmeri* throughout the growing season leads to season-long crop interference (Jha and Norsworthy, 2009). *Amaranthus palmeri* interference can cause substantial yield losses in agronomic crops (Fig. 2.2). Uncontrolled *A. palmeri* has caused yield losses as

FIGURE 2.2 Herbicide-resistant *A. palmeri* and *A. tuberculatus* competing with soybean in a production field in Nebraska, United States.

high as 91% in corn (*Zea mays* L.) (Massinga et al., 2001), 54% in cotton (Morgan et al., 2001), and 68% in soybean (Klingaman and Oliver, 1994). Bensch et al. (2003) reported 79% soybean yield loss at a density of 8 plants/m^2 in Kansas, United States. Early emergence (0–1 week after soybean emergence) and establishment of *A. palmeri* have been shown to reduce soybean yield compared to late emergence (2–8 weeks after soybean emergence) (Korres et al., 2019). *Amaranthus palmeri* at 1.58 plants/m^2 resulted in 38%–63% grain yield loss in sorghum (*Sorghum bicolor* L.) (Moore et al., 2004). In a study conducted at North Carolina, United States, Burke et al. (2007) reported that *A. palmeri* decreased the canopy diameter of peanut (*Arachis hypogaea* L.) and caused 28% and 68% yield loss at densities of 1 and 5 plants/m of peanut row, respectively, and interfere with the peanut harvesting.

A recent multilocation/year study conducted in Nebraska, United States, reported that the critical period of weed (including *A. palmeri*) removal in soybean was delayed from V1-V2 to V4-R5 using preemergence (PRE) herbicides such as saflufenacil + imazethapyr + pyroxasulfone and saflufenacil + imazethapyr (Knezevic et al., 2019). However, in the absence of a PRE herbicide, the critical time of *A. palmeri* removal evaluated at 5% soybean yield loss occurred at V1 and V6 soybean growth stages that was equivalent to the 194 and 480 growing degrees days (GDD) in 2018 and 2019, respectively (de Sanctis et al., 2021).

Reproductive biology

Amaranthus palmeri is a dioecious species; male (pistillate) and female (staminate) flowers are produced on separate plants (Mosyakin and Robertson, 2008). Male inflorescences can be distinguished from the female flower as they lack the sharp, stiff bracts that subtend female flowers and thus have a softer feel (Mosyakin and Robertson, 2008). Flowering in *A. palmeri* usually starts in summer and continues into fall. Keeley et al. (1987) reported that flowering of *A. palmeri* is influenced by day length; shorter days hasten flowering, which in turn, reduces plant size and number of seeds. Male *A. palmeri* plants produce wind-dispersed pollen grains (Borsch, 1998; Franssen et al., 2001a) and pollen-mediated gene flow from herbicide-resistant to susceptible *A. palmeri* has been reported under field conditions (Sosnoskie et al., 2012) thus the transfer of herbicide resistance alleles (Jhala et al., 2021).

Herbicide resistance

Glyphosate-resistant *A. palmeri* was first reported in Georgia (Culpepper et al., 2006), and since then, 27 other states in the United States have documented the presence of this weed as of 2020 (Heap, 2020). *Amaranthus palmeri* has evolved resistance to at least eight herbicide sites of action: microtubule-, ALS-, EPSPS-, PS II-, HPPD-, PPO-, and very long-chain fatty acid (VLCFA) inhibitors and synthetic auxins (Heap, 2020). *Amaranthus palmeri* has also been found to have multiple herbicide resistance, such as ALS- and EPSPS inhibitor resistance (Chahal et al., 2017); PS II/HPPD inhibitor resistance in Nebraska, United States (Jhala et al., 2014), as well as 2,4-D/glyphosate/atrazine/ALS inhibitor /HPPD inhibitor resistance in Kansas, United States (Kumar et al., 2019). The evolution of herbicide-resistant *A. palmeri* populations and their widespread occurrence has made the control of this weed difficult in agronomic crops.

Amaranthus tuberculatus

Amaranthus tuberculatus, a summer annual broadleaf weed, is a native plant of North America and considered to be one of the most troublesome agricultural weeds in the midwestern United States (Waselkov et al., 2020). Historically, this weed was commonly found along the riverbanks and pond margins (Sauer, 1957), but the changing cultivation practices in the 20th century, especially the adoption of conservation tillage and intensification of agriculture are believed to promote the invasion of *A. tuberculatus* in the agricultural lands in the United States (Buhler, 1995). Research conducted in southeast Texas, United States, reported that 37-years of no-till management increased *A. tuberculatus* density by four-fold compared to the conventional-tillage system (Govindasamy et al., 2020). Since its first report in 1830 in Oklahoma, this weed is continuously moving toward the northern and eastern parts of the United States and Canada (Korres et al., 2018; Sarangi and Jhala, 2018; Schryver et al., 2017).

Seed germination

Amaranthus tuberculatus is a small-seeded weed with high initial seed viability ($>89\%$) and likely to form a persistent seedbank (Korres et al., 2018; Leon et al., 2004). Fig. 2.3 is showing the relative seed size of *A. tuberculatus* in comparison with corn and soybean. In a study conducted in Nebraska, United States, Burnside et al. (1996) reported that 3% *A. tuberculatus* seeds were recovered after 17 years of burial at 20 cm soil depth. In another study, Buhler and Hartzler (2001) reported that 12% of *A. tuberculatus* seeds persisted after 4 years of burial. The size of seedbank, persistence, and emergence of *A. tuberculatus* can be impacted by several factors including tillage, weed management practices, cropping systems, and other environmental factors. For example, a study conducted in Illinois, United States showed that tillage increased the persistence of *A. tuberculatus* seedbank in the top 6 cm soil profile (Steckel et al., 2007). Refsell and Hartzler (2009) reported that in the no-till system, the seeds of *A. tuberculatus* were concentrated near the soil surface, but the majority of the seeds were found between 9 and 15 cm in the chisel-till system.

FIGURE 2.3 Relative seed size of *A. tuberculatus* in comparison with soybean and corn seeds. Provided scale is in centimeter.

Amaranthus tuberculatus is considered a late-emerging summer annual weed, and it starts emerging from mid-May in the midwestern United States (Werle et al., 2014). In a study conducted in Iowa, United States, *A. tuberculatus* continued to emerge until mid-August in no-till systems (Refsell and Hartzler, 2009). Leon and Owen (2006) reported that seedling emergence in no-till was at least four times higher than chisel and moldboard plow systems. Tillage can also promote the early emergence of *A. tuberculatus*; Govindasamy et al. (2020) reported that in conventional-tillage system, 63 GDD was required for 50% emergence of this weed, whereas 75 GDD was needed under no-till system in Texas, United States. The minimum temperature required for *A. tuberculatus* germination was 10°C, and a greater germination was observed with alternating temperatures than constant temperature (Leon et al., 2004). Similarly, Guo and Al-Khatib (2003) reported that *A. tuberculatus* germination was highest at 25/20°C day/night temperature. The dormancy of *A. tuberculatus* was phytochrome regulated and germination was promoted under continuous red light, whereas far-red light inhibited its germination (Leon and Owen, 2003).

Growth, flowering, and seed production

Amaranthus tuberculatus has an indeterminate growth habit; however, the plant morphology depends on the environmental conditions. Sarangi et al. (2016) reported that water stress can reduce plant height, number of leaves, growth index, and seed production of *A. tuberculatus* substantially. The 50% field capacity of water treatment showed 1.4- and 2.9-times less plant height and growth index, respectively, than the no water stress (100% field capacity). Guo and Al-Khatib (2003) reported that *A. tuberculatus* showed a higher heat tolerance than *A. retroflexus* and survived at 45/40°C day/night temperature for 12 days. Being a C4 weed, *A. tuberculatus* shows some physiological characteristics, including the high photosynthetic rate at high temperature, low CO_2 compensation point, and a greater water and nitrogen use efficiency, which are proven to be beneficial for this weed to survive (Costea et al., 2005).

Similar to *A. palmeri*, *A. tuberculatus* is also a dioecious and wind-pollinated species. The shape of the pollen grains is spherical, and the pollens are very small in size (18.5 μm diameter), which has favored the long-distance travel of *A. tubeculatus* pollen by wind (Franssen et al., 2001a). Also, the pollens of this weed showed lower settling velocity (0.0185–0.021 m/s) and can remain viable for 120 hours (Costea et al., 2005; Liu et al., 2012). Sarangi et al. (2017b) estimated the pollen-mediated gene flow of *A. tuberculatus* under field conditions and concluded that 50% gene flow occurred within 3 m from the pollen source, but 10% outcrossing was recorded at 88 m from the source.

Similar to other *Amaranthus* spp., *A. tubeculatus* shows a short-day flowering response. The flowering is initiated earlier under short-day conditions (Costea et al., 2005). Shading can also negatively impact *A. tubeculatus* growth and seed production. A study conducted in Illinois, United States, showed that 40%, 68%, and 99% shading reduced the seed production by 51%, 75%, and >99%, respectively, and the trend was similar to the biomass and plant height data (Steckel et al., 2003). The fecundity of *A. tubeculatus* is highly variable and depends on many factors including competition, emergence time, management practices, soil type, and other environmental factors. *Amaranthus tubeculatus* emerging with soybean can produce 300,000 to 2.3 million seeds per female plant (Hartzler et al., 2004). Emergence timing can also impact the fecundity; Wu and Owen (2014) reported that *A. tubeculatus* plants that emerged in early-July produced 75% fewer seeds than the plants emerged in May in Iowa, United States. Seed production of *A. tuberculatus* was also reduced by 70% when the soil water content was reduced by 50% (Sarangi et al., 2016).

Amaranthus tubeculatus seeds can disperse by water, farm machinery, manure and compost, wind, birds, and other animals. In a 2-year study in the United States, the viable seeds of *A. palmeri* and *A. tuberculatus* were recovered from the digestive tracts of the migratory waterfowls (birds) after 38 hours of feeding and this study also concluded that the seeds can potentially be dispersed up to 2900 km from the source by these birds (Farmer et al., 2017). Therefore it is considered that the seed dispersal of *A. tubeculatus* played an important role in spreading herbicide-resistant populations in North America.

Competitiveness and yield reduction

Similar to *A. palmeri*, *A. tubeculatus* is also considered a competitive weed that can cause total crop loss when growing at a high density. Interference of

A. tubeculatus began at the V6 corn stage caused 4%–23% yield reduction in a study conducted in Illinois, United States (Steckel et al., 2004). The same study also reported that the season-long competition of *A. tubeculatus* at a density of 310 plants/m^2 caused 74% yield loss in corn. Allowing *A. tubeculatus* to interfere with soybean at an average density of 199 plants/m^2 for 10 weeks after emergence caused 43% yield loss (Hager et al., 2002). Bensch et al. (2003) reported that *A. tubeculatus* at a density of 0.25–4 plants/m soybean row was less competitive than *A. palmeri* but caused greater yield reduction than *A. retroflexus* at the same density.

Herbicide resistance

As of 2020, the presence of herbicide-resistant *A. tuberculatus* has been reported from 19 states in the United States and Ontario, Canada (Heap 2020). Population resistant to ALS-, EPSPS-, and PS II-inhibiting herbicides are widespread in the midwestern and southern United States (Murphy et al., 2019; Rosenbaum and Bradley, 2013; Sarangi et al., 2015; Schultz et al., 2015; Singh et al., 2020; Vieira et al., 2018). Moreover, HPPD-, and PPO-inhibiting herbicide-resistant *A. tuberculatus* populations are spreading quickly across the midwestern United States (Hausman et al., 2011; Lee et al., 2008; McMullan and Green, 2011; Oliveira et al., 2017; Sarangi et al., 2019; Thinglum et al., 2011). An *A. tuberculatus* population from Missouri, United States, showed resistance to six herbicides (2,4-D, atrazine, chlorimuron, fomesafen, and mesotrione) from six different sites-of-action groups (Shergill et al., 2018). The use of soil residual herbicides from VLCFA inhibitor group was recommended for season-long control of herbicide-resistant *A. tuberculatus* population. As these herbicides interact with many primary target sites, the chance of evolving resistance to VLCFA inhibitors was low to none (Busi, 2014); however, some of the recent research reported that *A. tuberculatus* populations in the United States are also showing resistance to these herbicides (Evans et al., 2019).

Other weedy *Amaranthus* spp.

Amaranthus retroflexus, *A. spinosus*, and *A. hybridus* are the other important weedy *Amaranthus* species that can compete with the summer crops and reduce yield substantially. *Amaranthus retroflexus* is found throughout the northern, eastern, and central United States, northeastern Mexico, and southeastern Canada (Sauer, 1967). It is also present in Africa, Asia, and Europe (Brenan, 1961; Holm et al., 1991a,b; Wells et al., 1986). Moreover, it has naturalized to the newer environments of northern and southern hemispheres including a temperate climate. The tropical America is believed to be the native place for *A. spinosus*, although it was introduced later to other places in the world (Holm et al., 1991b). *Amaranthus hybridus* was historically found on the riverbanks in the eastern North America, northern South America, and Central America (Sauer, 1967); however, due to its use as a leafy vegetable, it is now found in the south-central Asia, Africa, and Australia (Costea et al., 2001).

Seed germination

Amaranthus retroflexus, *A. spinosus*, and *A. hybridus* are monoecious plants and have both male and female flowers on the same plant (Weaver and McWilliams, 1980). On an average, they can produce 100,000–300,000 seeds per plant (Sellers et al., 2003). Similar to *A. palmeri* and *A. tuberculatus*, these three species (*A. retroflexus*, *A. spinosus*, and *A. hybridus*) have the smaller seed size and they require 350 GDD (10°C base temperature) for germination and can emerge throughout the growing season (Sellers et al., 2003). Moreover, these weeds start emerging when the soil temperature reaches 20°C (Barton, 1962), however, emergence pattern may vary depending on the species and environmental condition (Sellers et al., 2003). Studies reported that optimum temperature for *A. retroflexus* germination ranged from 25°C–40°C (Kępczyński and Bihun, 2002; Guo and Al-Khatib, 2003), whereas it was 30°C–35°C for *A. spinosus* (Steckel et al., 2004) and 32°C–34°C for *A. hybridus* (Kępczyński and Bihun, 2002).

Soil type and burial depth also impacted the seed germination of these weeds. For example, the germination of *A. retroflexus* was better in the lighter soils (sandy loam, sandy clay loam, and loamy sand) than the heavier soils (silty clay, and sandy clay) (Ghorbani et al., 1999). Burial depth below 5 cm soil can reduce the germination percentage. Ghorbani et al. (1999) observed no seed germination for *A. retroflexus* beyond 3 cm soil depth. Similarly, *A. spinosus* germinated well up to 4 cm burial depth, but no seed was germinated when placed below 6 cm depth (Chauhan and Johnson, 2009). The seedbank persistence and size can be influenced by the cropping system and tillage practice. Cardina et al. (2002) reported that in the no-till system, the size of *A. retroflexus* seedbank was two times greater than the moldboard plow. The same study also reported that the corn-oats-hay rotation had greater seed density (260–630 seeds/m^2) than the corn-soybean rotation (10–270 seeds/m^2).

Herbicide resistance

Amaranthus retroflexus, *A. spinosus*, and *A. hybridus* have developed resistance to multiple herbicide sites-of-action, causing their management difficult. *Amaranthus retroflexus* has evolved resistance to ALS-, PS II-, and PPO-inhibiting herbicides (Heap, 2020; Wang et al., 2017; Woodyard et al., 2009). Likewise, *A. hybridus* has evolved resistance to ALS-, EPSPS-, PS II-, and PPO-inhibiting herbicides and synthetic auxins (Fuerst et al., 1986; Heap, 2020; Manley et al., 1999; Nie et al., 2019).

Competitiveness and yield loss

Amaranthus retroflexus, *A. spinosus*, and *A. hybridus* are troublesome weeds in more than 25 crops worldwide (Berry et al., 2006; Holm et al., 1991b). These weeds can emerge before or at the same time of the crop emergence and compete for resources such as space, water, and nutrients. Depending on the crop, weed density, emergence time, and duration of competition, these weeds can cause 5%–100% yield loss. *Amaranthus hybridus* at 6 plants/m^2 density caused 100% yield reduction in watermelon (*Citrullus vulgaris* Schrad.) and muskmelon (*Cucumis melo* L.) when allowed to compete for the whole season (Terry et al., 1997). In another study, Stuart et al. (1984) observed 43% cotton yield loss when the density of *A. hybridus* was 0.7 plants/m^2. Morales-Payan and Stall (2002a, 2002b) reported 30% and 61% yield loss in bell pepper (*Capsicum annuum* L.) at 1 and 6 plants/m^2 of *A. hybridus* density, respectively. A study conducted in Kansas, United States, reported that *A. retroflexus* at 12 plants/m^2 density caused 40% yield loss in sorghum (Knezevic et al., 1997). Dielman et al. (1995) also reported a 12.3% yield loss of soybean with two *A. retroflexus* plants/m^2. Depending on the weed density and emergence timing, *A. retroflexus* reduced corn yield up to 34% in a study conducted in Ontario, Canada (Knezevic et al., 1994).

Management of weedy *Amaranthus*

Chemical control of *Amaranthus* weeds

Due to having a wide window of emergence, a single herbicide application would not provide season-long control of *Amaranthus* spp. including *A. palmeri* and *A. tuberculatus*. Therefore management programs in agronomic crops should focus on clean start of the growing season with the application of a PRE residual herbicide with multiple sites of action followed by an effective postemergence (POST) herbicide application. Sarangi et al. (2017a) in a field experiment conducted in Nebraska, United States, reported that PRE followed by POST herbicide programs that include saflufenacil plus imazethapyr plus dimethenamid-P, sulfentrazone plus cloransulam, or *S*-metolachlor plus metribuzin as PRE, followed by fomesafen plus glyphosate, *S*-metolachlor plus fomesafen, or acifluorfen plus glyphosate as POST resulted in more than 90% control of glyphosate-resistant *A. tuberculatus* throughout the growing season, whereas, POST-only herbicide treatments had less than 70% control of glyphosate-resistant *A. tuberculatus*.

Mixing foliar-active POST herbicides with a residual herbicide would provide an overlapping residual activity to prevent *Amaranthus* spp. emergence later in the growing season; however, the presence of sufficient soil moisture is critical for the activity of soil residual herbicides. Sarangi and Jhala (2019) conducted a field experiment to compare PRE, PRE followed by POST, and PRE followed by POST plus residual herbicide programs for *A. palmeri* control and soybean yield, and reported that PRE herbicide showed 66% control of *A. palmeri*, whereas 86% and 97% control was obtained with the PRE followed by POST, and PRE followed by POST plus residual herbicide programs, respectively, at 28 days after POST herbicide application, and PRE followed by POST plus residual herbicide program resulted in higher soybean yield than other treatments. Aulakh and Jhala (2015) reported similar results with PRE followed by POST herbicide programs, where POST program includes foliar-active herbicides tank-mixed with acetochlor, pyroxasulfone, or *S*-metolachlor in glufosinate-resistant soybean. Multiple herbicide-tolerant corn, soybean, and cotton were launched recently, and they will provide additional herbicide choices for controlling the glyphosate-resistant *Amaranthus* species; however, growers should not rely only on 2,4-D choline, dicamba, or glufosinate or their sequential applications for POST weed control program (Shyam et al., 2021).

Need for integrated weed management

"If herbicide-resistant-weed problems are addressed only with herbicides, evolution will most likely win"

by Mortensen et al. (2012)

Overreliance on a single herbicide site of action selected the resistant populations of *Amaranthus*. The introduction of the genetically modified glyphosate-tolerant crops promoted the rapid spread of glyphosate-resistant weed populations in the world. Biotechnology companies are currently promoting multiple herbicide-tolerant crops as the solution to glyphosate-resistant weed issues; however, this can create new herbicide-resistant weed challenges if proper precautions are not taken. Herbicide-resistant weeds are not only a threat to global food security but also can negatively impact environmental quality by increasing the herbicide application. Therefore the integrated weed management approach should be taken to reduce the selection pressure, decrease herbicide use, and preserve the finite herbicide choices.

Herbicide-dependent weed control treatments sometime fall short of providing satisfactory control of *Amaranthus* spp. A late-season survey of cotton fields in Texas, United States, revealed that *A. palmeri* and *A. tuberculatus* that escaped the standard herbicide programs could add up to 14 million seeds/ha to seedbank (Werner et al., 2020). Therefore the successful adoption of a multitactic weed management approach that includes physical, mechanical, and biological control of weeds along with the herbicides is the need for better weed management (Bagavathiannan and Davis, 2018). For example, Price et al. (2016) concluded that *A. palmeri* density reduced exponentially as a function of cover crop biomass, and a deep tillage along with cover crops increased *A. palmeri* control substantially. Similarly, a study conducted in Georgia, United States, showed that winter cover crops with high biomass production ability suppressed *A. palmeri* germination early in the season, providing a slightly longer window for POST herbicide application (Webster et al., 2013).

Harvest weed seed control (HWSC) comprises a set of nonchemical weed management tactics that can destroy or remove weed seeds at crop harvest and prevent the addition of seeds to the seedbank. Some of these HWSC tactics are proven to be effective in depleting the *Amaranthus* spp. seedbank (Shergill et al., 2020b). Studies conducted in the United States reported that the impact mill can destroy 96.2% and 92.4% *A. tuberculatus* and *A. hybridus* seeds, respectively (Shergill et al., 2020a). Another research conducted in Arkansas, United States, showed that a heat index of 22,600 in narrow-windrow burning was enough to kill *A. palmeri* seeds by 100% and narrow-windrow burning or chaff removal integrated with an effective herbicide program reduced *A. palmeri* density and seedbank size significantly (Norsworthy et al., 2016, 2020).

References

Assad, R., Reshi, Z.A., Jan, S., Rashid, I., 2017. Biology of Amaranths. Bot. Rev. 83, 382–436.

Aulakh, J.S., Jhala, A.J., 2015. Comparison of glufosinate-based herbicide programs for broad-spectrum weed control in glufosinate-resistant soybean. Weed Technol. 29, 419–430.

Bagavathiannan, M.V., Davis, A.S., 2018. An ecological perspective on managing weeds during the great selection for herbicide resistance. Pest. Manag. Sci. 74, 2277–2286.

Barton, L.V., 1962. The germination of weed seeds. Weeds. 10, 174–182.

Becker, R.E., Wheeler, E.L., Lorentz, K., Stafford, A.E., Grosjean, O.K., Betschart, A.A., et al., 1981. A compositional study of Amaranth grain. J. Food Sci. 46, 1175–1180.

Bell, M.S., Hager, A.G., Tranel, P.J., 2013. Multiple resistance to herbicides from four site-of-action groups in waterhemp (*Amaranthus tuberculatus*). Weed Sci. 61, 460–468.

Bensch, C.N., Horak, M.J., Peterson, D., 2003. Interference of redroot pigweed (*Amaranthus retroflexus*), Palmer amaranth (*A. palmeri*), and common waterhemp (*A. rudis*) in soybean. Weed Sci. 51, 37–43.

Berry, A.D., Stall, W.M., Rathinasabapathi, B., Macdonald, G.E., Charudattan, R., 2006. Smooth pigweed (*Amaranthus hybridus* L.) and livid amaranth (*Amaranthus lividus*) interference with cucumber (*Cucumis sativus*). Weed Technol. 20, 227–231.

Borsch, T., 1998. Pollen types in the Amaranthaceae: morphology and evolutionary significance. Grana. 37, 129–142.

Brenan, J.P.M., 1961. *Amaranthus* in Britain. Watsonia. 4, 261–280.

Buhler, D.D., 1995. Influence of tillage systems on weed population dynamics and management in corn and soybean in the Central USA. Crop. Sci. 35, 1247–1258.

Buhler, D.D., Hartzler, R.G., 2001. Emergence and persistence of seed of velvetleaf, common waterhemp, woolly cupgrass, and giant foxtail. Weed Sci. 49, 230–235.

Burke, I.C., Schroeder, M., Thomas, W.E., Wilcut, J.W., 2007. Palmer amaranth interference and seed production in peanut. Weed Technol. 21, 367–371.

Burnside, O.C., Wilson, R.G., Weisberg, S., Hubbard, K.G., 1996. Seed longevity of 41 weed species buried 17 years in eastern and western Nebraska. Weed Sci. 44, 74–86.

Busi, R., 2014. Resistance to herbicides inhibiting the biosynthesis of very-long-chain fatty acids. Pest. Manag. Sci. 70, 1378–1384.

Cardina, J., Herms, C.P., Doohan, D.J., 2002. Crop rotation and tillage system effects on weed seedbanks. Weed Sci. 50, 448–460.

Chahal, P.S., Aulakh, J.S., Jugulam, M., Jhala, A.J., 2015. Herbicide-resistant Palmer amaranth (*Amaranthus palmeri S. Wats.*) in the United States—mechanisms of resistance, impact, and management. In: Price, A.J., Kelton, J.A., Sarunaite, L. (Eds.), Herbicides, Agronomic Crops, and Weed Biology. InTechOpen, pp. 1–40.

Chahal, P.S., Varanasi, V.K., Jugulam, M., Jhala, A.J., 2017. Glyphosate-resistant Palmer amaranth (*Amaranthus palmeri*) in Nebraska: Confirmation, *EPSPS* gene amplification, and response to post corn and soybean herbicides. Weed Technol. 31, 80–93.

Chahal, P.S., Irmak, S., Jugulam, M., Jhala, A.J., 2018. Evaluating effect of degree of water stress on growth and fecundity of Palmer amaranth (*Amaranthus palmeri*) using soil moisture sensors. Weed Sci. 66, 738–745.

Chauhan, B.S., Johnson, D.E., 2009. Germination ecology of spiny (*Amaranthus spinosus*) and slender amaranth (*A. viridis*): troublesome weeds of direct-seeded rice. Weed Sci. 57, 379–385.

Clouse, J.W., Adhikary, D., Page, J.T., Ramaraj, T., Deyholos, M.K., Udall, J.A., et al., 2016. The amaranth genome: genome, transcriptome, and physical map assembly. Plant. Genome. 9, 1–14.

Costea, M., Sanders, A., Waines, G., 2001. Priliminary results toward a revision of the *Amaranthus hybridus* species complex (Amaranthaceae). SIDA. 19, 931–974.

Costea, M., Weaver, S.E., Tardif, F.J., 2004. The biology of Canadian weeds. 130. *Amaranthus retroflexus* L., *A. powellii* S. Watson and *A. hybridus* L. Can. J. Plant. Sci. 84, 631–668.

Costea, M., Weaver, S.E., Tardif, F.J., 2005. The biology of invasive alien plants in Canada. 3. *Amaranthus tuberculatus* (Moq.) Sauer var. *rudis* (Sauer) Costea & Tardif. Can. J. Plant. Sci. 85, 507–522.

Culpepper, A.S., Grey, T.L., Vencill, W.K., Kichler, J.M., Webster, T.M., Brown, S.M., et al., 2006. Glyphosate-resistant Palmer amaranth (*Amaranthus palmeri*) confirmed in Georgia. Weed Sci. 54, 620–626.

de Sanctis, J.H., Knezevic, S.Z., Kumar, V., Jhala, A.J., 2021. Residual herbicides affects critical time of Palmer amaranth (*Amaranthus palmeri* S. Watson) removal in soybean in southcentral Nebraska. Agron. J. (in press).

Dielman, A., Hamill, A.S., Weise, S.F., Swanton, C.J., 1995. Empirical models of redroot redroot pigweed (*Amaranthus* spp.) interference in soybean (*Glycine max*). Weed Sci. 43, 612–618.

Dominguez-Valenzuela, J.A., Gherekhloo, J., Fernández-Moreno, P.T., Cruz-Hipolito, H.E., Alcántara-de la Cruz, R., Eduardo Sánchez-González, E., et al., 2017. First confirmation and characterization of target and non-target site resistance to glyphosate in Palmer amaranth (*Amaranthus palmeri*) from Mexico. Plant. Physiol. Bioch. 115, 212–218.

Ehleringer, J., 1983. Ecophysiology of *Amaranthus palmeri*, a Sonoran Desert summer annual. Oecologia. 57, 107–112.

Ehleringer, J., Forseth, I., 1980. Solar tracking by plants. Science. 210, 1094–1098.

Evans, C.M., Strom, S.A., Riechers, D.E., Davis, A.S., Tranel, P.J., Hager, A.G., 2019. Characterization of a waterhemp (*Amaranthus tuberculatus*) population from Illinois resistant to herbicides from five site-of-action groups. Weed Technol. 33, 400–410.

Farmer, J.A., Webb, E.B., Pierce, R.A., Bradley, K.W., 2017. Evaluating the potential for weed seed dispersal based on waterfowl consumption and seed viability. Pest. Manag. Sci. 73, 2592–2603.

Figueiredo, M.R.A., Leibhart, L.J., Reicher, Z.J., Tranel, P.J., Nissen, S.J., Westra, P., et al., 2018. Enhanced 2,4-D metabolism in *A. tuberculatus*. Pest. Manag. Sci. 74, 2356–2362.

Franssen, A.S., Skinner, D.Z., Al-Khatib, K., Horak, M.J., 2001a. Pollen morphological differences in *Amaranthus* species and interspecific hybrids. Weed Sci. 49, 732–737.

Franssen, A.S., Skinner, D.Z., Al-Khatib, K., Horak, M.J., Kulakow, P.A., 2001b. Interspecific hybridization and gene flow of ALS resistance in *Amaranthus* species. Weed Sci. 49, 598–606.

Fuerst, E.P., Arntzen, C.J., Pfister, K., Penner, D., 1986. Herbicide cross-resistance in triazine-resistant biotypes of four species. Weed Sci. 34, 344–353.

Gaines, T.A., Zhang, W., Wang, D., Bukun, B., Chisholm, S.T., Shaner, D.L., et al., 2010. Gene amplification confers glyphosate resistance in *Amaranthus palmeri*. Proc. Natl. Acad. Sci. 107, 1029–1034.

Gaines, T.A., Ward, S.M., Bekun, B., Preston, C., Leach, J.E., Westra, P., 2012. Interspecific hybridization transfers a previously unknown glyphosate resistance mechanism in *Amaranthus* species. Evol. Applic. 5, 29–38.

Ghorbani, R., Seel, W., Leifert, C., 1999. Effects of environmental factors on germination and emergence of *Amaranthus retroflexus*. Weed Sci. 47, 505–510.

Govindasamy, P., Sarangi, D., Provin, T., Hons, F., Bagavathiannan, M., 2020. No-tillage altered weed species dynamics in a long-term (36-year) grain sorghum experiment in southeast Texas. Weed Sci. 68, 476–484.

Guo, P., Al-Khatib, K., 2003. Temperature effects on germination and growth of redroot pigweed (*Amaranthus retroflexus*), Palmer amaranth (*A. palmeri*), and common waterhemp (*A. rudis*). Weed Sci. 51, 869–875.

Hager, A.G., Wax, L.M., Stoller, E.W., Bollero, G.A., 2002. Common waterhemp (*Amaranthus rudis*) interference in soybean. Weed Sci. 50, 607–610.

Hartzler, R.G., Battles, B.A., Nordby, D., 2004. Effect of common waterhemp (*Amaranthus rudis*) emergence date on growth and fecundity in soybean. Weed Sci. 52, 242–245.

Hausman, N.E., Singh, S., Tranel, P.J., Riechers, D.E., Kaundun, S.S., Polge, N.D., et al., 2011. Resistance to HPPD-inhibiting herbicides in a population of waterhemp (*Amaranthus tuberculatus*) from Illinois, United States. Pest. Manag. Sci. 67, 258–261.

Heap, I., 2020. The international herbicide-resistant weed database. http://weedscience.org/Home.aspx (Accessed December 15, 2020).

Holm, L.G., Pancho, J.V., Herberger, J.P., Plucknett, D.L., 1991a. A geographic atlas of world weeds. Krieger Publishing Co, Malabar, Florida, USA, p. 391.

Holm, L.G., Plucknett, D.L., Pancho, J.V., Herberger, J.P., 1991b. The World's Worst Weeds: Distribution and Biology. The University Press of Hawaii, Malabar, FL, p. 609.

Huffman, J., Hausman, N.E., Hager, A.G., Riechers, D.E., Tranel, P.J., 2015. Genetics and inheritance of nontarget-site resistances to atrazine and mesotrione in a waterhemp (*Amaranthus tuberculatus*) population from Illinois. Weed Sci. 63, 799–809.

Jha, P., Norsworthy, J.K., 2009. Soybean canopy and tillage effects on emergence of Palmer amaranth (*Amaranthus palmeri*) from a natural seed bank. Weed Sci. 57, 644–651.

Jhala, A.J., Sandell, L.D., Rana, N., Kruger, G.R., Knezevic, S.Z., 2014. Confirmation and control of triazine and 4-hydroxyphenylpyruvate dioxygenase-inhibiting herbicide-resistant Palmer amaranth (*Amaranthus palmeri*) in Nebraska. Weed Technol. 28, 28–38.

Jhala, A.J., Norsworthy, J.K., Ganie, Z.A., Sosnoskie, L.M., Beckie, H.J., Mallory-Smith, C.A., et al., 2021. Pollen-mediated gene flow and transfer of resistance alleles from herbicide-resistant broadleaf weeds. Weed Technol . Available from: https://doi.org/10.1017/wet.2020.101.

Kaundun, S.S., Hutchings, S.J., Dale, R.P., Howell, A., Morris, J.A., Kramer, V.C., et al., 2017. Mechanism of resistance to mesotrione in an *Amaranthus tuberculatus* population from Nebraska, USA. PLoS One. 12, e0180095.

Keeley, P.E., Carter, C.H., Thullen, R.J., 1987. Influence of planting date on growth of Palmer amaranth (*Amaranthus palmeri*). Weed Sci. 35, 199–204.

Kępczyński, J., Bihun, M., 2002. Induction of secondary dormancy in *Amaranthus caudatus* seeds. Plant. Growth Regul. 38, 135–140.

Klingaman, T.E., Oliver, L.R., 1994. Palmer amaranth (*Amaranthus palmeri*) interference in soybeans (*Glycine max*). Weed Sci. 42, 523–527.

Knezevic, S.Z., Weise, S.F., Swanton, C.J., 1994. Interference of redroot pigweed (*Amaranthus retroflexus*) in corn (*Zea mays*). Weed Sci. 42, 568–573.

Knezevic, S.Z., Horak, M.J., Vanderlip, R.L., 1997. Relative time of redroot pigweed (*Amaranthus retroflexus* L.) emergence is critical in pigweed-sorghum [*Sorghum bicolor* (L.) Moench] competition. Weed Sci. 45, 502–508.

Knezevic, S.Z., Pavlovic, P., Osipitan, O.A., Barnes, E.R., Beiermann, C., Oliveira, M.C., et al., 2019. Critical time for weed removal in glyphosate-resistant soybean as influenced by pre-emergence herbicides. Weed Technol. 33, 393–399.

Kohrt, J.R., Sprague, C.L., 2017. Response of a multiple-resistant Palmer amaranth (*Amaranthus palmeri*) population to four HPPD-inhibiting herbicides applied alone and with atrazine. Weed Sci. 65, 534–545.

Korres, N.E., Norsworthy, J.K., Young, B.G., Reynolds, D.B., Johnson, W.G., Conley, S.P., et al., 2018. Seedbank persistence of Palmer amaranth (*Amaranthus palmeri*) and waterhemp (*Amaranthus tuberculatus*) across diverse geographical regions in the United States. Weed Sci. 66, 446–456.

Korres, N.E., Norsworthy, J.K., Mauromoustakos, A., 2019. Effects of Palmer amaranth (*Amaranthus palmeri*) establishment time and distance from the crop row on biological and phenological characteristics of the weed: Implications on soybean yield. Weed Sci. 67, 126–135.

Kreiner, J.M., Giacomini, D., Bemm, F., Waithaka, B., Regalado, J., Lanz, C., et al., 2019. Multiple modes of convergent adaptation in the spread of glyphosate-resistant *Amaranthus tuberculatus*. Proc. Natl. Acad. Sci. USA. 116, 21076–21084.

Kumar, V., Liu, R., Boyer, G., Stahlman, P.W., 2019. Confirmation of 2,4-D resistance and identification of multiple resistance in a Kansas Palmer amaranth (*Amaranthus palmeri*) population. Pest. Manag. Sci. 75, 2925–2933.

Küpper, A., Borgato, E.A., Patterson, E.L., Netto, A.G., Nicolai, M., Carvalho, S.J.P., et al., 2017. Multiple resistance to glyphosate and acetolactate synthase inhibitors in Palmer amaranth (*amaranthus palmeri*) identified in Brazil. Weed Sci. 65, 317–326.

Lee, R.M., Hager, A.G., Tranel, P.J., 2008. Prevalence of a novel resistance mechanism to PPO-inhibiting herbicides in waterhemp (*Amaranthus tuberculatus*). Weed Sci. 56, 371–375.

Leon, R.G., Owen, M.D.K., 2003. Regulation of weed seed dormancy through light and temperature interactions. Weed Sci. 51, 752–758.

Leon, R.G., Owen, M.D.K., 2006. Tillage systems and seed dormancy effects on common waterhemp (*Amaranthus tuberculatus*) seedling emergence. Weed Sci. 54, 1037–1044.

Leon, R.G., Knapp, A.D., Owen, M.D.K., 2004. Effect of temperature on the germination of common waterhemp (*Amaranthus tuberculatus*), giant foxtail (*Setaria faberi*), and velvetleaf (*Abutilon theophrasti*). Weed Sci. 52, 67–73.

Lillie, K.J., Giacomini, D.A., Green, J.D., Tranel, P.J., 2019. Coevolution of resistance to PPO inhibitors in waterhemp (*Amaranthus tuberculatus*) and Palmer amaranth (*Amaranthus palmeri*). Weed Sci. 67, 521–526.

Liu, J., Davis, A.S., Tranel, P.J., 2012. Pollen biology and dispersal dynamics in waterhemp (*Amaranthus tuberculatus*). Weed Sci. 60, 416–422.

Lorentz, L., Gaines, T.A., Nissen, S.J., Westra, P., Strek, H.J., Dehne, H.W., et al., 2014. Characterization of glyphosate resistance in *Amaranthus tuberculatus* populations. J. Agric. Food Chem. 62, 8134–8142.

Manley, B.S., Singh, B.K., Shaner, D.L., Wilson, H.P., 1999. Imidazolinone resistance in smooth pigweed (*Amaranthus hybridus*) is due to an altered acetolactate synthase. Weed Technol. 13, 697–705.

Massinga, R.A., Currie, R.S., Horak, M.J., Boyer, J., 2001. Interference of Palmer amaranth in corn. Weed Sci. 49, 202–208.

McMullan, P.M., Green, J.M., 2011. Identification of a tall waterhemp (*Amaranthus tuberculatus*) biotype resistant to HPPD-inhibiting herbicides, atrazine, and thifensulfuron in Iowa. Weed Technol. 25, 514–518.

Molin, W.T., Nandula, V.K., Wright, A.A., Bond, J.A., 2016. Transfer and expression of ALS inhibitor resistance from Palmer amaranth (*Amaranthus palmeri*) to an *A. spinosus* × *A. palmeri* hybrid. Weed Sci. 64, 240–247.

Montgomery, J.S., Giacomini, D., Waithaka, B., Lanz, C., Murphy, B.P., Campe, R., et al., 2020. Draft genomes of *Amaranthus tuberculatus, Amaranthus hybridus*, and *Amaranthus palmeri*. Genome Biol. Evol. 12, 1988–1993.

Moore, J.W., Murray, D.S., Westerman, R.B., 2004. Palmer amaranth (*Amaranthus palmeri*) effects on the harvest and yield of grain sorghum (*Sorghum bicolor*). Weed Technol. 18, 23–29.

Morales-Payan, J.P., Stall, W.M., 2002a. Time of removal and population density effects of livid amaranth on bell pepper (*Capsicum annum*). Hortsci. 37, 747–748.

Morales-Payan, J.P., Stall, W.M., 2002b. Yield of polyethylene-mulched bell pepper (*Capsicum annuum*) as affected by time of emergence and population density of smooth pigweed (*Amaranthus hybridus* L.). Proc. Fla. State Soc. 115, 200–202.

Morgan, G.D., Baumann, P.A., Chandler, J.M., 2001. Competitive impact of Palmer amaranth (*Amaranthus palmeri*) on cotton (*Gossypium hirsutum*) development and yield. Weed Technol. 15, 408–412.

Mortensen, D.A., Egan, J.F., Maxwell, B.D., Ryan, M.R., Smith, R.G., 2012. Navigating a critical juncture for sustainable weed management. BioSci. 62, 75–84.

Mosyakin, S.L., Robertson, K.R., 2008. 39. Amaranthaceae. 3. Amaranthus. Flora North. Am. 4, 405–418.

Murphy, B.P., Larran, A.S., Ackley, B., Loux, M.M., Tranel, P.J., 2019. Survey of glyphosate-, atrazine- and lactofen-resistance mechanisms in Ohio waterhemp (*Amaranthus tuberculatus*) populations. Weed Sci. 67, 296–302.

Murray, M.J., 1940. The genetics of sex determination in the family Amaranthaceae. Genetics. 25, 409–431.

Nandula, V.K., Reddy, K.N., Kroger, C.H., Poston, D.H., Rimando, A.M., Duke, S.O., et al., 2012. Multiple resistance to glyphosate and pyrithiobac in Palmer amaranth (*Amaranthus palmeri*) from Mississippi and response to flumiclorac. Weed Sci. 60, 179–188.

Nandula, V.K., Ray, J.D., Ribeiro, D.N., Pan, Z., Reddy, K.N., 2013. Glyphosate resistance in tall waterhemp (*Amaranthus tuberculatus*) from Mississippi is due to both altered target-site and nontarget-site mechanisms. Weed Sci. 61, 374–383.

Nandula, V.K., Giacomini, D.A., Ray, J.D., 2020. Resistance to acetolactate synthase inhibitors is due to a W 574 to L amino acid substitution in the ALS gene of redroot pigweed and tall waterhemp. PLoS One. 15, e0235394.

National Research Council, 1984. Amaranth: Modern Prospects for an Ancient Crop. The National Academies Press, Washington, DC.

Nie, H., Mansfield, B.C., Harre, N.T., Young, J.M., Steppig, N.R., Young, B.G., 2019. Investigating target-site resistance mechanism to the PPO-inhibiting herbicide fomesafen in waterhemp and interspecific hybridization of *Amaranthus* species using next generation sequencing. Pest. Man. Sci. 75, 3235–3244.

Norsworthy, J.K., Korres, N.E., Walsh, M.J., Powles, S.B., 2016. Integrating herbicide programs with harvest weed seed control and other fall management practices for the control of glyphosate-resistant Palmer Amaranth (*Amaranthus palmeri*). Weed Sci. 64, 540–550.

Norsworthy, J.K., Green, J.K., Barber, T., Roberts, T.L., Walsh, M.J., 2020. Seed destruction of weeds in southern US crops using heat and narrow-windrow burning. Weed Technol. 34, 589–596.

Oliveira, M.C., Jhala, A.J., Gaines, T., Irmak, S., Amundsen, K., Scott, J.E., et al., 2017. Confirmation and control of HPPD-inhibiting herbicide–resistant waterhemp (*Amaranthus tuberculatus*) in Nebraska. Weed Technol. 31, 67–79.

Oliveira, M.C., Gaines, T.A., Dayan, F.E., Patterson, E.L., Jhala, A.J., Knezevic, S.Z., 2018a. Reversing resistance to tembotrione in an *Amaranthus tuberculatus (var. rudis)* population from Nebraska, USA with cytochrome P450 inhibitors. Pest. Manag. Sci. 74, 2296–2305.

Oliveira, M.C., Ganines, T., Patterson, E.L., Jhala, A.J., Irmak, S., Amundsen, K., et al., 2018b. Interspecific and intraspecific transference of metabolism based mesotrione resistance in dioecious weedy *Amaranthus*. Plant. J. 96, 1051–1063.

Pederson, B., Bach Knudsen, K.E., Eggum, B.O., 1990. The nutritive value of amaranth grain (*Amaranthus caudatus*). Plant. Foods Hum. Nutr. 40, 61–71.

Peterson, D.E., Nakka, S., Jugulam, M., Thompson, C.R., Godar, A.S., 2017. Rapid detoxification via glutathione S-transferase (GST) conjugation confers a high level of atrazine resistance in Palmer amaranth (*Amaranthus palmeri*). Pest. Manag. Sci. 73, 2236–2243.

Price, A.J., Monks, C.D., Culpepper, A.S., Duzy, L.M., Kelton, J.A., Marshall, M.W., et al., 2016. High-residue cover crops alone or with strategic tillage to manage glyphosate- resistant Palmer amaranth (*Amaranthus palmeri*) in southeastern cotton (*Gossypium hirsutum*). J. Soil. Water Conserv. 71, 1–11.

Rayburn, A.L., McCloskey, R., Tatum, T.C., Bollero, G.A., Jeschke, M.R., Tranel, P.J., 2005. Genome size analysis of weedy *Amaranthus* species. Crop. Sci. 45, 2557–2562.

Refsell, D.E., Hartzler, R.G., 2009. Effect of tillage on common waterhemp (*Amaranthus rudis*) emergence and vertical distribution of seed in the soil. Weed Technol. 23, 129–133.

Rosenbaum, K.K., Bradley, K.W., 2013. A survey of glyphosate-resistant waterhemp (*Amaranthus rudis*) in Missouri soybean fields and prediction of glyphosate resistance in future waterhemp populations based on in-field observations and management practices. Weed Technol. 27, 656–663.

Salas, R.A., Burgos, N.R., Tranel, P.J., Singh, S., Glasgow, L., Scott, R.C., et al., 2016. Resistance to PPO-inhibiting herbicide in Palmer amaranth from Arkansas. Pest. Manag. Sci. 72, 864–869.

Salas-Perez, R.A., Burgos, N.R., Rangani, G., Singh, S., Refatti, J.P., Piveta, L., et al., 2017. Frequency of Gly-210 deletion mutation among Protoporphyrinogen Oxidase inhibitor-resistant Palmer Amaranth (*Amaranthus palmeri*) populations. Weed Sci. 65, 718–731.

Sarangi, D., Jhala, A.J., 2018. A statewide survey of stakeholders to assess the problem weeds and weed management practices in Nebraska. Weed Technol. 32, 642–655.

Sarangi, D., Jhala, A.J., 2019. Palmer amaranth (*Amaranthus palmeri*) and velvetleaf (*Abutilon theophrasti*) control in no-tillage conventional (non–genetically engineered) soybean using overlapping residual herbicide programs. Weed Technol. 33, 95–105.

Sarangi, D., Sandell, L.D., Knezevic, S.Z., Aulakh, J.S., Lindquist, J.L., Irmak, S., et al., 2015. Confirmation and control of glyphosate-resistant common waterhemp (*Amaranthus rudis*) in Nebraska. Weed Technol. 29, 82–92.

Sarangi, D., Irmak, S., Lindquist, J., Knezevic, S., Jhala, A.J., 2016. Effect of water stress on the growth and fecundity of common waterhemp (*Amaranthus rudis*). Weed Sci. 64, 42–52.

Sarangi, D., Sandell, L.D., Kruger, G.R., Knezevic, S.Z., Irmak, S., Jhala, A.J., 2017a. Comparison of herbicide programs for season-long control of glyphosate-resistant common waterhemp (*Amaranthus rudis*) in soybean. Weed Technol. 31, 53–66.

Sarangi, D., Tyre, A.J., Patterson, E.L., Gaines, T.A., Irmak, S., Knezevic, S.Z., et al., 2017b. Pollen-mediated gene flow from glyphosate-resistant common waterhemp (*Amaranthus rudis* Sauer): consequences for the dispersal of resistance genes. Sci. Rep. 7, 44913.

Sarangi, D., Stephens, T., Barker, A.L., Patterson, E.L., Gaines, T.A., Jhala, A.J., 2019. Protoporphyrinogen oxidase (PPO) inhibitor-resistant waterhemp (*Amaranthus tuberculatus*) from Nebraska is multiple herbicide resistant: confirmation, mechanism of resistance, and management. Weed Sci. 67, 510–520.

Sauer, J.D., 1957. Recent migration and evolution of the dioecious amaranths. Evolution. 11, 11–31.

Sauer, J.D., 1967. The grain amaranths and their relatives: a revised taxonomic and geographic survey. Ann. Mo. Bot. Gard. 54, 103–137.

Schryver, M.G., Soltani, N., Hooker, D.C., Robinson, D.E., Tranel, P.J., Sikkema, P.H., 2017. Glyphosate-resistant waterhemp (*Amaranthus tuberculatus* var. *rudis*) in Ontario, Canada. Can. J. Plant. Sci. 97, 1057–1067.

Schultz, J.L., Chatham, L.A., Riggins, C.W., Tranel, P.J., Bradley, K.W., 2015. Distribution of herbicide resistances and molecular mechanisms conferring resistance in Missouri waterhemp (*Amaranthus rudis* Sauer) populations. Weed Sci. 63, 336–345.

Schwartz, L.M., Norsworthy, J.K., Young, B.G., Bradley, K.W., Kruger, G.R., Davis, V.M., et al., 2016. Tall waterhemp (*Amaranthus tuberculatus*) and Palmer amaranth (*Amaranthus palmeri*) seed production and retention at soybean maturity. Weed Technol. 30, 284–290.

Sellers, B.A., Smeda, R.J., Johnson, W.G., Kendig, J.A., Ellersieck, M.R., 2003. Comparative growth of six Amaranthus species in Missouri. Weed Sci. 51, 329–333.

Shergill, L.S., Barlow, B.R., Bish, M.D., Bradley, K.W., 2018. Investigations of 2,4-D and multiple herbicide resistance in a Missouri waterhemp (*Amaranthus tuberculatus*) population. Weed Sci. 66, 386–394.

Shergill, L.S., Bejleri, K., Davis, A., Mirsky, S.B., 2020a. Fate of weed seeds after impact mill processing in midwestern and mid-Atlantic United States. Weed Sci. 68, 92–97.

Shergill, L.S., Schwartz-lazaro, L.M., Leon, R., Ackroyd, V.J., Flessner, M.L., Bagavathiannan, M., et al., 2020b. Current outlook and future research needs for harvest weed seed control in North American cropping systems. Pest. Manag. Sci. 76, 3887–3895.

Shyam, C., Chahal, P.S., Jhala, A.J., Jugulam, M., 2021. Management of glyphosate-resistant Palmer amaranth (*Amaranthus palmeri*) in 2,4-D choline, glufosinate, and glyphosate-resistant soybean. Weed Technol. (in press) DOI: https://doi.org/10.1017/wet.2020.91.

Shyam, C., Jhala, A.J., Kruger, G., Jugulam, M., 2019. Rapid metabolism increases the level of 2,4-D resistance at high temperature in common waterhemp (*Amaranthus tuberculatus*). Sci. Rep. 9, 1–9.

Singh, V., Garetson, R., McGinty, J., Dotray, P., Morgan, G., Nolte, S., et al., 2020. Distribution of herbicide-resistant waterhemp (*Amaranthus tuberculatus*) across row crop production systems in Texas. Weed Technol. 34, 129–139.

Sosnoskie, L.M., Webster, T.M., MacRae, A.W., Grey, T.L., Culpepper, A.S., 2012. Pollen-mediated dispersal of glyphosate-resistance in Palmer amaranth under field conditions. Weed Sci. 60, 366–373.

Steckel, L.E., 2007. The dioecious *Amaranthus* spp.: here to stay. Weed Tech. 21, 567–570.

Steckel, L.E., Sprague, C.L., Hager, A.G., Simmons, F.W., Bollero, G.A., 2003. Effects of shading on common waterhemp (*Amaranthus rudis*) growth and development. Weed Sci. 51, 898–903.

Steckel, L.E., Sprague, C.L., Stoller, E.W, Wax, L.M., 2004. Temperature effects on germination of nine *Amaranthus* species. Weed Sci. 52, 217–221.

Steckel, L.E., Sprague, C.L., Stoller, E.W., Wax, L.M., Simmons, F.W., 2007. Tillage, cropping system, and soil depth effects on common waterhemp (*Amaranthus rudis*) seed-bank persistence. Weed Sci. 55, 235–239.

Steinau, A.N., Skinner, D.Z., Steinau, M., 2003. Mechanism of extreme genetic recombination in weedy *Amaranthus* hybrids. Weed Sci. 51, 696–701.

Stuart, B.L., Harrison, S.K., Abernathy, J.R., Krieg, D.R., Wendt, C.W., 1984. The response of cotton (*Gossypium hirsutum*) water relations to smooth pigweed (*Amaranthus hybridus*) competition. Weed Sci. 32, 126–132.

Terry Jr, R.E., Stall, W.M., Shilling, D.G., Bewick, T.A., Kostewitz, S.R., 1997. Smooth amaranth interference with watermelon and muskmelon production. Hortsci. 32, 630–632.

Thinglum, K.A., Riggins, C.W., Davis, A.S., Bradley, K.W., Al-Khatib, K., Tranel, P.J., 2011. Wide distribution of the waterhemp (*Amaranthus tuberculatus*) ΔG210 PPX2 mutation, which confers resistance to PPO-inhibiting herbicides. Weed Sci. 59, 22–27.

Tranel, P.J., Wassom, J.J., Jeschke, M.R., Rayburn, A.L., 2002. Transmission of herbicide resistance from a monoecious to a dioecious weedy *Amaranthus* species. Theor. Appl. Genet. 105, 674–679.

Trucco, F., Jeschke, M.R., Rayburn, A.L., Tranel, P.J., 2005a. *Amaranthus hybridus* can be pollinated frequently by *A. tuberculatus* under field conditions. Heredity. 94, 64–70.

Trucco, F., Jeschke, M.R., Rayburn, A.L., Tranel, P.J., 2005b. Promiscuity in weedy amaranths: high frequency of female tall waterhemp (*Amaranthus tuberculatus*) × smooth pigweed (*A. hybridus*) hybridization under field conditions. Weed Sci. 53, 46–54.

Trucco, F., Zheng, D., Woodyard, A.J., Walter, J.R., Tatum, T.C., Lane Rayburn, A., et al., 2007. Nonhybrid progeny from crosses of dioecious amaranths: implications for gene-flow research. Weed Sci. 55, 119–122.

Varanasi, V.K., Brabham, C., Norsworthy, J.K., Nie, H., Young, B.G., Houston, M., et al., 2018. A statewide Survey of PPO-inhibitor resistance and the prevalent target-site mechanisms in Palmer amaranth (*Amaranthus palmeri*) accessions from Arkansas. Weed Sci. 66, 149–158.

Vennapusa, A.R., Faleco, F., Vieira, B., Samuelson, S., Kruger, G.R., Werle, R., et al., 2018. Prevalence and mechanism of atrazine resistance in waterhemp (*Amaranthus tuberculatus*) from Nebraska. Weed Sci. 66, 595–602.

Vieira, B.C., Samuelson, S.L., Alves, G.S., Gaines, T.A., Werle, R., Kruger, G.R., 2018. Distribution of glyphosate-resistant *Amaranthus* spp. in Nebraska. Pest. Manag. Sci. 74, 2316–2324.

Wang, H., Guo, W., Zhang, L., Zhao, K., Ge, L., Lv, X., et al., 2017. Multiple resistance to thifensulfuron-methyl and fomesafen in redroot pigweed (*Amaranthus retroflexus* L.) from China. Chilean. J. Agric. Res. 77, 311–317.

Ward, S.M., Webster, T.M., Steckel, L.E., 2013. Palmer amaranth (*Amaranthus palmeri*): a review. Weed Technol. 27, 12–27.

Waselkov, K.E., Regenold, N.D., Lum, R.C., Olsen, K.M., 2020. Agricultural adaptation in the native North American weed waterhemp, *Amaranthus tuberculatus* (Amaranthaceae). PLoS One. 15, e0238861.

Wassom, J.J., Tranel, P.J., 2005. Amplified fragment length polymorphism-based genetic relationships among weedy *Amaranthus* species. J. Hered. 96, 410–416.

Weaver, S.E., McWilliams, E.L., 1980. The Biology of Canadian Weeds.: 44. *Amaranthus retroflexus* L., *A. powellii* S. Wats. and *A. hybridus* L. Can. J. Plant. Sci. 60, 1215–1234.

Webster, T.M., Scully, B.T., Grey, T.L., Culpepper, A.S., 2013. Winter cover crops influence *Amaranthus palmeri* establishment. Crop. Prot. 52, 130–135.

Wells, M.J., Balsinhas, A.A., Joffe, H., Engelbrecht, V.M., Harding, G., Stirton, C.H., 1986. A catalogue of problem plants in southern Africa incorporating the national weed list of South Africa. Memoirs, Botanical Surv. South. Afr. 658.

Werle, R., Hartzler, R.G., Buhler, D.D., Lindquist, J.L., Sandell, L.D., 2014. Predicting emergence of 23 summer annual weed species. Weed Sci. 62, 267–279.

Werner, K., Sarangi, D., Nolte, S., Dotray, P., Bagavathiannan, M., 2020. Late-season surveys to document seed rain potential of Palmer amaranth (*Amaranthus palmeri*) and waterhemp (*Amaranthus tuberculatus*) in Texas cotton. PLoS One. 15, e0226054.

Wetzel, D.K., Horak, M.J., Skinner, D.Z., Kulakow, P.A., 1999. Transferal of herbicide resistance traits from *Amaranthus palmeri* to *Amaranthus rudis*. Weed Sci. 47, 538–543.

Woodyard, A.J., Hugie, J.A., Riechers, D.E., 2009. Interactions of mesotrione and atrazine in two weed species with different mechanisms for atrazine resistance. Weed Sci. 57, 369–378.

Wu, C., Owen, M.D.K., 2014. When is the best time to emerge: reproductive phenology and success of natural common waterhemp (*Amaranthus rudis*) cohorts in the midwest United States? Weed Sci. 62, 107–117.

Chapter 3

Avena fatua and *Avena sterilis*

Breanne D. Tidemann[1], Charles M. Geddes[2] and Hugh J. Beckie[3]
[1]*Agriculture and Agri-Food Canada Research Centre, Lacombe, AB, Canada,* [2]*Agriculture and Agri-Food Canada Research Centre, Lethbridge, AB, Canada,* [3]*School of Agriculture and Environment, The University of Western Australia, Perth, WA, Australia*

Introduction

Avena fatua L. and *Avena sterilis* L. belong to the Poaceae/Gramineae—grass family. Synonyms for *A. fatua* include *Avena sativa* subsp. *fatua* (L.) Fiori. Common names are wild oat, wild oats, common wild oat, spring wild oat, folle avoine, oatgrass, poor oats, and wheatgrass (Beckie et al., 2012a; Sharma and Vanden Born, 1978). Synonyms for *A. sterilis* include the following: *Avena ludoviciana* Durieu; *Avena sterilis* subsp. *sterilis*; *Avena sterilis* subsp. *ludoviciana*; and *Avena macrocarpa* Moench. Common names are sterile oat, winter wild oat, animated oat, and wild oat (Anonymous, 2019). In this review, we follow the Weed Science Society of America (WSSA) and European and Mediterranean Plant Protection Organization nomenclature of *A. sterilis* being synonymous with *A. ludoviciana* Durieu, which is a subspecies of *A. sterilis* and not a separate species (Anonymous, 2019; Sáez et al., 2017; WSSA, 2020).

The botanical description for *A. fatua* is based on Beckie et al. (2012a). It is an annual grass with fibrous roots and counter-clockwise twisted juvenile leaves, that reaches up to 150 cm high. Leaves are linear-lanceolate, between 20 and 30 cm long and 0.4–1 cm wide. The leaf sheaths are open (no auricles) with slightly hairy edges and a 0.2–0.5 cm ligule that is grayish-white, papery, and irregularly lacerate at the top. The open, loose panicle is equilateral, between 10 and 40 cm long and drooping each with two glumes that contain three florets. The lemma is hairy or glabrous with a range of colors between light yellow to black, and the first floret is usually covered with dense hairs. The florets separate through disarticulation of the rachillas segments, and each floret has a geniculate, hygroscopic awn 3–4 cm long, twisted along the bottom portion. The seeds are elliptical, 1 cm long with two flowering scales enclosing the caryopsis with a depressed scar at the base (Fig. 3.1).

Biology and Management of Problematic Crop Weed Species. DOI: https://doi.org/10.1016/B978-0-12-822917-0.00015-X

FIGURE 3.1 Global distribution of (A) *Avena fatua* and (B) *Avena sterilis*. *Dataset from Anonymous (2019) and USDA (2020).*

The botanical description for *A. sterilis* outlined below is derived from Edgar (1980), Ivens (1989), Stace (1997), and Anonymous (2019). It is an annual that grows up to 150 cm high; juvenile growth is prostrate to erect. Stem is tufted, rarely geniculate, not branching, not rooting at the nodes, with nodes sometimes hairy. Leaf blades are up to 60 cm long, 0.6–1.4 cm wide, 30–40 times as long as wide, linear, and not hairy. Ligule is 2–5 mm long, membranous, acute to truncate, and sheath often on lower leaves. Inflorescence is an equilateral, pyramidal, or slightly one-sided loose panicle, 15–45 cm long, and 8–25 cm wide. Spikelets are 1.7–4.5 cm long without the awns, each with 2–5 florets of which only the lowest has a basal scar, pedicelled, disarticulating above the glumes, but not between the florets. Florets tend to be reddish color with straw yellow tips, but can be solid straw yellow. Glumes are equal or subequal, 2.4–5.0 cm long, 7–11 nerved. Lemma is narrow-lanceolate, usually densely hairy or glabrous, 1.5–4.0 cm long, 7 nerved, bidentate, the uppermost (third seed) being awnless, the lower two (first and second seeds) with a 3–9 cm long bent and twisted dorsal awn. Caryopsis length is 10 mm, width is 0.2–0.25 cm (Fig. 3.1).

The subspecies *sterilis* differs from subspecies *ludoviciana* by the larger size of some reproductive parts. In comparison with *A. fatua*, the stem of *A. sterilis* is more prostrate with more tillers at the maximum tillering stage, the panicles are lighter in weight because the spikelets bear fewer and smaller florets, the glume is longer, the awn is absent for the third seed, the spikelets are hard and do not break easily, and the seeds remain longer within the spikelet at maturity and are shed in units of 2 or 3 (Edgar, 1980; Moss, 2015; Stace, 1997). *Avena fatua* and *A. sterilis* are hexaploid with a chromosome count of $2n = 42$ and genome AACCDD.

Avena fatua originated in the western Mediterranean (Loskutov, 2008) and has since been introduced to most temperate world regions. It is found in Europe, the Middle East, western Asia, Canada, and the United States (Beckie et al., 2012a; USDA, 2020) (Fig. 3.2). *Avena sterilis* likely spread as a weed in wheat (*Triticum aestivum* L.) and barley (*Hordeum vulgare* L.) crops from the Fertile Crescent (Middle East) 3000 to 4000 years ago to its present native range throughout the Mediterranean region and southern Europe (Rosentrater and Evers, 2018). It was subsequently introduced into central and eastern Europe and other continents via seed contamination of wool and grain (Stace, 1997; Torner et al., 1991). The center of diversification of *A. sterilis* is present-day Turkey (Phillips et al., 1993) and possibly eastward in the Caucasus (Azerbaijan) (Soldatov and Loskutov, 1991).

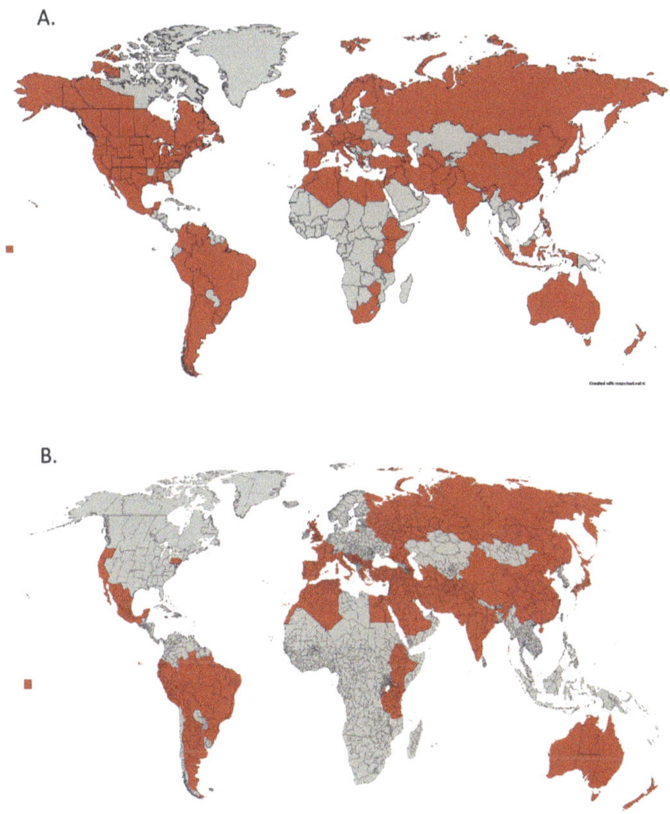

FIGURE 3.2 (A) (Clockwise from top left) *Avena fatua* collar region, hairs on leaf margin, three florets/seeds and mature plants in a crop canopy (bottom picture credit: K.N. Harker, used with permission). (B) *Avena sterilis*: mature plant (left; http://www.plantillustrations.org) and inflorescence (right; http://www.weedscience.org).

Avena sterilis is commonly found in semiarid, temperate, or subtropical regions globally; it occurs on all continents except Antarctica (O'Donnell et al., 2002). It is a naturalized species in parts of North, Central and South America, southern, central, and eastern Europe, most of Africa, Indian subcontinent, the Middle East, Asia, and Oceania (Anonymous, 2019; Floc'h, 1991; Holm et al., 1979). In Australia, both *A. fatua* and *A sterilis* cooccur; however, *A. sterilis* tends to be more prevalent in warmer areas of northern New South Wales and southern Queensland, whereas *A. fatua* dominates in southeastern areas of the country (McGillion and Storrie, 2006). An examination of the genetic diversity of 369 accessions from 26 countries using microsatellite (simple sequence repeat, SSR) markers indicated that accessions from Greece, Liberia, and Italy were most diverse, whereas those from Egypt, Georgia, Ethiopia, Gibralter, and Kenya were most distinct (Fu et al., 2007) (Fig. 3.2).

Avena fatua is found in a variety of noncrop disturbed (ruderal) and cultivated habitats and prefers clay or clay loam soils (Beckie et al., 2012a). Similarly, *A. sterilis* can occur in cropland, pastures, vineyards, and ruderal areas; it tends to prefer heavier-textured soils compared with that of *A. fatua* (Stace, 1997). In addition, the species can grow under drier conditions than *A. fatua* (Fernandez-Quintanilla et al., 1990; Thurston, 1951). The species is well adapted to a wide range of climatic conditions, as evident from its wide geographic distribution across semiarid, temperate, and Mediterranean (subtropical) regions globally. With a warming climate, the range of *A. sterilis* is expected to advance into northeastern Europe and contract in southern Europe (Castellanos-Frías et al., 2014). Climate has little impact on the distribution of *A. fatua* on the Canadian Prairies (Leeson and Beckie, 2012). However, O'Donnell and Adkins (2001) showed increased biomass and fecundity in plants grown in higher CO_2 environments with interactions between CO_2, temperature, moisture, and wild oat accession. Sharma and Vanden Born (1978) cited a number of investigations indicating *A. fatua* adaptation to environment and agronomic practices, resulting in biotypes which may respond differently to climate change. In addition, Ziska (2017) suggested that modern *A. fatua* accessions have increased competitive ability at early growth stages when grown in higher CO_2 environments than older *A. fatua* accessions.

Biology and population demography

Seed germination of *A. sterilis* is favoured by low temperatures more than that of *A. fatua* (Fernandez-Quintanilla et al., 1990). Both *A. sterilis* and *A. fatua* can germinate over a wide range of temperatures (5°C–30°C) and solute potentials (−0.025 to −1.4 MPa); germination rate is similar between species between 10°C and 20°C, with a higher germination rate of *A. fatua* above 20°C and a lower germination rate below 10°C relative to *A. sterilis* (Fernandez-Quintanilla et al., 1990; Quail and Carter, 1968). Minimum, maximum, and optimum temperatures for the germination of *A. sterilis* seeds were

2°C, 30°C, and 10°C, respectively (Uremis and Uygur, 1999), whereas those temperatures for buried *A. fatua* seeds were 3°C, 17°C, and 4°C–8°C, respectively (Murdoch, 1983). Other studies have reported 15°C (Mennan and Uygur, 1996) and 25°C (Hassanein et al., 1996) as the optimum germination temperatures for *A. sterilis*. In a Petri dish assay, *A. fatua* germination increased with higher osmotic potential, particularly in light compared with dark conditions, but was less impacted by oxygen concentration (Boyd and Van Acker, 2004). Light can both stimulate and inhibit *A. fatua* germination (Hou and Simpson, 1990; Hsiao and Simpson, 1971; Sharma and Vanden Born, 1978), and effects vary depending on the dormancy state of the population (Hou and Simpson, 1991, 1993). Gibberellic acid, potassium nitrate, smoke water, and some compounds found in smoke (e.g., butenolide) can stimulate germination (Adkins and Peters, 2001; Hilton, 1984; Kępczyński et al., 2006; Mennan and Uygur, 1996 Stevens et al., 2007). Piercing seeds with a needle can also promote germination (Foley, 1987; Hsiao et al., 1983).

Avena fatua or *A. sterilis* seed banks are considered moderately persistent (4–5 years), but a small proportion of seeds may persist longer (Terry, 1984; Van Acker, 2009). *Avena sterilis* seed persisted for a maximum of 2 years in Australia and Spain (Quail and Carter, 1968; Sanchez del Arco et al., 1995), whereas *A. fatua* seed persisted for >6.7 years in Alaska (Conn and Deck, 1995). *Avena fatua* seed longevity tends to increase with burial depth, although the degree of dormancy introduces some variability (Zorner et al., 1984); seeds at a soil depth of 34 cm persisted for up to 9 years (Miller and Nalewaja, 1990). Seed self-burial is aided by long geniculate hygroscopic awns (Somody et al., 1985).

Avena fatua seed dormancy varies widely among seed accessions, and it is postulated that this variability is due equally to genetic heterogeneity and phenotypic plasticity (Jana and Naylor, 1980; Jana and Thai, 1987). Primary seed dormancy imposed during *A. fatua* seed maturation reduces precocious germination while seeds remain on the mother plant (Baskin and Baskin, 1985, 2014). High temperatures or water stress during seed maturation can reduce primary dormancy (Adkins et al., 1987; Peters, 1982; Sawhney and Naylor, 1980). Often a period of warm, dry, after-ripening is required to alleviate primary dormancy (Foley, 1994). Much less is known about secondary dormancy in *A. fatua* seed, but it can be induced by greater light intensity, high temperature, and anoxia (Sawhney et al., 1986; Symons et al., 1987; Tilsner and Upadhyaya, 1985). Germination stimulants that are effective on *A. fatua* can be less effective on seeds that are in a secondary dormancy state (Kępczyński, 2018)

Emergence time of the species may differ spatially and temporally. For example, in northern and central Europe, emergence of *A. fatua* mainly occurs in spring, whereas *A. sterilis* emerges in winter and autumn; in southern Europe, both species often emerge in August (Aibar et al., 1991; Thurston, 1957). *Avena sterilis* exhibits prolonged emergence periodicity (Qasem, 2007). It grows rapidly, establishes extensive and deep root systems, and efficiently responds to high levels of nitrogen (Balyan et al., 1991). In greenhouse studies comparing several weed species, *A. fatua* was among the species with the greatest response to added nitrogen, but intermediate response to added phosphorous (Blackshaw et al., 2003, 2004).

Both species are highly selfing. Outcrossing rates for *A. fatua* range between 0% and 12.3% in noncrop environments and between 0.05% and 0.16% in cropped environments in Canada (Murray et al., 2002). Flowering in *A. sterilis* occurs earlier than *A. fatua* (Holm et al., 1979; Stace, 1997). In addition to hybridization between *A. sterilis* subspecies, *A. sterilis* hybridizes with other *Avena* species such as *A. fatua* (Aujas and Darmency, 1984; Cavan et al., 1998), *A. sativa* L. (Mariot et al., 1999; Sereno-Tavares et al., 1995), and *A. nuda* L. (Yu et al., 1998). *Avena fatua* can create fertile hybrids when crossed with *A. sterilis* (Aujas and Darmency, 1984; Cavan et al., 1998). Sterile hybrids have been recorded between *A. fatua* and *A. strigose* Schreb, *A. barbata* Pott ex Link, *A. magna* H.C. Murphy & Terrell, and *A. murphyi* Ladiz. (Rajhathy and Thomas, 1974).

Avena fatua produce up to 1070 seeds/plant in noncrop conditions; however, in competitive environments, 20–150 seeds/plant are more common (Beckie et al., 2012a). In spring wheat and barley crops, fecundity averages 20–70 seeds/plant (Belles et al., 2000; Thill and Mallory-Smith, 1997; Van Wychen et al., 2004). Early and late emerging *A. fatua* cohorts in the United States produced 130 and 60 seeds/plant, respectively (Dai et al., 2012). *Avena fatua* seed shatters before most crops mature. About 30%–70% of *A. fatua* seed shatters before wheat harvest in western Canada (Burton et al., 2016, 2017; Shirtliffe et al., 2000; Tidemann et al., 2017b). This variability is likely affected by genetic heterogeneity and phenotypic plasticity. Seed dispersal primarily occurs near the parent plant (Beckie et al., 2012a); however, combine harvesters can spread seeds over long distances (up to 145 m) (Shirtliffe and Entz, 2005). Planting of contaminated seedlots is another mechanism of dispersal (Michael et al., 2010; Thill and Mallory-Smith, 1997). *Avena fatua* seeds survive rumen digestion in some instances (Blackshaw and Rode, 1991) and can be dispersed along with cattle manure. This may be mitigated by ensiling (Blackshaw and Rode, 1991) or by composting manure for 20–30 days (Larney and Blackshaw, 2003).

Avena sterilis produce up to 200 seeds/plant, although substantially less seed is produced under interspecific or intra-specific competition (Fernandez-Quintanilla et al., 1986). In Spain, a population with 300 panicles/m^2 resulted in 2830 seeds/m^2 (Sanchez del Arco et al., 1995). Seed shatter from panicles may begin 2–3 weeks before typical wheat harvest (Balyan et al., 1991). Natural seed dispersal is mainly by birds or other animals, although the scale of anthropogenic dispersal via grain seedlot admixture, hay, or farm equipment is considerably greater (Stace, 1997; Terry, 1984).

Avena fatua is susceptible to predation by mammals, birds, or insects (Beckie et al., 2012a). Insect predation rates fall between 0%–100%, depending on spatial location and closeness to field edges (Jacob et al.,

2006). *Avena fatua* is also susceptible to various nematodes in India and Australia and snails in California, United States (Beckie et al., 2012a). Many diseases have been reported on *A. fatua* in Canada (Beckie et al., 2012a).

Avena sterilis exhibits variable susceptibility to barley yellow dwarf virus, oat crown rusts (*Puccinia coronata* f. sp. avenae) and stem rusts (*Puccinia graminis* f. sp. avenae) (Niekerk et al., 2001). More than 35 rust resistance genes have been identified from germplasm, and many of these genes were introgressed into *A. sativa* breeding programs (McKenzie and Harder, 1995). The species is a susceptible host for stem nematode (*Ditylenchus dipsaci*), a pest of vegetable crops in the Mediterranean region (Andaloussi and Bachikh, 2001), as well as root-lesion nematode (*Pratylenchus neglectus*) (Vanstone and Russ, 2001).

Herbicidal control and herbicide resistance

Herbicidal control of *A. fatua* and *A. sterilis* in-crop is heavily reliant on postemergence-applied acetyl-CoA carboxylase (ACCase)- and acetolactate synthase (ALS)-inhibiting herbicides [groups 1 (WSSA)/A (HRAC)/A (Australia) and 2/B/B, respectively], which were introduced in the late 1970s and early 1980s (Heap, 2020). Foliar-applied glyphosate (a 5-enolpyruvylshikimate-3-phosphate synthase inhibitor; group 9/G/M) is often used as a nonresidual preplant burnoff (or postemergence in glyphosate-resistant crops) to control these species. Other foliar-applied herbicides include photosystem-II inhibitors (group 5/C1/C), photosystem I electron diverters (group 22/D/L), and the glutamine synthetase inhibitor (group 10/H/N), glufosinate-ammonium (in glufosinate-resistant crops) (Beckie et al., 2012a). Preemergence control or suppression of *Avena* species rely on lipid synthesis inhibitors (group 8/N/J), alone or mixed with a microtubule assembly inhibitor (group 3/K1/D), and more recently a very long-chain fatty acid elongase (VLCFAE) inhibitor (group 15/K3/K) applied alone or with a protoporphyrinogen oxidase (PPO) inhibitor (group 14/E/G). Use of these residual preemergence herbicides for *Avena* species management varies globally because their efficacy and persistence is dependent on climatic and edaphic factors (Atienza et al., 2001; Tidemann et al., 2014). Glyphosate applied preharvest 0–15 days after *A. fatua* anthesis can reduce seed production and viability (Shuma et al., 1995), although preharvest application timing must adhere to the herbicide label to avoid risk of above-threshold residues in grain shipments (Beckie et al., 2020a). Generally, *A. fatua* and *A. sterilis* respond similarly to herbicides of different modes of action. However, in Australia, differential sensitivity between species to group 27/F2/H (4-hydroxyphenol-pyruvate dioxygenase inhibitors) has been observed and is being investigated by BASF Australia (BASF, 2019).

Avena fatua herbicide resistance was first reported in Australia in 1985 and has since been reported in 17 countries, whereas resistance in *A. sterilis* has been documented in nine countries since first reported in 1989 (Table 3.1) (Heap, 2020). Most cases are resistant to ACCase- or ALS-inhibiting

TABLE 3.1 Reported herbicide-resistant biotypes of *Avena sterilis* and *A. fatua*.

Species	Biotype	Country	Crop	Year
A. sterilis	ACCase inhibitor	Australia	Wheat, canola, pulses	1989
		Italy	Wheat	1992
		France	Wheat	1996
		Turkey	Wheat	1997
		Israel	Wheat	2006
		Iran	Wheat, canola	2006
		Greece	Wheat	2009
		Cyprus	Barley	2010
	ALS inhibitor	Australia	Wheat	2005
		France	Wheat	2006
		Italy	Wheat	2007
		Iran	Wheat	2009
	Antimicrotubule mitotic disrupter	Australia	Wheat	2002
	EPSP synthase inhibitor	Australia	Chickpea	2018
	ACCase + ALS inhibitor	Italy	Wheat	2004
		Turkey	Wheat	2008
		Iran	Wheat	2010
	ACCase + ALS inhibitor + antimicrotubule mitotic disrupter	United Kingdom	Wheat, cereals	1993
A. fatua	ACCase inhibitor	Australia (WA)	Wheat (1985), lupin, canola (1988), barley, chickpea, faba bean lentil (1991)	1985
		Canada	Barley, cropland, wheat, canola (1990), winter wheat (2012)	1990
		United States	Wheat (1990), cereals, sugar beet (1991), barley (1997), cropland (2002)	1990
		Belgum	Winter wheat	1996
		France	Wheat	1996

		Chile	Wheat	1998
		Mexico	Wheat	1998
		Iran	Wheat	2007
		Germany	Sugar beet	2009
		Argentina	Barley, wheat	2010
		Poland	Barley, wheat	2011
		Turkey	Wheat	2011
		Germany	Sugar beet	2012
		New Zealand	Wheat, clover	2014
		Syria	Wheat, potato, faba bean	2015
	ALS inhibitor	Canada	Cereals, cropland (1996), wheat (1997), canola, pea (1998)	1996
		United States	Cereals, wheat	1996
		France	Wheat	2006
		Poland	Barley, wheat	2011
	Cell elongation inhibitor	United States	Barley, wheat	1996
	Lipid inhibitor	Canada	Wheat	1996
	Microtubule inhibitor	United states	Cropland	1990
	EPSP synthase inhibitor	Australia	Chickpea	2018
	ACCase + ALS inhibitor	South Africa	Wheat	1986
		Germany	Sugar beet	2009
		Poland	Barley, wheat	2011
		United States	Wheat	2012
		Canada	Barley, cropland, wheat, canola	1994
		United Kingdom	Cereals, wheat, canola	1994
		Australia	Barley, wheat, chickpea	2006
	Cell elongation + lipid inhibitor	Canada	Barley, cereals	1989
		United States	Barley (1990), cereals (1993)	1990

(*Continued*)

TABLE 3.1 (Continued)

Species	Biotype	Country	Crop	Year
	ACCase + ALS + lipid inhibitor	Canada	Barley, lentil, wheat, canola (1996), cereals, peas (1998)	1998
	ACCase + ALS + lipid inhibitor + antimicrotubule mitotic disrupter	Canada	Wheat	1997
	ACCase + ALS + lipid + long-chain fatty acid + PPO inhibitor	Canada	Wheat	2015

Source: Adapted from Heap, I.M., 2020. International Herbicide-Resistant Weed Database. http://www.weedscience.org (Accessed 1 July 2020).

herbicides in small-grain cereals (wheat, barley), followed distantly by oilseeds and pulses.

The extent of herbicide resistance in *A. sterilis* is significantly less than that of *A. fatua* because of its reduced relative abundance and global distribution. While herbicide resistance in *A. sterilis* remains limited to ACCase and ALS-inhibiting herbicides, and the antimicrotubule mitotic disrupter (group 25/Z) flamprop-methyl, *A. fatua* populations in Canada or the United States have exhibited resistance to lipid synthesis inhibitors, cell elongation inhibitors (group 8/Z/Z), microtubule inhibitors, a VLCFAE inhibitor, and a PPO inhibitor (Heap, 2020; Mangin et al., 2017a; O'Donovan et al., 1994). Five-way herbicide resistance (WSSA groups 1, 2, 8, 14, and 15) was found in an *A. fatua* population from Manitoba, Canada (Mangin et al., 2017a), which is the only *Avena* spp. population reported to exhibit resistance to VLCFAE- or PPO-inhibiting herbicides. In 2018 the first and only cases of glyphosate resistance in *A. fatua* and *A. sterilis* were identified in a chickpea (*Cicer arietinum* L.) field in New South Wales, Australia (Heap, 2020). Although this biotype is not reported elsewhere, *A. fatua* was predicted to be at high risk of selection for glyphosate resistance in western Canada (Beckie et al., 2013a).

Herbicide resistance in *Avena* spp. is a global issue, but cross- or multiple-herbicide resistance is more extensive in Canada than other countries. In western Canada, routine surveys have documented herbicide resistance in *A. fatua* since the mid-1990s (Beckie et al., 2008, 2013b, 2020b). The most recent 2014–17 survey identified herbicide-resistant *A. fatua* in 69% of the 578/798 fields where the species was collected (Beckie et al., 2020b). This compared with 44% of fields with *A. fatua* in the previous survey (2007–09) (Beckie et al., 2013b). In the 2014–17 survey, *A. fatua* populations with resistance to ACCase inhibitors, ALS inhibitors, and cross- or multiple-resistance to both ACCase and ALS inhibitors were found in 62, 34, and 27% of fields where *A. fatua* was sampled, respectively (Beckie et al., 2020b). There was a marked increase in *A. fatua* resistance to ALS-inhibiting herbicides and resistance to both ACCase and ALS inhibitors in the most recent survey of western Canada due to increased use of ALS inhibitors to manage ACCase inhibitor-resistant *A. fatua* populations (Beckie et al., 2020b). Microtubule inhibitor-resistant *A. fatua* was present in 8% of fields where the species was collected in the 2007–09 survey (Beckie et al., 2013b). In contrast, herbicide resistance in *Avena* spp. remained stable in Western Australia between 2005 and 2010, with the most recent survey of 128 populations confirming resistance to ACCase inhibitors, ALS inhibitors, and both ACCase and ALS inhibitors in 48%, 2%, and 7% of populations, respectively (Owen and Powles, 2009, 2016). A 2007 survey of 113 *Avena* spp. populations from New South Wales, Australia, found that 38% of populations were resistant to ACCase inhibitors, whereas no populations exhibited resistance to ALS inhibitors (Broster et al., 2011). In Greece, 81% of 104 *A. sterilis* populations were ACCase inhibitor-resistant in 2009, and only 3% were ALS inhibitor-resistant (Travlos et al., 2011).

Resistance to ACCase- or ALS-inhibiting herbicides in *A. fatua* can be conferred by target site or nontarget site mechanisms (Beckie and Tardif, 2012). In a study of 16 Canadian *A. fatua* populations, seven ACC and two ALS gene mutations conferred resistance to ACCase and ALS inhibitors, respectively (Beckie et al., 2012b). In addition to these mutations, resistance was conferred also by enhanced metabolism by cytochrome P450 monoxygenases in some of the populations. Keith et al. (2017) showed that multiple resistance to these modes of action in some *A. fatua* populations in the United States was associated with elevated levels of mRNAs and proteins related to abiotic stress tolerance. Nontarget site resistance to flucarbazone, imazamethabenz, and pinoxaden in these populations was controlled by three separate, but closely linked, nuclear genes (Burns et al., 2018). Five *ACC* gene mutations are known to confer ACCase inhibitor resistance in *A. sterilis* in Australia (Liu et al., 2007). Both species are at risk of selection for nontarget site resistance through the intentional or unintentional use of low doses of ACCase- or ALS-inhibiting herbicides (Busi et al., 2016; Darmency and Uludag, 2018).

Weed management

Both *Avena* species are economically damaging, particularly in cereal crops, and rank high in terms of competitive ability. *Avena fatua* competition has reduced cereal crop yields by up to 70% (Beckie et al., 2012a) (Table 3.2). Recent results showed up to 78% yield loss in wheat in Pakistan with 40 *A. fatua* seeds/m^2 (Kulsoom et al., 2018). The impact of *A. fatua* is reduced in winter cereals compared to spring cereals with yield losses of ~40% in

TABLE 3.2 Crop yield loss due to *Avena fatua* and *Avena sterilis* interference.

Crop	Weed density (plants/m²)	Yield loss (%)	Reference
Avena fatua			
Spring wheat	64	28	Kirkland (1993)
	118	39	Kirkland (1993)
	100	40–63	Kirkland and Hunter (1991)
	≤500	54	Stougaard and Xue (2004, 2005)
	≤70	40–60	Carlson and Hill (1985)
	15–25	17	Rola (1984)
	100–150	50	Rola (1984)
	15	30	Khan et al. (2008)
	8	14	Wimschneider et al. (1990)
	3–54	5	Necajeva et al. (2017)
Winter wheat	120	28	Fast et al. (2009)
	15–25	6	Rola (1984)
	100–150	34	Rola (1984)
	250	≤40	Jäck et al. (2017)
	40	78	Kulsoom et al. (2018)
Barley	150	29	O'Donovan et al. (1985)
	170	42	Morishita and Thill (1988)
	70	25	Scursoni and Satorre (2005)
	+1[a]	1[a]	Wilson et al. (1990)
Maize	27[b]	25	Castillo and Ahrens (1986)
	9[b]	14	Castillo and Ahrens (1986)
Oat	>300	≤70	Willenborg et al. (2005)
Avena sterilis			
Wheat	3	Up to 15	Walia et al. (2001)
	40	16	Balyan and Malik (1989)
	10	30–40	Walia et al. (2001)
	160	46	Balyan and Malik (1989)
	146–162	27–60	Balyan et al. (1991)
	10–100	23–64	Walia and Singh (2005)
	110	58–65	Dhima and Eleftherohorinos (2001)

Barley	20–80	10	Torner et al. (1991)
	110	6–12	Dhima and Eleftherohorinos (2001)
	300	50	Torner et al. (1991)
	120	Up to 67	Dhima et al. (2000)
Triticale	110	55–73	Dhima and Eleftherohorinos (2001)

[a]*Each additional* A. fatua *plant resulted in a 1% increase in yield loss.*
[b]*Per meter of row (75 cm row spacing).*

winter cereals in Germany from *A. fatua* densities around 250 plants/m^2 (Jäck et al., 2017). For the greatest economical returns, *A. fatua* should be controlled by the tillering stage (Gillespie and Nalewaja, 1988); however, some studies suggest removal by 2 weeks after crop emergence (Harker et al., 2001, 2007). Other crops have a suggested critical period for weed control between those two extremes (Beckie et al., 2012a) (Table 3.2). Plant density of *A. sterilis* up to 160 plants/m^2 resulted in wheat yield losses of up to 60% (Balyan et al., 1991; Dhima and Eleftherohorinos, 2001; Walia et al., 2001); in barley, densities up to 300 plants/m^2 reduced yields by up to 65% (Dhima et al., 2000; Torner et al., 1991). Similar to wheat or barley, *A. sterilis* is highly resource (water and nutrients) competitive (Dhima et al., 2000; Gonzalez-Ponce and Santin, 2001). Therefore the control of *A. sterilis* at pre-tillering seedling stages is recommended to mitigate crop interference and yield loss (Table 3.2).

A large number of allelochemicals or plant residues containing unidentified compounds can suppress *A. fatua* germination and early growth under controlled conditions (Beckie et al., 2012a; Batish et al., 2002; Geddes et al., 2015; Jabran et al., 2010). Some aqueous extracts are selective to *A. fatua* and do not affect wheat in controlled studies (Bajwa et al., 2013). Recent research highlighted rhizobacteria and fungi, such as *Pseduomonas* species, that reduce *A. fatua* growth and increase mortality while having positive effects on a wheat crop (Dar et al., 2020; Mustafa et al., 2019). Solarization can reduce *A. fatua* germination, particularly after 30 days in moist soil (Arora and Yaduraju, 1998).

Effective integrated weed management (IWM) systems to manage *Avena* spp. include diverse crop types and sequences, use of competitive cultivars, high crop seeding rates or narrow row spacing, and fertilizer banded close to the seed row (Blackshaw et al., 2008; Jones and Medd, 1997). A competitive crop stand is the best strategy in mitigating the adverse impact of *Avena* spp. on crop yield or quality (Bajwa et al., 2017). Crop species that compete with *A. fatua* include barley, hybrid canola (*Brassica napus* L.), and yellow mustard (*Sinapis alba* L.) (Beckie et al., 2012a). Wheat, including winter varieties, is not as competitive as barley (Lanning et al., 1997, Scursoni et al., 1999); however, increased seeding rates of wheat improves competitive ability (Beckie et al., 2012a). Increased competition from wheat decreased *Avena* spp. density, and shifted biomass and seed production to higher in the canopy (Walsh, 2019). Increased seeding rates of other crops including barley and oats (*Avena sativa* L.) also reduced *A. fatua* biomass and seed production (Beckie et al., 2012a, May, 2018).

Varieties and growth habits of crops can also affect competitiveness with *A. fatua* [i.e., tall barley > short (Harker et al., 2009)]. Hybrid canola was more competitive than open-pollinated canola (Zand and Beckie, 2002). Taller and earlier maturing varieties may affect *A. fatua* in the year of growth, as well as population demography in subsequent years (Kieloch and Golebiowska, 2018; Lehnhoff et al., 2013). Perennial forages or annual crops cut for silage or hay manage *A. fatua* effectively, particularly when ensiling or manure composting renders *A. fatua* seed nonviable (Beckie et al., 2012a). Clipping of *A. fatua* panicles above the crop canopy prior to viable seed production particularly in short-statured crops, provides another opportunity to manage *A. fatua* (Tidemann et al., 2020).

Delayed crop seeding can help manage *A. fatua*, resulting in reduced population abundance by 70% and 96% in Canada and Japan, respectively (Asai et al., 2010; Darwent and Smith, 1985). However, Martin and Scursoni (2018) showed genetic differences between *A. fatua* in cultivated fields compared to nonfield populations that may allow field populations to adapt to delayed seeding techniques by also delaying their emergence through enhanced dormancy. In addition, short growing seasons can limit the implementation of delayed seeding, which may result in reduced crop yield potential or the inability to harvest a late maturing crop.

IWM becomes most successful when multiple strategies are combined. Utilizing a combination of crop rotation, seeding rate, and competitive varieties with limited herbicide use was as effective or better at managing *A. fatua* populations than full herbicide use alone (Harker et al., 2009). Increasing cropping system diversity combined with increased seeding rates and mechanical control strategies like silaging allowed *A. fatua* herbicide applications to be eliminated for 3 years while maintaining equivalent population abundance compared with full herbicide treatments alone (Harker et al., 2016). These treatments were often as profitable as conventional annual cropping systems with full herbicide usage (Smith et al., 2018). IWM strategies can also be incorporated with strategic herbicide applications to achieve effective control of *A. fatua*. (Mangin et al., 2017b)

Harvest weed seed control (HWSC) is a method of targeting weed seeds contained in the chaff fraction at harvest to prevent their entry into the soil weed seed bank (Walsh et al., 2018). *Avena fatua* seeds can be effectively destroyed by HWSC methods (Tidemann et al., 2017a; Walsh et al., 2012); however, seed retention at harvest may be too low in some cropping systems to effectively manage the population and is environment-dependent (Burton et al., 2016, 2017; Tidemann et al., 2016, 2017b; Walsh and Powles., 2014).

Increased crop seeding rate not only reduced *A. fatua* biomass and seed production but also increased the proportion of seed retained higher in the canopy (Walsh, 2019). Therefore agronomic strategies may be useful to increase the proportion of seeds available for management with HWSC.

IWM methods are also effective on *A. sterilis*. In Australia, Walker et al. (2002) found that seed production of *A. sterilis* was reduced in barley compared with wheat at reduced herbicide inputs. Increased planting density of wheat suppressed the growth, biomass production, and fecundity of *A. sterilis* (Radford et al., 1980). In a study in Spain, *A. sterilis* distribution in wheat fields was patchy and patches persisted over time (Castillejo-González et al., 2019). Therefore site-specific or precision management of this species based on precision maps developed in the previous growing season could be achieved with reduced herbicide inputs.

Herbicide-resistant *Avena* spp. warrant effective IWM strategies. It is clear that *Avena* spp. evade current weed control in contemporary cropping systems globally. A comprehensive understanding of *Avena* spp. biology and ecology are required to develop truly effective IWM programs. It is only by integrating knowledge of weed phenology and demography with that of management principles that these species will be managed effectively in contemporary cropping systems.

References

Adkins, S.W., Peters, N.C.B., 2001. Smoke derived from burn vegetation stimulates germination of arable weed seeds. Seed Sci. Res. 11, 213–222.

Adkins, S.W., Lowen, M., Symons, S.K., 1987. Variation within pure lines of wild oats (*Avena fatua*) in relation to temperature of development. Weed Sci. 35, 169–172.

Aibar, J., Ochoa, M.J., Zaragoza, C., 1991. Field emergence of *Avena fatua* L. and *A. sterilis* ssp. *ludoviciana* (Dur.) Nym. in Aragon, Spain. Weed Res. 31, 29–32.

Andaloussi, A., Bachikh, J., 2001. Studies on the host range of *Ditylenchus dipsaci* in Morocco. Nematol. Mediterr. 29, 51–57.

Anonymous, 2019. Invasive Species Compendium. CABI, Switzerland. Available from: http://www.cabi.org/isc, Accessed July 1, 2020.

Arora, A., Yaduraju, N.T., 1998. High-temperature effects on germination and viability of weed seeds. J. Agron. Crop. Sci. 181, 35–43.

Asai, M., Nakamura, N., Yogo, Y., 2010. Effects of delayed seeding and trifluralin application rate in controlling wild oat (*Avena fatua*) growing with winter wheat crops. J. Weed Sci. Technol. 55, 8–15.

Atienza, J., Tabernero, M.T., Álvarez-Benedí, J., Sanz, M., 2001. Volatilisation of triallate as affected by soil texture and air velocity. Chemosphere. 42, 257–261.

Aujas, C., Darmency, H., 1984. The concept of species in wild oats: *Avena fatua* L. and *A. sterilis*. In: Proc. 7th Int. Colloquium on Weed Ecology, Biology and Systematics (Vol. 1), European Weed Research Society, Paris, France, pp. 219–227.

Bajwa, A.A., Khalid, S., Sadia, S., Nabeel, M., Nafees, W., 2013. Influence of combinations of allelopathic water extracts of different plants on wheat and wild oat. Pak. J. Weed Sci. Res. 19, 157–166.

Bajwa, A.A., Akhter, M.J., Iqbal, N., Peerzada, A.M., Hanif, Z., Manalil, S., et al., 2017. Biology and management of *Avena fatua* and *Avena ludoviciana*: two noxious weed species of agro- ecosystems. Environ. Sci. Pollut. Res. 24, 19465–19479.

Balyan, R.S., Malik, R.K. 1989. Wild oat (Avena ludoviciana) competition with wheat (Triticum aestivum): effect of nitrogen fertilization. Season report of research projects of agronomy 2 and 4 on weed control, 31 pp.

Balyan, R.S., Malik, R.K., Panwar, R.S., Singh, S., 1991. Competitive ability of winter wheat cultivars with wild oat (*Avena ludoviciana*). Weed Sci. 39, 154–158.

BASF, 2019. Frequency Herbicide Label. BASF Australia, Victoria, Australia. Accessed November 15, 2020. https://crop-solutions.basf.com.au/products/frequency-herbicide.

Baskin, C.C., Baskin, J.M., 1985. The annual dormancy cycle in buried weed seeds. A continuum. BioScience. 35, 492–498.

Baskin, C.C., Baskin, J.M., 2014. Seeds: Ecology, Biogeography, and, Evolution of Dormancy and Germination, 2nd edn. Academic Press, San Diego, California, USA, p. 1600.

Batish, D.R., Singh, H.P., Kohli, R.K., Saxena, D.B., Kaur, S., 2002. Allelopathic effects of parthenin against two weedy species, *Avena fatua* and *Bidens pilosa*. Env. Exp. Bot. 47, 149–155.

Beckie, H.J., Tardif, F.J., 2012. Herbicide cross resistance in weeds. Crop. Prot. 35, 15–28.

Beckie, H.J., Leeson, J.Y., Thomas, G., Brenzil, C.A., Hall, L.M., Holzgang, G., et al., 2008. Weed resistance monitoring in the Canadian prairies. Weed Technol. 22, 530–543.

Beckie, H.J., Francis, A., Hall, L.M., 2012a. The biology of Canadian weeds. 27. *Avena fatua* L. (updated). Can. J. Plant. Sci. 92, 1329–1357.

Beckie, H.J., Warwick, S.I., Sauder, C.A., 2012b. Basis for herbicide resistance in Canadian populations of wild oat (*Avena fatua*). Weed Sci. 60, 10–18.

Beckie, H.J., Blackshaw, R.W., Low, R., Hall, L.M., Sauder, C.A., Martin, S., et al., 2013a. Glyphosate- and acetolactate synthase inhibitor-resistant kochia (*Kochia scoparia*) in western Canada. Weed Sci. 61, 310–318.

Beckie, H.J., Lozinski, C., Shirriff, S., Brenzil, C.A., 2013b. Herbicide-resistant weeds in the Canadian prairies: 2007-2011. Weed Technol. 27, 171–183.

Beckie, H.J., Flower, K.C., Ashworth, M.B., 2020a. Farming without glyphosate? Plants 9, e9010096.

Beckie, H.J., Shirriff, S.W., Leeson, J.Y., Hall, L.M., Harker, K.N., Dokken-Bouchard, F., et al., 2020b. Herbicide-resistant weeds in the Canadian prairies: 2012 to 2017. Weed Technol. 34, 461–474.

Belles, D.S., Thill, D.C., Shafii, B., 2000. PP-604 rate and *Avena fatua* density effects on seed production and viability in *Hordeum vulgare*. Weed Sci. 48, 378–384.

Blackshaw, R.E., Rode, L.M., 1991. Effect of ensiling and rumen digestion by cattle on weed seed viability. Weed Sci. 39, 104–108.

Blackshaw, R.E., Brandt, R.N., Janzen, H.H., Entz, T., Grant, C.A., Derksen, D.A., 2003. Differential response of weed speies to added nitrogen. Weed Sci. 51, 532–539.

Blackshaw, R.E., Brandt, R.N., Janzen, H.H., Entz, T., 2004. Weed species response to phosphorus fertilization. Weed Sci. 52, 406–412.

Blackshaw, R.E., Harker, K.N., O'Donovan, J.T., Beckie, H.J., Smith, E.G., 2008. Ongoing development of integrated weed management systems on the Canadian prairies. Weed Sci. 56, 146–150.

Boyd, N., Van Acker, R., 2004. Seed germination of common weed species as affected by oxygen concentration, light, and osmotic potential. Weed Sci. 52, 589–596.

Broster, J.C., Koetz, E.A., Wu, H., 2011. Herbicide resistance in wild oats (*Avena spp.*) in southern New South Wales. Plant. Prot. Q. 26, 106–107.

Burns, E.E., Keith, B.K., Talbert, L.W., Dyer, W.E., 2018. Non-target site resistance to flucarbazone, imazamethabenz and pinoxaden is controlled by three linked genes in *Avena fatua*. Weed Res. 58, 8–16.

Burton, N.R., Beckie, H.J., Willenborg, C.J., Shirtliffe, S.J., Schoenau, J.J., Johnson, E.N., 2016. Evaluating seed shatter of economically important weed species. Weed Sci. 64, 673–682.

Burton, N.R., Beckie, H.J., Willenborg, C.J., Shirtliffe, S.J., Schoenau, J.J., Johnson, E.N., 2017. Seed shatter of six economically important weed species in producer fields in Saskatchewan. Can. J. Plant. Sci. 97, 266–276.

Busi, R., Girotto, M., Powles, S.B., 2016. Response to low-dose herbicide selection in self-pollinated *Avena fatua*. Pest. Manag. Sci. 72, 603–608.

Carlson, H.L., Hill, J.E., 1985. Wild oat (*Avena fatua*) competition with spring wheat: plant density effects. Weed Sci. 33, 176–181.

Castellanos-Frías, E., de León, D.G., Pujadas-Salva, A., Dorado, J., Gonzalez-Andujar, J.L., 2014. Potential distribution of *Avena sterilis* L. in Europe under climate change. Ann. Appl. Biol. 165, 53–61.

Castillejo-González, I.L., de Castro, A.I., Jurado-Expísito, M., Peña, J.-M., García-Ferrer, A., López-Granados, F., 2019. Assessment of the persistence of *Avena sterilis* L. patches in wheat fields for site-specific sustainable management. Agronomy 9, e9010030.

Castillo, J., Ahrens, W.H., 1986. Wild oat competition in no-till and conventional tillage corn. In: Proceedings of the North-Central Weed Control Conference, Milwaee, Wisconsin, USA, 2–4 Dec 1986, p. 83.

Cavan, G., Biss, P., Moss, S.R., 1998. Herbicide resistance and gene flow in wild-oats (*Avena fatua* and *Avena sterilis* ssp. *ludoviciana*). Ann. Appl. Biol. 133, 207–217.

Conn, J.S., Deck, R.E., 1995. Seed viability and dormancy of 17 weed species after 9.7 years of burial in Alaska. Weed Sci. 43, 583–585.

Dai, J., Wiersma, J.J., Martinson, K.L., Durgan, B.R., 2012. Influence of time of emergence on the growth and development of wild oat (*Avena fatua*). Weed Sci. 60, 389–393.

Dar, A., Zahir, Z.A., Ashgar, H.N., Ahmad, R., 2020. Preliminary screening of rhizobacteria for biocontrol of little seed canary grass (*Phalaris minor* Retz.) and wild oat (*Avena fatua* L.) in wheat. Can. J. Microbiol. 66, 368–376.

Darmency, H., Uludag, A., 2018. A quantitative genetic examination of non-target-site resistance applied to *Avena* species. Weed Res. 58, 69–75.

Darwent, A.L., Smith, J.H., 1985. Delayed seeding for wild oat control in rapeseed in northwest Alberta. Can. J. Plant. Sci. 65, 1101–1106.

Dhima, K.V., Eleftherohorinos, I.G., 2001. Influence of nitrogen on competition between winter cereals and sterile oat. Weed Sci. 49, 77–82.

Dhima, K.V., Eleftherohorinos, I.G., Vasilakoglou, I.G., 2000. Interference between *Avena sterilis*, *Phalaris minor* and five barley cultivars. Weed Res. 40, 549–559.

Edgar, E., 1980. The identification of oats growing wild in New Zealand. In: Proceedings of the 33rd New Zealand Weed and Pest Control Conference, pp. 230–236.

Fast, B.J., Medlin, C.R., Murray, D.S., 2009. Five cool-season annual grass weeds reduce hard red winter wheat grain yield and price. Weed Technol. 23, 206–213.

Fernandez-Quintanilla, C., Navarrete, L., Andujar, J.L.G., Fernandez, A., Sanchez, M.J., 1986. Seedling recruitment and age-specific survivorship and reproduction in populations of *Avena sterilis* L. ssp. *ludoviciana* (Durieu) Nyman. J. Appl. Ecol. 23, 945–955.

Fernandez-Quintanilla, C., Gonzalez, A.J.L., Appleby, A.P., 1990. Characterization of the germination and emergence response to temperature and soil moisture of *Avena fatua* and *A. sterilis*. Weed Res. 30, 289–295.

Floc'h, E., 1991. Invasive plants of the Mediterranean basin. In: Groves, R.H., Castri, F. (Eds.), Biogeography of Mediterranean Invasions. Cambridge University Press, Cambridge, UK, pp. 67–80.

Foley, M.E., 1987. The effect of wounding on primary dormancy in wild oat (*Avena fatua*) caryopses. Weed Sci. 35, 180–184.

Foley, M.E., 1994. Temperature and water status of seed affect afterripening in wild oat (*Avena fatua*). Weed Sci. 42, 200–204.

Fu, Y.-B., Chong, J., Fetch, T., Wang, M.-L., 2007. Microsatellite variation in *Avena sterilis* germplasm. Theor. Appl. Genet. 114, 1029–1038.

Geddes, C.M., Cavaliere, A., Daayf, F., Gulden, R.H., 2015. The allelopathic potential of hairy vetch (*Vicia villosa* Roth.) mulch. Am. J. Plant. Sci. 6, 2651–2663.

Gillespie, G.R., Nalewaja, J.D., 1988. Economic control of weeds in wheat, *Triticum aestivum*. Weed Technol. 2, 257–261.

Gonzalez-Ponce, R., Santin, I., 2001. Competitive ability of wheat cultivars with wild oats depending on nitrogen fertilization. Agronomie 21, 119–125.

Harker, K.N., Blackshaw, R.E., Clayton, G.W., 2001. Timing weed removal in field pea (*Pisum sativum*). Weed Technol. 15, 277–283.

Harker, K.N., Blackshaw, R.E., Clayton, G.W., 2007. Wild oat (*Avena fatua*) vs. redstem filaree (*Erodium cicutarium*) interference in dry pea. Weed Technol. 21, 235–240.

Harker, K.N., O'Donovan, J.T., Irvine, R.B., Turkington, T.K., Clayton, G.W., 2009. Integrating cropping systems with cultural techniques augments wild oat (*Avena fatua*) management in barley. Weed Sci. 57, 326–337.

Harker, K.N., O'Donovan, J.T., Turkington, T.K., Blackshaw, R.E., Lupwayi, N.Z., Smith, E.G., et al., 2016. Diverse rotation and optimal cultural practices control wild oat (*Avena fatua*). Weed Sci. 64, 170–180.

Hassanein, E.E., Kholosy, A.S., Abd-Alla, M.M.S., Ibrahim, H.M., 1996. Effect of temperature degrees on seed germination and seedling vigour of different wild oat species. Ann. Agric. Sci. 34, 1373–1380.

Heap, I.M., 2020. International Herbicide-Resistant Weed Database. http://www.weedscience.org (Accessed 1 July 2020).

Hilton, J.R., 1984. The influence of light and potassium nitrate on the dormancy and germination of *Avena fatua* L. (wild oat) seed and its ecological significance. N. Phytol. 96, 31–34.

Holm, L., Pancho, J.V., Herberger, J.P., Plucknett, D.L., 1979. A Geographical Atlas of Worst Weeds. John Wiley and Sons Inc., Toronto, ON, Canada.

Hou, J.Q., Simpson, G.M., 1990. Phytochrome action and water status in seed germination of wild oats (*Avena fatua*). Can. J. Bot. 68, 1722–1723.

Hou, J.Q., Simpson, G.M., 1991. Effects of prolonged light on germination of six lines of wild oat (*Avena fatua*). Can. J. Bot. 69, 1414–1417.

Hou, J.Q., Simpson, G.M., 1993. Germination response to phytochrome depends on specific dormancy states in wild oat (*Avena fatua*). Can. J. Bot. 71, 1528–1532.

Hsiao, A.I., Simpson, G.M., 1971. Dormancy studies in seed of *Avena fatua*. 7. The effects of light and variation in water regime on germination. Can. J. Bot. 49, 1347–1357.

Hsiao, A.I., McIntyre, G.I., Haines, J.A., 1983. Seed dormancy in *Avena fatua*. I. Induction of germination by mechanical injury. Bot. Gaz. 144, 217–222.

Ivens, G.W., 1989. East African Weeds and their Control, 2nd edn. Oxford University Press, Tanzania Ltd., Nairobi, Kenya, p. 303.

Jabran, K., Farooq, M., Hussain, M., Rehman, H., Ali, M.A., 2010. Wild oat (*Avena fatua* L.) and canary grass (*Phalaris minor* Ritz.) management through allelopathy. J. Plant. Prot. Res. 50, 41–44.

Jäck, O., Menegat, A., Gerhards, R., 2017. Winter wheat yield loss in response to *Avena fatua* competition and effect of reduced herbicide dose rates on seed production of this species. J. Plant. Dis. Prot. 124, 371–382.

Jacob, H.S., Minkey, D.M., Gallagher, R.S., Borger, C.P., 2006. Variation in postdispersal weed seed predation in a crop field. Weed Sci. 54, 148–155.

Jana, S., Naylor, J.M., 1980. Dormancy studies in seed of *Avena fatua*. 11. Heritability for seed dormancy. Can. J. Bot. 58, 91–93.

Jana, S., Thai, K.M., 1987. Patterns of changes of dormant genotypes in *Avena fatua* populations under different agricultural conditions. Can. J. Bot. 65, 1741–1745.

Jones, R., Medd, R., 1997. Economic analysis of integrated management of wild oats involving fallow, herbicide and crop rotational options. Aust. J. Exp. Agric. 37, 683–691.

Keith, B.K., Burns, E.E., Bothner, B., Carey, C.C., Mazurie, A.J., Hilmer, J.K., et al., 2017. Intensive herbicide use has selected for constitutively elevated levels of stress-responsive mRNAs and proteins in multiple herbicide-resistant *Avena fatua* L. Pest. Manag. Sci. 73, 2267–2281.

Kępczyński, J., 2018. Induction of agricultural weed seed germination by smoke and smoke-derived karrikin (KAR_1), with a particular reference to *Avena fatua* L. Acta Physiol. Plant. 40, 87–96.

Kępczyński, J., Bialecka, B., Light, M.E., van Staden, J., 2006. Regulation of *Avena fatua* seed germination by smoke solutions, gibberellin A_3 and ethylene. Plant. Growth Regul. 49, 9–16.

Khan, I., Hassan, G., Khan, M.I., Gul, M., 2008. Effect of wild oats (*Avena fatua* L.) population and nitrogen levels on some agronomic and physiological traits of wheat. Electron. J. Environ. Agric. Food Chem. 7, 2723–2734.

Kieloch, R., Golebiowska, H., 2018. Influence of environmental conditions and crop competition on morphological and biological diversity of *Avena fatua* L. and *Solanum nigrum* L. Acta Agrobotanica 71, 1740.

Kirkland, K.J., 1993. Spring wheat (*Triticum aestivum*) growth and yield as influenced by duration of wild oat (*Avena fatua*) competition. Weed Technol. 7, 890–893.

Kirkland, K.J., Hunter, J.H., 1991. Competitiveness of Canada prairie spring wheats with wild oat (*Avena fatua* L.). Can. J. Plant. Sci. 71, 1089–1092.

Kulsoom, U., Gul, J., Bakht, T., Kanwal, S., 2018. Increasing densities of *Avena fatua* and *Rumex dentatus* reduce the yield of wheat under field conditions. Pak. J. Weed Sci. Res. 24, 155–164.

Lanning, S.P., Talbert, L.E., Martin, J.M., Blake, T.K., Bruckner, P.L., 1997. Genotype of wheat and barley affects light penetration and wild oat growth. Agron. J. 89, 100–103.

Larney, F.J., Blackshaw, R.E., 2003. Weed seed viability in composted beef cattle feedlot manure. J. Environ. Qual 32, 1105–1113.

Leeson, J.Y., Beckie, H.J., 2012. Influence of climate on weed species distribution in the Canadian Prairies. In: Ivany, J.A., Blackshaw, R.E. (Eds.), Climate Change and the Canadian Agricultural Environment. Topics in Canadian Weed Science, Vol. 8. Canadian Weed Science Society, Pinawa, Manitoba, Canada, p. 98.

Lehnhoff, E.A., Miller, Z.J., Brelsford, M.J., White, S., Maxwell, B.D., 2013. Relative canopy height influences wild oat (*Avena fatua*) seed viability, dormancy, and germination. Weed Sci. 61, 564–569.

Liu, W.J., Harrison, D.K., Chalupska, D., Gornicki, P., O'Donnell, C.C., Adkins, S.W., et al., 2007. Single-site mutations in the carboxyl-transferase domain of plastid acetyl-CoA carboxylase confer resistance to grass-specific herbicides. Proc. Natl. Acad. Sci. U.S.A. 104, 3627–3632.

Loskutov, I.G., 2008. On evolutionary pathways of *Avena* species. Genet. Resour. Crop. Ecol. 55, 211–220.

Mangin, A.R., Hall, L.M., Schoenau, J.J., Beckie, H.J., 2017a. Influence of tillage, vertical seed distribution, and pyroxasulfone application timing and rate on control of wild oat (*Avena fatua* L.). Can. J. Plant. Sci. 98, 601–608.

Mangin, A.R., Hall, L.M., Beckie, H.J., 2017b. Triallate-resistant wild oat (*Avena fatua* L.): unexpected resistance to pyroxasulfone and sulfentrazone. Can. J. Plant. Sci. 97, 20–25.

Mariot, M.P., Sereno, M.J.C., Federizzi, L.C., Carvalho, F.I.F., 1999. Inheritance of plant height and main panicle length for crosses between *Avena sativa* L. and *Avena sterilis* L. Pesq. Agro. Bras. 34, 77–82.

Martin, A.N., Scursoni, J.A., 2018. *Avena fatua* L. escapes and delayed emergence in wheat (*Triticum aestivum* L.) crops of Argentina. Crop. Prot. 103, 30–38.

May, W.E., 2018. Altering the competitiveness of tame oat (*Avena sativa* L.) versus wild oat (*Avena fatua* L.) with phosphorous and seeding rate. Can. J. Plant. Sci. 98, 582–590.

McGillion, T., Storrie, A. (Eds.), 2006. Integrated Weed Management in Australian Cropping Systems. CRC for Australian Weed Management, Adelaide, SA, Australia, pp. 199–201.

McKenzie, R.I.H., Harder, D.E., 1995. Oat. In: Slinkard, A.E., Knott, D.R. (Eds.), Harvest of Gold: The History of Field Crop Breeding in Canada. University of Saskatchewan, Saskatoon, SK, Canada, pp. 98–112.

Mennan, H., Uygur, F.N., 1996. The germination and growth biology of some important weeds in wheat fields. Ondokuz Mayis Uni Fac. Agric. J. 11, 153–166.

Michael, P.J., Owen, M.J., Powles, S.B., 2010. Herbicide-resistant weed seed contaminate grain sown in the Western Australia grainbelt. Weed Sci. 58, 466–472.

Miller, S.D., Nalewaja, J.D., 1990. Influence of burial depth on wild oats (*Avena fatua*) seed longevity. Weed Technol. 4, 514–517.

Morishita, D.W., Thill, D.C., 1988. Factors of wild oat (*Avena fatua*) interference on spring barley (*Hordeum vulgare*) growth and yield. Weed Sci. 36, 37–42.

Moss, S., 2015. Identification of Wild Oats. Rothamsted Technical Publication, Rothamsted Research, Harpenden, UK. Available from: http://www.rothamsted.ac.uk/tools, Accessed 1 July 2020.

Murdoch, A.J., 1983. Environmental control of germination and emergence in *Avena fatua*. Asp. Appl. Biol. 4, 63–69.

Murray, B.G., Morrison, I.N., Friesen, L.F., 2002. Pollen-mediated gene flow in wild oat. Weed Sci. 50, 321–325.

Mustafa, A., Muhammad, N., Abbas, T., Saeed, Q., Hussain, A., Ashraf, M.N., et al., 2019. Growth response of wheat and associated weeds to plant antagonistic rhizobacteria and fungi. Italian J. Agron. 14, 191–198.

Necajeva, J., Erdmane, Z., Isoda-Krasovska, A., Curiske, J., Dudele, I., Gaile, L., et al., 2017. Influence of wild oat plant density on spring wheat yield. Zemdirbyste-Agriculture 104, 209–218.

Niekerk, B.D., Pretorius, Z.A., Boshoff, W.H.P., 2001. Pathogenic variability of *Puccinia coronata* f. sp. avenae and *P. graminis* f. sp. avenae on oat in South Africa. Plant. Dis. 85, 1085–1090.

O'Donnell, C.C., Adkins, S.W., 2001. Wild oat and climate change: the effect of CO_2 concentration, temperature and water deficit on the growth and development of wild oat in monoculture. Weed Sci. 49, 694–702.

O'Donnell, C.C., Adkins, S.W., Walker, S.R., 2002. Herbicide resistance in the northern grain region of Australia: Developments in research from 1993 to present. In: Jacob, H.S., Dodd, J., Moore, J.H. (Eds.). Proceedings of the 13th Australian Weeds Conference, Perth, Western Australia, pp. 617–619.

O'Donovan, J.T., de St Remy, E.A., O'Sullivan, P.A., Dew, D.A., Sharma, A.K., 1985. Influence of the relative time of emergence of wild oat (*Avena fatua*) on yield loss of barley (*Hordeum vulgare*) and wheat (*Triticum aestivum*) yield and profitability. Can. J. Plant. Sci. 83, 977–985.

O'Donovan, J.T., Sharma, M.P., Harker, K.N., Maurice, D., Baig, M.N., Blackshaw, R.E., 1994. Wild oat (*Avena fatua*) populations resistant to triallate are also resistant to difenzoquat. Weed Sci. 42, 195–199.

Owen, M.J., Powles, S.B., 2009. Distribution and frequency of herbicide-resistant wild oat (*Avena* spp.) across the Western Australian grain belt. Crop. Pasture Sci. 60, 25–31.

Owen, M.J., Powles, S.B., 2016. The frequency of herbicide-resistant wild oat (*Avena* spp.) populations remains stable in Western Australian cropping fields. Crop. Pasture Sci. 67, 520–527.

Peters, N.C.B., 1982. The dormancy of wild oat seed (*Avena fatua* L.) from plants grown under various temperature and soil moisture conditions. Weed Res. 22, 205–212.

Phillips, T.D., Murphy, J.P., Goodman, M.M., 1993. Isozyme variation in germplasm accessions of the wild oat *Avena sterilis* L. Theor. Appl. Genet. 86, 54–64.

Qasem, J.R., 2007. Chemical control of wild-oat (*Avena sterilis* L.) and other weeds in wheat (*Triticum durum* Desf.) in Jordan. Crop. Prot. 26, 1315–1324.

Quail, P.H., Carter, O.G., 1968. Survival and seasonal germination of seeds of *Avena fatua* and *A. ludoviciana*. Aust. J. Agric. Res. 19, 721–729.

Radford, B.J., Wilson, B.J., Cartledge, O., Watkins, F.B., 1980. Effect of wheat seeding rate on wild oat competition. Aust. J. Exp. Agric. Anim. Husb. 20, 77–81.

Rajhathy, T., Thomas, H., 1974. Cytogenetics of Oats (*Avena* L.). The Genetics Society of Canada, Ottawa, Ontario, Canada, p. 90.

Rola, H., 1984. Influence of wild oat (*Avena fatua* L.) on the yields of winter wheat and spring wheat. Pamietnik Pulawski 84, 133–144.

Rosentrater, K.A., Evers, A.D., 2018. Introduction to cereals and pseudocereals and their production, Kent's Technology of Cereals, 5th ed. Science Direct, London, UK, pp. 1–76.

Sáez, L., Aedo, C., Romero-Zarco, C., 2017. Proposal to conserve the name *Avena sterilis* (*Poaceae*) with a conserved type. Taxon 66, 203–204.

Sanchez del Arco, M.J., Torner, C., Quintanilla, C.F., 1995. Seed dynamics in populations of *Avena sterilis* ssp. *ludoviciana*. Weed Res. 35, 477–487.

Sawhney, E., Naylor, J.M., 1980. Dormancy studies in seeds of *Avena fatua*. 12. Influence of temperature on germination behavior of nondormant families. Can. J. Bot. 58, 579–581.

Sawhney, R., Hsiao, A.I., Quick, W.Q., 1986. The influence of diffuse light and temperature on seed germination of three genetically nondormant lines of wild oats (*Avena fatua*) and its adaptive significance. Can. J. Bot. 54, 1910–1915.

Scursoni, J.A., Satorre, E.H., 2005. Barley (*Hordeum vulgare*) and wild oat (*Avena fatua*) is affected by crop and weed density. Weed Technol. 19, 790–795.

Scursoni, J.A., Benech-Arnold, R., Hirchoren, H., 1999. Demography of wild oats in barley crops: effect of crop, sowing rate and herbicide treatment. Agron. J. 91, 478–485.

Sereno-Tavares, M.J.C.M., Bodanese-Zanettini, M.H., de Carvalho, F.I.Fl, 1995. Hybridization among Brazilian oat (*Avena sativa* L.) cultivars and the wild oat *A. sterilis* L. Pesq. Agro. Bras. 30, 383–388.

Sharma, M.P., Vanden Born, W.H., 1978. The biology of Canadian weeds. 27. *Avena fatua* L. Can. J. Plant. Sci. 58, 141–157.

Shirtliffe, S.J., Entz, M.H., 2005. Chaff collection reduces seed dispersal of wild oat (*Avena fatua*) by a combine harvester. Weed Sci. 53, 465–470.

Shirtliffe, S.J., Entz, M.H., Van Acker, R.C., 2000. *Avena fatua* development and seed shatter as related to thermal time. Weed Sci. 48, 555–560.

Shuma, J.M., Quick, W.A., Raju, M.V.S., Hsiao, A.I., 1995. Germination of seeds from plants of *Avena fatua* L. treated with glyphosate. Weed Res. 35, 249–255.

Smith, E.G., Harker, K.N., O'Donovan, J.T., Turkington, T.K., Blackshaw, R.E., Lupwayi, N.Z., et al., 2018. The profitability of diverse crop rotations and other cultural methods that reduce wild oat (*Avena fatua*). Can. J. Plant. Sci. 98, 1094–1101.

Soldatov, V.N., Loskutov, I.G., 1991. Distribution of wild and weedy oat species in the Caucasus, USSR. Plant. Genet. Resour. Newsl. 86, 38.

Somody, C.N., Nalewaja, J.D., Miller, S.D., 1985. Self-burial of wild oat florets. Agron. J. 77, 359–362.

Stace, C., 1997. New Flora of the British Isles, 2nd edn. Cambridge University Press, Cambridge, UK.

Stevens, J.C., Merritt, D.J., Flematti, G.R., Ghisalberti, E.L., Dixon, K.W., 2007. Seed germination of agricultural weeds is promoted by the butenolide 3-methyl-2*H*-furo[2,3-*c*]pyran-2-one under laboratory and field conditions. Plant. Soil. 298, 113–124.

Stougaard, R.N., Xue, Q., 2004. Spring wheat seed size and seeding rate effects on yield loss due to wild oat (*Avena fauta*) interference. Weed Sci. 52, 133–141.

Stougaard, R.N., Xue, Q., 2005. Quality versus quantity: spring wheat seed size and seeding rate effects on *Avena fatua* interference, economic returns and economic thresholds. Weed Res. 45, 351–360.

Symons, S.J., Simpson, G.N., Adkins, S.W., 1987. Secondary dormancy in *Avena fatua*: effect of temperature and after-ripening. Physiol. Plant. 70, 419–426.

Terry, P.J., 1984. A Guide to Weed Control in East African Crops. Kenya Literature Bureau, Nairobi, Kenya, p. 186.

Thill, D.C., Mallory-Smith, C.A., 1997. The nature and consequence of weed spread in cropping systems. Weed Sci. 45, 337–342.

Thurston, J.M., 1951. Some experiments and field observations on the germination of wild oat (*Avena fatua* and *A. ludoviciana*) seeds in soil and the emergence of seedlings. Ann. Appl. Biol. 38, 812–823.

Thurston, J.M., 1957. Morphological and physiological variation in wild oats (*Avena fatua* L. and *A. ludoviciana* Dur.) and in hybrids between wild and cultivated oats. J. Agric. Sci. 49, 259–274.

Tidemann, B.D., Hall, L.M., Johnson, E.N., Beckie, H.J., Sapsford, K.L., Raatz, L.L., 2014. Efficacy of fall- and spring-applied pyroxasulfone for herbicide-resistant weeds in field pea. Weed Technol. 28, 351–360.

Tidemann, B.D., Hall, L.M., Harker, K.N., Alexander, B.C.S., 2016. Identifying the crtical control points in the wild oat (*Avena fatua*) life cycle and the potential effects of harvest weed seed control. Weed Sci. 64, 463–473.

Tidemann, B.D., Hall, L.M., Harker, K.N., Beckie, H.J., 2017a. Factors affecting weed seed devitalization with the Harrington Seed Destructor. Weed Sci. 65, 650–658.

Tidemann, B.D., Hall, L.M., Harker, K.N., Beckie, H.J., Johnson, E.N., Stevenson, F.C., 2017b. Suitability of wild oat (*Avena fatua*), false cleavers (*Galium spurium*), and volunteer canola (*Brassica napus*) for harvest weed seed control in western Canada. Weed Sci. 65, 769–777.

Tidemann, B.D., Harker, K.N., Johnson, E.N., Willenborg, C.J., Shirtliffe, S.J., 2020. Time of wild oat (*Avena fatua*) panicle clipping influences seed viability. Weed Sci. 68, 260–267.

Tilsner, H.R., Upadhyaya, M.K., 1985. Induction and release of secondary seed dormancy in genetically pure lines of *Avena fatua*. Physiol. Plant. 64, 377–382.

Torner, C., Gonzalez-Andujar, J.L., Fernandez-Quintanilla, C., 1991. Wild oat (*Avena sterilis*) competition with winter barley: plant density effects. Weed Res. 31, 301–307.

Travlos, I.S., Giannopolitis, C.N., Economou, G., 2011. Diclofop resistance in sterile wild oat (*Avena sterilis* L.) in wheat fields in Greece and its management by other post-emergence herbicides. Crop. Prot. 30, 1449–1454.

Uremis, I., Uygur, F.N., 1999. Minimum, optim and maximum germination temperatures of some important weed species in the Çukurova region of Turkey. Turk. J. Weed Sci. 2, 1–12.

USDA, 2020. Plants profile for *Avena fatua*. [online]. Accessed August 7, 2020. Available at: https://plants.usda.gov/core/profile?symbol = AVFA.

Van Acker, R.C., 2009. Weed biology serves practical weed management. Weed Res. 49, 1–5.

Van Wychen, L.R., Maxwell, B.D., Bussan, A.J., Miller, P.R., Luschei, E.C., 2004. Wild oat (*Avena fatua*) habitat and water use in cereal grain cropping systems. Weed Sci. 52, 352–358.

Vanstone, V.A., Russ, M.H., 2001. Ability of weeds to host the root legion nematodes*Pratylenchus neglectus* and *P. thornei*. I. Grass weeds. Aust. Plant. Pathol. 30, 245–250.

Walia, U.S., Singh, M., 2005. Studies on the threshold values of *Avena ludoviciana* and *Rumex spinosus* in wheat. Ind. J. Weed Sci. 37, 91–92.

Walia, U.S., Seema, J., Brar, L.S., Singh, M., 2001. Competitive ability of wheat with variable population of wild oats (*Avena ludoviciana* Dur.). Ind. J. Weed Sci. 33, 171–173.

Walker, S.R., Medd, R.W., Robinson, G.R., Cullis, B.R., 2002. Improved management of *Avena ludoviciana* and *Phalaris paradoxa* with more densely sown wheat and less herbicide. Weed Res. 42, 257–270.

Walsh, M.J., 2019. Enhanced wheat competition effects on the growth, seed production, and seed retention of major weeds of Australian cropping systems. Weed Sci. 67, 657–665.

Walsh, M.J., Powles, S.B., 2014. High seed retention at maturity of annual weeds infesting crop fields highlights the potential for harvest weed seed control. Weed Technol. 28, 486–493.

Walsh, M.J., Harrington, R.B., Powles, S.B., 2012. Harrington Seed Destructor: a new nonchemical weed control tool for global grain crops. Crop. Sci. 52, 1343–1347.

Walsh, M.J., Broster, J.C., Schwartz-Lazaro, L.M., Norsworthy, J.K., Davis, A.S., Tidemann, B. D., et al., 2018. Opportunities and challenges for harvest weed seed control in global cropping systems. Pest. Manag. Sci. 74, 2235–2245.

Willenborg, C.J., Rossnagel, B.G., May, W.E., Lafond, G.P., Shirtliffe, S.J., 2005. Influence of wild oat (*Avena fatua*) relative time of emergence and density on cultivated oat yield, wild oat seed production, and wild oat contamination. Weed Sci. 53, 342–352.

Wilson, B.J., Cousens, R., Wright, K.J., 1990. The response of spring barley and winter wheat to *Avena fatua* population density. Ann. Appl. Biol. 116, 601–609.

Wimschneider, W., Bachthaler, G., Fischbeck, G., 1990. Competitive effects of *Avena fatua* L. (wild oats) on wheat (*Triticum aestivum* L.) as a basis for effective weed control. Weed Res. 30, 43–52.

WSSA, 2020. Weed Science Society of America. Composite List of Weeds. http://wssa.net/wssa/weed/composite-list-of-weeds (Accessed 1 July 2020).

Yu, Y., Chen, P., Liu, D., 1998. Cytogenetic study on cross between *A. nuda* and *A. sterilis*. J. Nanjing Agric. Univ. 21, 1–6.

Zand, E., Beckie, H.J., 2002. Competitive ability of hybrid and open-pollinated canola (*Brassica napus*) with wild oat (*Avena fatua*). Can. J. Plant. Sci. 82, 473–480.

Ziska, L.H., 2017. Could recent increases in atmospheric CO_2 have acted as a selection factor in *Avena fatua* populations? A case study of cultivated and wild oat competition. Weed Res. 57, 399–405.

Zorner, P.S., Zimdahl, R.L., Schweizer, E.E., 1984. Sources of variable seed loss in buried dormant and non-dormant populations of wild oat (*Avena fatua* L.) seed in Colorado. Weed Res. 24, 143–150.

Chapter 4

Bromus diandrus and *Bromus rigidus*

Catherine P.D. Borger[1], Joel Torra[2], Aritz Royo-Esnal[2], Laura Davies[3] and George Newcombe[4]

[1]Department of Primary Industries and Regional Development, Northam, WA, Australia, [2]Department d'Hortofructicultura, Botànica i Jardineria, Agrotecnio, Universitat de Lleida, Lleida, Spain, [3]ADAS, Battlegate Road, Boxworth, Cambridgeshire, United Kingdom, [4]University of Idaho, College of Natural Resources, Mowcow, Idaho

Introduction

Bromus diandrus Roth and *Bromus rigidus* Roth are native to the Mediterranean region and Eurasia. *Bromus diandrus* is now found throughout Europe, Asia, Africa, North America, and Oceania (Clapham et al., 1952; Kon and Blacklow, 1989). *Bromus rigidus* is found in similar areas, but in the United States the same species is considered as *B. diandrus* so there are fewer records specifically attributed to *B. rigidus* (Sales, 1993). Likewise, in Australia, various species of the *Bromus* genus including *B. diandrus* were widespread by the mid-1800's. However, *B. rigidus* was not reported until 1986, although it was probably present before this date (CABI, 2020). In western and northern Europe, both species are regarded as being introduced, with *B. diandrus* more prevalent than *B. rigidus* (Royal Botanic Garden Edinburgh, 2020). Additionally, in these areas, both species are often confused with the common, native *Bromus sterilis* L. and are possibly more widespread than indicated by current records (Davies and Hull, 2018).

Both species are increasingly prevalent in agricultural systems. Throughout Europe and Australia, this has been attributed to the spread of conservation tillage and low diversity of crop rotations (Kleemann and Gill, 2009a; Garcia et al., 2014). In southern Europe and Australia the introduction of conservation tillage increases moisture availability and crop water use efficiency, vital to the nonirrigated, low yielding, cereal dominant cropping systems (D'Emden et al., 2008; Garcia et al., 2014). In northern and central Europe, conservation tillage is adopted to improve soil health and reduce costs (Petersen, 2006; Morris et al., 2010; Davies and Hull, 2018).

Chauhen-BMPCWS-1632754. DOI: https://doi.org/10.1016/B978-0-12-822917-0.00004-5

Diversity of crop rotations has reduced, due to climatic limitations or market forces, encouraging a cereal monoculture, highly dependent on a small number of herbicides (Kleemann and Gill, 2009a; Morris et al., 2010; Garcia et al., 2014).

There are few effective in-crop herbicides for the control of *Bromus* spp. in cereals, and most are applicable to wheat (*Triticum aestivum* L.) rather than barley (*Hordeum vulgare* L.) (Kleemann and Gill, 2009a; Garcia et al., 2014; Royo-Esnal et al., 2018). Further, the herbicides that are available have variable performance depending on environmental conditions and weed density (Kleemann and Gill, 2008; Garcia et al., 2014; Kleemann et al., 2016). The highly competitive nature of these species and staggered emergence times, combined with variable efficiency of control tactics makes integrated weed management (IWM) necessary for effective control in agronomic systems (Kleemann and Gill, 2008; Garcia et al., 2014; Kleemann et al., 2016).

Taxonomy and description

Bromus diandrus and *B. rigidus* are found within tribe Bromeae, subfamily Pooideae. These species are closely related, with the exact taxonomic classification still open to international debate. Both species are polyploids; *B. diandrus* is $2n = 28$, 42, and 56, and *B. rigidus* is $2n = 28$, 42, 56, 70 (Oja and Laarmann, 2002). *Bromus diandrus* and *B. rigidus* are annual plants, 15–90 cm tall. They are found in cropped fields or other disturbed sites, and it is common to find both species in the same field (Clapham et al., 1952; Kon and Blacklow, 1989).

For both species, leaves are wide (10 mm) and generally hairy (hairs of 0–0.8 mm, with 0–750 hairs/cm^2), with a membranous, prominent ligule (Kon and Blacklow, 1989). The panicles are 9–20 cm long and can be loose and nodding (spikelet branches longer than the spikelet) or more compact and erect (spikelet branches shorter than the spikelet, Fig. 4.1). Tetraploid, hexaploid, and octoploid ecotypes within the *B. diandrus–rigidus* complex may have the spreading panicles typically associated with *B. diandrus* or the contracted, erect panicles often associated with *B. rigidus* (Kon and Blacklow, 1989; Oja and Laarmann, 2002). The spikelet is 24–44 mm long with 4–8 florets. Awn length is variable (29–78 mm long), and the lemmas are elongated (20–35 mm long). Lemmas either have a short ($\leq$1 mm), spherical callus with a rounded tip or an elongated callus ($>$ 1 mm) with a pointed tip. The scar from the rachilla is circular in tetraploids or elliptical (more pointed) in hexaploids or octoploids (Fig. 4.2, Clapham et al., 1952; Kon and Blacklow, 1989; Oja and Laarmann, 2002).

Glumes are unequal, and the flowers of *B. diandrus* are hermaphrodite and self-pollinated (cleistogamous); outcrossing is possible by wind pollination (anemogamy) but infrequent ($<$1%, Kon and Blacklow, 1990). It is

FIGURE 4.1 A loose panicle from a *Bromus diandrus* plant (left) and a compact panicle from a *B. rigidus* plant (right) from South Australia. *Photos courtesy of Dr. Sam Kleemann, Plant Science Consulting, Australia.*

FIGURE 4.2 A seed with the circular rachilla scar observed in tetraploids (left) and the pointed scar observed in hexa or octoploids (right), from *Bromus* species ecotypes in Western Australia. *Photos courtesy of Mr Andrew Storrie, AGRONOMO Australia.*

likely that the breeding system of *B. rigidus* is similar, based on the high proportion of seed set in cleistogamous florets and similar floral structures (Kon and Blacklow, 1990). The panicles and leaves may remain green as the plants mature or turn to a reddish-purple prior to senescence (Kon and Blacklow, 1989).

In the United States, *B. diandrus* and *B. rigidus* are considered a single species, or *B. rigidus* is treated as a subspecies of *B. diandrus*, although the two species are considered separately in the record of potential biocontrol agents (Tsvelev, 1984; O'Connor, 1990; Farr and Rossman, 2020). More recently, Sales (1993) classified *B. diandrus*, *B. rigidus*, and *B. sterilis* as varieties of one species. In Europe, *B. diandrus* and *B. rigidus* are considered separate species (Stace, 1991). Although *Bromus* is the most widespread term, the synonyms *Anisantha diandra* (Roth) Tutin ex Tzvelev. or *A. rigida* (Roth) Hyl. are also used (Stace, 1991). Likewise, in Australia the two have been treated as separate species, mainly because populations lack the genetic diversity of international ecotypes. Australian species are not polyploids; *B. diandrus* $2n = 56$ and *B. rigidus* $2n = 42$ (Kon and Blacklow, 1989).

An international genetic analysis is required to fully understand the relationships within the *Bromus* genus; the *B. diandrus–rigidus–sterilis* complex may be in the process of speciation. However, there is too much genetic variability and shared or variable morphological traits to allow a final taxonomic decision without further research (Kon and Blacklow, 1989; Oja and Laarmann, 2002). The current review includes research on both *B. diandrus* and *B. rigidus* (or *A. diandra* and *A. rigida*) and accepts the name assigned by the authors of the research under discussion. Research referring to "*Bromus* sp." is also included, where papers originate from countries where either *B. diandrus* or *B. rigidus* are the most common representatives of the *Bromus* genus.

Seed biology, seed bank dynamics, and germination ecology

Under controlled conditions, *B. diandrus* populations from Australia, France, Israel, New Zealand, and Spain produced 1196–2729 seed/plant and 59–134 seed/panicle (Burghardt and Froud-Williams, 1997). In controlled conditions in Australia, populations of both *B. diandrus* and *B. rigidus* from Western and South Australia had similar seed production; 1156–2908 seed/plant for *B. rigidus* and 661–3380 seed/plant for *B. diandrus* (Kon and Blacklow, 1989). At harvest, *B. rigidus* seed generally have some level of innate dormancy, and they needed a long period to allow after-ripening of all seed (Table 4.1, Kon and Blacklow, 1989). By comparison, the initial level of dormancy in *B. diandrus* seed is variable. Gill and Carstairs (1988) noted less innate dormancy (higher initial germinability) in *B. diandrus* compared to *B. rigidus* in southern Australia. By comparison, Gill and Blacklow (1985) found genetic variation in the initial dormancy requirements of

TABLE 4.1 A selection of the parameters used to describe biological characteristics of *Bromus diandrus* and *B. rigidus* in the Weed Seed Wizard decision support software (Agriculture and Food Western Australia, 2020).

Species	*Bromus diandrus*	*Bromus rigidus*
Degree days after-ripening for any germination	100	100
Degree days after-ripening for 10% germination	280	480
Degree days after-ripening for 90% germination	2000	9000
Minimum and maximum germination temperature (°C)	5–22	5–22
Mass of 1000 seed (g)	9	8.3
Minimum degree days for seed set	2200	1400
Degree days to maximum seed set	2700	2000
Degree days to initial shedding	2500	1800
Degree days to senescence	2800	2300
Persistence (annual loss on surface, at 1 cm, at 10 cm)	40%, 50%, 20%	40%, 60%, 20%

B. diandrus populations from Western Australia. Seed from most populations were fully dormant at harvest in late spring, although seed from one population had over 20% germinability. Of these populations with high initial dormancy, some had 100% germinability by mid-summer, and some retained some level of dormancy by May (end of autumn) (Gill and Blacklow, 1985). This same behavior was observed in Spain for *B. diandrus* (Torra and Royo-Esnal, personal observations). Likewise, Chapman et al. (1999) and Cheam (1986) also found *B. diandrus* seed from Western Australia was fully dormant at harvest in late spring, although these populations lost dormancy rapidly (Chapman et al., 1999). It is clear that genetic or environmental factors influence the degree of innate dormancy for *B. diandrus* seed.

Following the release of innate dormancy, seed of *B. diandrus* was still inhibited from germinating by high temperature and low soil moisture content (Cheam, 1986; Chapman et al., 1999). Germination of *B. diandrus* seed in South Australia was stimulated by cold stratification, as the cold temperature triggered the synthesis of gibberellic acid within the seed. This ensured seed emerged once temperatures dropped below 4°C in autumn and winter (Kleemann and Gill, 2013). Likewise, cereal crops in Spain usually experience peak emergence of *B. diandrus* after cold temperatures in early to mid-November in autumn and a major rainfall event (Garcia et al., 2014). Despite the loss of dormancy, seed will not germinate if autumn climatic conditions are too dry and in northeast Spain, the autumn emergence flush was delayed to winter in one out of three observed seasons due to a prolonged summer-autumn drought (Royo-Esnal et al., 2018). The inhibition of germination in arid climates, as found over the summer of Mediterranean climates, and initial innate dormancy observed in some populations, would protect *B. diandrus* seed from germination following rainfall over the summer-autumn fallow, where seedlings would die due to subsequent arid conditions. In northern Europe, *B. diandrus* emerges in the late summer to autumn. Although a small proportion of a population may germinate in the spring, these plants are unlikely to set viable seed (Peters et al., 1993; AHDB, 2014).

Both *B. diandrus* and *B. rigidus* seed is highly sensitive to burial. The after-ripening process of *B. diandrus* seed was accelerated by burial at 2 cm. By late autumn, 95%–99% of seed germinated in darkness, whereas germination in the light varied from 55% to 95% (Cheam, 1986). Del Monte and Dorado (2011) demonstrated that seed of *B. diandrus* in Madrid, Central Spain have photoinhibition of germination, with optimal emergence once seed were in dark conditions. If seed were exposed to light, initial dormancy could be extended 2–4 months. When in dark conditions, seed germinated easily, and only very low temperatures or severe drought can prevent germination. By comparison, the germination of *B. diandrus* seed in South Australia was not inhibited by light (Kleemann and Gill, 2006). However, the *B. diandrus* population from Del Monte and Dorado (2011) came from a cropped field. While the research by Kleemann and Gill (2006) used seed from both field and roadside populations, their investigation on sensitivity of *B. diandrus* seed to light was only conducted on roadside populations. Weeds in cropped fields may evolve delayed emergence to avoid preseeding herbicides, as Fleet and Gill (2012) found for *Hordeum murinum* ssp. *glaucum*. Thus it would be necessary to compare the populations collected from a similar habitat to confirm the differences in photoinhibition of germination. Similarly to the Spanish *B. diandrus* population, *B. rigidus* in Australia showed low germinability (0%–40%) from April to September (over the winter cropping season) but high germinability of up to 80%–90% in June-July, when buried (Kleemann and Gill, 2006; Del Monte and Dorado, 2011).

Seed bank longevity is affected by burial and is similar between the species. Harradine (1986) found that although *B. diandrus* seed were 95% germinable 27 days after shedding from the parent plant, seed placed on the soil surface emerged over 18 months (Harradine, 1986). Seed buried at 5 or 15 cm germinated or degraded over 6 months. For seed buried at 5 cm, 97%

emerged within 1 month, whereas 99% of seed buried at 15 cm germinated but failed to emerge. These results contrast with Dastgheib and Poole (2010), who did not observe any difference in the percentage of emergence following burial from 0 to 20 cm depth, although seedling growth was affected when seedlings emerged from depth. Likewise, *B. rigidus* seed on the soil surface were depleted within 15 months in field conditions, in Oregon, United States (Gleichsner and Appleby, 1989). Seed viability was 83%, 62%, and 23% after 1, 9, and 12 months, respectively. The surface seed that persisted over 1 year had a high proportion of enforced dormancy. Seed buried at 1–30 cm germinated rapidly, with over 90% germination within the first month (Gleichsner and Appleby, 1989).

Seed bank longevity beyond 2 years has not been investigated in detail. Harradine (1986) found that a proportion of the ungerminated *B. diandrus* seed on the soil surface remained viable for over 2 years. At four sites in Western Australia, *B. diandrus* seed that were sown in autumn and buried when the soil was agitated to mimic the seeding system had over 86% emergence (as a percent of total emergence) in year one (Fig. 4.3, C. Borger

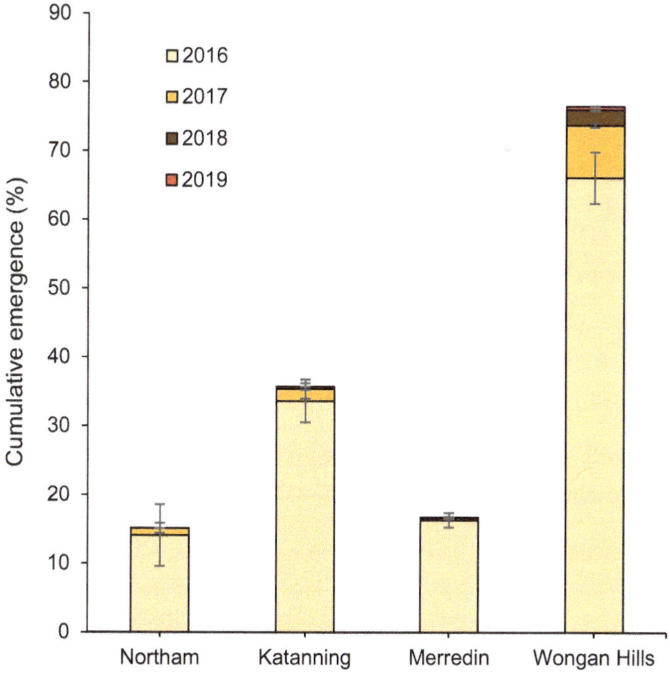

FIGURE 4.3 Cumulative emergence of *Bromus diandrus* seed sown in micro-plots in May 2016 at four sites in Western Australia, with seedlings counted and removed for four years. Note that in May of each year the soil was disturbed to a depth of 5 cm to simulate the seeding system. Vertical bars indicate the standard error of three replications. *Unpublished data from C. Borger.*

unpublished data). All sites had low emergence (0.5%–2%) by year three, and only Wongan Hills Western Australia had emergence (0.5%) in year four. Gill and Carstairs (1988) concluded that *B. rigidus* seed was more likely to have dormant seed surviving between years than *B. diandrus*. Clearly, both *B. diandrus* and *B. rigidus* can lose dormancy and germinate rapidly when buried, although some seed persist for multiple years.

While there is considerable variation in innate dormancy and after-ripening requirements, it is clear that the management of these species will depend on dormancy release and seed burial. Crop seeding is likely to bury a proportion of seed, ensuring that both *B. diandrus* and *B. rigidus* will germinate and emerge at a similar time to the crop. How this affects management, and how seeding systems can be manipulated to affect emergence needs to be considered in IWM systems.

Population dynamics

As discussed, emergence of *B. diandrus* is closely correlated to soil temperature and moisture, and these factors have been used to successfully develop a hydrothermal time (HTT) based model for seedling emergence under different soil management regimes in Lleida, Spain (Garcia et al., 2013). In this semiarid Mediterranean region, emergence occurred from approximately 150–200 HTT to 400 HTT (Royo-Esnal et al., 2018). In southern Spain (Huelva), *B. diandrus* seed planted in late autumn (to a depth of 5 cm) emerged during winter, with over 90% of seedlings emerged by mid-winter (2 February, Cao et al., 2012). This emergence period was longer in north-east Spain, where cohorts of *B. diandrus* in autumn sown crops emerged throughout winter and spring (Garcia et al., 2014). Likewise, *B. diandrus* cohorts in eastern Australia had greatest emergence in autumn, but cohorts continued to emerge up to 4 months after crop seeding (Heenan et al., 1990).

The staggered emergence of this weed will make management more difficult. For example, Heenan et al. (1990) attempted to control *B. diandrus* with initial residue burning in autumn to destroy seed on the soil surface, full cultivation at crop seeding to kill existing seedlings, preseeding and post-emergent herbicide application. Yet the late emerging cohorts escaped all weed control attempts and a low level weed density remained at crop harvest (Heenan et al., 1990). However, the first cohort of a weed can compete most effectively with the crop and produce the greatest number of seed with the highest survivorship, compared to later weed cohorts. In Madrid, Spain, 85% of *B. diandrus* seed emerged in the first cohort (Mokhtassi-Bidgoli et al., 2013). Four cohorts emerged over the cropping season, but average seed per plant for cohorts 1–4 were 25, 11, 2, and 1, respectively. Survivorship (%) of the successive cohorts declined from 74, 39, 37, to 5. In the absence of control, the population increased until reaching an equilibrium with a density of 17,324 seeds/m^2 in the soil seed bank (and 11,487 mature plants/m).

However, where postemergent herbicide controlled of 90% of plants in the initial cohort, the seed bank was reduced to <500 seed/m^2.

Bromus rigidus matures earlier than *B. diandrus*, with the minimum degree days for seed production estimated at 1400 and 2200 in Australia (Table 4.1). In two studies in grain crops in Spain, *B. diandrus* densities from 2008 to 2011 at harvest ranged from 0.1 to 563 plants/m^2, with 1.8–27,782 seed/m^2 (14–223 seed/plant), or 1–266 plants m^{-2} at harvest, with 15–5434 seed/m^2 (12–56 seed/plant) (Garcia et al., 2014; Recasens et al., 2016). In nine fields with cereal crops in Western Australia (in 2008), *Bromus* species (most likely *B. diandrus* and *B. rigidus*) at harvest grew at densities of 12–41 plants/m^2, with 2374–4059 seed/m^2 (or 55–192 seed/plant[1]) (Walsh and Powles, 2014). *Bromus rigidus* may also start shedding seed earlier than *B. diandrus* (Table 4.1).

A decision support tool developed in Australia, the Weed Seed Wizard (WSW), describes a range of weed species, and has been successfully utilized to model IWM programs for *Lolium rigidum* Gaud. harvest weed seed destruction (Borger et al., 2018; Borger et al., 2020a). The model summarizes population ecology of both *B. diandrus* and *B. rigidus* in Australia (Table 4.1). It highlights that seed size, optimal temperature for germination and seed degradation on the soil surface or at depth is similar between species, and the largest differences between the two species is the longer period of after-ripening to achieve germination of *B. rigidus* and the earlier maturity and earlier seed shedding and senescence of this species.

Interference with agronomic production

Bromus species are closely related to tribe Triticeae, which includes economically important genera like wheat, barley, and rye (*Secale cereale* L.) (Kon and Blacklow, 1989; O'Connor, 1990). Therefore it is not surprising that these species are favored by and highly competitive within those systems and environments managed to favor crop growth. In Tunisia, field surveys have indicated that *B. diandrus* at a density of 120 plants/m^2 can reduce wheat yield by up to 80% (Mejri et al., 2010). In Spain, *B. diandrus* at densities of 12–500 plants/m^2 reduced cereal yield (barley–wheat–wheat rotation) by 22%–71% (Garcia et al., 2014). In New Zealand, *B. diandrus* reduced cereal grain (wheat and barley) yield by 25%–30% (Dastgheib et al., 2003). In the United Kingdom, a population density as low as three *B. diandrus* plants/m^2 caused a 2% yield loss in wheat (AHDB, 2014). Competition of *B. diandrus* in lupin (*Lupinus angustifolius* L.) in Western Australia occurred within the first 16 days of crop development, and lupin seed yield was reduced from 475 to 195 g/m^2 as *B. diandrus* density increased from 0 to 507 plants/m^2 (65,910 seed/m^2) in the high fertilizer treatments (Supasilapa et al., 1992). By comparison, yield only reduced to 240 g/m^2 at 507 *B. diandrus* plants/m^2 in treatments with lower fertilizer applications (with 30,927 seed/m^2). Therefore the fertilizer favored growth and competitive ability of *B. diandrus*. Further, both *B. diandrus* and *B. rigidus* contaminate grain (Gleichsner and Appleby, 1989; Michael et al., 2010). Farmers seeding contaminated grain may introduce the weeds to a new area, and those seed in grain are frequently herbicide resistant (Michael et al., 2010). The sharp awns on *B. diandrus* seed allow them to get caught in wool and penetrate the skin, mouth, eyes, and intestines of livestock, causing animal welfare issues and reduced revenue from contaminated wool or carcass (Edgar and Connor, 2000).

Management

Bromus diandrus has been the subject of extensive research into weed control, with less research into the control of *B. rigidus* (Table 4.2). Management includes herbicide application, soil cultivation or amelioration, biological control, grazing, and harvest weed seed control (HWSC). It is notable that Royo-Esnal et al. (2018) found no value in higher crop sowing densities for *B. diandrus* control, possibly due to the high competitive ability of this species. The efficiency of many control techniques was either low or highly variable depending on site and seasonal conditions (Table 4.2).

Chemical control

Nonselective herbicides (glyphosate, etc.) are usually applied before crop seeding and/or emergence and are more effective when combined with delayed seeding (Royo-Esnal et al., 2018). As discussed, *B. diandrus* seed rapidly lose dormancy and germinate in cold temperatures. In a cereal rotation in Spain, *B. diandrus* emergence prior to crop seeding (i.e., those plants that could be killed by nonselective herbicide) ranged from less than 1% to over 90% in the first year (2014/15), depending on time of sowing (with the experiment investigating early, medium and late sowing times). The autumn of the following year (2015/16) was very dry, with less than 10% emergence prior to all sowing times (Royo-Esnal et al., 2018). Aside from seasonal variability in effectiveness of delayed sowing, it has the added disadvantage of reduced yield. The baseline sensitivity of *B. diandrus* to nonselective herbicides may also be lower than that of other grass weed species. In Spain, LD_{50} values of sensitive *B. diandrus* populations exposed to glyphosate in plant assays ranged from 85–116 g a.e./ha, higher than sensitive *L. rigidum* populations from the same region (17–45 g a.e. glyphosate/ha) (Barroso et al., 2010). Nonselective herbicides may also be used for spray-topping pasture to prevent seed set of *Bromus* species, but efficiency depends on weed development stage. This technique requires (a) multiple cohorts of the *Bromus* species reaching anthesis and seed set at a uniform time across the

TABLE 4.2 Weed control strategies applied to *Bromus diandrus, B. rigidus*, or both species in various countries and the percentage weed control or range in control achieved by each strategy.

Strategy	Country	Weed control (and reference)
Burning residue	Australia, USA	65%*** (Sweet et al., 2008; Bowcher et al., 2019), 96%* (Heenan et al., 1990)
Shallow cultivation in autumn	Australia	50%*** (Bowcher et al., 2019)
Full soil inversion	Spain	97%–99%* (Garcia et al., 2013; Recasens et al., 2016).
Biological control	Tunisia, United Kingdom	30%* (Mejri et al., 2010), >85% dry weight reduction (Lawrie et al., 1998)
Crop allellopathy	Tunisia	Variable suppression*** (Bouhaouel et al., 2019; Bouhaouel et al., 2020)
Delayed sowing	Spain, Australia	55% ≥ 90%* (Royo-Esnal et al., 2018), 30%–90%*** (Bowcher et al., 2019)
Nonselective herbicide	Spain	<1%* (dry seeding) to >90%* (Royo-Esnal et al., 2018)
Preemergence herbicide	Australia, New Zealand	8%–95%* (Kleemann et al., 2016), 82%–93%** (Kleemann and Gill, 2008), 60%* (Dastgheib et al., 2003)
Postemergence herbicide	Australia, Spain	>87%** (Kleemann and Gill, 2009b), 90%*** (Gill and Bowran, 1990), 55%–100%* (Garcia et al., 2014), >90%* (Royo-Esnal et al, 2018)
Spray-topping	Australia	50%–90%*** (Bowcher et al., 2019)
Grazing	Australia, New Zealand	30%* (Tozer et al., 2007), 20%–80%*** (Bowcher et al., 2019)
Harvest weed seed control	Australia	36%–72%* (Borger et al., 2020b), 77%*** (Walsh and Powles, 2014)

The symbol * is used to indicate research on *B. diandrus*, ** for research on *B. rigidus*, or *** for research on both species.

field, and (b) the weed reaching maturity at a different time to the desirable pasture species (Bowcher et al., 2019).

Efficacy of a range of preemergent herbicides in South Australia and Victoria ranged from 8% to 95% (Kleemann et al., 2016). Effective control was provided by metribuzin plus trifluralin at three sites (69%–89% control), but there was only a 33% and 51% reduction in plant and panicle density at the fourth site, probably due to dry conditions (Kleemann et al., 2016). Pyroxasulfone plus triallate provided reliable control of *B. diandrus*, with an 89%–95% reduction in plant density, but the high cost of this mixture will ensure low adoption in rain-fed (low yielding), dryland wheat crops (Kleemann et al., 2016). In New Zealand, *B. diandrus* was controlled by preemergent applications of cyanazine or chlorsulfuron plus terbuthylazine and metribuzin. However, in spite of high herbicide efficiency, only chlorsulfuron is registered for use in New Zealand (Dastgheib et al., 2003). Preemergent (incorporated by sowing) metribuzin controlled 82%–90% of *B. rigidus* in barley, but crop density was reduced 23% at one site in South Australia. Metribuzin plus pendimethalin controlled 89%–93% of *B. rigidus* without damaging the crop (Kleemann and Gill, 2008).

Postemergent herbicide applications also have varying efficiency. In grain crops in Spain, mesosulfuron-methyl plus iodosulfuron-methyl-sodium controlled 55%–59% (Garcia et al., 2014) or 91%–100% of *B. diandrus* (Royo-Esnal et al., 2018). Further, seed production on the surviving plants was reduced by 91%, and seed weight reduced by 31% (Garcia et al., 2015). Pyroxsulam plus florasulam or mesosulfuron-methyl plus propoxycarbazone-sodium controlled over 95% of *B. diandrus* in wheat (Royo-Esnal et al., 2018). In recent years an expanding range of crops tolerant to herbicides (like imazapic/imazapyr or metribuzin) have offered growers more flexibility for selective weed control. Imidazolinone herbicides provided high levels of control of *B. rigidus* in South Australia (> 87% control, with <6 seed produced/m^2) compared to postemergent mesosulfuron-methyl (11%–67% control with 461–3983 seed/m^2) (Kleemann and Gill, 2009b). Use of diverse crop species in rotations also allows variation in herbicide use. Royo-Esnal et al. (2018) obtained 90%–100% control of *B. diandrus* when clethodim was sprayed in field pea (*Pisum sativum* L.), winter oilseed rape (*Brassica napus* L.), and camelina (*Camelina sativa* L.).

It is clear that chemical control is highly variable, but is also threatened by herbicide resistance. *Bromus diandrus* and *B. diandrus* ssp. *rigidus* (or *B. rigidus*) in multiple states in Australia have developed resistance to ACCase inhibitors and ALS inhibitors, and *B. diandrus* has developed resistance to EPSP synthase inhibitors (glyphosate, Malone et al., 2015; Heap, 2020). Likewise, populations in Spain are showing resistance to ALS and EPSP inhibitors (Escorial et al., 2011). Populations of *A. diandra* are resistant to ALS inhibitors in the United Kingdom (Davies et al., 2020). Resistance development for ACCase or ALS inhibitors occurs over a short period, and it is likely that other countries will rapidly develop resistance (Heap, 2020).

Allelopathy

Research has indicated *B. diandrus* Roth. (syn. *B. rigidus* Roth. subsp. gussonii Parl.) growth may be suppressed by the phenolic acids released by the roots of barley crops (Bouhaouel et al., 2019; Bouhaouel et al., 2020). *Bromus diandrus* root length was reduced by 9%–42% and coleoptile length by 10%–36%. However, the experiments were in controlled conditions, and growth suppression depended on barley genotypes and soil properties like sodium content, microbial communities, and drainage. By comparison, aqueous extracts of *B. diandrus* or *B. rigidus* reduced the root length of barley and wheat by 20%–40% (Lehoczky et al., 2011). It is clear that the impact of allellopathy from both crops and weeds requires further research.

Harvest weed seed control

HWSC is a potential control technique for *Bromus* species. Annual burning of crop residue in 1985–87 in New South Wales, Australia, reduced *B. diandrus* density by 96% compared to no burning (Heenan et al., 1990). Burning was applied in February, at the end of the summer/autumn fallow, prior to seeding winter annual crops. Despite the effectiveness of burning stubble residue, this practice is no longer a common weed management technique in Australia due to concerns regarding environmental impact (Borger et al., 2018). In the European Union, crop residue burning for weed control is banned through the common agricultural policy (European Union, 2013).

Other forms of HWSC include burning windrows or capturing seed mechanically at harvest (Walsh et al., 2018b). In Western Australia, 77% of *Bromus* sp. seed was retained at the beginning of harvest (i.e., at crop maturity) above the 15 cm harvest height, although seed retention dropped to 41% by the end of the harvest period (28 days after crop maturity, Walsh and Powles, 2014). This research was conducted in a single season, on unspecified *Bromus* species. There is both environmental and genetic variation in shedding time of *B. diandrus*. Borger et al. (2020b) found that *B. diandrus* seed retention at the end of spring (i.e., at maturity of the winter annual crop) for a single population in Western Australia ranged from 0% to 54% over 3 years. In Spain, it was noted that the crop camelina matured earlier than cereal crops. Camelina maturity was prior to seed shedding of *B. diandrus*, which occurred in mid-June, at the same time cereal crops reached maturity (Westerman et al., 2012; Royo-Esnal et al., 2018). It is possible HWSC for *Bromus* sp. would be improved though use of early maturing crops or cultivars.

There is no field research to assess how effectively the seed of *Bromus* sp. are captured at harvest, which depends on plant and seed head height, lodging, and time of shedding (Walsh and Powles, 2014; Walsh et al., 2018a). However, modeling has indicated that HWSC is a highly effective control technique even if it's only practiced in alternate years (for *L. rigidum*) or if it only removes a very small proportion of *B. diandrus* seed each year (Borger et al., 2018; Borger et al., 2020a). Therefore while more research is needed, it is likely that HWSC will be a key part of the management of *Bromus* sp. in future.

Cultivation

As discussed, burial can stimulate germination of both *B. diandrus* and *B. rigidus*, although there is conflicting results on how well the germinated seed emerge from depths of up to 20 cm (Harradine, 1986; Gleichsner and Appleby, 1989; Dastgheib et al., 2003). Consequently, cultivation is a highly effective control method. Shallow burial of seed of both species (dark conditions) prior to seeding stimulates emergence by removing photoinhibition of germination. Seedlings are easier to control if they all emerge early in the season. In the 1980's, prior to the introduction of conservation tillage, cultivation prior to seeding reduced *B. diandrus* densities by 64% in New South Wales, Australia (Heenan et al., 1990). Shallow cultivation to 5 cm ("autumn tickle") is occasionally used in the current Australian farming system to simulate emergence and buries 70% of surface weed seed (Agriculture and Food Western Australia, 2020). Soil amelioration to address soil constraints may be applied, and a full soil inversion (burial of the surface soil at a depth of 10–30 cm, i.e., moldboard plow) can bury 96% of surface seed beyond 10 cm deep (Agriculture and Food Western Australia, 2020). While conservation tillage systems with minimal soil disturbance are increasingly popular in Australia and Europe, cultivation may occasionally be applicable for weed control or soil amelioration (D'Emden et al., 2008; Murphy et al., 2017).

A series of trials in Spain highlighted varying emergence of *B. diandrus* following different methods of cultivation (Table 4.3). A full soil inversion provided highly effective weed control due to deep burial of seed. Tillage methods that resulted in shallow seed burial increased seedling emergence (due to removal of photoinhibition). However, density was also reduced in the no-tillage system (Table 4.3). This is similar to the findings of Hernandez-Plaza et al. (2011), but in direct contrast to Kleemann and Gill (2006). However, the method of no-tillage varies between studies (no tillage, zero tillage, minimum tillage, etc.). Further, the environmental changes resulting from no-tillage, in terms of crop residue remaining on the soil surface, soil temperature, soil moisture retention, and so on, also varies between systems. For choice of crop seeding method to become a reliable weed control technique, more information is required to compare tillage techniques in different environments.

Biological control

For the US investigations of biological control, *B. diandrus* and *B. rigidus* have been treated as closely related but distinct species in the basic reporting of their pathogens (Farr and Rossman, 2020). There are 30 global pathogen records for *B. diandrus* and 68 for *B. rigidus*, but interestingly, the majority of records were from the invaded range for these species (i.e., North America, Australia, New Zealand, Japan, and South Africa) rather than the native ranges (Mediterranean Basin) (Farr and Rossman, 2020). This may

TABLE 4.3 Reduction in cumulative emergence of *Bromus diandrus* following a range of cultivation techniques in two studies from Spain: Garcia et al. (2013), A, and Recasens et al. (2016), B.

Cultivation technique	Description	Cumulative emergence
Chisel plow	Disturb to 10–40 cm and superficially bury the surface seed.	Highest plant density. A: >600, >1100 and >500 plants/m^2 in year 1–3. B: 266, 241 and 75 plants/m^2 in year 1–3.
Subsoiler	Disturb to a depth of 20–25 cm, with 4 cm wide shanks at 35 cm apart.	Compared to the chisel plow, reduced density by: A: 59%, 56%, and 48% in year 1–3. B: 22%, 44%, and 45% in year 1–3.
Full inversion	Moldboard with skimmers to bury topsoil and 96% of weed seed to below 10 cm depth.	Compared to the chisel plow, reduced density by: A: 97%, 98%, and 99% in year 1–3. B: 98%, 98%, and 99% in year 1–3.
No tillage	Disc drill, where flat discs create a furrow to 5 cm, followed by a tyne to place seed and fertilizer in the row, 95% of seed remains on the soil surface and 5% buried to 5 cm.	Compared to the chisel plow, reduced density by: B: 83%, 99%, and 89% in year 1–3.

mean that greater research is needed to identify pathogens in the native range. However, it may also relate to the previously discussed close relationship of *Bromus* species to many crop species, making it more susceptible to crop disease. The recorded pathogens of *Bromus* species include common crop diseases including "take-all" disease common to cereal crops (*Gaeumannomyces graminis*), "net blotch" disease of barley (*Pyrenophora teres*), and "blotch" disease of wheat (*Parastagonospora nodorum*) (Cook and Dubé, 1989). Host specificity is a key aspect of identifying a successful biological control agent (Meyer et al., 2016). The close relationship between the *Bromus* genus and the tribe Triticeae complicates the search for potential biological control agents.

Possible biological control agents include a rhizobacterium, *Pseudomonas trivialis* strain X33d, investigated against *B. diandrus* in Tunisia (Table 4.2,

Mejri et al., 2010). *Pyrenophora chaetomioides* (Speg) strains were investigated in the United Kingdom for control of several *Bromus* species. Although dry weight of *B. diandrus* plants were reduced by 85%, no plants were killed by the fungus and regrowth occurred (Lawrie et al., 1998). The seed pathogen *Pyrenophora semeniperda* has been investigated as a potential bio-herbicide, as applications of inoculum around mid-anthesis in wheat substantially reduced emergence and seedling vigor of *B. diandrus* and a range of other grass weeds in Australia (Medd and Campbell, 2005). The advantage of this biological control agent is that it is already widely distributed in North and South America, Australia, South Africa, New Zealand, Argentina, and Egypt. However, there are numerous factors that will affect the likelihood of bio-herbicide development, including the staggered emergence of *Bromus* species, which may make it difficult to target plants at anthesis, and the dry climate of Australia that does not always favor the growth of fungi (Medd and Campbell, 2005).

Integrated weed management

A consistent message is that no control technique consistently removes 100% of plants. Obviously, the emergence of multiple cohorts may have confused assessment of weed control in some of the studies within this review, but it is the multiple cohorts of *B. diandrus* and *B. rigidus* that make complete control difficult. Successful management programs in the field have in common the use of multiple weed control strategies over multiple years (Heenan et al., 1990; Kleemann and Gill, 2009a; Recasens et al., 2016; Royo-Esnal et al., 2018). This conclusion was confirmed by modeling of *B. diandrus* control in Madrid, Spain, in a dryland cereal rotation (Mokhtassi-Bidgoli et al., 2013). Modeling concluded that control tactics that reduce plants by 92% will eventually result in the seed bank reaching zero, but no single control tactic could provide this level of control (Mokhtassi-Bidgoli et al., 2013). It was likewise the finding of this review that no individual control tactic reliably gave over 90% control (with the exception of a full soil inversion) in all seasonal conditions (Table 4.2). Moreover, the use of chemicals is restricted in Europe and New Zealand, which highlights the necessity of combining nonchemical control tactics to achieve long-term control.

A key determinate of the success of IWM practices was initial weed density. Royo-Esnal et al. (2018) compared 12 IWM programs and found that at a starting density of 139 *B. diandrus* plants/m^2, multiple IWM programs had close to 100% control in the first year. By comparison, an initial density of 812–2105 plants/m^2 took 3–4 years to control. Likewise, Garcia et al. (2014) noted in-crop selective herbicide efficiency was related to weed density, and 100% control only occurred where weed density was low. Also, it is worth noting that those treatments with 100% control from the herbicide still had 1.9–17.5 seed/m^2 at harvest, as the staggered emergence of *B. diandrus* ensure that it is not possible to apply in-crop herbicide to all seedlings (Garcia et al., 2014). Both Royo-Esnal et al. (2018) and Montull and Taberner (2013) observed that the most effective cultural control methods were a cereal rotation with a dicotyledonous crop (winter oilseed rape, field pea, or camelina), allowing use of selective grass herbicides, and delayed sowing. These strategies could be complemented with HWSC, or residue management prior to seeding the following crop, to achieve control of seed from late cohorts.

It is clear that a long-term IWM program is required for effective control of *B. diandrus* and *B. rigidus*, and once the seed bank is reduced, it is feasible to maintain the populations at sufficiently low levels to avoid yield loss. There are multiple decision support tools to assist growers or advisors in the optimal combination of IWM tactics. Examples include the WSW or the Brome RIM (Ryegrass Integrated Management) model (Borger et al., 2018; Torra and Monjardino, 2020). While the Brome RIM model gives a better assessment of economic performance of IWM strategies for *Bromus* species, WSW focusses on specific weed ecology and seasonal variation (Borger et al., 2018; Torra and Monjardino, 2020).

Conclusion

Bromus diandrus and *B. rigidus* are widespread and difficult to control weeds, but a specific IWM strategy over 2–4 years (depending on initial weed density) can reduce this weed to very low densities. However, the varying biological and ecological traits both within and between species indicate that management strategies will need to be adapted to each particular population. Growers or advisors will need to assess populations and adjust control strategies based on the ecological traits of the population in question and seasonal climatic conditions. There are a range of control tactics that would benefit from more research, including cultivation and no-tillage seeding systems, allelopathic crops, biological control, and HWSC. There is little research on increasing the competitive ability of crops against *Bromus* species, but these techniques may have limited efficiency due to the very high competitive ability of this species compared to other common winter annual grasses like *L. rigidum* or *Hordeum* species (Borger et al., 2020c). There is less research on the control of *B. rigidus* than *B. diandrus*, but it is likely that the control tactics that successfully control *B. diandrus* will also be effective against *B. rigidus*. However, the potentially longer dormant seed bank of *B. rigidus* indicates that IWM programs may need to continue for more than 2–4 years. The earlier maturity and shedding times of *B. rigidus* may also reduce the efficiency of HWSC. For this reason, more research is needed to clarify the relationship between *B. diandrus* and *B. rigidus* species, to allow control strategies for *B. diandrus* to be successfully adapted to *B. rigidus*.

References

Agriculture and Food Western Australia, 2020. Weed Seed Wizard. Version 7.1.7. Department of Primary Industries and Regional Development Western Australia, Grains Research and Development Corporation, University of Western Australia, University of Adelaide, NSW Department of Primary Industries, Queensland Department of Agriculture, Fisheries and Forestry, Australia, Perth, Western Australia. https://www.agric.wa.gov.au/weed-seed-wizard-0 (Accessed 29 April 2020).

AHDB, 2014. Identification and control of brome grasses. Information sheet 31. https://media.ahdb.org.uk/media/Default/Imported%20Publication%20Docs/Identification%20and%20control%20of%20brome%20grasses.pdf (Accessed 17 June 2020).

Barroso, J., Loureiro, I., Escorial, M.C., Chueca, M.C., 2010. The response of *Bromus diandrus* and *Lolium rigidum* to dalapon and plyphosate 1: baseline sensitivity. Weed Res. 50, 312–319.

Borger, C.P.D., Riethmuller, G., Renton, M., 2018. Weed Seed Wizard: a tool that demonstrates the value of integrated weed management tactics such as harvest weed seed destruction. Comput. Electron. Agric. 127, 27–33.

Borger, C., Petersen, D., Gill, G., 2020a. Great brome and barley grass—modelling the long term value of harvest weed seed control. GRDC Grains Research Updates. Grains Industry Association of Western Australia, Perth, Western Australia.

Borger, C.P.D., Hashem, A., Gill, G.S., 2020b. Comparison of growth, survivorship, seed production and shedding of eight weed species in a wheat crop in Western Australia. Weed Res. 60, 415–424.

Borger, C.P.D., Hashem, A., Van Burgel, A., Gill, G., 2020c. Invasiveness of agronomic weed species in wheat in Western Australia. Weed Res. 60, 251–258.

Bouhaouel, I., Richard, G., Fauconnier, M.L., Ongena, M., Franzil, L., Gfeller, A., et al., 2019. Identification of barley (*Hordeum vulgare* L. subsp. vulgare) root exudates allelochemicals, their autoallelopathic activity and against *Bromus diandrus* Roth. germination. Agron 9, 345.

Bouhaouel, I., Gfellera, A., Boudabous, K.H., Fauconnier, M.L., Slama Ayed, L., Slim Amara, H., et al., 2020. Effects of physico-chemical and biological properties of soil on the allelopathic activity of barley (*Hordeum vulgare* L. subsp. vulgare) root exudates against *Bromus diandrus* Roth. and *Stelleria media* L. weeds. Allelopathy J. 49, 17–34.

Bowcher, A., Cheam, A., Gill, G., Moore, J., Kleemann, S., 2019. Weed 6. Brome grass (*Bromus* spp.). In: Storrie, A.E. (Ed.), Integrated Weed Management Manual. Grains Research and Development Corporation, Australia, pp. 32–35.

Burghardt, G., Froud-Williams, R.J., 1997. Phenology and reproductive allocation of *Bromus sterilis, B. diandus, B. hordeaceus* and *B. commutatus*. The 1997 Brighton Crop Protection Conference: Weeds. British Crop Protection Council, Brighton, UK, pp. 479–484.

CABI, 2020. *Bromus rigidus* (ripgut brome). CAB International, Wallingford, UK. Available from: https://www.cabi.org/isc/datasheet/31167, last (Accessed 24 June 2020).

Cao, R., Francisco-Fernandez, M., Anand, A., Bastida, F., Gonzalez-Andujar, J.L., 2012. Modelling *Bromus diandrus* seedling emergence using nonparametric estimation. J. Agr. Biol. Envir. St. 18, 64–86.

Chapman, R., Bolger, T.P., Le Coultre, I.F., 1999. Seed dormancy release in three common pasture grasses from a Mediterranean-type environment under contrastion conditions. Aust. J. Exp. Agric. 39, 143–147.

Cheam, A.H., 1986. Patterns of change in seed dormancy and persistence of *Bromus diandrus* Roth. (great brome) in the field. Aust. J. Agric. Res. 37, 471–481.

Clapham, A.R., Tutin, T.G., Warburg, E.F., 1952. Flora of the British Isles. Cambridge University Press, Cambridge, UK.

Cook, R.P., Dubé, A.J., 1989. Host-Pathogen Index of Plant Diseases in South Australia. South Australian Department of Agriculture, South Australia.

D'Emden, F.H., Llewellyn, R.S., Burton, M.P., 2008. Factors influencing adoption of conservation tillage in Australian cropping regions. Aust. J. Agr. Resour. Ec. 52, 169–182.

Dastgheib, F., Poole, N., 2010. Seed biology of brome grass weeds (*Bromus diandrus* and *B. hordeaceus*) and effects of land management. N. Z. Plant. Prot. 63, 78–83.

Dastgheib, F., Rolston, M.P., Archie, W.J., 2003. Chemical control of brome grasses (*Bromus* spp.) in cereals. N. Z. Plant. Prot. 56, 227–232.

Davies, L.R., Hull, R., 2018. Presence and distribution of brome weeds in UK arable farming. Asp. Appl. Biol. 141, 37–46.

Davies, L.R., Onkokesungn, N., Brazier-Hicks, M., Edwards, R., Moss, S., 2020. Detection and characterization of resistance to acetolactate synthase inhibiting herbicides in *Anisantha* and *Bromus* species in the United Kingdom. Pest. Manag. Sci. 76, 2473–2482.

Del Monte, J.P., Dorado, J., 2011. Effect of light conditions and after ripening time on seed dormancy loss of *Bromus diandrus*. Weed Res. 51, 581–590.

Edgar, E., Connor, H.E., 2000. Flora of New Zealand. Volume 5. Gramineae. Manaaki Whenua Press, Lincoln, New Zealand.

Escorial, C., Loureiro, I., Rodriguez-Garcia, E., Chueca, C., 2011. Population variability in the response of ripgut brome (*Bromus diandrus*) to sulfosulfuron and plyphosate herbicides. Weed Sci. 59, 107–112.

European Union, 2013. Regulation (EU) No 1306/2013. On the financing, management and monitoring of the common agricultural policy and repealing Council Regulations (EEC) No 352/78, (EC) No 165/94, (EC) No 2799/98, (EC) No 814/2000, (EC) No 1290/2005 and (EC) No 485/2008. https://eur-lex.europa.eu/legal-content/EN/TXT/HTML/?uri = CELEX:32013R1306&from = en (Accessed 17 June 2020).

Farr, D.F., Rossman, A.Y., 2020. Fungal Databases. ARS, USDA, USA. https://nt.ars-grin.gov/fungaldatabases/ (Accessed 4 May 2020).

Fleet, B., Gill, G., 2012. Seed dormancy and seedling recruitment in smooth barley (*Hordeum murinum* ssp. *glaucum*) populations in southern Australia. Weed Sci. 60, 394–400.

Garcia, A.L., Recasens, J., Forcella, F., Torra, J., Royo-Esnal, A., 2013. Hydrothermal emergence model for ripgut brome (*Bromus diandrus*). Weed Sci. 61, 146–153.

Garcia, A.L., Royo-Esnal, A., Torra, J., Cantero-Martinez, C., Recasens, J., 2014. Integrated management of *Bromus diandrus* in dryland cereal fields under no-till. Weed Res. 54, 408–417.

Garcia, A.L., Royo-Esnal, A., Torra, J., Recasens, J., 2015. Integrated effect of crop sowing date and herbicide stress on fitness of *Bromus diandrus* Roth. Span. J. Agric. Res. 13, e09-e001.

Gill, G.S., Blacklow, W.M., 1985. Variations in seed dormancy and rates of development of great brome, *Bromus diandrus* Roth., as adaptations to the climates of southern Australia and implications for weed control. Aust. J. Agric. Res. 36, 295–304.

Gill, G.S., Bowran, D.G., 1990. Tolerance of wheat cultuvars to metribuzin and implicaitons for the control of *Bromus diandrus* and *B. rigidus* in Western Australia. Aust. J. Exp. Agric. 30, 373–378.

Gill, G.S., Carstairs, S.A., 1988. Morphological, cytological and ecological discrimination of *Bromus rigidus* from *Bromus diandrus*. Weed Res. 28, 399–405.

Gleichsner, J.A., Appleby, A.P., 1989. Effect of depth and duration of seed burial on ripgut brome (*Bromus rigidus*). Weed Sci. 37, 68–72.

Harradine, A.R., 1986. Seed longevity and seedling establishment of *Bromus diandrus* Roth. Weed Res. 26, 173–180.

Heap, I., 2020. The International Survey of Herbicide Resistant Weeds. www.weedscience.org (Accessed 23 April 2020).

Heenan, D.P., Taylor, A.C., Leys, A.R., 1990. The influence of tillage, stubble management and crop rotaiton on the persistence of great brome (*Bromus diandrus* Roth). Aust. J. Exp. Agric. 30, 227–230.

Hernandez-Plaza, E., Kozak, M., Navarrete, L., Gonzalex-Andujar, J.L., 2011. Tillage system did not affect weed diversity in a 23-year experiment in Mediterranean dryland. Agric. Ecosyst. Environ. 140, 102–105.

Kleemann, S.G.L., Gill, G.S., 2006. Differences in the distribution and seed germination behaviour of populations of *Bromus rigidus* and *Bromus diandrus* in South Australia: adaptations to habitat and implications for weed management. Aust. J. Agric. Res. 57, 213–219.

Kleemann, S.G.L., Gill, G.S., 2008. Application of metribuzin for the control of rigid brome (*Bromus rigidus*) in no-till barley crops of southern Australia. Weed Technol. 22, 34–37.

Kleemann, S.G.L., Gill, G.S., 2009a. Population ecology and management of rigid brome (*Bromus rigidus*) in Australian cropping systems. Weed Sci. 57, 202–207.

Kleemann, S.G.L., Gill, G.S., 2009b. The role of imidazolinone herbicides for the control of *Bromus rigidus* (rigid brome) in wheat in southern Australia. Crop. Prot. 28, 913–916.

Kleemann, S.G.L., Gill, G.S., 2013. Seed dormancy and seedling emergence in ripgut brome (*Bromus diandrus*) populations in southern Australia. Weed Sci. 61, 222–229.

Kleemann, S.G.L., Boutsalis, P., Gill, G.S., Preston, C., 2016. Applications of pre-emergent pyroxasulfone, flufenacet and their mixtures with triallate for the control of *Bromus diandrus* (ripgut brome) in no-till wheat (*Triticum aestivum*) crops of southern Australia. Crop. Prot. 80, 144–148.

Kon, K.F., Blacklow, W.M., 1989. The biology of Australian weeds 19. *Bromus diandrus* Roth and *B. rigidus* Roth. Plant. Prot. Q. 4, 51–60.

Kon, K.F., Blacklow, W.M., 1990. Polymorphism, outcrossing and polyploidy in *Bromus diandrus* and *B. rigidus*. Aust. J. Bot. 38, 609–618.

Lawrie, J., Greaves, M.P., Down, V.M., 1998. Drechslera sp. (*Pyrenophora chaetomioides* (Speg.)), a potential biocontrol agent for *Bromus sterilis* and other *Bromus* spp. Biocontrol Sci. Technol. 8, 479–484.

Lehoczky, E., Nelima, M.O., Szabo, R., Szalai, A., Nagy, P., 2011. Allelopathic effect of *Bromus* spp. and *Lolium* spp. shoot extracts on some crops. Commun. Agric. Appl. Biol. Sci. 76, 537–544.

Malone, J.M., Morran, S., Shirley, N., Boutsalis, P., Preston, C., 2015. EPSPS gene amplification in glyphosate-resistant *Bromus diandrus*. Pest. Manag. Sci. 72, 81–88.

Medd, R.W., Campbell, M.A., 2005. Grass seed infection following inundation with *Pyrenophora semeniperda*. Biocontrol Sci. Technol. 15, 21–36.

Mejri, D., Gamalero, E., Tombolini, R., Musso, C., Massa, N., Berta, G., et al., 2010. Biological control of great brome (*Bromus diandrus*) in durum wheat (*Triticum durum*): specificity, physiological traits and impact on plant growth and root architecture of the fluorescent pseudomonad strain X33d. BioControl 55, 561–572.

Meyer, S.E., Beckstead, J., Pearce, J.L., 2016. Community ecology of fungal pathogens on *Bromus tectorum*. In: Germino, M.J., Chambers, J.C., Brown, C.S. (Eds.), Exotic Brome-Grasses in Arid and Semiarid Ecosystems of the Western US: Causes, Consequences, and Management Implications. Springer International Publishing, Switzerland, pp. 193–223.

Michael, P.J., Owen, M.J., Powles, S.B., 2010. Herbicide-resistant weed seeds contaminate grain sown in the Western Australian grainbelt. Weed Sci. 58, 466–472.

Mokhtassi-Bidgoli, A., Navarrete, L., AghaAlikhani, M., Gonzalex-Andujar, J.L., 2013. Modelling the population dynamic and management of *Bromus diandrus* in a non-tillage system. Crop. Prot. 43, 128–133.

Montull, J.M., Taberner, A., 2013. *Bromus diandrus* integrated management. Results from second year of trials. Proceedings of the XIV Congress of the Spanish Weed Science Society, pp. 313–318.

Morris, N.L., Miller, P.C.H., Orson, J.H., Froud-Williams, R.J., 2010. The adoption of non-inversion tillage systems in the United Kingdom and the agronomic impact on soil, crops and the environment—a review. Soil. Till. Res. 108, 1–15.

Murphy, D., Leopold, M., van Gool, D., Hoyle, F., Stockdale, E., 2017. Soil quality: 1 constraints to plant production. SoilsWest, Perth, Western Australia.

O'Connor, P., 1990. Poaceae. In: Wagner, W.L., Herbst, D.R., Sohmer, S.H. (Eds.), Manual of the Flowering Plants of Hawaii. University of Hawaii Press & Bishop Museum Press, Honolulu, Hawaii, pp. 1481–1604.

Oja, T., Laarmann, H., 2002. Comparative study of the ploidy series *Bromus sterilis*, *B. diandrus* and *B. rigidus* (Poaceae) based on chromosome numbers, morphology and isozymes. Plant. Biol. (Stuttg.) 4, 484–491.

Peters, N.C.B., Froud-Williams, R.J., Orson, J.H., 1993. The rise of barren brome, *Bromus sterilis* in UK cereal crops Brighton Crop Protection Conference, weeds. British Crop Protection Council (BCPC Farnham UK), Brighton, UK, pp. 773–780.

Petersen, J., 2006. Distribution, importance and control of *Bromus* species in the middle west of Germany. J. Plant. Dis. Protect. Spec. Issue 20, 289–296.

Recasens, J., Garcia, A.L., Cantero-Martinez, C., Torra, J., Royo-Esnal, A., 2016. Long-term effect of different tillage systems on the emergence and demography of *Bromus diandrus* in rainfed cereal fields. Weed Res. 56, 31–40.

Royal Botanic Garden Edinburgh, 2020. Flora Europaea Database. Royal Botanic Garden, Edinburgh, UK. http://rbg-web2.rbge.org.uk/FE/fe.html (Accessed 17 July 2020).

Royo-Esnal, A., Recasens, J., Garrido, J., Torra, J., 2018. Rigput brome (*Bromus diandrus* Roth.) management in a no-till field in Spain. Agron 8, 251.

Sales, F., 1993. Taxonomy and nomenclature of *Bromus* sect. Genea. Taxonomy and nomenclature of *Bromus* sect. Genea 50, 1–31.

Stace, C., 1991. New Flora of the British Isles. Cambridge University Press, Cambridge, UK.

Supasilapa, S., Steer, B.T., Milroy, S.P., 1992. Competition between lupin (*Lupinus angustifolius* L.) and great brome (*Bromus diandrus* Roth.): development of leaf area, light interception and yields. J. Exp. Agric. 32, 71–81.

Sweet, S.B., Kyser, G.B., DiTomaso, J.M., 2008. Susceptibility of exotic annual grass seeds to fire. Invas. Plant. Sci. Mana. 1, 158–167.

Torra, J., Monjardino, M., 2020. Chapter II: Ryegrass integrated management (RIM)-based DSS. In: Chantre, G.R., Gonzalez-Andujar, J.L. (Eds.), Decision Support Systems for Weed Management. Springer International Publishing, Switzerland, p. 341.

Tozer, K.N., Marshall, A.J., Sedcole, J.R., Edwards, G.R., 2007. Ripgut brome (*Bromus diandrus*) distribution in relation to topography and management on seven high country properties in the south island. N. Z. Plant. Protect. 60, 168–173.

Tsvelev, N., 1984. Grasses of the Soviet Union. Part 1. Balkema, Rotterdam, Netherlands.

Walsh, M.J., Powles, S.B., 2014. High seed retention at maturity of annual weeds infesting crop fields highlights the potential for harvest weed seed control. Weed Technol. 28, 486–493.

Walsh, M.J., Broster, J.C., Aves, C., Powles, S.B., 2018a. Influence of crop competition and harvest weed seed control on rigid ryegrass (*Lolium rigidum*) seed retention in wheat crop canopies. Weed Sci. 66, 627–633.

Walsh, M.J., Broster, J.C., Schwartz-Lazaro, L.M., Norsworthy, J.K., Davis, A.S., Tidemann, B. D., et al., 2018b. Opportunities and challenges for harvest weed seed control in global cropping systems. Pest. Manag. Sci. 74, 2235–2245.

Westerman, P.R., Atanackovic, V., Royo-Esnal, A., Torra, J., 2012. Differential weed seed predation in dryland cereals. Arthropod-Plant Interact. 6, 591–599.

Chapter 5

Chenopodium album and *Chenopodium murale*

Seyed Vahid Eslami[1] and Sarah Ward[2]
[1]Associate Professor, Department of Agronomy, University of Birjand, Birjand, Iran,
[2]Associate Professor, Department of Soil and Crop Sciences, Colorado State University, Fort Collins, CO, United States

Origins and taxonomy

The genus *Chenopodium*, formerly in the temperate family Chenopodiaceae but more recently placed within the larger and more widely distributed Amaranthaceae, consists of approximately 250 species worldwide (Prasad and Shivay, 2019; Christenhusz and Byng, 2016). Several species in this genus have a long history of human association. Plants utilized for food or medical purposes include *C. quinoa* Willd. (quinoa) originally domesticated as a grain crop in the Andes and now grown in cool temperate zones worldwide, and *C. ambrosoides* L. (wormseed or epazote) native to Central and South America, used as a culinary and medicinal herb. The examination of the stomach contents of Iron Age human bodies preserved in peat bogs in northern Europe reveal that *C. album* leaves and seeds were harvested and consumed over 3000 years ago (Behre, 2008; Renfrew, 1973). Two partially domesticated subspecies of *Chenopodium berlandieri* Moq. (pitseed or netseed goosefoot) have been utilized for human food: *C. berlandieri* subsp. *jonesianum* was widely cultivated as a pseudocereal in eastern North America prior to the introduction of maize (*Zea mays* L.) (Kistler and Shapiro, 2011; Wilson, 1981), whereas the leaves and immature inflorescences of *C. berlandieri* subsp. *berlandieri* are still harvested and consumed in parts of Central America. In Mexico a domesticated form of *C. berlandieri* utilized in this way has been classified separately as *C. nuttaliae* Safford (Wilson and Heiser, 1979), known locally as huazontle. Partially domesticated forms of *Chenopodium album* L. are cultivated throughout the Indian Himalayas as a green vegetable known as *bathua* (Prasad and Shivay, 2019; Partap and Kapoor, 1985). *Chenopodium album* is also used in parts of southeast Asia as an antihelminthic and medicinal herb (Poonia and Upadhayay, 2015).

Biology and Management of Problematic Weed Species. DOI: https://doi.org/10.1016/B978-0-12-822917-0.00009-4

This long history of utilization and sporadic domestication has created a complex association between crops and wild *Chenopodium* species that coevolved in shared agroecosystems. Crop-weed gene flow between quinoa and a closely related companion weed, *C. hircinum* Schrad., has been documented throughout the Andes (Wilson, 1988) and was also observed between *C. berlandieri* and quinoa when the latter was introduced as a novel crop to North America (Ward, 1998). Human transport of *Chenopodium* propagules, whether intentional for cultivation and utilization, or unintentional as contaminants of agricultural products, has dispersed a number of *Chenopodium* species beyond their geographical centers of origin. This chapter focuses on two globally distributed and economically important weed species: *C. album* (common lambsquarters, white goosefoot) and *Chenopodium murale* L. (nettleleaf goosefoot, sowbane).

The taxon *C. album* is a species aggregate consisting of wild and cultivated forms, reflecting its history of multiple interactions with human agriculture (Krak et al., 2019; Mandák et al., 2012; Wilson, 1990). In addition to variable morphotypes, multiple ploidy levels have been identified within this aggregate (Mandak et al., 2018). The globally distributed weed most commonly identified as *C. album* s. str. is an allohexaploid ($2n = 6x = 54$), probably the result of hybridization between unidentified diploid and tetraploid parents followed by chromosome doubling (Walsh et al., 2015; Mandák et al., 2012). Chloroplast DNA haplotype diversity analyses suggest that this *C. album* polyploid complex originated in Central Asia and human-mediated dispersal into Europe occurred when agriculture spread west during the Neolithic period (Krak et al., 2019). It is possible that the allopolyploid genome of weedy *C. album* facilitates its competitiveness in diverse agricultural environments and its ability to evade chemical management: various studies have identified associations between polyploidy, weediness, and the evolution of herbicide resistance (e.g., Cheng et al., 2020; Yu et al., 2013; Bennett et al., 1998).

The origins and taxonomy of *C. murale* are less well studied than those of *C. album*. This species is native to southern Eurasia and north Africa but is now naturalized globally (Bajwa et al., 2019). Populations of *C. murale* are reported to be diploid in India (Gangopadhyay et al., 2002).

Chenopodium album and *C. murale* as global weeds

The lengthy association between *C. album* and agriculture, with humans acting as intentional or accidental vectors for seed, has led to *C. album* becoming one of the world's most widely distributed weeds, occurring from sea level to 3600 m and from latitude 70°N to latitude 50°S (Holm et al., 1991). *Chenopodium album* and *C. murale* are opportunistic colonizers of disturbed areas, hence their success as agronomic weeds in annual crops. These weeds are also found along roadsides and in areas disturbed by overgrazing or other human or animal activities. Neither species is a significant invader of intact plant communities in undisturbed natural areas, and they are generally less problematic in perennial crops, although *C. murale* has been reported as a noxious weed of alfalfa (*Medicago sativa* L.) in India (Mseddi et al., 2017). This species occurs as far north as Canada and Scotland (Holm et al., 1991; Hulteen and Fries, 1986); however, it is less cold-tolerant than *C. album* and therefore more prevalent as a weed in subtropical and tropical latitudes.

Chenopodium album and *C. murale* are efficient resource competitors that reduce crop yields by interference (Qasem, 2018; Bhowmik and Reddy, 1988). Both species also produce allelopathic compounds, primarily as root exudates that enhance their competitive abilities by inhibiting crop seedling germination and growth. Inhibitory compounds identified in leaf extracts and root exudates from *C. album* and *C. murale* include oxalic acid, alkaloids, aldehydes, flavonoids, and saponins (Rezaie and Yarnia, 2009; Batish et al., 2007a,b; Reinhardt et al., 1994). Globally, these species are problematic weeds across a diverse range of crops. *Chenopodium album* is the most commonly observed and most troublesome weed of sugar beet (*Beta vulgaris* L.), tomato (*Solanum lycopersicum* L.) and other vegetable crops in North America, where it is also ranked among the top five most damaging weeds in soybean (*Glycine max* L.) and other pulse crops (Van Wychen, 2019). In Europe, *C. album* is a major weed of spring-planted crops, especially maize (Keller et al., 2014), sugar beet (Aper et al., 2012a,b) and soybean (Weber et al., 2015). This species has also been reported as reducing yields of potatoes (*Solanum tuberosum* L.) and sunflower (*Helianthus annuus* L.) in several countries (Holm et al., 1991). In China, *C. album* has become a serious weed of maize and soybean (Huang et al., 2019). *Chenopodium murale* occurs in multiple crops across its range (Bajwa et al., 2019) but published reports of crop losses have centered on warm-season vegetables such as tomatoes, garlic (*Allium sativum* L.) and onions (*Allium cepa* L.) (Qasem, 2018; Edim, 2009; Qasem, 1995).

Biology and ecology

Chenopodium album and *C. murale* are erect annual herbs with the characteristic goosefoot-shaped leaves for which the genus *Chenopodium* is named. *Chenopodium album* can grow up to 2 m in height, with a strong taproot and ridged branching stems that often have red stripes. *Chenopodium murale* has a similar growth form but is bushier and more compact, up to 70 cm in height, with a reddish stem. Both species have alternate simple green to dark green leaves, farinose in the seedlings and on the abaxial surface in mature plants. Greenish perfect flowers are small, radially symmetric, enclosed within five tepals and form dense sessile clusters at the tips of stems and stem branches. The seeds of *C. album* are typically disc-shaped, approximately 1.5 mm in diameter with a smooth shiny black surface. Seeds of *C. murale* are similar in size and shape but dull black with a pitted testa. In both species the pericarp forms a thin papery covering that encloses the mature seed.

Both *C. album* and *C. murale* are predominantly self-pollinating, although up to 3% wind-mediated outcrossing at distances up to 2 m has been reported in *C. album* (Yerka et al., 2012). Plants of both species have high rates of seed production: up to 24,000 seeds per plant for *C. murale* (Holm et al., 1991) and over 150,000 seeds per plant for *C. album* (Colquhoun et al., 2001). Seed heteromorphism occurs in *C. album*, resulting in variable numbers of brown seeds that are non-dormant and larger than the typical dormant black seeds (Williams and Harper, 1965). The proportion of brown seeds produced may increase when the plant is stressed, facilitating rapid germination and establishment in less favorable environments, including saline soils (Yao et al., 2010). Under favorable conditions, dormant black *C. album* seeds can remain viable for more than 30 years (Conn and Deck, 1995; Toole and Brown, 1946), allowing this species to become a major component of the soil seedbank. The persistence of *C. murale* seed in various soil conditions has been less extensively studied; more information on the behavior of this weed in the seedbank is needed for effective management.

Chenopodium album and *C. murale* are summer annuals. *Chenopodium album* seeds germinate at temperatures ranging from 5°C to 30°C (Bajwa et al., 2019; Bhowmik, 1982). This allows seedlings to emerge at cooler temperatures earlier in the season than many other summer annual weeds, enhancing competitiveness (DeGreeff et al., 2018). Germination continues throughout the growing season, with small later-germinating plants flowering in early fall at higher latitudes (Bassett and Crompton, 1978). The flexible phenology of *C. album*, with accelerated time to flowering in shorter photoperiods, contributes to its capacity for interference with multiple crops and cropping systems. It has been suggested that despite its near-global distribution, *C. album* is more important as a weed in temperate zones because cooler temperatures and longer photoperiods at higher latitudes allow greater vegetative growth before flowering, resulting in larger more competitive plants with increased seed production (Holm et al., 1991). The phenology of *C. murale* appears to be similarly flexible: for example, Qasem (2018) reports *C. murale* germinates in fall in Jordan and grows throughout the winter, flowering from spring onwards.

Chenopodium album thrives in diverse soil types and in soil pH from acid to alkaline (Holm et al., 1991) and tolerates both dry and saline conditions (Yao et al., 2010). The wide global distribution of this species and of *C. murale* reflects their phenological plasticity and capacity to adapt to a range of environments and edaphic conditions. It is also worth noting that the genus *Chenopodium* was thought to contain only C_3 plants, but intermediate C_3–C_4 forms in several species, including *C. album*, have recently been reported (Yorimitsu et al., 2019), possibly an evolutionary step that will facilitate range expansion under climate change.

Herbicide resistance

In common with an increasing number of weed species, *C. album* worldwide has responded to chemical management relying on repeated application of the same herbicide modes of action (MOAs) by evolving resistance. *Chenopodium album* populations resistant to photosystem II inhibitors are widespread in Europe and North America: this was the first type of herbicide resistance to appear in this species, with reports dating back to the 1970s (Heap, 2020). In New Zealand a population of *C. album* resistant to atrazine was the first reported case of any herbicide-resistant weed in that country (Rahman, 1982). Resistance to auxinic herbicides including aminopyralid, clopyralid, and dicamba has also been reported in *C. album* in New Zealand (Rahman et al., 2014), where multiple resistance is now emerging in *C. album* populations resistant to both triazines and auxinics (Ghanizadeh and Harrington, 2017). Resistance to acetolactase synthase-inhibiting herbicides has been reported in *C. album* in the United States, Canada, Europe, and most recently China (Heap, 2020; Huang et al., 2019; Konstantinovic et al., 2015), and *C. album* populations with moderate levels of glyphosate resistance have been sporadically reported in North America over the past decade (e.g., Yerka et al., 2013; Westhoven et al., 2008a).

To date, there have been few investigations of the molecular basis for herbicide resistance or associated fitness costs in *C. album*. Photosystem II inhibitor resistance in a German population was mediated by a target-site mutation in the chloroplast D1 protein (Thiel and Varrelmann, 2014). Glyphosate resistance in a US *C. album* population appeared to depend on reduced translocation, not target-site mutation or gene amplification (Yerka et al., 2013). As with other weed species, triazine and dicamba resistance is associated with reduced fitness in some *C. album* populations (Ghanizadeh and Harrington, 2019a,b) which could limit the spread of these forms of resistance, especially if frequent applications of the herbicides in question can be avoided.

As yet there are no confirmed reports of herbicide resistance for *C. murale*. It is unclear whether this reflects less intense selection pressure from herbicide use, or lack of investigation and reporting of *C. murale* populations that fail to respond to chemical management.

Management

Preventing seed spread

Once a weed has colonized an area and become established, eradication is far more expensive and greater resources will be required to control further spread and reduce its impact. Given the history of human vectors moving *Chenopodium* species to new locations, preventing seed spread is important when managing *C. album* and *C. murale*. Harvesting, mowing, hay baling, tilling, earth moving, and manure spreading equipment all have the potential to disseminate seeds of *C. album* and *C. murale*, and cleaning agricultural equipment is a vital strategy to prevent dispersal of weed seeds over large distances (Boyd and White, 2009). Studies in Belguim showed that seed transport by agricultural machinery and manure can spread metamitron-resistant biotypes of *C. album* (Aper et al., 2012a,b). Goldwasser et al. (2011) reported that spreading viable weed seeds with manure is a major contributor to infesting fields with different weed species, including *C. album. Chenopodium album* seeds were found in a range from 5 to 413 viable seeds/ Mg of manure in 12 of 20 surveyed farms in New York state in the United States. (Mt Pleasant and Schlather, 1994). After unsuccessful weed control, seeds of surviving weeds may also contaminate harvested produce and feed. This is particularly risky for fields with herbicide-resistant *C. album* plants that escaped weed control in fields of forage maize, newly sown grassland, fodder beet and other tuber and root crops used as feed. Ensiling a fodder crop after harvesting is an effective strategy for reducing *C. album* and *C. murale* seed viability: exposure of seeds to low pH (2–4), high temperatures, high pressure and the absence of oxygen during ensiling considerably reduce the germination and viability of *C. album* seeds (Seglar, 2003). Aper et al (2013) concluded that keeping a fresh maize silo closed for at least 4 weeks and heaping farmyard manure to raise the temperature are excellent preventive measures to limit the spread of resistant *C. album* seeds between fields.

Chenopodium album and *C. murale* shed seeds during and after harvest, and combine harvesting facilitates their seed dispersal (Donald and Nalewaja, 1991). Mechanical scarification by machinery further aids germination: *C. album* seeds germinated better when collected by combines than when harvested by hand in the same field (Currie and Peeper, 1988). However, mechanical milling to destroy weed seeds in chaff at harvest is a promising management option: in tests using a stationary impact mill, between 96% and 99% of *C. album* seeds mixed with soybean chaff were destroyed (Shergill et al., 2020).

Cultural methods

Cover crops and crop residue

Various studies in Europe and North America have documented suppressive effects of cover crops on *C. album*. A living cover crop of spring-planted rye (*Secale cereale* L.) reduced early-season biomass of *C. album* 98% compared to control plots with no rye (Zimdahl, 2007). Kruidhof et al. (2008) tested various cover crop species and reported that weed biomass consisting of 98% *C. album* was reduced more than 70% by fodder radish (*Raphanus sativus* L.), winter oilseed rape (*Brassica napus* L.), and winter rye. Gallandt (2006) found that alternate years of vegetable and cover crops with a summer fallow reduced *C. album* seed rain in a vegetable rotation of broccoli (*Brassica oleracea* L.) and winter squash (*Cucurbita moschata* Zhit.) more than consecutive years of red clover (*Trifolium pretense* L.) when compared to a no cover crop control, most likely because the fallow periods depleted the weed seedbank. Mennan et al. (2009) found that ryegrass (*Festuca perennis* Lam.) and hairy vetch (*Vicia villosa* L.) as winter cover crops, and grain sorghum (*Sorghum bicolor* L.) and sudangrass (*Sorghum x drummondi* Millsp & Chase) as summer cover crops suppressed *C. album* in early-season organic tomato, pepper (*Capsicum annuum* L.), kale (*Brassica oleracea* L.) and lettuce (*Lactuca sativa* L.) production. Mirsky et al. (2010) reported that a yellow mustard (*Sinapis alba* L.)/ buckwheat (*Fagopyrum esculentum* Moench)/winter rape (*Brassica napus* L.) mixture combined with summer fallow reduced the *C. album* seedbank by 85%; losses to germination were highest when soil disturbance was associated with cover cropping

The effects of cover crop residue on *C. album* have also been investigated. Dyck and Liebman (1994) reported that incorporation of crimson clover (*Trifolium incarnatum* L.) residue into soil reduced the emergence rate, density, and biomass production of *C. album*, with negligible effects on the emergence and growth of sweet corn (maize). In a related study, these authors found that *C. album* plants were smaller and inflicted less damage on sweet corn when clover residue as green manure, rather than synthetic fertilizer, was used as the major N source. They attributed these results to two factors: (1) lower levels of soil inorganic N present in the green manure treatment during the early portion of the growing season, which retarded growth of the weed but had little effect on the crop, and (2) selective phytotoxic effects of clover residue on the weed (Dyck and Liebman, 1995). Boydston and Hang (1995) found that *C. album* density and biomass were significantly reduced in potato when rapeseed green manure was tilled into the soil the following spring before planting; this treatment was more effective than bare fallow or sudangrass tilled in the fall. Soil incorporation of sorghum residues alone and when mixed with sunflower, rice (*Oryza sativa* L.), and *Brassica* spp. trashes provided effective control against *C. album* in wheat (*Triticum aestivum* L.), cotton (*Gossypium hirsutum* L.), and maize fields (Bajwa et al., 2015). Occasionally, more weeds will emerge in low levels of cover crop residue (1–2 Mg/ha) than in uncovered control plots (Teasdale and Mohler, 2000), probably due to retarding evaporation of soil moisture and providing more uniform moisture conditions, and stimulating germination of selected weed species through releasing nitrogenous compounds particularly from legume cover crops.

Soil disturbance and mechanical management

Current trends toward conservation agriculture minimize soil disturbance in cropping systems allowing weed seeds to remain at or near the soil surface, which may lead to changes in weed composition. Although it has been shown that annual small-seeded broad-leaved weeds like *C. album* and *C. murale* often appear in low densities in conservation tillage systems (Eslami, 2014), as noted earlier in this chapter, *C. album* and *C. murale* establishment is favored by disturbance, and field research in North America and elsewhere indicates these weed species are more prevalent in cropping systems that include soil cultivation. For example, Tuesca et al. (2001) recorded major changes in the spectrum of weeds present under different tillage systems and attributed the lower densities of *C. album* under no-till systems to the inhibitory effect of crop residues on light interception. Malidza et al. (2009) determined the influence of different soil cultivations consisting of plowing, disking, chiseling and no-till over a 10-year period on weed populations in soybeans and found that *C. album* recruitment was greater in tilled than in no-till treatments. Buhler (1992) compared the effects of moldboard plowing, chisel plowing, ridge tillage, and no-tillage on weeds and found that *C. album* density was highest with chisel plowing. Mulugeta and Stoltenberg (1997) reported increased emergence of *C. album* following soil disturbance in a no-tillage system, and O'Donovan and McAndrew (2000) in Alberta observed that *C. album* density increased in the soil seedbank as tillage increased. Bernstein et al. (2014) found that *C. album* recruitment in organic soybean was greater in tilled than no-till treatments and was less effectively suppressed by a mowed rye cover crop than by roller-crimping. Mirsky et al. (2010) reported a cumulative 85% decrease in *C. album* in the weed seedbank when germination was stimulated by repeated soil tilling throughout the growing season; these authors also noted, however, that adjusting tillage timing to optimize cover crop biomass could increase *C. album* seedbank recruitment.

The longevity of seeds tends to increase with depth of burial and to decline as soil disturbance increases. In one field study, *C. album* seedling emergence was 9% after 6 years of burial in cultivated soil and 53% after 6 years of burial in undisturbed soil. Since environmental conditions near the soil surface are usually warmer and dryer than deeper in the soil profile, the inability of seeds to remain in a fully imbibed state may account for greater seed losses at shallow depths and after soil disturbance (Radosevich et al., 2007). Tillage and other soil disturbances stimulate decomposition of organic matter and nitrification of ammonium released by decomposition. The presence of nitrate indicates not only enhanced availability of mineral nutrients, but also the elimination of competing vegetation, and emergence and density of *C. album* therefore increases in response to elevated nitrate concentrations (Liebman et al., 2004). Tillage that alters seed position in the soil profile, together with irrigation, may change the moisture condition of seeds, breaking dormancy and affecting germination behavior. Hu et al. (2018) found that the germination of *C. album* and *C. hybridum* initially increased but then decreased as soil moisture content increased. No-till maintains seeds on the soil surface where soil moisture content is relatively low during winter, decreasing seed dormancy release efficiency; moreover, seeds on the soil surface experience greater seed predation as most seed predators are surface foragers (Fenner and Thompson, 2005).

Cultivators are effective management tools against weeds, killing them through uprooting, dismemberment, or burial (Mohler, 2001). Mohler et al. (2016) found that no *C. album* seedlings recovered from 4 cm of burial, but seedlings showed 57%–67% recovery when part of a single leaf was left exposed, suggesting the importance of throwing as much soil into the row as the crop will tolerate to completely bury weed seedlings. Withholding irrigation for several days after hilling-up operations minimizes weed recovery from burial. Perron and Legere (2000) in a maize/soybean rotation found that *C. album* seed production was greater under mechanical weed control compared with chemical weed control, with 73,000 and 766,000 seeds/m^2 produced under chemical and mechanical weed control measures, respectively. Lundkvist (2009) in a field study in Sweden concluded that the late emerging annual weed species including *C. album* were most effectively controlled when preemergence (PRE) weed harrowing was combined with one or two weed harrowing treatments after crop emergence. However, continuous use of heavy machinery in fields may result in soil compaction that reduces oxygen concentration in soil capillaries and lowers the O_2 availability for germinating seeds. Yasin and Andreasen (2019) concluded that reduced O_2 concentrations due to soil compaction may be one reason why *C. album* has increased in some area as its germination is favored by lower oxygen levels. Flushes of early-emerging weeds such as *C. album* can be managed with preplant tillage as long as it destroys weeds rather than merely uprooting them. Singh and Singh (2004) found that presowing irrigation reduced the dry weight of C. *album* and C. *murale* by 21% and 25%, respectively; subsequent wheat yield was 12% higher compared to postsowing irrigation. This can be ascribed to the destruction of the first flush of weeds during land preparation and sowing.

Thermal weeding techniques including flaming and solarization are promising approaches for controlling *C. album* and *C. murale* (Bajwa et al., 2019; Ravangard et al., 2015). Ravangard et al. (2015) buried *C. album* seeds at depths from 0 to 15 cm and covered the field soil with a two-layer clear plastic sheet for 45 days. They reported this completely prevented seed germination of *C. album* at 0–5 cm burial depth and only 9% of seeds germinated at 10 cm burial depth, whereas 95% of seeds buried at 15 cm germinated. *Chenopodium album* and *C. murale* are sensitive to flaming, as they possess unprotected growing points and thin leaves. These species can be killed at early seedling stages (0–4 true leaves) using 20–50 kg/ha propane; at later stages (4–12 leaves) higher rates of 50–200 kg/ha propane, are required (Upadhyaya and Blackshaw, 2007). Predispersal flaming did not affect viability of *C. album* seeds, it is, however, a promising method to reduce density of seeds following dispersal (Gallandt, 2014). Steam application can also control *C. album*: in a study using superheated steam (400°C), 2–4 s of exposure to steam killed seedlings of *C. album*, whereas seed production of plants that survived was significantly reduced (Upadhyaya and Blackshaw, 2007). Lasers may also be an effective management tool: Heisel et al. (2001) found that when stems were cut below the meristems, 0.9 and 2.3 J/mm of CO_2 laser energy was sufficient to cause 90% biomass reduction in *C. album.*

Fertilization strategies

Chenopodium album and *C. murale* are efficient accumulators of mineral nutrients and will absorb them as much or more than crops. High demand and nutrient accumulation of *C. album* under low and high fertility levels has been noted (Qasem, 2010). Qasem and Hill (1994) reported that N, P, K, and Mg concentrations were higher in shoots of *C. album* compared to many associated crops. Greenhouse and outdoor trials confirmed that the dry weight of *C. album* growing in competition with crops increased with increasing N application rates (Sønderskov et al., 2012), whereas N uptake and assimilation rates were reported to be higher in *C. album* compared with crop plants, making it more competitive (Lindsey et al., 2013). Nitrate has a stimulatory effect on germination of many small-seeded weed species like *C. album* and *C. murale*. It has been noted that mineralization of N from N-enriched residues of alfalfa may enhance emergence of *C. album* (Teasdale et al., 2019). *Chenopodium murale* favors a high level of fertilization and is more abundant in N-rich habitats in the Mediterranean region (Shimwell, 2006). This species is a superior accumulator of N, P, K, Mg, and Se compared to associated vegetable crops and many other weed species under field conditions (Qasem, 2010).

Nitrogen management strategies that limit N assimilation by weeds and favor crop growth should be considered to reduce crop–weed competition (DiTomaso, 1995). Banding N or P is one potential approach to reduce the negative effect of *C. album* and *C. murale* on crops (Blackshaw et al., 2004a,b). Timing fertilizer applications when crop assimilation is rapid may also reduce weed interference (Lindsey et al., 2013). Paolini et al. (1999) found that time of N fertilization did not influence biomass, yield and yield quality of the weed-free sugar beet (*Beta vulgaris* L.), but early N fertilization reduced crop biomass in the presence of *C. album*. Blackshaw et al. (2004a,b) found that density and biomass of *C. album* plants were sometimes lower with spring-applied N than with fall-applied N. Delayed fertilizer application increased maize biomass as much as 70% and reduced *C. album* biomass as much as 50%, compared with early application of the same total quantity of fertilizer (Liebman et al., 2004).

Competitive interference for P and, to some extent, for nitrogen was noted between wheat and *C. album* (Zimdahl, 2004). Nutrient interference played a major part in limitation of growth of wheat plants by *C. album*, whereas wheat exhibited greater noncompetitive interference in restricting potassium uptake by *C. album*. Blackshaw et al. (2004a,b) found that *C. album* was more responsive to P addition than canola and wheat in a range of 5 to 60 mg P/kg soil. *Chenopodium album* plants have been shown to respond strongly to P amendment as conventional fertilizers such as composted poultry manure (Little et al., 2015). *Chenopodium murale* requires high K soil level for maximum competitiveness (Qasem, 2018). Reduced K levels decreased *C. murale* competition effects on tomato and cauliflower (Qasem, 2010), suggesting control of the weed could be improved by managing K levels in the field. Manipulation of fertilizer rate or level of a single mineral element applied under field conditions may be integrated with other methods of weed control for better management of *C. murale* in field crops of low K demand.

Crop rotation and crop competition

Weeds typically associate with crops that have similar life cycles, so rotating crops with different life cycles can disrupt the development of weed–crop associations (Eslami, 2014). Crop rotations also allow different herbicides to be applied: Manley et al. (2001) reported this management strategy successfully controlled populations of *C. album*. Germination of *C. album* is favored by cool temperatures. Teasdale et al. (2019) found that *C. album* presence had a strong negative association with early-season temperature, but this relationship was modified by rotation length, with longer rotations increasing the probability of occurrence particularly at higher temperatures. An effective crop rotation program to control *C. album* and *C. murale* should alternate fall and spring sown crops, rotate broadleaf crops with grass or cereal crops, and include crops with different planting dates in different years to prevent weed adaptation and buildup. Planting later in such a rotation may allow the use of a preplanting burndown herbicide or tillage to control early-emerging weeds such as *C. album* and *C. murale*.

Maximizing crop competition against weeds by manipulating row spacing and planting competitive crop cultivars can be an important tool for an integrated weed management program (Watson et al., 2006). Increased crop yield and improved control of *C. album* through planting crops in narrow rows have been reported for cotton, maize, and soybeans (Bernstein et al., 2014; Tharp and Kells, 2001; Culpepper and York, 2000). Qasem (2018) investigated competition of 31 tomato cultivars with *C. murale* over different growing periods under natural conditions and found considerable variation in cultivar competitiveness and weed impact on growth and yield. Crop genotypes with early canopy closure, high competitive ability, and/or allelopathic suppressive potential against *C. album* and *C. murale* may control these species more efficiently (Qasem, 2018). Rotations incorporating fodder crops that are planted at higher densities and in narrower rows compared to those sown for their seeds is a sound strategy for the control of these species.

Allelopathic weed management

The suppressive potential of allelopathic compounds released by plants against their neighbors has been investigated for control of *C. album* and *C. murale* in various cropping systems. The application of rye cover crop mulch and root residues reduced *C. album* by 90% in tobacco, sunflower, and soybean in no-till systems (Barnes et al., 1986; Shilling et al., 1985). Ahmad et al. (1995) reported that sorghum residues suppressed emergence and growth of *C. album* plants. Abdul Raoof and Siddiqui (2013) found that allelochemicals present in leaves of *Tinospora cordifolia* Miers had suppressive impact on chlorophyll and protein content of *C. album* and *C. murale*. Exudates from different oat cultivars reduced the germination of *C. album* from 10% to 86% (Grimmer and Masiunas, 2005).

The efficacy of crop allelopathic water extracts for control of *Chenopodium* species has been explored in several studies. Sorghum water extract ("sorghaab") suppressed *C. album* in wheat (Cheema et al., 1997). Use of reduced doses of pendimethalin (413 g ai/ha) in combination with sorghum/sunflower water extract (15–18 l/ha each) was effective in complete suppression of *C. album* (Awan et al., 2009). Batish et al. (2002) found that parthenin, a sesquiterpene lactone extracted from *Parthenium hysterophorus* L., significantly reduced germination, seedling growth, dry weight accumulation and chlorophyll content of *C. murale*, suggesting that parthenin could be a useful natural herbicide for weed management. Ailanthone, a phytotoxic compound isolated from tree of heaven (*Ailanthus altissima* Mill.), showed both PRE and postemergence (POST) herbicidal activity and reduced *C. album* biomass (Heisey and Heisey, 2003). Qasem (2002) reported that aqueous extracts and dried shoots of several medicinal plant species were toxic to *C. murale* germination and seedling growth. Stokłosa et al. (2012) confirmed the herbicidal activity of clove oil (*Syzygium aromaticum* L.) against *C. album*. Al-Harbi (2018) found that allelopathic effects on the germination of *C. murale* seeds was obtained with leaf extracts of *Rhanterium epapposum* and *Salsola imbricata* especially at concentration of 40% and 60%. Results also showed that the root and shoot lengths of *C. murale* were more sensitive to allelopathic impact of *R. epapposum* leaf extract compared to *S. imbricata* leaf extract. Qasem and Issa (2018) showed that residues of cabbage, cauliflower, and bean (*Phaseolus vulgaris* L.) were remarkably toxic against germination and growth of *C. murale*. Crops with confirmed allelopathic effects can be exploited as smother crops, planted in a rotational sequence, left as residue or mulch especially in low or zero-tillage systems, and bred for allelopathic varieties to include promising allelochemicals in a new generation of herbicides to control *C. murale* and *C. album*.

Biological control

Some specialized natural enemies have been successfully tested as potential biological control agents for *C. album*, but none for *C. murale* (Bajwa et al., 2019). Kempenaar et al. (1996) found that application of *Ascochyta caulina* spore suspensions improved the performance of maize and sugar beet grown with *C. album*, reducing *C. album* density up to 65% without harming the crops. Scheepens et al. (1997) suggested that a microbial herbicide based on *A. caulina* could be used to control this weed. Vurro et al. (2001) achieved improved control of *C. album* using low rates of herbicides (metribuzin and rimsulfuron at 20% of the labeled rate) in combination with toxins from *A. caulina*. If other weeds are present, the fungus may be tank mixed with half the regular dose of a sulphonylurea herbicide.

It should be ensured that any biological control is compatible with other measures in a cropping system. For instance, in a potato crop the application of fungicides could interfere with the control of *C. album* by *A. caulina*. Timing of fungicide application or the use of selective fungicides that do not inhibit *A. caulina* could solve the incompatibility problem. In maize and sugar beet crops, no fungicides are usually applied, so no compatibility problems with mycoherbicides will arise. The use of N fertilizer is an important management consideration in biological control of weeds by fungi, because high N supply may stimulate the growth of both weed and pathogen (Ghorbani et al., 2002). Some genera of deleterious bacteria, such as *Achromobacter*, *Alcaligenes*, *Arthrobacter*, *Citrobacter*, *Enterobacter*, *Erwinia*, *Flavobacterium*, *Klebsiella*, and *Pseudomonas*, may be useful for *Chenopodium* species suppression, whereas they have benign or neutral effects on crops (Liebman et al., 2004).

Weed seeds are lost from the seedbank through aging, attacks by seed predators, and seed decay as well as germination (Eslami, 2014). Nearly all (99.8%) of the current year's seed rain of *C. album* was eliminated under a canopy of alfalfa (Radosevich et al., 2007). The seed predator in this case was a small field mouse (*Peromyscus* sp.). Mice also consumed up to 20% of seeds of *C. album* in no-till soybean and maize in Canada (Bajwa et al., 2015). Plant residues can also promote the occurrence of generalist weed seed feeders such as beetles, ants, crickets, and small rodents (Liebman and Mohler, 2001). Weed seed predation by invertebrates such as field crickets and ground beetles can play an important role in reducing the soil seedbanks of weed species with small seeds: ground-dwelling invertebrates alone can consume 80%–90% of the postdispersal seeds of *C. album*. For example, the ground beetle *Harpalus rufipes* Degeer is capable of consuming up to 90% of the postdispersal seeds of certain weed species. As the density of *H. rufipes* increases, a corresponding increase in seed predation of preferred species, such as *C. album* has been observed (Eslami, 2014). Blubaugh and Kaplan (2016) found that seed predators, predominantly invertebrates, were responsible for a 38% reduction in seedling emergence and 81% reduction in *C. album* biomass in fallow plots following simulated seed rain, suggesting that granivory can effectively suppress weeds. van der Laat et al. (2015) quantified postdispersal seed predation of *C. album* in maize and soybean managed with conventional, reduced, and zero-tillage systems. They showed that zero-tillage might limit weed seed predation because abundant plant residue may be a barrier for the mobility of insect seed predators even if their populations are higher than in other tillage regimes.

Chemical management

The importance of herbicides in modern weed management is underscored by estimates that losses in the agricultural sector would increase about 500% without their use (Eslami, 2014). Both PRE and POST herbicides have been shown to be effective against *C. album* and *C. murale*, although—as previously noted—the evolution of resistance to multiple MOAs in *C. album* threatens future use if herbicides are not carefully managed. Sprague (1999) found that isoxaflutole applied PRE controls *C. album* under conventional and no-till maize. Maize injury was most commonly observed with both tillage systems in coarse textured soils and was more severe with higher rates of isoxaflutole. Isoxaflutole applied postemergence also controlled *C. album*. Singh and Singh (2004) in a study of control of *C. album* and *C. murale* weeds found that PRE application of pendimethalin at 0.75 kg/ha supplemented by one hand weeding or 2,4-D, 0.50 kg/ha at 30 days after sowing (DAS) significantly increased wheat yield due to better weed control. Qasem (2005) in a study on an onion (*Allium cepa* L.) field highly infested with *C. murale* (206 pl/m^2) found that oxyfluorfen and oxadiazon were the best options for weed control and improved onion yield, followed by pendimethalin, methabenzthiazuron, and linuron as preplanting treatments, and pendimethalin, linuron, and paraquat as postplanting applications. Field studies on garlic (*Allium sativum* L.) showed that pendimethalin in combination with manual hoeing can successfully control *C. album*, and *C. murale* and also raise garlic bulb yield (Khatam et al., 2012; Mehmood et al., 2007). Mehriya et al. (2008) in a field study on cumin (*Cuminum cyminum* L.) infested mostly by *C. murale* (70.0%) and *C. album* (16.3%) found that oxyfluorfen or oxadiargyl 50 g/ha each applied at 20 DAS followed by one hand weeding at 35 DAS controlled weeds effectively with higher cumin seed yield. Nagar et al. (2009) also found that hand weeding twice at 30 and 45 DAS and PRE application of pendimethalin at 1.0 kg/ha + hand weeding at 45 DAS were at par in controlling *C. murale* and *C. album* in coriander (*Coriandrum sativum* L.). Alebrahim et al. (2012) observed that the efficacy ranking of four herbicides studied for *C. album* control in potatoes was as ethalfluralin, metribuzin, pendimethalin and trifluralin, whereas ranking of herbicides for potato safety was metribuzin, trifluralin, pendimethalin, and ethalfluralin. Among newer herbicides, the synthetic auxin halauxifen acid is effective for preplant burndown control of *C. album* when applied 14 days prior to planting for maize and soybeans.

Tank mixes of two or more herbicides are increasingly used in many agronomic crops to control a wide spectrum of weeds, reduce production costs, and to delay the development of herbicide-resistant weeds (Zand et al., 2008). A synergistic response of *C. album* to mixtures of halosulfuron-methyl and 2,4-D was reported, with weed biomass for all combinations less than expected based on independent herbicide action (Isaacs et al., 2006). Deveikyte et al. (2015) concluded the PRE combination of metamitron with phenmedipham plus desmedipham plus ethofumesate plus rapseed oil resulted in excellent (>96%) control of *C. album* and other broad-leaved weeds. Assadollahi et al. (2017) reported that a combination of equal doses of foramsulfuron and nicosulfuron optimized *C. album* damage compared to other ratios. Moreover, combining these two herbicides with ammonium sulfate had synergistic effects on *C. album* control. Smith et al. (2019) found that where there was sufficient rainfall for herbicide activation, isoxaflutole + metribuzin at all rates achieved better than 90% control of *C. album*. Tank mixes of metribuzin, flumioxazin, and imazethapyr are effective on many annual broadleaf weeds such as *C. album* and *C. murale* (Owen, 2017).

Herbicide application with low quality water reduces efficacy. Water hardness caused by increased calcium carbonate content has been reported to limit the efficacy of dicamba, 2,4-D, and quinclorac used for the control of broad-leaved species like *C. album* and *C. murale* (Izadi Darbandi et al., 2011, 2012). Roskamp et al. (2013) found that control of *C. album* in field experiments was reduced by up to 56% when saflufenacil was applied in water with a pH of 4.0 compared with water with a pH of 7.7.

As *C. album* is currently the fourth most globally prevalent herbicide-resistant weed (Heap 2020), management of resistant *C. album* populations is increasingly important. Westhoven et al. (2008b) found that glyphosate-resistant *C. album* was controlled by an integrated strategy that included a combination of preplant glyphosate, 2,4-D, and residual herbicides, with POST application of glyphosate. As with any weed, the effectiveness of glyphosate on *C. album* can be reduced by drought, short photoperiods, or insect infestation reducing plant growth rate; below-label application rates; rainfall too soon after application; and dust.

Triazine-resistant *C. album* has been a global problem since the 1980s. Studies have reported satisfactory control of these populations with PRE herbicides. Chomas and Kells (2004) examined metolachlor, pendimethalin, and acetochlor applied PRE for control of triazine-resistant *C. album* in maize and found that pendimethalin provided greater (98%) and more consistent control than metolachlor (66%) or acetochlor (86%). They also found that isoxaflutole (35 g ai/ha), flumetsulam (35 g ai/ha), and rimsulfuron plus thifensulfuron (26 g ai/ha) provided 98% or greater control of triazine-resistant *C. album*.

Arslan et al. (2016) found that treatments with tembotrione POST either with or without interrow cultivation consistently resulted in weed control and crop yield comparable to standard treatments with atrazine POST. Additional atrazine-free treatments with topramezone applied POST worked well in Oregon where small-seeded weed species such as *C. album* were prevalent.

Integrated weed management

Awareness of the dangers associated with a sole reliance on herbicides for weed control is growing both in the agricultural and public sectors, and interest is shifting toward a more integrated approach to weed management based on ecological principles (Dillard et al., 2004). A best approach to manage weeds is to integrate two or more available and effective technologies (Sanyal et al., 2008). For example, Boydston and Vaughn (2002) in a 2-year study in potatoes found that the use of a combination of cover crops, cultivation, and herbicide controlled weeds including *C. album*, reduced PRE-applied herbicide inputs 66%, and maintained potato yield. Dillard et al. (2004) found that crop rotation combined with nonchemical measures such as mulching can be effective in managing *C. album* populations. The inclusion of PRE herbicides with tillage increased control of *C. album* compared to tillage used without a herbicide (Chauhan et al., 2012), whereas Mohammadduost Chamanabad et al. (2011) found that combining the application of metribuzin and leaving barley residue on the soil surface significantly increased *C. album* control compared to either alone. Given the competitive efficiency of nutrient uptake and utilization in *C. album* and *C. murale*, consideration must also be given to fertilizer type, dose, and application timing when devising integrated management strategies. Additional integrated strategies such as stale seedbed, where irrigation stimulates the germination and emergence of weeds that can then be killed using nonselective herbicides, are promising approaches for managing early-emerging weeds such as *C. album* and *C. murale* before planting (Cloutier et al., 2007).

Conclusion

The ecological amplitude, adaptive capacity, competitiveness, high rates of reproduction, and genetic diversity of *C. album* and *C. murale* have enabled them to become globally important weeds that impact a diverse range of temperate and subtropical cropping systems. The accumulation of resistance to different herbicide MOAs by *C. album* indicates that continued dependence on chemical control is not feasible: integrating additional strategies will be necessary for long-term sustainable management. The more recent emergence of *C. murale* as a weed of international importance is reflected in the relative lack of published research on this species compared to *C. album*. This needs to be addressed: in particular, more information is needed on effective chemical and nonchemical management of *C. murale* in different agroenvironments.

References

Abdul Raoof, K.M., Siddiqui, M.B., 2013. Evaluation of allelopathic effect of *Tinospora cordifolia* on biochemical activity of some weeds. Int. J. Agric. Technol 9, 237–244.

Ahmad, S., Rehman, A., Cheema, Z.A., Tanveer, A., Khaliq, A., 1995. Evaluation of some crop residues for their allelopathic effects on germination and growth of cotton and cotton weeds. in: 4th Pakistan Weed Science Conference, Faisalabad, Pakistan, pp. 63–71.

Alebrahim, M.T., Rashed-Mohassel, M.H., Wilcokson, S., Baghestani, M.A., Ghorbani, R., 2012. Evaluating of some preemergence herbicides for lambsquarter and redroot pigweed control in potato fields. J. Plant Prot 25, 358–367.

Al-Harbi, N.A., 2018. Allelopathic effect of leaf extract of two wild plants on seed germination, shoot and root length of two weed species; *Portulaca oleracea* and *Chenopodium murale*. Biosci. Biotechnol. Res. Asia 15, 929–935.

Aper, J., De Riek, J., De Cauwer, B., Bulke, R., Rehuel, D., 2012a. The origin of metamitron resistant Chenopodium album populations in sugar beet. Commun. Agric. Appl. Biol. Sci 77 (3), 335–342.

Aper, J., Mechant, E., De Riek, J., Van Laere, K., Bulcke, R., Reheul, D., 2012b. Analysis of local spread of metamitron-resistant *Chenopodium album* patches in Belgium. Weed Res 52, 421–429.

Aper, J., De Cauwer, B., De Roo, S., Lourenc, M., Fievez, V., Bulcke, R., et al., 2013. Seed germination and viability of herbicide resistant and susceptible *Chenopodium album* populations after ensiling, digestion by cattle and manure storage. Weed Res 54, 169–177.

Arslan, Z.F., Williams II, M.M., Becker, R., Fritz, V.A., Ed Peachey, R., Rabaey, T.L., 2016. Alternatives to atrazine for weed management in processing sweet corn. Weed Sci 64, 531–539.

Assadollahi, H., Rastgoo, M., Izadi Darbandi, E., Ghanbari, A., 2017. Evaluation of foramsulfuron and nicosulfuron combination with ammonium sulfate on lambsquarters. Iranian J. Field Crops Res 15, 19–30.

Awan, I.U., Khan, M.A., Zareef, M., Khan, E.A., 2009. Weed management in sunflower with allelopathic water extract and reduced doses of a herbicide. Pak. J. Weed Sci. Res 15, 19–30.

Bajwa, A.A., Mahajan, G., Chauhan, B.S., 2015. Nonconventional weed management strategies for modern agriculture. Weed Sci 63, 723–747.

Bajwa, A.A., Zulfiqar, U., Sadia, S., Bhowmik, P., Chauhan, B., 2019. A global perspective on the biology, impact and management of *Chenopodium album* and *Chenopodium murale*: two troublesome agricultural and environmental weeds. Environ. Sci. Pollut. Res 26, 5357–5371.

Barnes, J.P., Putnam, A.R., Burke, B.A., 1986. Allelopathic activity of rye. In: Putnam, A.R., Tang, C.S. (Eds.), The Science of Allelopathy. Willey Interscience, New York, pp. 271–286.

Bassett, I.J., Crompton, C.W., 1978. The biology of Canadian weeds. 32. *Chenopodium album* L. Can. J. Plant Sci 58, 1061–1072.

Batish, D.R., Singh, H.P., Saxena, D.B., Kohli, R.K., 2002. Weed suppressing ability of Parthenin—a sesquiterpene lactone from *Parthenium hysterophorus*. N. Z. Plant Prot 55, 218–221.

Batish, D.R., Lavanya, K., Singh, H.P., Kohli, R.K., 2007a. Phenolic allelochemicals released by *Chenopodium murale* affect the growth, nodulation and macromolecule content in chickpea and pea. Plant Growth Regul 51, 119–128.

Batish, D.R., Lavanya, K., Singh, H.P., Kohli, R.K., 2007b. Root mediated allelopathic interference of nettle-leaved goosefoot (*Chenopodium murale*) on wheat (*Triticum aestivum*). J. Agron. Crop Sci 193, 37–44.

Behre, K.E., 2008. Collected seeds and fruits from herbs as prehistoric food. Veg. History Archaeobotany 17, 65–73.

Bennett, M.D., Leitch, I.J., Hanson, L., 1998. DNA amounts in two samples of angiosperm weeds. Ann. Botany 82 (Supplement A), 121–134.

Bernstein, E.R., Stoltenberg, D.E., Posner, J.L., Hedtcke, J.L., 2014. Weed community dynamics and suppression in tilled and no-tillage transitional organic winter rye–soybean systems. Weed Sci 62, 125–137.

Bhowmik, P.C., 1982. Differential growth and response to temperature between two biotypes of *Chenopodium album*. Proc. North. Weed Sci. Soc 36, 89–93.

Bhowmik, P.C., Reddy, K.N., 1988. Interference of common lambsquarters (*Chenopodium album*) in transplanted tomatoes (*Lycopersicon esculentum*). Weed Technol 2, 505–508.

Blackshaw, R.E., Brandt, R.N., Janzen, H.H., Entz, T., 2004a. Weed species response to phosphorus fertilization. Weed Sci 52, 406–412.

Blackshaw, R.E., Molnar, L.J., Janzen, H.H., 2004b. Nitrogen fertilizer timing and application method affect weed growth and competition with spring wheat. Weed Sci 52, 614–622.

Blubaugh, C.K., Kaplan, I., 2016. Invertebrate seed predators reduce weed emergence following seed rain. Weed Sci 64, 80–86.

Boyd, N.S., White, S., 2009. Impact of wild blueberry harvesters on weed seed dispersal within and between fields. Weed Sci 57, 541–546.

Boydston, R.A., Hang, A., 1995. Rapeseed green manure crop suppresses weeds in potato. Weed Technol 9, 669–675.

Boydston, R.A., Vaughn, S.F., 2002. Alternative weed management systems control weeds in potato. Weed Technol 16, 23–28.

Buhler, D.D., 1992. Population dynamics and control of annual weeds in corn as influenced by tillage systems. Weed Sci 40, 241–248.

Chauhan, B.S., Singh, R.G., Mahajan, G., 2012. Ecology and management of weeds under conservation agriculture: a review. Crop Prot 38, 57–65.

Cheema, Z.A., Luqman, M., Khaliq, A., 1997. Use of allelopathic extracts of sorghum and sunflower herbage for weed control in wheat. J. Anim. Plant Sci 7, 91–93.

Cheng, J., Yang, X., Lifang, X., 2020. Polyploidization contributes to evolution of competitive ability: a long-term common garden study on the invasive *Solidago canadensis* in China. Oikos 129, 700–713.

Chomas, A.J., Kells, J.J., 2004. Triazine-resistant common lambsquarters control in corn with preemergence herbicides. Weed Technol 18, 551–554.

Christenhusz, M.J.M., Byng, J.W., 2016. The number of known plant species in the world and its annual increase. Phytotaxa 261, 201–217.

Cloutier, D.C., van der Weide, R.Y., Peruzzi, A., Leblanc, M.L., 2007. Mechanical weed management. In: Upadhyaya, M.K., Blackshaw, R.E. (Eds.), Non-Chemical Weed Management: Principles, Concepts and Technology. CABI, Oxfordshire, pp. 111–134.

Colquhoun, J., Stoltenberg, D.E., Binning, L.K., Boerboom, C.M., 2001. Phenology of common lambsquarters growth parameters. Weed Sci 49, 177–183.

Conn, J.S., Deck, R.E., 1995. Seed viability and dormancy of 17 weed species after 9.7 years of burial in Alaska. Weed Sci 43, 583–585.

Culpepper, A.S., York, A.C., 2000. Weed management in ultra-narrow row cotton. Weed Technol 14, 19–29.

Currie, R.S., Peeper, T.F., 1988. Combine harvesting affects weed seed germination. Weed Technol 2, 499–504.

DeGreeff, R.D., Varanasi, A.V., Dille, J.A., Peterson, D.E., Jugulam, M., 2018. Influence of plant growth stage and temperature on glyphosate efficacy in common lambsquarters (*Chenopodium album*). Weed Technol 32, 448–453.

Deveikyte, I., Sarunaite, L., Seibutis, V., 2015. Evaluation of pre- and postemergence herbicide combinations for broadleaved weeds in sugar beet. In: Price, A., Kelton, J., Sarunaite, L. (Eds.), Herbicides, Agronomic Crops and Weed Biology. IntechOpen, pp. 53–72.

Dillard, H.R., Bellinder, R.R., Shah, D.A., 2004. Integrated management of weeds and diseases in a cabbage cropping system. Crop Prot 23, 163–168.

DiTomaso, J.M., 1995. Approaches for improving crop competitiveness through the manipulation of fertilization strategies. Weed Sci 43, 491–497.

Donald, W.W., Nalewaja, J.D., 1991. Northern great plains. In: Donald, W.W. (Ed.), Systems of Weed Control in Wheat in North America. Monograph #6 Weed Sci. Soc. Am, Champaign, IL, pp. 90–126.

Dyck, E., Liebman, M., 1994. Soil fertility management as a factor in weed control: the effect of crimson clover residue, synthetic nitrogen fertilizer, and their interaction on emergence and early growth of lambsquarters and sweet corn. Plant Soil 167, 227–237.

Dyck, E., Liebman, M., 1995. Crop–weed interference as influenced by a leguminous or synthetic fertilizer nitrogen source. II. Rotation experiments with crimson clover, field corn, and lambsquarters. Agr. Ecosyst. Environ 56, 109–120.

Edim, H., 2009. Effect of nettle-leaf goosefoot (*Chenopodium murale*) on bulb onion (*Allium cepa*) yield. Appl. Entomol. Phytopathol 76, 55–64.

Eslami, S.V., 2014. Weed management in conservation agriculture systems. In: Chauhan, B.S., Mahajan, G. (Eds.), Recent Advances in Weed Management. Springer, New York, pp. 87–124.

Fenner, M., Thompson, K., 2005. The Ecology of Seeds. Cambridge University Press, New York.

Gallandt, E.R., 2006. How can we target the weed seedbank? Weed Sci 54, 588–596.

Gallandt, E.R., 2014. Weed management in organic farming. In: Chauhan, B.S., Mahajan, G. (Eds.), Recent Advances in Weed Management. Springer, New York, pp. 63–86.

Gangopadhyay, G., Das, S., Mukherjee, K.K., 2002. Speciation in *Chenopodium* in West Bengal. India. Genet. Res. Crop Evol 49, 503–510.

Ghanizadeh, H., Harrington, K.C., 2017. Cross-resistance to auxinic herbicides in dicamba-resistant *Chenopodium album*. N. Z. J. Agric. Res 60, 45–53.

Ghanizadeh, H., Harrington, K.C., 2019a. Ecological evidence for the fitness trade-off in triazine resistant *Chenopodium* album L.: Can we exploit the cost of resistance? Agronomy-Basel 9. Available from: https://doi.org/10.3390/agronomy9090523.

Ghanizadeh, H., Harrington, K.C., 2019b. Fitness costs associated with multiple resistance to dicamba and atrazine in *Chenopodium album*. Planta 249, 787–797.

Ghorbani, R., Scheepens, P.C., Zweerde, W.V.D., Leifert, C., McDonald, A.J.C., Seel, W., 2002. Effects of nitrogen availability and spore concentration on the biocontrol activity of *Ascochyta caulina* in common lambsquarters. Weed Sci 50, 628–633.

Goldwasser, Y., Miron, J., Sibony, M., Sorek, P., Yosef, E., Rubin, B., 2011. Spread of weed seeds by cattle manure—is there a danger of infesting farmers' fields with weeds? in: Proceedings of the 3rd International Conference on Novel and Sustainable Weed Management in Arid and Semi-Arid Ecosystems. Huesca, Spain.

Grimmer, O.P., Masiunas, J.B., 2005. The weed control potential of oat cultivars. Hort. Technol. 15, 140–144.

Heap, I., 2020. The International Survey of Herbicide Resistant Weeds. http://www.weedscience.org (accessed on May 20, 2020).

Heisel, T., Schou, J., Christensen, S., Andreasen, C., 2001. Cutting weeds with a CO2 laser. Weed Res 41, 19–29.

Heisey, R.M., Heisey, T.K., 2003. Herbicidal effects under field conditions of *Ailanthus altissima* bark extract, which contains ailantheone. Plant Soil 256, 85–99.

Holm, L.G., Plucknett, D.L., Pancho, J.V., Herberger, J.P., 1991. *Chenopodium album* L. The World's Worst Weeds: Distribution and Biology. Krieger Publishing, Malabar, FL, pp. 84–91.

Hu, X.W., Ding, X.Y., Baskin, C.C., Wang, Y.R., 2018. Effect of soil moisture during stratification on dormancy release in seeds of five common weed species. Weed Res 58, 210–220.

Huang, Z., Zhou, X., Zhang, C., Jiang, C., Huang, H., Wei, S., 2019. First report of molecular basis of resistance to imazethapyr in common lambsquarters. Weed Sci 68, 63–68.

Hulteen, E., Fries, E.M., 1986. Atlas of North European vascular plants north of the Tropic of Cancer. Lubrecht & Cramer, Port Jervis, NY, p. 335.

Isaacs, M.A., Hatzios, K.K., Wilson, H.P., Toler, J., 2006. Halosulfuron and 2,4-D mixtures effects on common lambsquarters. Weed Technol 20, 137–142.

Izadi Darbandi, E., Nessari, N., Azarian, F., 2011. Investigating the effect of water hardness on 2,4-D amine efficacy on redroot pigweed and common lambsquarters control. J. Plant Prot 25, 258–265.

Izadi Darbandi, E., Azarian, F., Nessari, N., 2012. Investigating the effect of water calcium carbonate on sinsulfuron efficacy on redroot pigweed and common lambsquarters control. Iranian J. Field Crops Res 10, 84–90.

Keller, M., Böhringer, N., Möhring, J., Rueda-Ayala, V., Gutjahr, C., Gerhards, R., 2014. Long-term changes in weed occurrence, yield and use of herbicides in maize in south-western Germany, with implications for the determination of economic thresholds. Weed Res. 54, 457–466.

Kempenaar, C., Horsten, P.J.F.M., Scheepens, P.C., 1996. Growth and competitiveness of common lambsquarters after foliar application of *Ascochyta caulina* as a mycoherbicide. Weed Sci 44, 609–614.

Khatam, A., Ahmad, A.W., Zafarullah Khan, M., 2012. Perception of farmers regarding effect of various weed management practices in onion crop. Pak. J. Weed Sci. Res 18, 553–559.

Kistler, L., Shapiro, B., 2011. Ancient DNA confirms a local origin of domesticated chenopod ieastern North America. J. Archaeol. Sci 38, 3549–3554.

Konstantinovic, B., Blagojevic, M., Konstantinovic, B., Samardzic, N., 2015. Resistance of weed species *Chenopodium album* L. to ALS-inhibitors. Roman. Agricult. Res 32, 253–261.

Krak, K., Habibi, F., Douda, J., Vit, P., Lomonosova, M.N., Wang, L., et al., 2019. Human-mediated dispersal of weed species during the Holocene: a case study of *Chenopodium album* agg. J. Biogeog 46, 1007–1019.

Kruidhof, H.M., Bastiaans, L., Kropff, M.J., 2008. Ecological weed management by cover cropping: effects on weed growth in autumn and weed establishment in spring. Weed Res 48, 492–502.

Liebman, M., Mohler, C.M., 2001. Weeds and the soil environment. In: Liebman, M., Mohler, C.M., Staver, C.P. (Eds.), Ecological Management of Agricultural Weeds. Cambridge University Press, New York, pp. 210–268.

Liebman, M., Mohler, C.L., Staver, C.P., 2004. Ecological Management of Agricultural Weeds. Cambridge University Press, UK.

Lindsey, L.E., Darryl, D., Warncke, K.S., Wesley, J.E., 2013. Fertilizer and population affect nitrogen assimilation of common lambsquarters and redroot pigweed. Weed Sci 61, 131–135.

Little, N.G., Mohler, C.L., Ketterings, Q.M., DiTommaso, A., 2015. Effects of organic nutrient amendments on weed and crop growth. Weed Sci 63, 710–722.

Lundkvist, A., 2009. Effects of pre- and post-emergence weed harrowing on annual weeds in peas and spring cereals. Weed Res 49, 409–416.

Malidza, G., Vrbnicanin, S., Kurjacki, I., Pavlovic, D., 2009. Influence of different soil cultivation systems on weed population in soybeans. in: Proceedings of 8th EWRS Workshop on Physical and Cultural Weed Control, Zaragoza, Spain, pp. 9–10

Mandák, B., Trávníček, P., Paštová, L., Kořínková, D., 2012. Is hybridization involved in the evolution of the *Chenopodium album* aggregate? An analysis based on chromosome counts and genome size estimation. Flora 207, 530–540.

Mandak, B., Krak, K., Vit, P., Lomonosova, M.N., Belyayev, A., Habibi, F., et al., 2018. Hybridization and polyploidization within the *Chenopodium album* aggregate analyzed by means of cytological and molecular markers. Mol. Phyl. Evol 129, 189–201.

Manley, B.S., Wilson, H.P., Hines, T.E., 2001. Weed management and crop rotations influence populations of several broadleaf weeds. Weed Sci 49, 106–122.

Mehmood, T., Khokhar, K.M., Shakeel, M., 2007. Integrated weed management practices in garlic crop in Pakistan. Crop Prot 26, 1031–1035.

Mehriya, M.L., Yadav, R.S., Jangir, R.P., Poonia, B.L., 2008. Effect of different weed management practices on weeds and yield of cumin. Ann. Arid Zone 47, 139–144.

Mennan, H., Ngouajio, M., Isık, D., Kaya, E., 2009. Integrated winter/summer cover crops into weed management program in organically grown vegetables. in: Proceedings of 8th EWRS Workshop on Physical and Cultural Weed Control, Zaragoza, Spain, pp. 9–10.

Mirsky, S.B., Gallandt, E.R., Mortensen, D.A., Curran, W.S., Shumway, D.L., 2010. Reducing the germinable weed seedbank with soil disturbance and cover crops. Weed Res 50, 341–352.

Mohammadduost Chamanabad, H.R., Asghari, A., Habibi, Gh.R., Pourmorad Kaleibar, B., 2011. Effect of herbicides and crop residue on integrated weed control in potato. Electron. J. Crop Prod 4, 171–185.

Mohler, C.L., 2001. Mechanical management of weeds. In: Liebman, M., Mohler, C.L., Staver, C.P. (Eds.), Ecological Management of Agricultural Weeds. Cambridge University Press, UK, Cambridge, pp. 139–209.

Mohler, C.L., Iqbal, J., Shen, J., DiTommaso, A., 2016. Effects of water on recovery of weed seedlings following burial. Weed Sci 64, 285–293.

Mseddi, K., Alghamdi, A., Sharawy, S., Ibrahim, N., 2017. Screening of weeds and their effect on alfalfa (*Medicago sativa*). Indian J. Agricult. Sci 87, 1565–1571.

Mt Pleasant, J.M., Schlather, K.J., 1994. Incidence of weed seed in cow manure and its importance as a weed source for cropland. Weed Technol 8, 304–310.

Mulugeta, D., Stoltenberg, D.E., 1997. Increased weed emergence and seed bank depletion by soil disturbance in a no-tillage system. Weed Sci 45, 234–241.

Nagar, R.K., Meena, B.S., Dadheech, R.C., 2009. Effect of integrated weed and nutrient management on weed density, productivity and economics of coriander (*Coriandrum sativum*). Indian J. Weed Sci 41, 71–75.

O'Donovan, J.T., McAndrew, D.W., 2000. Effect of tillage on weed populations in continuous barley. Weed Technol 14, 726–733.

Owen, M.D.K., 2017. Weed management update for 2018 and beyond: The more things change (2017). Proceedings of the Integrated Crop Management Conference 13, 77–88. Available from: https://lib.dr.iastate.edu/icm/2017/proceedings/13.

Partap, T., Kapoor, P., 1985. The Himalayan grain chenopods. I. Distribution and ethnobotany. Agric. Ecosystems Environ. 14, 185–199.

Paolini, R., Principi, M., Froud-Williams, R.J., Del Puglia, S., Biancardi, E., 1999. Competition between sugarbeet and *Sinapis arvensis* and *Chenopodium album*, as affected by timing of nitrogen fertilization. Weed Res 39, 425–440.

Perron, F., Legere, A., 2000. Effects of crop management practices on *Echinochloa crus-galli* and *Chenopodium album* seed production in a maize/soyabean rotation. Weed Res 40, 535–547.

Poonia, A., Upadhayay, A., 2015. *Chenopodium album* Linn: review of nutritive value and biological properties. J. Food Sci. Technol 52, 3977–3985.

Prasad, R., Shivay, Y.S., 2019. *Chenopodium* species: from weeds to a healthy food grain 'quinoa'. Current Sci 116, 171–173.

Qasem, J.R., 1995. Allelopathic effect of *Amaranthus retroflexus* and *Chenopodium murale* on vegetable crops. Allelopathy J 2, 49–66.

Qasem, J.R., 2002. Allelopathic effects of selected medicinal plants on *Amaranthus retroflexus* and *Chenopodium murale*. Allelopathy J 10, 105–122.

Qasem, J.R., 2005. Chemical control of weeds in onion. J. Hortic. Sci. Biotech 80, 721–726.

Qasem, J.R., 2010. Competitive relationship of cauliflower and nettle-leaved goosefoot and their differential responses to potassium level. J. Plant Nutr. 33, 2135–2147.

Qasem, J.R., 2018. Competition of redroot pigweed (*Amaranthus retroflexus* L.) and nettle-leaved goosefoot (*Chenopodium murale* L.) with tomato (*Lycopersicon esculentum* Mill.) cultivars. J. Hortic. Sci. Biotech 93. Available from: https://doi.org/10.1080/14620316.2017.1421052.

Qasem, J.R., Hill, T.A., 1994. Inter- and intraspecific competition of fat-hen and groundsel. Weed Res 34, 109–118.

Qasem, J.R., Issa, N.N., 2018. Germination and growth management of some common annual weeds by phytotoxicity of selected vegetable crops. Sci. Hortic 233, 431–445.

Radosevich, S.R., Holt, J.S., Ghersa, C.M., 2007. Ecology of Weeds and Invasive Plants: Relationship to Agriculture and Natural Resource Management. John Wiley & Sons, Inc, Hoboken, New Jersey.

Rahman, A., 1982. Resistance of weeds to herbicides in New Zealand. In: Geelen, J.A., Esson, M.J. (Eds.), Pesticide Resistance Seminar. Hawkes Bay Branch of New Zealand Weed and Pest Control Society, Hastings, New Zealand, pp. 13–36.

Rahman, A.T., James, T., Trolove, M., 2014. Characteristics and control of dicamba-resistant common lambsquarters (*Chenopodium album*). Weed Biol. Manag 14, 88–98.

Ravangard, S.E., Eslami, S.V., Mahmudi, S., 2015. Effect of soil solarization on common lambsquarters control in Birjand region. J. Plant Prot. 29, 250–258.

Reinhardt, C.F., Meissner, R., Labuschagne, N., 1994. Allelopathic interactions between *Chenopodium album* L. and certain crop species. South Afr. J. Plant Soil 11, 45–49.

Renfrew, J.M., 1973. Palaeoethnobotany. Methuen, London, p. 263.

Rezaie, F., Yarnia, M., 2009. Allelopathic effects of *Chenopodium album*, *Amaranthus retroflexus* and *Cynodon dactylon* on germination and growth of safflower. J. Food Agric. Environ 7, 516–521.

Roskamp, J.M., Turco, R.F., Bischoff, M., Johnson, W.G., 2013. The influence of carrier water pH and hardness on saflufenacil efficacy and solubility. Weed Technol 27, 527–533.

Sanyal, D., Bhowmik, P.C., Anderson, R.L., Shrestha, A., 2008. Revisiting the perspective and progress of integrated weed management. Weed Sci 56, 161–167.

Scheepens, P.C., Kempenaar, C., Andreasen, C., Eggers, T.H., Netland, J., Vurro, M., 1997. Biological control of the annual weed *Chenopodium album*, with emphasis on the application of *Ascochyta caulina* as a microbial herbicide. Integr. Pest Manag. Rev 2, 71–76.

Seglar, B., 2003. Fermentation analysis and silage quality testing, Proceedings of the Minnesota Dairy Health Conference, 18. University of Minnesota, USA.

Shergill, L.S., Kreshnik, B., Davis, A., Mirsky, S.B., 2020. Fate of weed seeds after impact mill processing in midwestern and mid-Atlantic United States. Weed Sci 68, 92–97.

Shilling, D.G., Liebl, R.A., Worsham, A.D., 1985. Rye and wheat mulch: the suppression of certain broad-leaves weeds and the isolation and identification of phytotoxins. In: Thompson, A.C. (Ed.), Chemistry of Allelopathy. ACS symposium series. American Chemical Society, Washington, pp. 243–271.

Shimwell, D.W., 2006. A shoddy tale: perspectives on the wool alien flora of West Yorkshire in the twenty-first century. Watsonia 26, 127–137.

Singh, R., Singh, B., 2004. Effect of irrigation time and weed management practices on weeds and wheat Yield. Indian J. Weed Sci 36, 25–27.

Smith, A., Soltani, N., Kaastra, A.C., Hooker, D.C., Robinson, D.E., Sikkema, P.H., 2019. Isoxaflutole and metribuzin interactions in isoxaflutole-resistant soybean. Weed Sci 67, 485–496.

Sønderskov, M., Swanton, C.J., Kudsk, P., 2012. Influence of nitrogen rate on the efficacy of herbicides with different modes of action. Weed Res 52, 169–177.

Sprague, C.L., 1999. Herbicide tolerance and weed control in corn with isoxaflutole (PhD dissertation). Michigan State University.

Stokłosa, A., Matraszek, R., Isman, M.B., Upadhyaya, M.K., 2012. Phytotoxic activity of clove oil, its constituents, and its modification by light intensity in broccoli and common lambsquarters. Weed Sci 60, 607–611.

Teasdale, J.R., Mohler, C.L., 2000. The quantitative relationship between weed emergence and the physical properties of mulches. Weed Sci 48, 385–392.

Teasdale, J.R., Mirsky, S.B., Cavigelli, M.A., 2019. Weed species and traits associated with organic grain crop rotations in the Mid-Atlantic region. Weed Sci 67, 595–604.

Tharp, B.E., Kells, J.J., 2001. Effect of glufosinate-resistant corn population and row spacing on light interception, corn yield, and common lambsquarters growth. Weed Technol 15, 413–418.

Thiel, H., Varrelmann, M., 2014. Identification of a new PSII target site *psbA* mutation leading to D1 amino acid $Leu_{218}Val$ exchange in the *Chenopodium album* D1 protein and comparison to cross-resistance profiles of known modifications at positions 251 and 264. Pest Manag. Sci 70, 278–285.

Toole, E.H., Brown, E., 1946. Final results of the Duvel buried seed experiments. J. Agric. Res 72, 201–210.

Tuesca, D., Puricelli, E., Papa, J.C., 2001. A long-term study of weed flora shifts in different tillage systems. Weed Res 41, 369–382.

Upadhyaya, M.K., Blackshaw, R.E., 2007. Non Chemical Weed Management: Principles, Concepts and Technology. Cabi Publishing, UK.

van der Laat, R., Owen, M.D.K., Liebman, M., Leon, R.G., 2015. Postdispersal weed seed predation and invertebrate activity density in three tillage regimes. Weed Sci 63, 828–838.

Van Wychen, L., 2019. 2019 Survey of the Most Common and Troublesome Weeds in Broadleaf Crops, Fruits & Vegetables in the United States and Canada. Weed Science Society of America National Weed Survey Dataset. Available. Available from: http://wssa.net/wp-content/uploads/2019-Weed-Survey_Broadleaf-crops.xlsx.

Vurro, M., Chiara Zonno, M., Evidente, A., Andolfi, A., Montemurro, P., 2001. Enhancement of efficacy of *Ascochyta caulina* to control *Chenopodium album* by use of phytotoxins and reduced rates of herbicides. Biol. Control 21, 182–190.

Walsh, B.M., Adhikary, D., Maughan, P.J., Emshwiller, E., Jellen, E.N., 2015. *Chenopodium* polyploidy inferred from salt overly sensitive (SOS1) data. Am. J. Bot 102, 533–543.

Ward, S.M., 1998. A new source of restorable cytoplasmic male sterility in quinoa. Euphytica 10, 157–163.

Watson, P.R., Derksen, D.A., Van Acker, R.C., 2006. The ability of 29 barley cultivars to compete and withstand competition. Weed Sci 54, 783–792.

Weber, J.F., Kunz, C., Peteinatos, G.G., Zikeli, S., Gerhards, R., 2015. Weed control using conventional tillage, reduced tillage, no-tillage, and cover crops in organic soybean. Agriculture 32, 253–261.

Westhoven, A.M., Kruger, G.R., Gerber, C.K., Stachler, J.M., Loux, M.M., Johnson, W.G., 2008a. Characterization of selected common lambsquarters (*Chenopodium album*) biotypes with tolerance to glyphosate. Weed Sci 56, 685–691.

Westhoven, A.M., Stachler, J.M., Loux, M.M., Johnson, W.G., 2008b. Management of glyphosate-tolerant common lambsquarters in glyphosate-resistant soybean. Weed Technol 22, 628–634.

Williams, J.T., Harper, J.L., 1965. Seed polymorphism and germination. I. The influence of nitrates and low temperatures on the germination of *Chenopodium album*. Weed Res 5, 141–150.

Wilson, H.D., 1981. Domesticated Chenopodium of the Ozark Bluff Dwellers. Econ. Bot 35, 233–239.

Wilson, H.D., 1988. Allozyme variation and morphological relationships of *Chenopodium hircinum*. Syst. Bot 13, 215–228.

Wilson, H.D., 1990. Quinua and relatives (*Chenopodium* sect. *Chenopodium* subsect. Cellulata). Econ. Bot 44, 92–110.

Wilson, H.D., Heiser, D.B., 1979. The origin and evolutionary relationships of 'Huauzontle' (*Chenopodium nuttalliae* Safford), domesticated chenopod of Mexico. Am. J. Bot 66, 198–206.

Yao, S., Lan, H., Zhang, F., 2010. Variation of seed heteromorphism in *Chenopodium album* and the effect of salinity stress on the descendants. Ann. Bot 105, 1015–1025.

Yasin, M., Andreasen, C., 2019. The effect of oxygen concentration on the germination of some weed species under control conditions. Weed Sci 67, 580–588.

Yerka, M.K., de Leon, N., Stoltenberg, D.E., 2012. Pollen-mediated gene flow in common lambsquarters (*Chenopodium album*). Weed Sci 60, 600–606.

Yerka, M.K., Wiersma, A.T., Lindenmayer, R.B., Westra, P., Johnson, W.G., de Leon, N., et al., 2013. Reduced translocation is associated with tolerance of common lambsquarters to glyphosate. Weed Sci 61, 353–360.

Yorimitsu, Y., Kadosono, A., Hatakeyama, Y., Yabiku, T., Ueno, O., 2019. Transition from C3 to proto-Kranz to C3-C4 intermediate type in the genus *Chenopodium* (Chenopodiaceae). J. Plant Res 132, 839–855.

Yu, Q., Ahmad-Hamdani, M.S., Han, H., Christoffers, M.J., Powles, S.B., 2013. Herbicide resistance-endowing ACCase gene mutations in hexaploid wild oat (*Avena fatua*): insights into resistance evolution in a hexaploid species. Heredity 110, 220–231.

Zand, E., Mousavi, S.K., Heidari, A., 2008. Herbicides and Their Application. Jehad daneshgahi Mashhad Press.

Zimdahl, R.L., 2004. Weed-Crop Competition: A Review. Blackwell Publishing Ltd, Iowa, USA.

Zimdahl, R.L., 2007. Fundamentals of Weed Science. Elsevier Inc, USA.

Chapter 6

Chloris truncata and *Chloris virgata*

Het Samir Desai[1] and Bhagirath Singh Chauhan[2]
[1]The University of Queensland, Gatton, QLD, Australia, [2]The Centre for Crop Science, Queensland Alliance for Agriculture and Food Innovation (QAAFI) and School of Agriculture and Food Sciences (SAFS), The University of Queensland, Gatton, QLD, Australia

Introduction

Chloris truncata R. Br. and *Chloris virgata* Sw. are the hardy grass weeds of the tropics. *Chloris truncata* is known to have the origin in Australia (Borger et al., 2011); but, there are mixed reviews regarding the exact origin of *C. virgata* as many researchers believe that it is native to either Eurasia, Africa, the United States, or other tropical parts of the world (Anderson, 1974). Despite originated in Australia, *C. truncata* is naturalized in southern Africa, North America, and New Zealand. The future distribution of this weed species could be influenced by climate change (Weller et al., 2019). *Chloris virgata* is present in its uncertain origins (Eurasia, Africa, and the United States) but also distributed as a naturalized species in Hawaii, Australia, Canary Islands, India, and some Arabic countries (Anderson, 1974). Both species are quite challenging to identify at an early growth stage (3–4 leaf stage). However, both can easily be distinguished due to the wide variation in their morphological characteristics. *Chloris virgata* could be confused with other nonweed *Chloris* spp. such as *Chloris gayana* Kunth and *Chloris pilosa* Schumach in their spikelet appearance but this species may formally be separated in being annual and later having relatively long awns (Anderson, 1974).

Both species were considered as roadside environmental weeds in many countries a decade ago; however, due to the major shift of agricultural systems such as traditional tillage to no-tillage or zero-tillage, these weeds species have widely established in warm regions' agricultural systems (Osten, 2012). *Chloris virgata* is considered a common weed species in alfalfa (*Madicago Sativa* L.) fields of southwest United States and sorghum [*Sorghum bicolor* (L.) Moench] and maize (*Zea maize* L.) plantations of Mexico (Barkworth, 2007; Vibrans, 2009a,b). Both species have been listed in the top 10 emerging weed species of Australian agricultural systems (Llewellyn et al., 2016).

Biology and Management of Problematic Weed Species. DOI: https://doi.org/10.1016/B978-0-12-822917-0.00006-9

Due to their morphological characteristics, these weed species tend to disperse rapidly from one agroecosystem to another, and the rapid dispersal ability might have worked as a catalyst in the establishment of these species as major cropping weeds from environmental weeds. Many agriculturally progressive countries have been adopting conservation tillage as it increases moisture availability and crop water use efficiency, particularly in low-yielding cereal cropping systems (Reicosky and Saxton, 2007). These changes influence weed management practices, attributed to overdependence on chemical weed control tactics, especially on glyphosate (Desai et al., 2020).

There are several reasons behind the introduction of conservational tillage, for instance, it was introduced in northern and central Europe to ameliorate soil health and reduce production costs (Busari et al., 2015), whereas in Australia, the conservation tillage was introduced to reduce soil erosion and moisture evaporation rate (D'Emden et al., 2006). Crop diversity around the world has largely decreased due to climatic restrictions and consumer preferences and this resulted in cereal monoculture, highly reliant on very few herbicide options (Desai et al., 2020). These facts therefore caused the immediate need to develop alternative control protocols for weed species that can proliferate in diverse weather conditions such as *C. truncata* and *C. virgata*.

History and distribution

Chloris truncata and *C. virgata* are emerging weed species in many tropical parts of the world. Both species were previously considered roadside environmental weeds but with an advent of the evolution in farming systems and the ability of these species to disperse rapidly; both have been registered as problematic cropping weeds in many countries (Anderson, 1974) and widely present in tropical and temperate parts of the globe (Table 6.1).

Chloris genus was first described by Swartz in 1788 (Anderson, 1974), and since then approximately, 65 species have been identified by different botanists and scientists. *Chloris truncata* and *C. virgata* were first described by Robert Brown in 1810 and Olof Peter Swartz in 1797, respectively. *Chloris virgata* is reported to have a vast amount of variation within biotypes among *Chloris* genus as it possesses various distinctive morphological features in different biotypes collected from across the globe (Anderson, 1974), whereas *C. truncata* exhibited similar morphological characteristics irrespective of biotypes.

Chloris truncata has a long and well-documented distribution history as a prevalent global invader. A model suggested that it is present in many European countries: Argentina, Belgium, Czech Republic, France, Germany, Netherlands, Poland, Spain, Sweden, and Switzerland; some parts of the

TABLE 6.1 Geographic description of *Chloris truncata* and *Chloris virgata* distribution and habitat.

Description	*C. truncata*	*C. virgata*	Reference
Preferred region	Temperate, tropics, and subtropics	Tropical and temperate	Michael et al. (2012), Clayton et al. (2015)
Elevation	<500 m	2500 m	Vibrans (2009a,b)
Mean annual rainfall	—	500–750 mm	Vibrans (2009a,b)
Soil type	Prefers gray cracking clays to light sandy soil	Prefers heavy soil but has a wide soil range including saline, alkaline, and dark clay soil	Michael et al. (2012), Clayton et al. (2015)
Habitat	Drylands, woodlands, agricultural lands, and road/railway sides	Cultivated lands, desert, road/railway sides, grazing lands, disturbed lands, riverbanks, and coastal area	—

United States: California, Georgia, Hawaiian Island, and South Carolina; Japan, the United Kingdom, Philippines, Tonga, Niue, New Zealand, Fiji, and Australia (Michael et al., 2012). The distribution history of *C. virgata* has been poorly recorded so far. It is present in tropical regions across the globe (Table 6.2).

Chloris virgata is present across the mainland of Australia infesting an area of 118,000 ha of Australian grain growing belts, contributing to considerable yield and revenue losses (39,300 tons of grain and AUD 7.7 million per year, respectively) (Llewellyn et al., 2016). Therefore it has been included in the top 20 weeds of Australia in terms of yield loss and revenue loss (Llewellyn et al., 2016). *Chloris truncata* is not as widespread as *C. virgata* in Australia; however, *C. truncata* has been listed the top 10 most important weed of summer fallow, considering the yield loss (36,871,000 kg) and revenue loss (AUD 9.1 million) potential per year (Llewellyn et al., 2016). *Chloris virgata* is however considered as a cropping and environmental weed in many tropical countries; but, accurate data on yield and revenue loss by *C. virgata* can only be found in Australia. Consequently, there is considerable research gap on nature of this species across the globe.

TABLE 6.2 Global distribution of *Chloris truncata* and *Chloris virgata*.

Continent/ region	Countries		Origin	Reference
	C. truncata	*C. virgata*		
Africa	South Africa	Algeria, Angola, Botswana, Burkina Faso, Cabo Verde, Cameroon, Central African Republic, Chad, Egypt, Eritrea, Eswatini, Ethiopia, Kenya, Lesotho, Libya, Madagascar, Malawi, Mali, Mauritania, Morocco, Mozambique, Namibia, Niger, Nigeria, Saint Helena, Seychelles, Somalia, South Africa, and South Sudan	Introduced	Michael et al. (2012), Clayton et al. (2015)
Asia	Japan and Philippines	Afghanistan, Azerbaijan, Bangladesh, Bhutan, China, India, Indonesia, Iran, Iraq, Israel, Japan, Kazakhstan, Kyrgyzstan, Lebanon, Mongolia, Myanmar, Nepal, North Korea, Oman, Pakistan, Palestine, Saudi Arabia, South Korea, Taiwan, Turkmenistan, Uzbekistan, Vietnam, and Yemen	Introduced	Michael et al. (2012), Clayton et al. (2015)
Europe	Belgium, Czech Republic, France, Germany, Netherlands, Poland, Spain, Sweden, Switzerland, and the United Kingdom	Belgium, Czechia, France, Portugal, Russia, Spain, and the United Kingdom	Introduced	Verloove (2005), Michael et al. (2012), Clayton et al. (2015)

North America	The United States	Anguilla, Antigua and Barbuda, Aruba, Barbados, Bonaire; Saint Eustatius, and the United States	Native or introduced	Michael et al. (2012), Acevedo-Rodríguez (2012), Clayton et al. (2015)
Oceania	Australia, Fiji, Niue, and Tonga	Australia, French Polynesia, Nauru, New Caledonia, Northern Mariana Islands, Palau, Papua New Guinea, the US minor outlying Islands, and Vanuatu	Native or Introduced	Fosberg et al. (1987), Michael et al. (2012), Florence et al. (2013), Clayton et al. (2015)
South America	Argentina	Argentina, Bolivia, Brazil, Chile, Columbia, Ecuador, Paraguay, Peru, and Venezuela	Native or Introduced	Zuloaga et al., (2008), Michael et al. (2012), Clayton et al. (2015)

Taxonomy and description

The genus *Chloris* is found in the tribe Chloridoideae and family Poaceae. The *Chloris* genus is characterized by the basic chromosome number of $n = 10$ (Anderson, 1974). *Chloris truncata* is a tetraploid with $2n = 40$, whereas *C. virgata* is a diploid with $2n = 20$. These two species are closely related but the exact taxonomic classification is still a matter of scrutiny and international debate (Anderson, 1974). *Chloris truncata* is a summer active perennial grass, and *C. virgata* is a summer annual grass species. Therefore, it is common to observe both species in the same field and time.

Chloris truncata is a short-lived, stoloniferous species, having a C_4 photosynthesis mechanism (Weller et al., 2019). It is 20–50 cm tall, sheaths glabrous; ligules short ciliate; the blade is glabrous to scabrous having dimensions of 9–19 cm length × 0.2–0.3 mm width; spike 5–15, 5–23 cm long, radiate. *Chloris truncata* is efficient in seed production as very small plants (less than 1 g dry matter, 3–5 cm long) can produce seed heads and an individual head can produce 300–500 seeds (Weller et al., 2019). *Chloris truncata* can produce maximum biomass of 1500 kg/ha and 60,000 seeds/m^2 (Borger et al., 2011). *Chloris truncata* can be confused with *C. virgata* at an early vegetative stage. However, both can easily be distinguished due to considerable height difference and appearance of panicles. The panicle of *C. truncata* is green at the emergence and later achieved dark gray to black color with seed maturity (Fig. 6.1). Each panicle usually possesses 10–12 spikes and 70–80 seeds per spike. The healthy plant of *C. truncata* can produce up to 20,000 seeds (Chauhan et al., 2016).

Chloris virgata is a summer annual grass, possessing a C_4 photosynthesis mechanism (Desai et al., 2020). It can achieve a maximum height of >1 m and produce an optimum of 600 g/m^2 (Ngo, 2017). However, the size is extremely variable within the biotypes (Anderson, 1974). *Chloris virgata* is halophyte; culms usually tufted, many proliferating forms one root system, sometimes

FIGURE 6.1 Mature panicle of *Chloris virgata* (A) and *Chloris truncata* (B).

stoloniferous with the capability of producing roots at the joints in case of direct contact of root nodes with soil; sheaths usually glabrous; ligules glabrous; blades up to 30 cm long and 1.5 cm wide, usually glabrous with scabrous margins (Anderson, 1974). Its panicles appear green in color immediately after emergence but later become shiny ivory (Osten, 2012) (Fig. 6.1). The seed production potential of *C. virgata* has greatly increased over time due to the strong adaptive traits of this species as a previous study on seed production of *C. virgata* suggested that it can produce >40,000 seeds/plant (Osten, 2012); however, a recent study revealed that the healthy plant of *C. virgata* can produce >1,00,000 seeds in its entire life cycle (Manalil et al., 2020).

Biology and ecology

Weed biology covers many aspects of plant attributes such as seed dormancy and germination, growth physiology, competitive ability, and reproductive strategies (Gill, 2017). It is important to understand weed biology as its enhanced knowledge can provide the necessary information to develop effective and sustainable weed control programs. Both species have different favorable conditions for germination and emergence (Table 6.3).

TABLE 6.3 Favorable conditions of *Chloris truncata* and *Chloris virgata* for germination and emergence.

Description	*C. truncata*	*C. virgata*
Optimum temperature	20°C–25°C	23°C–28°C
Soil type	Gray clays to light sandy	Wide soil type: heavy black to light sandy
Burial depth	<0.5 cm	<0.5 cm
Soil pH	6.5–7.5	6.4–8
Ability to tolerate moisture stress	−0.4 MPa osmotic potential	−0.8 MPa osmotic potential
Ability to tolerate salt stress	250 mM NaCl	250 mM NaCl
Growing degree days from emergence to seed production	748–786	1200

Source: From Lodge, G., Whalley, R., 1981. Establishment of warm-and cool-season native perennial grasses on the north-west slopes of New South Wales. I. Dormancy and germination. Aus. J. Botany 29, 111–119; Ngo, T., Boutsalis, P., Preston, C., Gill, G., 2014. Seed biology of *Chloris truncata* (windmill grass) and *Chloris virgata* (feathertop Rhodes grass). 19th Australasian Weeds Conference, Science, Community and Food Security: the Weed Challenge, Hobart, Tasmania, Australia, 1–4 September 2014. Tasmanian Weed Society, pp. 75–78; Fernando, N., Humphries, T., Florentine, S.K., Chauhan, B.S., 2016. Factors affecting seed germination of feather fingergrass (*Chloris virgata*). Weed Sci. 64, 605–612.; Ngo D., Boutsalis, P., Preston, C., Gill, G., 2017a. Growth, development, and seed biology of feather fingergrass (*Chloris virgata*) in Southern Australia. Weed Sci. 65, 413–425; Boutsalis et al., 2017b; Chauhan, B.S., Manalil, S., Florentine, S., Jha, P., 2018. Germination ecology of *Chloris truncata* and its implication for weed management. PLoS One 13, e0199949.

A study of germination response of south-east Queensland populations revealed that *C. truncata* seeds can germinate at wide alternating temperature regimes (from 15/5 to 30/20°C); however, germination was highest (>80%) at 25/15 and 30/20°C (Chauhan et al., 2018). The germination of *C. truncata* was stimulated by light but 50% of seeds were also able to germinate in completely dark conditions (Chauhan et al., 2018). These results suggested that *C. truncata* can germinate throughout the year (summer and winter) in Australia.

In a similar study, germination response of *C. truncata* to moisture stress and salinity level was also evaluated, in which germination was completely inhibited at 300 mM NaCl and −0.1 mM osmotic potential and 179 mM NaCl and −0.52 MPa osmotic potential were required to inhibit 50% germination (Chauhan et al., 2018). These results suggested that seeds of *C. truncata* can tolerate a high level of salinity and moisture stress conditions and even germinate in saline soil and drought conditions which are common in Australia due to low rainfall and high evaporation rate (Rengasamy, 2002). Seeds of *C. truncata* exhibited >70% germination at a wide range of pH (from 4 to 10), suggesting that seeds can germinate in a variety of soils (Chauhan et al., 2018).

The potential of seedlings to emerge from burial depth is an important aspect to choose tillage timing and develop nonchemical weed control programs. Seedlings emerged in the highest proportion (67%) when seeds were placed on the soil surface; however, emergence significantly declined to 45% at 0.5 cm depth and further inhibited completely at 3 cm burial depth (Chauhan et al., 2018). A study conducted with the south Australian population showed a similar response to burial depth as seedlings of *C. truncata* are highly responsive to burial regardless of depth as seedling emergence was zero at 0.5, 1.0, 2.0, and 5.0 cm burial depth, while more than 60% emergence was observed from the soil surface (Ngo et al., 2017b).

In South Australia, seedlings of *C. truncata* required 748−786 growing degree days to reach maturity from the day of emergence (Ngo et al., 2017b). In a similar study, freshly harvested seeds of *C. truncata* possessed a low level of dormancy compared to *C. virgata* with 16%−40% germination (Ngo et al., 2017b). This study revealed opposite results from a study conducted with Queensland populations, whereas seeds exhibited 50% germination in completely dark conditions; however, there was only 2% germination observed in a study carried out with South Australian populations. Therefore light significantly stimulated germination with negligible germination in completely dark conditions. The study carried out by Ngo et al., 2017b with South Australian populations of *C. truncata* suggests that seeds of *C. truncata* can germinate under a wide range of temperatures (from 10°C to 40°C) and the ideal temperature for optimum germination is 20°C−25°C, which is similar to the Queensland populations. Germination of South Australian populations was completely inhibited at 250 mM NaCl level and −0.8 MPa osmotic potential (Ngo et al., 2017b) which is similar to that of the Queensland population of *C. truncata*.

In Australia, several studies have been conducted on seed biology and germination ecology of both weed species. A population of *C. virgata* from south-east Queensland exhibited optimum germination (92%) at the 30/20°C alternating day/night temperature (Fernando et al., 2016). Germination of the same population was similarly optimum (29%) at 30/20°C in 24 h dark conditions (Fernando et al., 2016). The lowest germination (below 15%) was observed at the lowest temperature regime (17/7°C) in both light/dark and completely dark conditions (Fernando et al., 2016). A recent study with 10 different populations from the eastern part of Australia showed similar germination at five alternating temperature regimes (15/5, 20/10, 25/15, 30/20, and 35/25°C) (Chauhan and Johnson, 2010). These results suggested an adaptation in *C. virgata* germination to a wide range of temperature and the seeds can now germinate in winter months in Australia.

In-field conditions in South Australia, the seedlings of this species require 1200-degree days to attain optimum maturity from the day of emergence (Ngo et al., 2017a). Fresh seeds of *C. virgata* exhibit initial moderate dormancy for 2 months and require 5 months of an after-ripening period to break dormancy (Ngo et al., 2017a). *Chloris virgata* seeds are halophyte since they can germinate in high saline conditions [250 mM sodium chloride (NaCl)] (Ngo et al., 2017a) and high-water stress conditions (−0.8 MPa osmotic potential) (Fernando et al., 2016). Moreover, *C. virgata* seeds are highly vulnerable to burial depth as seedling emergence drastically declined at 1 cm and completely inhibited at 5 cm (Ngo et al., 2017a).

A study conducted in the Argentine Pampas region with three *C. virgata* populations showed that all populations possessed the highest germination at the constant temperature of 25°C (Rodriguez et al., 2020). In another study, germination response of *C. virgata* to temperature was tested to alternating (25/15 and 30/20°C) and constant temperatures (5, 20, 25, and 30°C) regimes with populations collected from western Jilin province of China (Zhang et al., 2012). In this study, seeds possessed more than 90% germination at all the temperature regimes.

Management

High competitiveness due to rapid vegetative growth, substantial dispersibility, germination ability under a variety of weather conditions, peculiar biology, natural glyphosate tolerance, and unpredictable emergence nature are the major characteristics of *C. truncata* and *C. virgata* in agroecosystems. These features enhance the difficulties to control *C. truncata* and *C. virgata*. Some management options are discussed below and are primarily categorized as nonchemical and chemical control.

Nonchemical control

Due to the aforementioned characteristics, few cultural control options have been identified for these species. *Chloris turncata* and *C. virgata* can be controlled up to a decent level by manual hand-weeding and manual weed-chipping in conventional cropping systems; however, these options are not economically sustainable in broad-scale farming (Davidson et al., 2019). As discussed earlier, these two weed species are sensitive to burial regardless of depth and possess positive photoblastic nature; therefore deep tillage can be used to bury their seeds to suppress the emergence and subsequent establishment (Fernando et al., 2016; Chauhan et al., 2018). However, deep tillage operations are only possible in conventional farming thus causing limitations in no-till farming systems to manage these species. Consequently, these weed species have become major threats in the areas where no-till systems have been in practice for a long period (Llewellyn et al., 2016). Some other cultural weed control practices can also be useful such as cover-cropping, crop rotation, mulching, and soil solarization; however, the economic sustainability of these weed control approaches still needs to be evaluated especially in broad-scale farming.

The density and seed production potential of *C. truncata* and *C. virgata* can be reduced by an increase in crop density and competition. Several studies have been conducted regarding the potential of crop competition to suppress major cropping weed species such as *Avena fatua* L., *Lolium* spp., *Raphanus raphanistrum* L., and *Conyza bonariensis* (L.) Cronq; however, little research emphasis has been given to the emerging weed species such as *C. truncata* and *C. virgata*. The biomass and seed production of *C. virgata* was minimized by 74% and 100%, respectively, when surrounded by 6 mungbean (*Vigna radiata* L. R. Wilczek) plants in-field conditions (Sharma, 2020). Similarly, 10 sorghum plants/m^2 reduced the biomass of *C. virgata* up to >90% (Sharma, 2020). However, increased crop competition cannot be followed as a standalone weed management practice due to the inability to demonstrate complete control, it should rather be integrated with different weed management approaches to achieve optimum control (Bajwa et al., 2017).

Over the past few decades, considerable attention has been associated with the cover crop to manage many weed species in conservation agricultural systems. Cover crops, which have rapid vegetative growth, have become a viable alternative for sustainable weed management, especially in no-till systems (Teasdale, 1996). Cover crops for weed control have been emphasized in conservation agriculture (e.g., no-till systems) due to the interaction with weed germination, emergence, and therefore with the subsequent establishment of weed populations. The primary focus of using cover crops is to replace undesirable weed populations with feasible cover crops. Cover crops are often considered an effective tool to suppress annual weeds. In Australia, farmers have been observing *C. virgata* throughout the year including in summer fallow conditions despite its summer annual life cycle (Chauhan, 2019), thus utilizing nutrients and moisture which would have been available for winter cereal crops (e.g., wheat and barley) if not absorbed by the weeds in summer fallow conditions (Hunt et al., 2013). Considering these facts, the use of cover crops for the management of C. *virgata* and *C. truncata* opens the door for future research direction on this aspect.

In conservation agricultural systems the options for weed management are limited compared to conventional farming systems (Nichols et al., 2015). The recent emphasis on the utilization of crop residues gained noticeable momentum when it comes to sustainable control of weeds in conservation agricultural systems. In a pot study, 6 tons/ha of sorghum residue suppressed 94% and 67% of the biomass of *C. truncata* and *C. virgata*, respectively (Mobli Chauhan, 2020). Similarly, the seedling emergence was inhibited by 92% and 67% for *C. truncata* and *C. virgata* by 6 tons/ha of sorghum residue treatment compared to the residue-free treatment (Mobli and Chauhan, 2020). The slow release of allelopathic chemicals such as ferulic, *p*-coumaric, *p*-hydroxybenzoic, vanillic, and syringic acids during the decomposition of sorghum residue might be responsible for suppressing seedling emergence and biomass (Sowiński et al., 2020). However, pot studies do not ensure successful results in-field conditions; therefore future research should focus on this specific aspect and in-field viability should be explored to achieve effective control of these two emerging weed species in conservation as well as conventional farming systems.

Some cultural weed control practices are not always sustainable for controlling fast-growing grass weed species as several studies suggested that some cultural weed control operations (e.g., mowing) could facilitate rigidity to the weed plants and promote side branches by chopping shoot apex, thus making weed control nearly impossible even with postemergence herbicide applications (Shrestha et al., 2008).

Seed production of weed species in standing crop situations enhance the weed seedbank and contribute to weed complication in future crops and fallow conditions. Curtailing weed seedbank replenishment is essential for the adequate management of annual weeds. Harvest weed seed control (HWSC) is a novel approach to target weed species by collecting and/or destructing seeds during crop harvest to minimize the weed seedbank restoration (Walsh et al., 2018a). Seed retention and height of weed species at crop maturity are two determining factors for the success of HWSC. A large proportion of *C. virgata* seeds was retained at crop maturity, for instance, >70% in chickpea and sorghum. Recent studies suggested that the seeds of *C. virgata* were retained >90% in mungbean (Manalil et al., 2020) and >65% in sorghum (Mahajan et al., 2020) at crop harvest. Similarly, the seeds of *C. truncata* were retained in a considerable quantity (>75%) at the time of harvest in mungbean (Manalil et al., 2020). These characteristics of *C. truncata* and *C. virgata* could enable the HWSC tactic to capture the seeds before dispersal and reduce weed seedbanks (Walsh et al., 2018b).

Although cultural strategies can control *C. truncata* and *C. virgata* to some extent in conventional farming systems, systematic adjustments are required between these options in conservation farming. A mindful integration of cultural methods with herbicide application can surely bring sustainability to control these two emerging weed species across a variety of cropping systems.

Chemical control

Chloris species are considered difficult to control with a single herbicide application due to their staggered and year-round germination. Owing to this, seedling emergence is not limited to one season. Both species can germinate throughout the year at varying proportions depending upon the environmental conditions (Fernando et al., 2016; Chauhan et al., 2018). Many herbicides have been evaluated for *C. truncata* and *C. virgata* control in Australia and provided a varying degree of control (Table 6.4). These two species are hard-to-control weeds with a single-knockdown herbicide application (Widderick and McLean, 2018). Therefore double-knock tactics are being used to manage these weed species in which different or same mode of action herbicides are applied sequentially at certain intervals (Desai et al., 2020). The knowledge of available chemical options is limited to control *C. truncata* and *C. virgata* compared to other major weed species. Consequently, the effect of many chemical herbicide formulations on these two weed species still needs to be evaluated.

Integrated weed management

As the consequences of altering crop production systems, an alarming risk of herbicide resistance development, and unpredictable biology and ecophysiology of weed species, single herbicide control methods without integration become unfeasible and ineffective (Bajwa, 2014). Therefore well-planned integration of weed control approaches needs to be followed to target predominantly herbicide-resistant weed species such as *C. truncata* and *C. virgata*. The integrated use of cultural weed control approaches such as crop competition, row adjustment, crop residue management, and allopathic interactions in conjunction with chemical approaches have been found effective against many annual grass weed species; however, the knowledge of synergistic effects of various weed control tactics are limited to achieve satisfactory control of emerging weed species such as *C. truncata* and *C. virgata*. Therefore, future research needs to be conducted to identify suitable integrated weed management tactics to optimize the control of these hard-to-control weed species.

Conclusion

Chloris truncata and *C. virgata* are two troublesome weed species, spanning a wide range of geographic and climatic conditions. Owing to their distinctive biological characteristics, highly adaptive ability, high competitiveness, and year-round germination ability, these weed species have been rapidly emerging in agroecosystems across the world. Due to rapid vegetative growth, rapid seed dispersal, and ability to act as a host for pests, these two weed species are responsible for huge yield, quality, and revenue loss of

TABLE 6.4 Chemical control of *Chloris truncata* and *Chloris virgata*.

Species	Herbicide	Dose (g a.i./ha)	Application type	Situation	Growth stage	Mode of action	Reference
C. truncata	Butroxydim	300	Single	Cotton, mungbean, peanut, and sunflower	4–5 LS	Inhibits acetyl coenzyme A carboxylase	Chauhan et al. (2016)
	Chlorthal-dimethyl	555	Single	Cotton, soybean, and mungbean	4–5 LS	Inhibits microtubule assembly	Chauhan et al. (2016)
	Glyphosate	800	Single	Fallow	4–5 LS	Inhibits EPSP synthase	Chauhan et al. (2016)
	Haloxyfop FB paraquat	156 FB 500	Sequential	Fallow	2–8 LS	Inhibits acetyl coenzyme A carboxylase FB Inhibits photosynthesis	Chauhan et al. (2016)
C. virgata	Haloxyfop	76	Single	Fallow	3–4 LS	Inhibits acetyl coenzyme A carboxylase	Daniel et al. (2020), Desai et al. (2020)
	Propaquizafop	500	Single	Fallow, cotton, peanut, and sunflower	3–5 LS	Inhibits acetyl coenzyme A carboxylase	Daniel et al. (2020)
	Clethodim	90	Single	Fallow	4–5 LS	Inhibits acetyl coenzyme A carboxylase	Daniel et al. (2020), Desai et al. (2020)
	Butroxydim	300	Single	Sunflower	4 LS	Inhibits acetyl coenzyme A carboxylase	Aisthorpe (2014)
	Paraquat	600	Single	Fallow	4–5 LS	Inhibits photosynthesis	Desai et al. (2020)
	Isoxaflutole	750	Single	Fallow	4–5 LS	Inhibits 4-hydroxyphenyl-pyruvate dioxygenase	Aisthorpe (2014)
	Chlorthal-dimethyl	555	Single	Cotton	Early tillering stage	Inhibits microtubule assembly	Aisthorpe (2014)

(*Continued*)

TABLE 6.4 (Continued)

Species	Herbicide	Dose (g a.i./ha)	Application type	Situation	Growth stage	Mode of action	Reference
	Glufosinate-ammonium	750	Single	Fallow	4–5 LS	Inhibits glutamine synthetase	Desai et al. (2020)
	Isoxaflutole FB Paraquat	75 fb 500	Sequential	Fallow	2–8 LS	Inhibits 4-hydroxyphenyl-pyruvate dioxygenase fb Inhibits photosynthesis	Davidson et al. (2019)
	Atrazine FB Paraquat	2997 FB 500	Sequential	Fallow	2–8 LS	Inhibits photosynthesis at photosystem 2 FB Inhibits photosynthesis	Davidson et al. (2019)
	Haloxyfop FB Paraquat	156 FB 500	Sequential	Fallow	2–8 LS	Inhibits acetyl coenzyme A carboxylase FB inhibits photosynthesis	Davidson et al. (2019)
	Glyphosate FB Paraquat	1140 FB 600	Sequential	Fallow	4–5 LS	Inhibits EPSP synthase FB inhibits photosynthesis	Desai et al. (2020)
	Glyphosate FB Paraquat	1140 FB 1200	Sequential	Fallow	Up to 10 leaf stage	Inhibits EPSP synthase FB inhibits photosynthesis	Desai et al. (2020)
	Glufosinate-ammonium FB Paraquat	750 FB 600	Sequential	Fallow	Up to 10 leaf stage	Inhibits glutamine synthetase FB inhibits photosynthesis	Desai et al. (2020)
	Glufosinate-ammonium FB Paraquat	750 FB 1200	Sequential	Fallow	Up to 10 leaf stage	Inhibits glutamine synthetase FB Inhibits photosynthesis	Desai et al. (2020)

FB, followed by treatment; *LS*, leaf stage.

agricultural commodities. A thorough understanding of weed seed biology would aid in the development of highly effective weed control programs and thus make growers proactive to manage highly aggressive weed species. *Chloris truncata* and *C. virgata* can germinate in a wide range of temperatures; therefore, both species could germinate throughout the year and thus competing with various crops grown in different seasons.

Considering the abovementioned features, no single-weed control tactic can provide satisfactory control of *C. truncata* and *C. virgata*; therefore the use of a suitable blend of cultural methods along with chemical control approaches is required for the management of these two problematic weed species. Simulation model-based research has suggested the highly effective nature of double-knock herbicide applications, which can delay herbicide resistance development. Consequently, more research should be carried out to optimize the efficacy of double-knock herbicide application tactics and integrated weed management to manage these weed species in different cropping systems.

References

Acevedo-Rodríguez, P.M.T.S., 2012. Catalogue of the Seed Plants of the West Indies. Smithsonian Institution, Washington, DC USA.

Aisthorpe, D., 2014. Integrated Weed Management of Feathertop Rhodes Grass. Department of Agriculture, Grain Research & Development Corporation Australia.

Anderson, D.E., 1974. Taxonomy of the genus *Chloris* (Gramineae). Brigham Young University Science Bulletin. Biol. Ser. 19, 1.

Bajwa, A.A., 2014. Sustainable weed management in conservation agriculture. Crop. Prot. 65, 105–113.

Bajwa, A.A., Walsh, M., Chauhan, B.S., 2017. Weed management using crop competition in Australia. Crop. Prot. 95, 8–13.

Barkworth, M., 2007. *Chloris* Sw. Manual of Grasses for North America. Utah State University Press, Logan, USA.

Borger, C., Riethmuller, G., Hashem, A., 2011. Emergence, survival, biomass production, and seed production of *Chloris truncata* (windmill grass) in the Western Australian wheatbelt. Crop. Pasture Sci. 62, 678–685.

Busari, M.A., Kukal, S.S., Kaur, A., Bhatt, R., Dulazi, A.A., 2015. Conservation tillage impacts on soil, crop and the environment. Int. Soil. Water Conserv. Res. 3, 119–129.

Chauhan, B.S., 2019. Weed biology insights to improve Feathertop Rhodes grass & Barnyard grass. www.weedsmart.org.au/video/management-of-feathertop-rhodes-grass-barnyard-grass/.

Chauhan, B.S., Johnson, D.E., 2010. Implications of narrow crop row spacing and delayed *Echinochloa colona* and *Echinochloa crus-galli* emergence for weed growth and crop yield loss in aerobic rice. Field Crop. Res. 117, 177–182.

Chauhan, B.S., Widderick, M., Werth, J., Street, M., Cook, T., Congreve, M., 2016. Northern IWM Fact Sheet: Windmill Grass. Department of Agriculture and Fisheries, Queensland, Australia.

Chauhan, B.S., Manalil, S., Florentine, S., Jha, P., 2018. Germination ecology of *Chloris truncata* and its implication for weed management. PLoS One 13, e0199949.

Clayton, W.D., Harman, K.T., Williamson, H., Vorontsova, M., 2015. World Checklist of Poaceae. Royal Botanic Garden, Richmond, UK.

Daniel, R., Wu, H., Congreve, M., 2020. Feathertop Rhodes grass ecology and management. What strategies are working best? Grains Res. Update 54.

Davidson, B., Cook, T., Chauhan, B.S., 2019. Alternative options to glyphosate for control of large *Echinochloa colona* and *Chloris virgata* plants in cropping fallows. Plants 8, 245.

D'Emden, F.H., Llewellyn, R.S., Burton, M.P., 2006. Adoption of conservation tillage in Australian cropping regions: an application of duration analysis. Technol. Forecast. Soc. Change 73, 630–647.

Desai, H.S., Thompson, M., Chauhan, B.S., 2020. Target-Site resistance to glyphosate in *Chloris virgata* biotypes and alternative herbicide options for its control. Agronomy 10, 1266.

Fernando, N., Humphries, T., Florentine, S.K., Chauhan, B.S., 2016. Factors affecting seed germination of feather fingergrass (*Chloris virgata*). Weed Sci. 64, 605–612.

Florence, J., Chevillotte, H., Ollier, C., Meyer, J., 2013. Base de données botaniques Nadeaud de l'Herbier de la Polynésie Française (PAP) (Botanical database of the Nadeaud Herbarium of French Polynesia).

Fosberg, F.R., Sachet, M.-H., Oliver, R., 1987. A geographical checklist of the Micronesian Monocotyledonae. Micronesia 20, 1–129.

Hunt, J., Browne, C., McBeath, T., Verburg, K., Craig, S., Whitbread, A., 2013. Summer fallow weed control and residue management impacts on winter crop yield though soil water and N accumulation in a winter-dominant, low rainfall region of southern Australia. Crop. Pasture Sci. 64, 922–934.

Llewellyn, R., Ronning, D., Clarke, M., Mayfield, A., Walker, S., Ouzman, J., 2016. Impact of Weeds in Australian Grain Production. Grains Research and Development Corporation, Canberra, ACT, Australia, pp. 24–108.

Mahajan, G., Walsh, M., Chauhan, B.S., 2020. Junglerice (*Echinochloa colona*) and feather fingergrass (*Chloris virgata*) seed production and retention at sorghum maturity. Weed Technol. 34, 272–276.

Manalil, S., Mobli, A., Chauhan, B.S., 2020. Interference of windmill grass [*Chloris turncata*] and feathertop Rhodes grass [*Chloris virgata*] in mungbean. Crop. Pasture Sci 71.

Michael, P.J., Yeoh, P.B., Scott, J.K., 2012. Potential distribution of the Australian native *Chloris truncata* based on modelling both the successful and failed global introductions. PLoS One 7, e42140.

Mobli, A., Chauhan, B.S., 2020. Crop residue retention suppresses seedling emergence and biomass of winter and summer Australian weed species. Weed Biol. Manag. 20, 118–128.

Ngo, D., 2017. Biology and glyphosate resistance in *Chloris truncata* (windmill grass) and *Chloris virgata* (feathertop Rhodes grass) in southern Australia (Doctoral dissertation).

Ngo, D., Boutsalis, P., Preston, C., Gill, G., 2017a. Growth, development, and seed biology of feather fingergrass (*Chloris virgata*) in Southern Australia. Weed Sci. 65, 413–425.

Ngo, D., Boutsalis, P., Preston, C., Gill, G., 2017b. Plant development and seed biology of windmillgrass (*Chloris truncata*) in Southern Australia. Weed Sci. 65, 395–405.

Nichols, V., Verhulst, N., Cox, R., Govaerts, B., 2015. Weed dynamics and conservation agriculture principles: a review. Field Crop. Res. 183, 56–68.

Osten, V., 2012. Feathertop Rhodes Grass: A Best Weed Management Guide. Department of Agriculture, Fisheries and Forestry, Queensland, Brisbane, Australia.

Reicosky, D.C., Saxton, K.E., 2007. The benefits of no-tillage. No-Tillage Seeding in Conservation Agriculture. CAB Publishing, pp. 11–20.

Rengasamy, P., 2002. Transient salinity and subsoil constraints to dryland farming in Australian sodic soils: an overview. Australian J. Exp. Agriculture 42, 351–361.

Rodriguez, S., Kruk, B.C., Satorre, E.H., 2020. Modelling *Chloris virgata* germination and emergence under different temperature and light quality conditions. Weed Res. 60, 287–296.

Sharma, A.D., 2020. Crop competition as a strategy to control glyphosate-resistant Chloris virgata and application of biochar to control glyphosate mobility in Australian Soils (Doctoral disscrtation).

Shrestha, A., Hembree, K., Wright, S., 2008. Biology and Management of Horseweed and Hairy Fleabane in California. University of California, Agriculture and Natural Resources, pp. 6–9.

Sowiński, J., Dayan, F.E., Głąb, L., Adamczewska-Sowińska, K., 2020. Sorghum allelopathy for sustainable weed management. In: Mérillon, J.-M., Ramawat, K.G. (Eds.), Plant Defence: Biological Control. Springer International Publishing, Cham, pp. 263–288.

Teasdale, J.R., 1996. Contribution of cover crops to weed management in sustainable agricultural systems. J. Prod. Agriculture 9, 475–479.

Verloove, F., 2005. New records of interesting xenophytes in Spain. Lazaroa 26, 141.

Vibrans, H., 2009a. *Chloris virgata* Sw.

Vibrans, H., 2009b. Ficha Informativa par la especie Chloris virgata. Malezas de. Mexico.

Walsh, M.J., Broster, J.C., Aves, C., Powles, S.B., 2018a. Influence of crop competition and harvest weed seed control on rigid ryegrass (*Lolium rigidum*) seed retention height in wheat crop canopies. Weed Sci. 66, 627–633.

Walsh, M.J., Broster, J.C., Schwartz-Lazaro, L.M., Norsworthy, J.K., Davis, A.S., Tidemann, B. D., et al., 2018b. Opportunities and challenges for harvest weed seed control in global cropping systems. Pest. Manag. Sci. 74, 2235–2245.

Weller, S., Florentine, S., Mutti, N., Jha, P., Chauhan, B.S., 2019. Response of Chloris truncata to moisture stress, elevated carbon dioxide and herbicide application. Sci. Rep. 9, 1–10.

Zhang, H., Irving, L., Tian, Y., Zhou, D., 2012. Influence of salinity and temperature on seed germination rate and the hydrotime model parameters for the halophyte, *Chloris virgata*, and the glycophyte, *Digitaria sanguinalis*. South. Afr. J. Botany 78, 203–210.

Chapter 7

Erigeron bonariensis, Erigeron canadensis, and *Erigeron sumatrensis*

Singarayer Florentine[1], Talia Humphries[1] and Bhagirath Singh Chauhan[2]

[1]Centre for Environmental Management, School of Health and Life Sciences, Federation University Australia, Ballarat, VIC, Australia, [2]The Centre for Crop Science, Queensland Alliance for Agriculture and Food Innovation (QAAFI) and School of Agriculture and Food Sciences (SAFS), The University of Queensland, Gatton, QLD, Australia

Introduction

As the global human population is projected to increase to an estimated 8–10 billion by 2050, the demand for essential foodstuffs will increase significantly by 59%–98% (Alexandratos and Bruinsma, 2012). Whilst farmers across the globe will be expected to meet this demand, it is recognized that the agricultural sector in general is already facing appreciable problems with a range of elements such as insect pests, climate change disruptions, reduction in the areas of fertile land, and herbicide-resistant weeds. Holm (1979) estimated that approximately 200,000 species of plants have been identified and recorded across the world, and of these, 4% (approximately 8000 species) have been categorized as agronomic weeds. Within the agricultural industry, these weed species compete with crops for sunlight, nutrients, space, and water. In addition to restricting the growth and quality of food producing plants, the co-maturation of weeds with food crops can lower the economic and nutritional worth of the product by contaminating the harvest. Studies have shown that weeds have the most significant impact on the agricultural industry when compared to other factors such as insect pests and global climate change. Farmers across the world have been forced to spend increasing amounts of money to combat weeds in an effort to maintain production yields. Llewellyn et al. (2016) reported that an estimated AU$3.3 billion per year has been taken from the grain industry in terms of lost yield and cost of mitigation management efforts. It was also reported that

Biology and Management of Problematic Crop Weed Species. DOI: https://doi.org/10.1016/B978-0-12-822917-0.00024-0

approximately $1.3 billion had been lost in winter cropping systems alone. Similarly, Williamson (2002) and Pimentel et al. (2005) calculated that farmers in the United Kingdom spend €150 million per year on weed control whilst in the USA spend approximately US$3 billion per year.

Although in recent years, a range of herbicides have been established as the preferred option to reduce the impacts of weeds on the crops, a plethora of studies have shown the evolution of herbicide resistance in agricultural weeds (Heap, 2017; Moss, 2017; Marwat et al., 2008; Owen et al., 2007). Recent studies show that approximately 88 weed species in Australia have developed resistance to selected herbicides (Heap, 2017). Therefore it is essential to not only determine the future weed impacts on the global agricultural sector, but to predict which weed species are going to be problematic in terms of herbicide resistance. Such knowledge will assist in the implementation of early intervention control approaches to reduce the impact of weeds on the growth of crops and therefore facilitate greater economic yields. In order to develop suitable management strategies for these invasive species, it is vital that we have a clear understanding of the biology of the species. This fundamental understanding of weed species ecology and biology is essential to develop sound integrated weed management strategies. Therefore the objective in this chapter is to synthesize the available literature on relevant species' biology, their impacts on crops, current effective control measures, and the management of these three global invasive species: *Erigeron bonariensis* (L.) Cronq. (synonym *Conyza bonariensis*), *Erigeron canadensis* (L.) Cronq. (synonym *Conyza canadensis*), and *Erigeron sumatrensis* (Retz.) (synonym *Conyza sumatrensis*).

Taxonomy

The *Erigeron* genus consists of approximately 200 species and are mostly native to North America (Noyes, 2000; Thebaud and Abbott, 1995; Everett, 1990). Historically, Cronquist (1943) attempted to distinguish between *Conyza* and *Erigeron* species based on the (1) ligules being either present or absent, (2) the ligule length, and (3) the total number of pistillate and complete flowers. Based on these parameters, these plants were reclassified from *Erigeron* species to the *Conyza* class, noting that Conyza belongs to the *Asteracea* family, tribe *Asterace*, and subtribe *Conyzinae*. However, a recent taxonomic revision has now been completed in Australia, and *Conyza* has been transferred back to *Erigeron*. As a result, the names of *Conyza bonariensis* (L.) Cronq., *C. canadensis* (L.) Cronq., and *C. sumatrensis* (Retz.) E. Walker have been changed to *E. bonariensis* (L.) Cronquist., *E. canadensis* (L.) Cronquist., and *E. sumatrensis*. Note that *E. sumatrensis* is often confused with *E. bonariensis* at early stages, but *E. bonariensis* is usually a more robust plant, generally more pale in color with even broader leaves (Fig. 7.1).

FIGURE 7.1 *Erigeron* species. *Erigeron bonariensis* (L.) Cronquist (A) mature plant, (D) leaf shape and arrangement and (G) seed; *E. canadensis* (L.) Cronquist (B) mature plant, (E) leaf shape and arrangement, and (H) seed; and *E. sumatrensis* (C) mature plant, (F) leaf shape and arrangement, and (I) seed. *Sources: Photographs (B, C, E, and H) were obtained from GBIF. Org (accesed 18th May 2020 from https://www.gbif.org/occurrence/2012991530); Photographs (A, D, F, G, and I) were obtained from the* Atlas of Living Australia.

Current global distribution

The genus *Erigeron* comprises over 50 species (Kissmann and Groth, 1999), and the current distribution of *E. bonariensis*, *E. canadensis*, and *E. sumatrensis* is provided in Fig. 7.2. These three species are known to be widespread, due to their aggressive weed traits, such as high seed production, quick germination, and long-distance seed dispersal (Hao et al., 2009).

E. bonariensis is a common agronomic weed species originally from the temperate part of South America (Michael, 1977). However, this species is

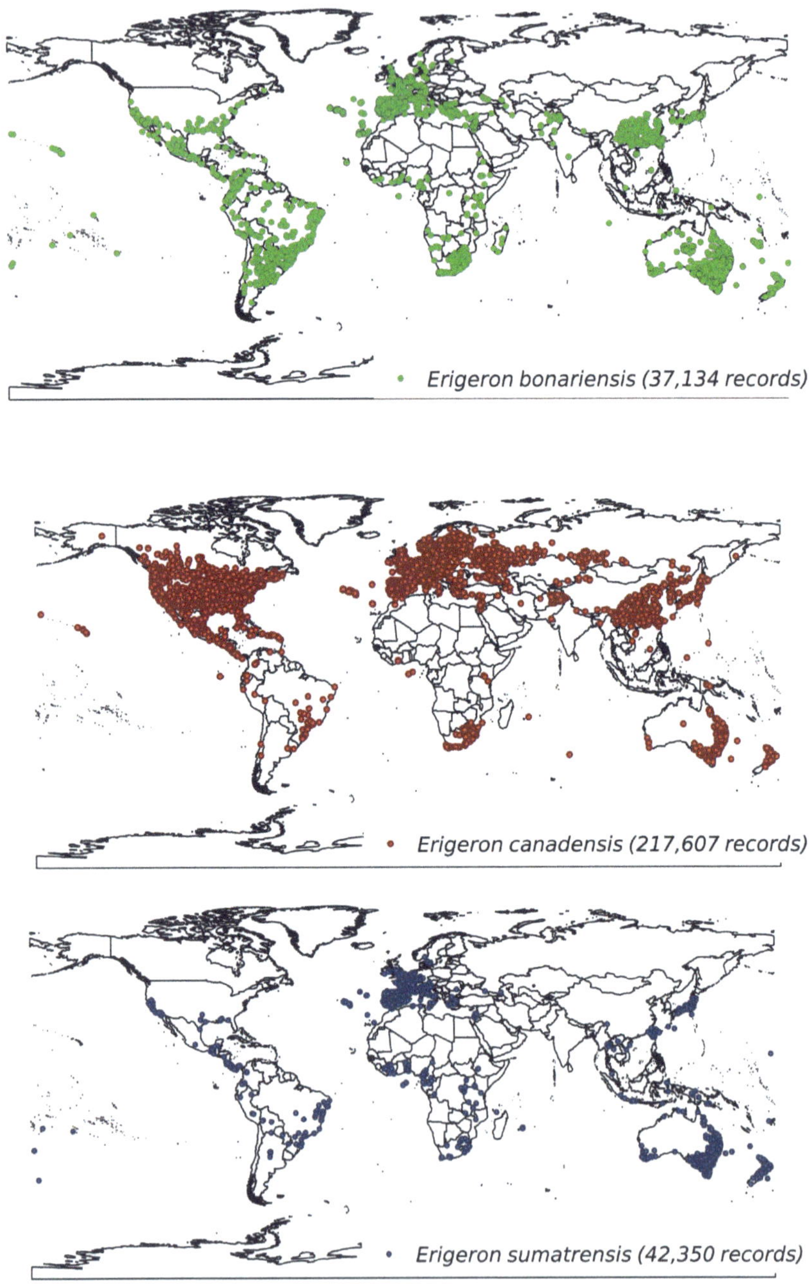

Data source: GBIF.org (24 March 2020) GBIF Occurrence Download https://doi.org/10.15468/dl.qqqoks7

FIGURE 7.2 Current distribution of *Erigeron bonariensis*, *E. canadensis*, and *E. sumatrensis*.

now naturalized in Africa, Asia, Australasia, Northern America, the Pacific region, Belgium, Germany, the Netherlands, France, Portugal, Spain, and the southern region of the USA (Negrean and Ciortan, 2012). *E. bonariensis* is cosmopolitan, and studies have shown that this species can invade various crops and disturbed lands, although it is more commonly found in orchards, vineyards, and abandoned industrial sites (Heap, 2017; Prieur-Richards et al., 2000; Burry and Kloot, 1982).

E. canadensis is native to North America and is presently found in approximately 70 countries (CABI, 2020; Holm et al., 1997). Michael (1977) reported that this species spread to Europe during the 1600s. *E. canadensis* is at present the most widespread invasive species of the three, and is found in almost all parts of China (Beres et al., 2018). In Africa, this species has invaded both the north and south parts of the tropical regions (Holm et al., 1997).

E. sumatrensis is a pan-tropical agronomic weed with American origin, and is currently present in 49 countries (USDA-ARS 2004; Thebaud and Abbott 1995; Holm et al., 1979). The species was first recorded in Java, Indonesia during 1860 (Kostermans et al., 1987). This species is now naturalized in Australia, and is present in most of Southeast Asia, Western Europe, and also some parts of the Mediterranean basin (Guillerm et al., 1990). Of these three species, *E. canadensis* and *E. sumatrensis* are most widespread across the globe (Thebaud and Abbott, 1995). In Australia, both *E. bonariensis* and *E. sumatrensis* are found in all states and territories (Wu, 2007), but *E. canadensis* is present only in a restricted range in Australia.

Reproduction

Kempen and Graf (1981) conducted a study on the phenology of *E. bonariensis* and reported that a single plant can produce up to 266,750 seeds. Similarly, a study conducted by Wu et al. (2007) in Australia found that a mature plant can produce up to 119,000 seeds. *E. bonariensis* prefers light for enhanced seed germination.

E. canadensis is a winter and summer annual plant, native to North America and is distributed worldwide in a range of temperate climates. This plant is considered an important agricultural weed because it can reduce agricultural yields by 90% at high densities and becomes problematic under low-tillage agriculture. *E. canadensie* can produce up to 70 seeds per flower. A study by Bhowmilk and Bekech (1993) found that a single plant can produce up to 200,000 seeds. *E. canadensis* seeds are not dormant at maturity (Buhler and Hoffman, 1999; Buhler and Owen, 1997). Seeds have the capability to germinate over a temperature range from 5°C to 30°C. However, the optimum temperature is 15°C under 12-h light and dark conditions (Ottavini et al., 2019). Seeds of these three species are attached to a pappus, which facilitate its spread by wind (anemochory) and water (hydrochory). Table 7.1

TABLE 7.1 *Erigeron bonariensis, Erigeron canadensis,* **and** *Erigeron sumatrensis* **species: nomenclature, origin, life form, and biological information.**

Parameters	*Erigeron bonariensis*	*Erigeron canadensis*	*Erigeron sumatrensis*	References
Origin	South America	North America	South America	Holm et al. (1979), Chaudhary et al. (1981), Thebaud and Abbott (1995)
Altitude (m)	Colder areas in altitudes between $\leq 400-500$	Warmer area in lower altitudes >1000	Sunny and warm climatic conditions Altitude is not a limiting factor, occurring from sea level to >3000	Vladimirov (2009)
Life cycle	Annual	Annual or biennial	Annual or short-lived perennial	Holm et al. (1979), Thebaud and Abbott (1995)
Height (m)	0.6–1.0	1.5–2.5	2–3	Thebaud et al. (1996)
Flowering season	Spring–Autumn	Autumn or winter	Summer–Autumn	
Seeds/plant	~119,000	~200,000	~60,000	Wu et al. (2007), Bhowmilk and Bekech (1993), Hao et al. (2009)
Seed dispersal	Mainly wind (anemochory) but secondary transport via water (hydrochory)	Mainly wind (anemochory) but secondary transport via water (hydrochory)	Mainly wind (anemochory) but secondary transport via water (hydrochory)	Green et al. (2008)
Seed viability	1–2 years	2–3 years	2–3 years	Wu et al. (2007), Weaver (2001), Hayashi (1979)

Seed dormancy	Non-deep physiological dormancy	Non-deep physiological dormancy	No dormancy period	Karlsson and Milberg (2007), Green et al. (2008), Wu et al. (2007)
Optimum temperature (°C) for germination	30/20	30/20	Data not available	Loura et al. (2020) Shontz and Oosting (1970)
Light required for germination	Yes	Yes	Yes	Michael (1977), Zinzolker et al. (1985), Wu et al. (2007)
Sodium chloride (mM) threshold. No germination occurs at or beyond	>200	>150	Unknown	Loura et al. (2020), Nandula et al. (2006)
Soil moisture required for germination (MPa)	< -0.8	< -0.8	Unknown	Loura et al. (2020), Nandula et al. (2006)
Seedling emergence reduced to zero when buried at depth (cm)	1.0	0.5	Unknown	Loura et al. (2020), Nandula et al. (2006)
Genetics	Hexaploid with chromosome number $2n = 54$	Diploid with chromosome number $2n = 18$	Hexaploid with chromosome number $2n = 54$	Thebaud and Abbott (1995)
Floret color	White to pink	Cream	Straw	Everett (1990)
Leaf dimensions	40–90 mm long 5–15 mm wide	20–100 mm long 3–8 mm wide	40–100 mm long 5–12 mm wide	Everett (1990)

shows the variations in growth preferences and reproduction between *E. bonariensis*, *E. canadensis*, and *E. sumatrensis*.

Impacts on crops

All three species cause significant impacts to agricultural industries across the world. A study conducted by Wu et al. (2010) found that *E. bonariensis* was able to reduce sorghum [*Sorghum bicolor* (L.) Moench.] yield by 65%–98%. Similarly, Trezzi et al. (2015) reported that *E. bonariensis*, at a density of 3 plants/m^2, reduced the soybean (Latin name with authority) yield by up to 36% when established at 0, 5, and 11 weeks before sowing of soybean. In Australia, *E. bonariensis* has been ranked as one of the top 10 weed species based on its contribution to significant losses of grain yield (Llewellyn et al., 2016). These authors also reported that revenue loss caused by *E. bonariensis* is worth AU$ 1.1 million per annum in the northern grain region of Australia. It has also been observed that these three *Erigeron* species were hosts of the tomato (Latin name with authority) yellow leaf curl virus in Spain (Jorda et al., 2001) and Turnip mosaic virus in Zimbabwe (Chivasa et al., 2002). Similarly, *E. bonariensis* tends to provide a habitat for various ant species, which are a serious crop pest in China (Xie and Yao, 1989). In Australia, this species is causing a significant impact to dry-land cropping systems. Specifically, *E. bonariensis* has invaded wheat (Latin name with authority), soybean, chickpea (Latin name with authority), cotton (Latin name with authority), and sorghum crops (Wu et al., 2007; Wu and Walker, 2004). Storrie (2007) indicated that *E. bonariensis* has become a problematic weed in cropping systems since Australian agricultural practices have changed from conventional to no-till farming practices, and has also been promoted by the reduction in application of herbicides.

E. canadensis is also a problematic agronomic weed across the world and it is found in 40 economically important crops (Holm et al., 1979). Similar to *E. bonariensis*, it is a major problem in conservation agriculture (Wiese et al., 1995). In the USA, this species causes a significant impact to corn, soybean, and cotton crops, as well as orchards and vineyards (Buhler, 1992; Kapusta 1979). Bruce and Kells (1990) conducted a study to examine the soybean yield losses from *E. canadensis* and suggested that losses run to about 83%. Similarly, Holm et al., (1979) pointed out that *E. canadensis* reduced sugar beet (Latin name with authority) yields by 64% due to heavy infestations in Germany. *E. canadensis* is the third most problematic weed in the USA (WSSA, 2017).

E. sumatrensis has been observed as a threat in a variety of minimum tillage cropping systems throughout the world; however, it has not received the same attention with regard to its effect on yield or economic losses (Ngow et al., 2020; Marambe et al., 2002). *E. sumatrensis* has become an agricultural weed more recently than *E. bonariensis* and *E. canadensis*; however, a competition study demonstrated *E. sumatrensis* outcompeted *E. canadensis* for water and nutrients in a Mediterranean field experiment (Thebaud et al., 1996). It has also developed multiple herbicide-resistant biotypes (Heap, 2017), and this, coupled with its aggressive nature, suggests that it has the potential to cause as much economic damage as *E. bonariensis* and *E. canadensis*.

Management

Tillage

Soil tillage is a widely used technique that can provide excellent weed control by burying surface seeds deeper into the soil profile and by uprooting establishing seedlings. Tillage can also soften the soil bed prior to seeding (Chauhan and Opena, 2012; Mohler and Galford, 1997), helping to deliver nutrients deeper into the soil profile (Issaka et al., 2019). Soil tillage is a very effective control option for *Erigeron* species, and can be used without chemical intervention, particularly if implemented when this weed is in its small seedling stage (Shresha et al., 2008). This method is effective as the majority of *Erigeron* species seeds germinate at the soil surface (Nandula et al., 2006), and light has been observed to be an important requirement for its seed germination (Wu et al., 2007). As a consequence, deeper buried seeds are not problematic. The majority of these species' seeds germinate in autumn; therefore tillage, which is conducted early winter (Buhler and Owen, 1997), is often most effective control strategy as the rosettes are still small and are easily removed (Shresha et al., 2008). In-row cultivation is available for a variety of annual and perennial systems, including cropping systems, vineyards, and orchids, making tillage a widely adaptable and effective technique.

Despite the weed control benefits that tillage provides, many land managers want to move away from tillage as a control option as it can (1) reduce the soil quality, (2) alter soil moisture, (3) increase soil erosion, (4) disturb beneficial soil biota (such as worms), and (5) can resurface deeper buried viable weed seeds (Bhowmilk and Bekech, 1993). Whilst no-till techniques can result in the faster decay of weed seeds, *Erigeron* species seeds tend to thrive as they are dynamic in terms of suitable germination conditions. No-till systems allow the seeds to remain close to the soil surface, which is favorable for light-sensitive seeds. *Erigeron* seed germination is most active in autumn, and its survival is high over winter, making them problematic to remove prior to spring sowing. Plants can also emerge during the spring, making them an ongoing challenge within the active growing season. In these no-till systems, the integrated use of other cultivation techniques and herbicides is required for the control of these three aggressive *Erigeron* species.

Herbicides

The broad-spectrum herbicide, glyphosate, is very effective for targeting broad-leaf weeds like *Erigeron* species. In most cases, this herbicide, when applied alone at a standard application of 1080 g ha (Tahmesebi et al., 2018) within a glyphosate-tolerant cropping system, will effectively control this weed. However, glyphosate resistance is becoming an issue of increasing concern to systems that have relied on this herbicide, and resistant *Erigeron* species biotypes are now globally widespread (Heap, 2017). *Erigeron* species were the first weed to adapt resistance to glyphosate, just three years after its introduction (VanGessel, 2001), and is currently one of the widest spread glyphosate-resistant weeds (Heap, 2017). Resistance to glyphosate has been confirmed to occur through the identification of lower shikimate accumulation, and even a dose twice the recommended rate has been shown to be ineffective at killing resistant biotypes (Tahmesebi et al., 2018). In addition to glyphosate resistance, *Erigeron* species have also adapted resistance to at least 20 other widely used herbicides, allowing this weed to tolerate photosystem II inhibitors, bipyridiliums, ureas, amides, and acetolactate synthase inhibitors (Heap, 2017). Multiple herbicide-resistant biotypes, such as resistance to both glyphosate and flazasulfuron, have also been observed (Tahmesebi et al., 2018).

The high level of herbicide resistance within *Erigeron* species has been overcome through various solutions, including novel herbicide mixtures and application timing, as summarized in Table 7.2. Targeting *Erigeron* species at an earlier growth stage, with combinations of herbicides, can greatly improve the success of the treatment, and broaden the options for suitable treatments. Cesco et al. (2019) explored the effect of 13 different herbicide combinations and sequential applications on 3–5 or 15 cm tall *Erigeron* species. It was found that eight of the tested treatments were effective at significantly reducing the shorter group, while only three of these treatments [(1) glyphosate + 2,4-D followed by saflufenacil, (2) 2,4-D followed by glyphosate + saflufenacil followed by paraquat, and (3) 2,4-D + glyphosate + saflufenacil followed by glyphosate + saflufenacil] were effective for the taller group. As *Erigeron* species germination occurs sporadically throughout the year, one herbicide treatment is unlikely to be effective for providing significant reduction. Byker et al. (2013) recommended the implementation of a preplant herbicide, followed by a postplant herbicide, for soybean cropping systems. The application of dicamba (600 g ha) pre- and postplanting provided the most consistent control of 86% reduction, and the herbicide combinations of glyphosate + dicamba or 2,4-D provided 78%–97% control 8 weeks after application, while the combination of glyphosate + chloransulam (17.5 g ha) provided 90% control. Dicamba was observed to be effective on *E. canadensis* rosettes up to 10 cm (Crose et al., 2020). Using only postplant herbicides did not provide adequate control on

TABLE 7.2 Overcoming herbicide resistance within *Erigeron* species using alternative solution combinations, including novel herbicide mixtures and application timing.

Mode of action	Site of action	Herbicides	Species	Effect	Reference
Amino acid synthesis inhibitor	EPSP synthetase Inhibitor	Glyphosate	*E. bonarensis*, *E. canadensis*, *E. sumatrensis*	Up to 60% control observed when applied to small plants (3–5 cm basal diameter)	Cesco et al. (2019), Byker et al. (2013)
Cell membrane inhibitor	PPO inhibitors	Carfentrazone-ethyl, oxyfluorfen, saflufenacil	*E. bonarensis*, *E. canadensis*, *E. sumatrensis*	All plants survived; however, growth was reduced	Tahmesebi et al. (2018)
Growth regulators	Synthetic auxins	2,4-D, 2,4-D amine, 2,4-D ester, dicamba, diglycolamine salt of dicamba, MCPA	*E. bonarensis*, *E. canadensis*, *E. sumatrensis*	All except MCPA provided over 90% control, and were more effective on younger plants. MCPA was not effective with over 90% survival rate	Tahmesebi et al. (2018), Cesco et al. (2019), Byker et al. (2013), Kruger et al. (2010)
Inhibits plant amino acid synthesis	Acetohydroxyacid synthase	Cloransulam, imazapyr, metsulfuron, chlorimuron-ethyl, chlorimuron-methyl,	*E. bonarensis*, *E. canadensis*,	Metsulfuron, chlorimuron-ethyl, chlorimuron-methyl, flazasulfuron, and thifensulfuron provided over 90% control,	Crose et al. (2020), Tahmesebi et al. (2018), Byker et al. (2013)

(*Continued*)

TABLE 7.2 (Continued)

Mode of action	Site of action	Herbicides	Species	Effect	Reference
		flazasulfuron, thifensulfuron	*E. sumatrensis*	while cloransulam had varied control between 28% and 90%	
	Inhibition of protoporphyrinogen oxidase (PPO)	Flumioxazin, pyraflufen-ethyl	*E. bonarensis, E. canadensis, E. sumatrensis*	Flumioxazin reduced growth by 50% and pyraflufen-ethyl reduced growth of *E. canadensis* by up to 95%	Tahmesebi et al. (2018)
	Inhibition of glutamine synthetase	Glufosinate	*E. bonarensis, E. canadensis, E. sumatrensis*	Reduced growth rate by up to 100%	Tahmesebi et al. (2018)
Photosynthesis inhibitors	Photosystem I	Diquat	*E. bonarensis, E. canadensis, E. sumatrensis*	Treatment provided 80%–90% growth reduction	Tahmesebi et al. (2018)
	Photosystem II	Atrazine	*E. bonariensis*	Provided over 90% control nine weeks after application, and between 65% and 87% control 21 weeks after application	Wu et al. (2008)

+ shows the herbicides that were combined and applied at the same time, → shows the second knock of herbicides that followed after the initial herbicide application.

their own. However, they can play an important role for the control of *E. canadensis* emerging later in the season (Byker et al., 2013).

Using the sequential application of herbicides is also referred to as the "double-knock technique." This method has been observed to improve *E. bonarensis* control, regardless of the density or age of the infestation (Walker et al., 2012). Double knock applications, using a combination of glyphosate + (2,4-D amine + picloram) followed by paraquat + diquat applied 7 days later, provided 92%–100% control, irrespective of density or age of the plants. Werth et al. (2010) found that following an initial glyphosate application with 2,4-D provided 85%–93% control, and when paraquat + diquat was used, control was improved to 88%–97%. Applying the sequential herbicide 7 days or more after the initial treatment reduced the level of control (Werth et al., 2010), suggesting that the time frame within which the sequential herbicide is applied, is critical. Further, the effect of applying herbicides at different times of the day have also provided different levels of efficacy (Montgomery et al., 2017). The herbicides 2,4-D, dicamba, glufosinate, and saflufenacil were found to be more effective when applied in the middle of the day, while paraquat was more effective when applied at sunset (Montgomery et al., 2017).

Recently, new herbicides, such as halauxifen-methyl from the new herbicide group called the arylpicolinates, have now become available. Zobiole et al. (2017) explored the effect of using these new herbicides to create novel combinations and found that the mixture of disclosulam + halauxifen-methyl + glyphosate (applied at a rate of 30.6, 38.2, and 960 g/ha, respectively), followed by a sequential treatment of either ammonium glufosinate, paraquat, or saflufenacil 10 days after the initial application, provided over 90% control of *E. canadensis*, *E. sumatrensis*, and *E. bonariensis*. Further, these combinations were effective for plants that were up to 62 cm tall, suggesting these new combinations could be effective for larger, more difficult plants.

Competition

As it has been previously mentioned, *Erigeron* species are not strong interspecific competitors, and control techniques such as cover crops, narrower row spacing, mulching, or crop residue can be implemented to target this weakness (Weaver, 2001). In addition to providing weed control services, cover crops are also beneficial for improving soil conditions in cropping systems, particularly when legumes are added to the cover crop mixture. When used as a monoculture, or in a mix, the crops cereal rye (Latin name with authority), hairy vetch (Latin name with authority), and crimson clover (Latin name with authority) were all observed to provide equal or better *E. canadensis* suppression than residual herbicides (flumioxazin + paraquat and metribuzin + chlorimurn-ethyl) (Pittman et al., 2019). Using cereal rye as a cover crop and leaving the crop residue resulted in delayed emergence of herbicide-resistant *E. canadensis* by 4 weeks within a soybean system, resulting in a significant reduction in this weed's density at the time of soybean planting (Sherman et al., 2020). Furthermore, residue was observed to reduce *E. canadensis* emergence by 80% in a no-tillage corn cropping-system, and, in addition, suppressed autumn germination (Bhowmilk and Bekech, 1993). In corn cropping systems, the most effective cover crop combination for consistent suppression, density reduction, and biomass reduction of glyphosate-resistant *E. canadensis* was crimson clover and annual ryegrass (Latin name with authority) (Cholette et al., 2018). Anderson (2014) showed that using cool season crops followed by cover crops in a 2-year sequence reduced *Erigeron* species density by 63%. However, it should be noted that the study suggested that this method would have to be ongoing to maintain control. Indeed, integrating cover crops with herbicides has been observed to improve control of glyphosate-resistant *E. canadensis* and *E. sumatrensis* biotypes in soybean systems, resulting in improved crop yields (Marochi et al., 2018; Sherman et al., 2020).

Grazing and mowing

Livestock are able to graze *Erigeron* without any associated health risks; however, it has been observed that after grazing, the plants can regrow quickly. Mowing these weeds had a similar regrowth effect (Bajwa et al., 2016; Shresha et al., 2008). Therefore it is suggested that grazing and mowing are not suitable treatments to use in isolation. However, grazing and mowing can be beneficial in order to reduce plant height prior to herbicide application to improve its efficacy (Wu and Koetz, 2012). Green et al. (2008) identified that *E. bonariensis* was more susceptible to grazing than *E. canadensis* and *E. sumatrensis*, and unlike the latter two, it was only found growing within grazing exclusions.

Fire

Fires of moderate to high severity can promote *E. bonariensis*, *E. canadensis*, and *E. sumatrensis* populations and stimulate recruitment from the seedbank (Pino et al., 2013). *E. canadensis* germination rate and total percentages were significantly improved when exposed to 110°C for up to 10 min; however, temperatures of 150°C and over devitalized the seeds (Aran et al., 2013). In addition to temperature, Aran et al. (2013) also observed that smoke exposure enhanced germination percentage. As *Erigeron* species can quickly regenerate after a fire (Pino et al., 2013), burning is not considered a suitable technique for its control in agricultural systems.

Management implications

E. bonarensis, *E. canadensis*, and *E. sumatrensis* are having a significant impact on cropping systems, globally. Each has high fecundity and the ability to disperse seeds great distances via hydrochory and anemochory. They have relaxed germination requirements, which makes their management an ongoing process. Establishing a more competitive environment through the use of denser row spacing and using crops can therefore assist with year-round management. Fire has been observed to stimulate germination, and therefore could be implemented to flush the soil seedbank; however, this technique is unlikely to devitalize the seeds. As many cropping systems are now moving away from regular cultivation, which is deemed the most effective management technique for these species, an overreliance on herbicides, particularly glyphosate, has been implemented for their control. This has resulted in glyphosate-resistant biotypes emerging throughout the world, as well as biotypes with multiple resistances to other herbicide groups. While new herbicides offer highly effective control, over time, it is likely that resistant biotypes will emerge if herbicide use is not integrated with other conventional control techniques. Grazing prior to herbicide application can improve its absorption, as it was demonstrated that herbicide treatments were more effective on shorter plants.

Extensive research into the biology, ecology and management for *E. bonarensis*, *E. canadensis*, and *E. sumatrensis* has been conducted, yet these species are still globally wide-spread and are having a significant impact on minimal tillage cropping systems. Future research should investigate the integration of control techniques that are effective under field conditions, whilst being economically viable for farmers.

References

Alexandratos, N., Bruinsma, J., 2012. World agriculture towards 2030/2050: the 2012 revision. FAO, Rome, ESA Working Paper No: 12–03.

Anderson, R.L., 2014. A cultural system to reduce weed interference in organic soybean. Renew. Agriculture Food Syst. 30 (4), 392–398.

Aran, D., Garcia-Duro, J., Reyes, O., Casal, M., 2013. Fire and invasive species: modifications in the germination potential of *Acacia melanoxylon, Conyza canadensis* and *Eucalyptus globulus*. For. Ecol. Manag. 302, 7–13.

Bajwa, A.A., Sadia, S., Ali, H.H., Jabran, K., Peerzada, A.M., Chauhan, B.S., 2016. Biology and management of two important *Conyza* weeds: a global review. Environ. Sci. Pollut. Res. 23, 24694–24710.

Beres, Z.T., Ernst, E.E., Ackley, B.A., Loux, M.M., Owen, M.D.K., Snow, O.A., 2018. High levels of glyphosate resistance in *Conyza canadensis* from agricultural and non-agricultural sites in Ohio and Iowa. Scientific Reports 8, 10483.

Bhowmilk, P.C., Bekech, M.M., 1993. Horseweed (*Conyza canadensis*) seed production, emergence, and distribution in no-tillage and conventional tillage corn (*Zea mays*). Agron. Trends Agric. Sci. 1, 67–71.

Bruce, J.A., Kells, J.J., 1990. Horseweed (*Conyza canadensis*) control in no-tillage soybeans (*Glycine max*) with pre-plant and pre-emergence herbicides. Weed Technol. 4, 642–647.
Buhler, D.D., 1992. Population dynamics and control of annual weeds in corn (*Zea mays*) as influenced by tillage systems. Weed Sci. 40, 241–248.
Buhler, D.D., Hoffman, M.L., 1999. Andersen's Guide to Practical Methods of Propagating Weeds & Other Plants. Weed Science Society of America, Lawrence, KS, p. 248.
Buhler, D.D., Owen, M.D.K., 1997. Emergence and survival of horseweed (*Conyza canadensis*). Weed Sci. 45, 98–101.
Burry, J.N., Kloot, P.M., 1982. The spread of Composite (*Compositae*) weeds in Australia. Contact Dermatitis 8, 410–413.
Byker, H., Soltani, N., Rbinson, D.E., Tardif, F.J., Lawton, M.B., Sikkema, P.H., 2013. Control of glyphosate-resistant horseweed (*Conyza canadensis*) with dicamba applied preplant and postemergence in dicamba-resistant soybean. Weed Technol. 27 (3), 492–496.
CABI, 2020. *Conyza canadensis* (Canadian fleabane). Accessed 17 May 2021. Available from: https://www.cabi.org/isc/datasheet/15251.
Cesco, V.J.S., Nardi, R., Krenchinski, F.H., Alberecht, A.J.P., Rodrigues, D.M., Albrecht, L.P., 2019. Management of resistant *Conyza* spp. during soybean pre-sowing. Planta Daninha 37, e019181064.
Chaudhary, S.A., Parker, C., Kasasian, L., 1981. Weeds of central, southern and eastern Arabian Peninsula. Tropical Pest Management 27, 181–190.
Chauhan, B.S., Opena, J., 2012. Effect of tillage systems and herbicides on weed emergence, weed growth, and grain yield in dry-seeded rice systems. Field Crop. Res. 137, 56–69.
Chivasa, S., Ekpo, E.J.A., Hicks, R.G.T., 2002. New hosts of Turnip mosaic virus in Zimbabwe. Plant Pathology 51, 386.
Cholette, T.B., Soltani, N., Hooker, D.C., Robinson, D.E., Sikkema, P.H., 2018. Suppression of glyphosate-resistant Canada fleabane (*Conyza canadensis*) in corn with cover crops seeded after wheat harvest the previous year. Weed Technol. 32, 244–250.
Cronquist, A., 1943. The separation of Erigeron from Conyza. Bulletin of the Torrey Botanical Club 70, 629–632.
Crose, J.A., Manuchehri, M.R., Baughman, T.A., 2020. Horseweed (*Conyza canadensis*) management in Oklahoma winter wheat. Weed Technol. 34, 229–234.
Everett, J., 1990. Asteraceae/Conyza. In: Harden, G.J. (Ed.), Flora of New South Wales. NSW University Press, Kensington.
Green, T.D., Sindel, B.M., Charles, G., Werth, J., 2008. A review of the ecology of fleabane (*Conyza* spp.). In: van Klinken, R.D., Osten, V.A., Panetta, F.D., Scanlan, J.C. (Eds.), Proceedings of the 16th Australian Weeds Conference. Queensland Weeds Society, Brisbane, pp. 171–173. Cairns, Queensland, 18–22 May 2008.
Guillerm, J.L., Floc'h, E., Maillet, J., Boulet, C., 1990. The invading weeds within the Mediterranean Basin. Biological invasions in Europe and the Mediterranean Basin Dordrecht. Kluwer Academic Publishers, Netherlands, pp. 61–84.
Hao, J.H., Qiang, S., Liu, Q.Q., Cao, F., 2009. Reproductive traits associated with invasiveness in *Conyza sumatrensis*. J. Syst. Evol. 47, 245–254.
Hayashi, I., 1979. Secondary succession of herbaceous communities in Japan: seed germination and shade tolerance of seedlings of the dominants. Bulletin of the Yokohama Phytosociological Society 16, 407–419.
Heap, I. 2017. The international survey of herbicide resistant weeds. Online. Heap, I., 2020. The International Herbicide-Resistant Weed Database. Online. Accessed: April 5, 2020. Available from: http://www.weedscience.org.

Holm, L.G., Pancho, J.V., Herberger, J.P., Plucknett, D.L., 1979. A geographical atlas of world weeds. John Wiley and Sons, New York, USA, pp. 391.

Holm, L., Pancho, J.V., Herberger, J.P., Plucknett, D.L., 1979. A geographical atlas of world weeds. John Wiley and Sons, New York, Chichester, Brisbane, Toronto, UK, p. 391.

Holm, L., Doll, J., Holm, E., Pancho, J.V., Herberger, J.P., 1997. World weeds: natural histories and distribution. Wiley, New York, 1129 p. In: No tillage and conventional-tillage corn (*Zea mays*). Agronomy 1993, 1, 67–71.

Issaka, F., Zhang, Z., Zhao, Z.Q., Asenso, E., Li, J.H., Li, Y.T., et al., 2019. Sustainable conservation tillage improves soil nutrients and reduces nitrogen and phosphorous losses in maize farmland in Southern China. Sustainability 11, 2397.

Jorda, C., Font, I., Martinez, P., Juarez, M., Ortega, A., Lacasa, A., 2001. Current status and new natural hosts of tomato yellow leaf curl virus (TYLCV) in Spain. Plant Disease 85, 445.

Kapusta, G., 1979. Seedbed tillage and herbicide influence on soybean (*Glycine max*) weed control and yield. Weed Sci. 27, 52–526.

Karlsson, L.M., Milberg, P., 2007. Comparing after-ripening response and germination requirements of *Conyza canadensis* and *C. bonariensis* (Asteraceae) through logistic functions. Weed Research 47, 433–441.

Kempen, H.M., Graf, J., 1981. Weed seed production. Proceedings of the Western Society of Weed Science 34, 78–81.

Kissmann, K., Groth, D., 1999. Plantas Infestantes e Nocivas, 2nd edn BASF, São Paulo.

Kostermans, A.J.H.H., Wirjahardja, S., Dekker, R.J., 1987. The weeds: description, ecology, and control. In: Soerjani, M., Kostermans, A.J.G.H., Tjitrosoepomo, G. (Eds.), Weeds of rice in Indonesia. Balai Pustaka, Jakarta, Indonesia, pp. 24–565.

Kruger, G.R., Davis, V.M., Weller, C.S., John, W.G., 2010. Control of horseweed (*Conyza canadensis*) with growth regulator herbicides. Weed Technology 24, 425–429.

Llewellyn, R.S., Ronning, D., Ouzman, J., Walker, S., Mayfield, A., Clarke, M., 2016. Impact of weeds on Australian grain production: the cost of weeds to Australian grain growers and the adoption of weed management and tillage practices Report for GRDC. CSIRO, Australia.

Loura, D., Sahil, Florentine, S., Chauhan, B.S., 2020. Germination ecology of hairy fleabane (*Conyza bonariensis*) and its implications for weed management. Weed Sci. 68, 411–417.

Marambe, B., Nissanka, S.P., De Sliva, L., Anandacoomaraswamy, A., Priyantha, M.G.D.L., 2002. Occurance of paraquat-resistant *Erigeron sumatrensis* (Retz.) in upcountry tea land of Sri Lanka. J. Plant. Dis. Prot. 18, 973–978.

Marochi, A., Ferreira, A., Takano, K., Oliveira, R.S., Ovejero, R.L.O., 2018. Managing glyphosate-resistant weeds with cover crop associated with herbicide rotation and mixture. Ciênc. Agrotecnol. 42 (4), 381–394.

Marwat, K.B., Muhammad, S., Zahid, H., Bakhtiar, G., 2008. Study of various herbicides for weed control in wheat under irrigated conditions. Pak. J. Weed Sci. Res. 14, 1–8.

Michael, P.W. 1977. Some weedy species of Amaranthus (amaranths) and Conyza/Erigeron (fleabanes) naturalized in the Asian-Pacific region. Proceedings of the 6th Asian-Pacific Weed Science Society Conference, Indonesia, 87–95.

Mohler, C.L., Galford, A.E., 1997. Weed seedling emergence and seed survival: separating the effects of seed position and soil modification by tillage. Weed Res. 37 (3), 147–155.

Montgomery, G.B., Treadway, J.A., Reeves, J.L., Steckel, L.E., 2017. Effect of time of day of application of 2,4-d, dicamba, glufosinate, paraquat, and saflufenacil on horseweed (*Conyza canadensis*) control. Weed Technol. 31, 550–556.

Moss, S. 2017. Herbicide resistance in weeds. In: Weed Research: Expanding Horizons, 181.

Nandula, V.K., Eubank, T.W., Poston, D.H., Koger, C.H., Reddy, K.N., 2006. Factors affecting germination of horseweed (*Conyza canadensis*). Weed Sci. 54, 898–902.

Negrean, G., Ciortan I., 2012. *Conyza bonariensis*, a new plant with invasive character in Romanian flora. Analele Universităţii din Craiova - Biologie, Horticultura, Tehnologia Prelucrarii Produselor Agricole, Ingineria Mediului 17, 743–748.

Ngow, Z., Chynoweth, R.J., Gunnarsson, M., Rolstn, P., Buddenhagen, C.E., 2020. A herbicide resistance risk assessment for weed in wheat and barley cropping systems in New Zealand. Plos One.

Noyes, R.D., 2000. Biogeographical and evolutionary insights on *Erigeron* and allies (Asteraceae) from ITS sequence data. Plant. Syst. Evol. 220, 93–114.

Ottavini, D., Pannacci, E., Onofri, A., Tei, F., Jensen, P.K., 2019. Effects of light, temperature, and soil depth on the germination and emergence of *Conyza canadensis* (l.) cronq. Agronomy 9, 533.

Owen, M.J., Walsh, M.J., Llewellyn, R.S., Powles, S.B., 2007. Widespread occurrence of multiple herbicide resistance in Western Australian annual ryegrass (*Lolium rigidum*) populations. Aust. J. Agric. Res. 58, 711–718.

Pimentel, D., Zuniga, R., Morrison, D., 2005. Update on the environmental and economic costs associated with alien-invasive species in the United States. Ecol. Econ. 52, 273–288.

Pino, J., Arnan, X., Rodrigo, A., Retana, J., 2013. Post-fire invasion and subsequent extinction of *Conyza* spp. in Mediterranean forests is mostly explained by local factors. Weed Res. 53, 470–478.

Pittman, K.B., Barney, J.N., Flessner, M.L., 2019. Horseweed (*Conyza canadensis*) suppression from cover crop mixtures and fall-applied residual herbicides. Weed Technol. 33, 303–311.

Prieur-Richards, A.H., Lavorel, S., Grigulis, K., Dos Santos, A., 2000. Plant community diversity and invasibility by exotics: invasion of Mediterranean old fields by *Conyza bonariensis* and *Conyza canadensis*. Ecological Letters 3, 412–422.

Sherman, A.D., Haramoto, E.R., Green, J.D., 2020. Integrating fall and spring herbicides with a cereal rye cover crop for horseweed (*Conyza canadensis*) management prior to soybean. Weed Technol. 34, 64–72.

Shontz, J.R., Oosting, H.J., 1970. Factors affecting interaction and distribution of *Haplopappus divaricatus* and *Conyza canadensis* in North Carolina old fields. Ecology 51, 780–793.

Shresha, A., Hembree, Write, S., 2008. Biology and management of horseweed and hairy fleabane in California. ANR Publications, The Regents of the University of California.

Storrie, A., 2007. Distribution and identification of *Conyza* species. In: Walker, S., Werth, J., Widderick, M. (Eds.), Proceedings of 2nd Fleabane workshop. Department of Primary Industries and Fisheries, Toowoomba, Australia, pp. 10–11.

Tahmesebi, B.K., Alebrahim, M.T., Roldan-Gmez, R.A., da Silveira, H.M., de Carvalho, L.B., la Cruz, R.A., et al., 2018. Effectiveness of alternative herbicides on three *Conyza* species from Europe with and without glyphosate resistance. Crop. Prot. 112, 350–355.

Thebaud, C., Abbott, R.J., 1995. Characterization of invasive *Conyza* species (Asteraceae) in Europe: quantitative trait and isozyme analysis. Am. J. Botany 82, 360–368.

Thebaud, C., Finzi, A.C., Affre, L., Debussche, M., Escarre, J., 1996. Assessing why two introduced *Conyza* differ in their ability to invade Mediterranean old fields. Ecology 77, 791–804.

Trezzi, M.M., Vidal, R.A., Patel, F., Miotto, E., Debastiani, F., Balbinot, A.A., Mosquen, R., 2015. Impact of *Conyza bonariensis* density and establishment period on soyabean grain yield, yield components and economic threshold. Weed Research 55, 34–41.

USDA-ARS, 2004. Germplasm Resources Information Network (GRIN). National Germplasm Resources Laboratory, Beltsville, Maryland, USA, Online Database.

VanGessel, M.J., 2001. Glyphosate-resistant horseweed from Delaware. Weed Sci. 49, 703–705.

Vladimirov, V., 2009. *Erigeron sumatrensis* (Asteraceae): a recently recognized alien species in the Bulgarian flora. Phytologia Balcanica 15, 361–365.

Walker, S., Boucher, L., Cook, T., Davidson, B., McLean, B., Widderick, M., 2012. Weed age affects chemical control of *Conyza bonariensis* in fallows. Crop. Prot. 38, 15–20.

Weaver, S.E., 2001. The biology of Canadian weeds. 115. *Conyza canadensis*. Can. J. Plant. Sci. 81, 867–875.

Weed Science Society of America (2017) WSSA survey ranks most common and most troublesome weeds in broadleaf crops, fruits, and vegetables. http://wssa.net/2017/05/wssa-survey-ranks-most-commonand-most-troublesome-weeds-in-broadleaf-crops-fruits-and-vegetables/. Accessed: November 11, 2018.

Werth, J., Walker, S., Boucher, L., Robinson, G., 2010. Applying the double knock technique to control *Conyza bonariensis*. Weed Biol. Manag. 10, 1–8.

Wiese, A.F., Salisbury, C.D., Bean, B.W., 1995. Downy brome (*Bromus tectorum*), jointed goatgrass (*Aegilops cylindrica*) and horseweed (*Conyza canadensis*) control in fallow. Weed Technol. 9, 249–254.

Williamson, M., 2002. "Alien plants in the British Isles". In: Pimentel, D. (Ed.), Biological invasions: economic and environmental costs of alien plant, animal and microbe species. CRC Press, pp. 91–112.

Wu, H., Walker, S. 2004. Flaxleaf fleabane: a problem weed in zero-tillage dryland cropping. Australian Cottongrower October-November, 68–71.

Wu, H., Walker, S., Robinson, G., 2008. Chemical control of flaxleaf fleabane (*Conyza bonariensis* (L.) Cronquist) in winter fallows. Plant Protection Quarterly 23, 162–165.

Wu, H., Walker, S., Robinson, G., Coombes, N., 2010. Control of flaxleaf fleabane (*Conyza bonariensis*) in wheat and sorghum. Weed Technology 24, 102–107.

Wu, H., Koetz, E., 2012. Management of flaxleaf fleabane (*Conyza bonariensis*) in lucerne pastures. In: Eighteenth Australasian Weeds Conference. 211–214.

Wu, H., Walker, S., Rollin, M.J., Tan, D.K.Y., Robinson, G., Werth, J., 2007. Germination, persistence, and emergence of flaxleaf fleabane (*Conyza bonariensis* [L.] Cronquist). Weed Biol. Manag. 7, 192–199.

Xie, F.Y., Yao, L.X., 1989. A study on *Dorylus orientalis* Westwood. Insect Knowl. 26, 291–293.

Zinzolker, A., Kigel, J., Rubin, B., 1985. Effects of environmental factors on the germination and flowering of *Conyza albida*, *C. bonariensis* and *C. canadensis*. Phytoparasitica 13, 229–230.

Zobiole, L.H.S., Krenchinski, F.H., Pereira, G.R., Rampazzo, P.E., Rubin, R.S., Lucio, F.R., 2017. Management programs to control *Conyza* spp. in pre-soybean sowing applications. Planta Daninha 36, e018175883.

Chapter 8

Cyperus esculentus and *Cyperus rotundus*

Gerald M. Henry[1], Matthew T. Elmore[2] and Travis W. Gannon[3]
[1]*University of Georgia, Athens, GA, United States,* [2]*Rutgers University, New Brunswick, NJ, United States,* [3]*North Carolina State University, Raleigh, NC, United States*

Biology and morphology

Cyperus esculentus L. and *Cyperus rotundus* L. are consistently ranked as two of the most troublesome weeds in the world (Peerzada, 2017; Van Wychen, 2019, 2020). These C_4 perennial sedge species are problematic due to prolific underground vegetative growth. A network of rhizomes, tubers, and basal bulbs are the primary mechanism by which both species proliferate. These sedges overwinter through tubers, which are the main reason for their weediness in cropping systems. A single plant can produce thousands of tubers annually, while the carrying capacity is estimated at 10–30 million tubers/ha without crop interference (Hauser, 1962; Horowitz, 1972; Tumbleson and Kommedahl, 1961). The competition with crops and subsequent yield reductions are thought to occur primarily through interference, but tuber extracts from both species have demonstrated allelopathic effects on crops (Drost and Doll, 1980; Javaid et al., 2007; Wills, 1987).

The life cycle of these perennial sedges begins with overwintering tubers. In temperate climates, first emergence originates from tubers in the spring. A quiescent bud on each tuber sprouts to produce a rhizome. Rhizome stems elongate, typically until they reach 1–4 cm of the soil surface. At this depth, likely in response to light and diurnal temperature fluctuations, internode elongation slows and leaves elongate to form a basal bulb (Stoller and Woolley, 1983). The basal bulb is the primary site of meristematic activity and the origin of all subsequent shoot, root, and rhizome growth (Gifford and Bayer, 1995; Jansen, 1971). Differentiation to basal bulbs usually marks the beginning of the rapid shoot and root growth as leaves emerge in the characteristic one-third phyllotaxy (Stoller et al., 1972). Rhizome development begins shortly thereafter. Within a few weeks, a single-basal bulb often produces several rhizomes that ascend or descend (Hauser, 1962).

Biology and Management of Problematic Crop Weed Species. DOI: https://doi.org/10.1016/B978-0-12-822917-0.00011-2

These rhizomes ultimately form new (secondary) basal bulbs and the process of shoot and root growth followed by rhizome production repeats, resulting in a rapidly expanding network of rhizomes. Rhizome and bulb networks expand from just one tuber to a patch of several square meters in just a few months in the presence of optimal growing conditions and without competition from other plants (Horowitz, 1972; Stoller and Sweet, 1987).

At some point in the life cycle, a proportion of rhizomes differentiate into tubers instead of basal bulbs. Commonly referred to as tubers among weed scientists and agronomists, these structures are botanically classified as corms and contain a large proportion of starch (Rodrigues and Estelita, 2009). Tubers contain four to seven nodes usually less than 1 cm long, with each node containing one bud. Typically, no more than two buds sprout at a time unless a sprout is removed, in which case the apical dominance is removed and several more buds sprout (Bendixen, 1973). Consequently, tillage and other strategies to physically remove their shoots are not effective for control. Apical dominance is also observed within *C. rotundus* tuber chains, as only the apical tuber in a chain sprouts. If the chain is broken, by cultivation, for example, apical dominance is released and each individual tuber will sprout from the terminal bud.

Cyperus esculentus tubers are terminal and solitary, whereas *C. rotundus* tubers often form in chains along the rhizome (Gifford and Bayer, 1995; Siriwardana and Nishimoto, 1987). Generally, tuber formation begins 4–6 weeks after emergence. In *C. esculentus*, tuberization predominates in shorter (<12 h) photoperiods and basal bulbs predominate with longer photoperiods (Garg et al., 1967; Stoller and Sweet, 1987). Increased nitrogen (N) fertilization also tends to promote shoot growth and reduce tuberization in *C. esculentus*, but implications for weed control in a field setting have not been explored (Garg et al., 1967). In *C. rotundus* the proportion of tuber to basal bulb production is not daylength-dependent, and tuber production can even increase as day length increases (Williams, 1978). Daylength-dependent tuberization is an advantageous trait for *C. esculentus* in temperate climates, because survival depends on tuber production as days shorten before freezing temperatures arrive.

Cyperus esculentus tubers can survive freezing soil temperatures, but *C. rotundus* tubers cannot. In laboratory experiments, *C. rotundus* tuber survival was <10% when chilled to −2°C, whereas *C. esculentus* required −9°C for significant survival reduction. Even in nonfreezing temperatures, *C. rotundus* tuber survival was <10% when exposed to 2°C for 12 weeks (Stoller, 1973). Tolerance to chilling temperatures is further reduced if *C. esculentus* tubers are subjected to desiccation (Thomas, 1969). At warmer temperatures ideal for growth, the literature suggests that *C. rotundus* is less tolerant to desiccation as the survival of tubers severed from their root system was just a few weeks in dry soil (Smith and Fick, 1937). Others suggested that this root system is important to maintain *C. rotundus* tuber moisture during dry periods (Andrews, 1940). Both species enjoy global tuber distribution as contaminants of root crops such as potato (*Solanum tuberosum* L.), peanut (*Arachis hypogaea* L.), and other feedstocks, but they are established as major weed problems in distinct regions worldwide due to tuber survival. *Cyperus rotundus* is found worldwide mostly in tropical and Mediterranean climates where soils do not freeze and moisture is not limiting (Bendixen and Nandihalli, 1987). *Cyperus esculentus* is established in colder climates and is especially problematic in southern Africa and North America where soils commonly freeze in winter. These tubers are most commonly found within 18 cm of the soil surface in heavier clay soils, but deeper in sandy soils (Nishimoto, 2001).

Tubers remain viable in the soil for at least a few years and population half-lives were reported to be 16 months for *C. rotundus* in a tropical climate and 4–6 months for *C. esculentus* in a temperate climate (Neeser et al., 1997; Stoller and Wax, 1973). These studies likely underestimate real dormancy observed under field conditions, as tubers were removed from the soil and detached from the tuber network before observation. This likely removed dormancy enforced by apical dominance and made tubers more prone to desiccation.

Apical dominance is not the only factor controlling these species tuber dormancy. As with seeds of annual weeds, a large proportion of *Cyperus* species tubers are dormant at maturity. When *C. esculentus* is grown in a temperate climate, tuber dormancy is high for tubers collected in the autumn and low for tubers collected in early spring, but dormancy can be reversed by 2 weeks of low-temperature exposure (Tumbleson and Kommedahl, 1962). Abscisic acid plays a major role in regulating *C. esculentus* tuber dormancy (Muñiz and Tames, 1982). *Cyperus rotundus* tubers also display some innate dormancy, although this is not well understood (Nishimoto, 2001). Interestingly, when tubers are separated from rhizome chains, immature tubers have less dormancy than mature tubers (Neeser et al., 1997; Shamsi et al., 1978). *Cyperus rotundus* dormancy may be governed by abscisic acid and possibly phenolic compounds, but more research is needed. Interestingly, one group reported that if one bud sprouts and is killed, it is common for *C. rotundus* tubers to enter a state of secondary dormancy where other buds on the tuber may not sprout immediately (Neeser et al., 1997). Much as seed dormancy complicates annual weed control with preemergent herbicides, tuber dormancy complicates control with herbicides as only a portion of the population will be nondormant and exposed to the herbicide.

It is important to understand conditions for tuber sprouting when planning herbicide applications. The minimum temperatures for tuber sprouting in *C. esculentus* and *C. rotundus* are 6°C and 10°C, respectively, and the optimal temperature range is 25°C–35°C (Holt and Orcutt, 1996). As with many annual weed seeds, alternating temperatures increase sprouting and bud elongation (Sun and Nishimoto, 1999). Fluctuations of just 6°C with the high temperature achieved for just a few minutes can dramatically increase sprouting percentage compared to constant temperature (Sun and Nishimoto, 1997). The emergence of disturbed tubers can also be predicted using accumulated growing degree-days (GDD), which may be especially useful when trying to time a second cultivation event to control these species prior to crop planting. If the first cultivation event occurs shortly after initial emergence, this GDD interval could be used to schedule the second cultivation after the second flush of emergence (Wilen et al., 1996).

Both *Cyperus* species are obligate outcrossers; so reproduction via seeds provides genetic variability to more rapidly evolve in changing environments (Horak and Holt, 1986). *Cyperus esculentus* seed viability was highly variable among seed lots, ranging from 1% to 78%, when seeds were collected at the appropriate time after anthesis (Hill et al., 1963; Thullen and Keeley, 1979). Comparatively, *C. rotundus* produces few seeds and viability is extremely poor (Thullen and Keeley, 1979). Regardless of seed viability, it is thought that seedling vigor is too poor to survive cultivation and other control strategies used in crop production, and thus propagation via seeds is unimportant, especially in dryland systems (Lapham and Drennan, 1990; Stoller and Sweet, 1987). However, in wet areas not frequently disturbed, it seems reasonable to assume that new populations can originate from seeds.

Research on *Cyperus* species population genetics has sought to help determine the contribution of sexual reproduction, but results are variable. Okoli et al. (1997) determined considerable variation in nine *C. esculentus* populations from various locations in California, United States, but considerably less variability among *C. rotundus* populations. A more extensive sampling with more modern molecular genetic techniques found a low level of genetic diversity among *C. esculentus* populations from cultivated fields in southwest France (Dodet et al., 2008). This suggests infestations spread by asexual means, likely through movement of tubers by agricultural machinery between fields in the region. Further supporting the idea that sexual reproduction is somewhat common, especially if it confers traits advantageous for survival, a *C. esculentus* and *C. rotundus* hybrid with resistance to acetolactate synthase-inhibiting herbicides was reported in rice production (Tehranchian et al., 2015). Given the advances in molecular genetics in recent years, more research on this subject would reveal how quickly this weed can adapt to invade new environmental conditions and management strategies.

Habitat, adaptation, and competition

Habitat

Cyperus esculentus and *C. rotundus* are commonly found throughout tropical, subtropical, and temperate areas of the world (Bendixen and Nandihalli, 1987;

Jordan-Molero and Stoller, 1978; Wills, 1987). Global species distribution may be a function of physical dissemination and ecological adaptation. *Cyperus* species propagules often contaminate crops during harvest, become digested by foraging animals, or adhere to agronomic equipment during cultivation, ultimately leading to short and long-range spread (Bendixen and Nandihalli, 1987; Lapham and Drennan, 1990; Schippers et al., 1993; Webster 2005). Buoyancy of tubers and seeds may also contribute to dissemination through surface irrigated crops (Felix et al., 2012). However, heterogeneous distribution of both species within climatic regions is greatly associated with a few edaphic variables.

Both species are favored by high light intensity and tolerate a wide range of soil types, but the most important factors affecting their prevalence are temperature, soil moisture, and photoperiod (Bendixen and Nandihalli, 1987; Stoller and Sweet, 1987). Cold soil temperatures can reduce tuber germination and survival; however, increases in starch, sugar, and lipid tuber content of *C. esculentus* may enhance tolerance to frozen soil conditions (Nishimoto, 2001; Stoller and Wax, 1973; Stoller and Weber, 1975). Therefore the presence of *C. rotundus* is typically restricted to warmer northern latitudes where average, mid-winter, low air temperatures are $> -1^{\circ}C$, whereas *C. esculentus* is capable of surviving $-17^{\circ}C$ (Glaze, 1987; Okoli et al., 1996; Stoller, 1981; Ueki, 1969).

Cyperus species infestations generally start in low-lying, poorly drained areas of fields where soil moisture levels are high. Dense patches tend to form in locations where competing annual weeds and crops are flooded out (Stoller, 1981). Therefore these sedges are only problematic in arid environments when present within irrigated fields (Stoller and Sweet, 1987). Although both species tolerate a wide range of moisture conditions (Keeley and Thullen, 1978; Nishimoto et al., 1978; Stoller and Weber, 1975), *C. esculentus* tolerates soil drying more than *C. rotundus* (Horak and Holt, 1986; Lapham and Drennan, 1990). Desiccation of *C. rotundus* tubers that result in the loss of approximately 85% water content resulted in tuber death, whereas even intermediate moisture levels still reduced tuber viability (Rao and Nagarajan, 1962; Smith and Fick, 1937).

Short photoperiods tend to stimulate *C. esculentus* reproductive growth, whereas long photoperiods begin to stimulate vegetative growth (Jansen, 1971); however, interactions with environmental and soil parameters can greatly influence the rate and magnitude of these responses (Garg et al., 1967). Typically, photoperiods of 8–12 h have led to the promotion of tuberization in *C. esculentus*, whereas16 h photoperiods maximized shoot formation and elongation (Bell et al., 1962; Garg et al., 1967). The influence of photoperiod on *C. rotundus* tuberization is poorly understood. Horowitz (1972) reported that natural day length of 10–14 h had little to no effect on *C. rotundus* tuberization, whereas Williams (1978) observed an increase in tuber production as photoperiod decreased.

Competition

Cyperus esculentus and *C. rotundus* have become two of the world's worst agricultural and horticultural weeds in an estimated 40 and 90 countries, respectively (Bendixen and Nandihalli, 1987). Both species are serious or principal weeds within rice (*Oryza sativa* L.), sugarcane (*Saccharum officinarum* L.), cotton (*Gossypium* spp.), corn (*Zea mays* L.), and soybeans (*Glycine max* L. Merr.) worldwide (Bendixen and Nandihalli, 1987). High growth rates and quick emergence of their seedlings create competition during early crop development that significantly impacts annual crop production (Edenfield et al., 2005; Horowitz, 1973). Infestations of these species can result in yield reductions up to and sometimes exceeding 50%, leading to significant economic losses (Keeley, 1987). Crop quality may be compromised while weed control strategies and harvest operations are often hindered (Grichar and Sestak, 2000; Moffett and McCloskey, 1998; Wilcut et al., 1994).

Competition studies conducted in Brazil concluded that season-long *C. rotundus* interference of 160 plants 0.1/m^2 in vegetable crops caused yield reductions in garlic (*Allium sativum* L.) of 89%, okra (*Hibiscus esculentus* L.) of 62%, green bean (*Phaseolus vulgaris* L.) of 41%, cucumber (*Cucumis sativus* L.) of 43%, cabbage (*Brassica oleracea* L.) of 35%, and tomato (*Lycopersicon esculentum* Mill.) of 53% (William and Warren, 1975). Bryson et al. (2003) reported seed cotton yield reductions of 62%–85% in response to *C. rotundus* interference, whereas Leon et al. (2003) reported that fresh weight of cotton was reduced by 9%–42% when grown in competition with *C. rotundus* tuber densities of more than 180 tubers/m^2. Reductions in soybean yield of 29% and 36% were observed in response to *C. rotundus* shoots/m^2 of 148 and 165, respectively (Dev et al., 1997), whereas shoots/m^2 of 58–246 and 675–1198 resulted in reductions in sugarcane yield of 14% and 45%, respectively (Durigan, 2005).

Similarly, season-long *C. esculentus* interference of 100 shoots/m of crop row resulted in a 34% reduction in seed cotton yield (Keeley and Thullen, 1975), whereas Patterson et al. (1980) observed linear decreases in seed cotton yield in response to *C. esculentus* densities approaching 93 shoots/m^2. *Cyperus esculentus* densities up to 13 plants/m^2 reduced soybean yields 9% to 34% (Nelson and Smoot, 2010).

Light/shade

Several environmental factors and management inputs may influence the interaction between *Cyperus* species and the crops they infest. This is directly related to weed-crop competition and the efficiency in which both utilize available resources. *Cyperus* species are photosynthetically efficient weeds with high light compensation points, so CO_2 fixation continues as light intensity approaches full sunlight conditions (Black et al., 1969; Boardman, 1977).

Patterson (1982) reported reductions in *C. esculentus* total biomass of 65%, 86%, and 96% in response to 40%, 70%, and 85% shade, whereas *C. rotundus* was reduced 67%, 81%, and 94% when grown under those same parameters. Wills (1975) observed maximum growth of *C. rotundus* grown in the greenhouse in response to 19 klux illumination compared to the least amount of growth in response to 9 klux illumination. Therefore crop species that exhibit quick growth or produce tall canopies can often shade out *Cyperus* species. Keeley and Thullen (1978) observed that ≥90% photosynthetically active radiation was intercepted by corn, potatoes (*Solanum tuberosum* L.), and safflower (*Carthamus tinctorius* L.) 8–9 weeks after planting. However, when competing against less competitive crops with slower growth habits, competition from *Cyperus* species may significantly reduce crop production (Cherry, 1973). Buker et al. (2003) reported over 40% watermelon [*Citrullus lanatus* (Thunb.) Matsum. & Nakai] yield loss when in direct competition with 12 *C. esculentus* plants/m^2. Transplanting watermelons versus direct-seeding did not increase canopy establishment and reduce *C. esculentus* interference. Similarly, Thullen and Keeley (1980) observed a reduction in Japanese millet (*Echinochloa esculenta* H. Scholz) dry biomass when plants were in competition with *C. esculentus* in the greenhouse, because tubers sprouted and emerged several days before the millet. This trend was not consistent with field trials where both species had similarly timed emergence.

Nitrogen

Competitive relationships between weeds and crops can be substantially influenced by N amounts and timings (DiTomaso, 1995; Iqbal and Wright, 1997). Weeds accumulate and utilize nutrients more efficiently than crops, which gives them a distinct advantage when it comes to competition for other limiting resources. Furthermore, the presence of weeds often depletes the soil nutrient status, leading to additional crop yield loss. Additive series studies conducted by Morales-Payan et al. (1998) revealed that the greatest amount of bell pepper (*Capsicum annuum* L.) fruit reduction (73%) was in response to 210 kg N/ha plus *C. rotundus* densities of 300 plants/m^2. Radish (*Raphanus sativus* L.) yield losses were 100% when grown in competition with *C. rotundus* at densities of 75 and 125 plants/m^2 at N rates of 200 and 100 kg/ha, respectively (Santos et al., 1998). The competitiveness of *C. rotundus* was enhanced almost 15 times that of cilantro (*Coriandrum sativum* L.) when N was increased to 72 and 108 kg/ha (Morales-Payan et al., 2000). However, N at 120 kg/ha resulted in an increase in corn shoot dry weight, leaf area, and plant height of 49%, 29%, and 27%, respectively, whereas only increasing *C. rotundus* shoot dry weight (Tuor and Froud-Williams, 2002). The application of 120 kg N/ha did not influence soybean or *C. rotundus* plant weight, leaf area, or height when grown in competition with one another (Tuor and Froud-Williams, 2002)

Control options

Rapid growth and vegetative reproduction make the control of *Cyperus* species very difficult. There is no single control tactic that provides season-long control of *C. esculentus* or *C. rotundus*. An integrated approach that involves biological, cultural, and chemical control options will be most effective. Complete eradication should not be the goal, but selection of control methods that focus on the prevention of significant crop yield loss.

Biological

Controlling weed infestations by promoting large numbers of natural enemies for their suppression in agronomic and horticultural crops is often referred to as bioherbicide applications or innundative biological control (Auld et al., 2003; Bailey, 2014; Eilenberg et al., 2001). Increased environmental awareness, expanded pesticide regulation, and herbicide resistance concerns have prompted the examination and use of alternative strategies for *Cyperus* species control. Several pathogens have shown efficacy in the field for *Cyperus* species control (Charudattan, 2001; Charudattan and Dinoor, 2000). Pathogenicity and host-range tests have confirmed that *Dactylaria higginsii* (Luttrell) M. B. Ellis shows potential as a bioherbicide on *C. esculentus* and *C. rotundus* (Kadir and Charudattan, 2000; Kadir et al., 2000a). Shoot number, shoot dry weight, and tuber dry weight were reduced by almost 90% in the field in response to three applications of *D. higginsii* (Kadir and Charudattan, 2000; Yandoc et al., 2006). Kadir et al. (2000b) noted that the disease developed in 5 days and resulted in heavy *C. rotundus* defoliation. Similarly, a rust fungi *Puccinia canaliculata* (Schw.) Lagerh. reduced *C. esculentus* stand density by 46%, tuber number by 66%, and inhibited inflorescence development (Phatak et al., 1983). Another pathogenic strain of *P. canaliculata* reduced *C. esculentus* shoots by 3% and tuber number by 44% (Bruckart et al., 1988). Gupta et al. (2002) described the pathogenicity of another *Puccinia* spp. on *C. rotundus* present within mulberry (*Morus alba* L.) fields. Shoot necrosis and tuber reduction was observed when symptoms of *P. romagnoliana* Marie and Sacc. became severe. Several insects have also been identified for use as a bioherbicide of nutsedge species.

The larval stage of the moth *Bactra verutana* Zeller commonly infests *C. esculentus* and can severely damage plants in the field, but greenhouse studies revealed that injured plants often survive to produce additional plants and tubers (Keeley et al., 1970). *Cyperus rotundus* appears to be a more suitable host for *B. verutana*, since 90% of the larvae reached maturity on *C. rotundus*, whereas only 65% survived to maturity on *C. esculentus* (Frick et al., 1979). *Cyperus rotundus* foliage was reduced 30%–68% in response to three to eight weekly releases of *B. verutana* in cotton fields (Frick and Chandler, 1978). In greenhouse studies, tuber dry weights and numbers were reduced 93% and 77% for *C. rotundus*, whereas *C. esculentus* was reduced 80% and 62% (Frick et al., 1979). Several other pathogens and insects have potential for use as bioherbicides on *Cyperus* species. However, associated costs, application frequency, application technique, and possible antagonism from crop management practices can limit the efficacy and use of bioherbicides for nutsedge control.

Cultural

Mulches

Polyethylene mulches are an efficacious weed management tool used in certain agronomic cropping systems. Mulches have been investigated for *Cyperus* species control in specialty cropping systems, but researchers noted differences in response between *C. esculentus* and *C. rotundus*. Webster (2005) concluded that *C. esculentus* growth was hindered by polyethylene mulches, whereas *C. rotundus* growth was similar to the nonmulched. Conventionally, mulches have only been used in specialty systems due to disadvantages on a larger scale including costs, specialty equipment, removal, and lack of control of some species.

Webster (2005) reported *C. esculentus* shoot production was reduced 46% and 72% by black-opaque and clear polyethylene mulch, respectively. Webster further concluded black-opaque and clear polyethylene mulch reduced emerged shoots by 96%. Webster (2005) also concluded black-opaque and clear polyethylene mulch reduced *C. esculentus* tuber production 49%. Johnson and Mullinix (2008) investigated clear, black, and white on black low-density polyethylene mulch and concluded each of the mulches installed to freshly prepared seedbeds within 1 week prior to cantaloupe (*Cucumis melo* L.) transplant effectively suppressed *C. esculentus* without the need for fumigation or herbicide inputs. Monday et al. (2015) investigated an integrated control approach of drip-applied herbicides under polyethylene mulched beds and concluded fomesafen, halosulfuron, and S-metolachlor provided similar *C. esculentus* control and tomato yield as the commercial standard (S-metolachlor sprayed to the bed surface prior to mulch application).

Previous research reported surface-applied and incorporated mulch reduced *C. rotundus* density and biomass by 45%–53% and 40%–50%, respectively (Mahmood and Cheema, 2004; Peerzada, 2017). *Cyperus rotundus* shoot and tuber growth in black-opaque polyethylene mulch was similar to the nonmulched, whereas clear polyethylene mulch reduced tuber biomass compared to the nonmulched (Webster, 2005). Roozkhosh et al. (2017) further concluded utilizing a two-layer black polyethylene mulch was effective for *C. rotundus* control. Hajjaj and El Oualkadi, 2019 evaluated tillage in combination with straw mulch for *C. rotundus* control and concluded tillage reduced dry biomass by 62%, whereas straw mulch reduced dry biomass by 35% 90 days after treatment. Marí et al. (2020) evaluated various biodegradable materials including biodegradable plastic films and paper mulches, and concluded each are suitable alternatives to polyethylene mulches for *C. rotundus*. In conclusion, mulches may be a viable option for *Cyperus* species control in select cropping systems, but do not scale up to large acreage crops.

Tilling/physical removal

In recent years with the adoption of no-till or reduced tillage systems, small-seeded annual grasses and perennial species such as *C. esculentus* and *C. rotundus* incidence has increased. Tillage has been investigated for its impact on *C. rotundus* control and was concluded to be the least expensive control option in organic production systems (Bangarwa et al., 2008; Peerzada, 2017). While there are disadvantages associated with tillage, it has been shown to effectively suppress select *Cyperus* species. Tillage brings tubers closer to the soil surface, which make them more prone to frost, desiccation, and carbohydrate starvation, ultimately reducing viability and regenerative capacity (Hershenhorn et al., 2015; Neeser et al., 1997; Riemens et al., 2008). In recent years, mechanical diggers have been investigated for their utility in managing *Cyperus* species in various cropping systems. These implements mimic a peanut digger and penetrate the soil, lifting tubers to the soil surface and causing 70%–100% reduction in tuber sprouting by exposing the tubers to heat and dehydration (Hershenhorn et al., 2015). Johnson et al. (2015) evaluated a cart attached to a peanut digger that collected the plant material and concluded that *C. esculentus* density was reduced by >99% 1 week later. Price et al. (2018) evaluated *C. esculentus* control with conservation tillage alone or in combination with polyethylene mulch compared to conventional tillage with bedded polyethylene mulch in watermelon production and concluded control was insufficient. Repeated hand weeding, hoeing, and stale-bed preparation have been reported to suppress *C. rotundus* by 74%–93% (Akbar et al., 2011; Islam et al., 2009; Peerzada, 2017; Wang et al., 2009). In conclusion, tillage followed by mulching during the season was considered an economical method for *C. rotundus* control in specialty cropping systems or in systems where broadcast herbicide applications are not a viable option.

Crop rotation and row spacing

Crop rotations impact weed species incidence and shifts for a number of reasons including herbicide inputs, cultural practices in various systems, and competitive ability of crops (Johnson and Coble, 1986; Vencill and Banks, 1994). Manley et al. (2002) investigated the effect of rotational crops on weed incidence including *C. esculentus* at two locations. Specifically, they evaluated continuous corn, continuous soybean, corn–soybean, and corn–tomato–soybean at two sites from 1991 to 1994 and concluded the interaction of crop rotation and evaluated herbicide programs was significant in most years. They further concluded rotating herbicides improved efficacy of select herbicides, regardless of crop rotation system.

Much research has investigated the effect of agronomic crop row spacing on weed incidence. Fundamentally, crop row spacing and density affects time required for canopy closure, which directly impacts the critical weed-free period with narrower row spacing reducing the time for canopy closure and may decrease reliance on herbicides. The impact of narrow row corn spacing on weed competition and critical weed-free period is not clear; however, previous research has shown some benefits of narrow cotton rows for increased competition with weed species. Tursun et al. (2015) investigated the effect of cotton row spacing on critical period for weed control in a naturally infested site including *Sorghum halepense* (L.) Pers., *Solanum nigrum* L., *Portulaca oleracea* L., *C. rotundus*, and *Xanthium strumarium* L. They concluded the critical period for weed control was significantly longer for 70 and 90 cm compared to 50 cm row spacing which has implications for competitiveness against weeds. Similarly, Daramola et al. (2020) concluded narrow row soybean is more competitive against select weed species and reduced the density and biomass compared to wider row soybean, although *Cyperus* species were not included in this research.

Cover crops/smother crops

Cover crops may also be included in integrated weed management plans and have proven to effectively suppress weeds after termination and during early stages of crop growth. Cover crops suppress weeds by several mechanisms including physical barriers left on the soil surface that reduce light. Furthermore, cover crops provide weed suppression through competition (Mirsky et al., 2013), smothering (Hutchinson and McGiffen, 2000), or allelopathic activity (Barnes et al., 1987; Kunz et al., 2016) and can also suppress weeds by increasing organic matter and improving soil health, tilth, nitrogen fixation, and soil structure (Bachie and McGiffen, 2013; Norsworthy et al., 2007; Osipitan et al., 2018; Teasdale and Mohler, 2000). Although previous research indicates cover crops may provide adequate suppression of selected weed species through winter and spring, they do not provide adequate control throughout the growing season in most summer cropping systems (Teasdale, 1996; Monday et al., 2015) indicating that they may be a useful tool in combination with other techniques for *Cyperus* control. Monday et al. (2015) concluded preherbicides and postherbicides utilized in combination with a cereal rye (*Secale cereale* L.) cover crop may improve *Amaranth* spp. and *C. esculentus* control in watermelon.

Stale seedbed

Stale seedbeds are another weed management technique that cultivates the soil to promote weed seed germination near the soil surface (germination zone).

After germination and emergence, weed seedlings are controlled (typically with an herbicide application and minimal soil disturbance) and the crop is planted or transplanted soon thereafter into a mostly weed-free seedbed. Kanatas et al. (2020) concluded employing a stale seedbed in combination with glyphosate application provided 72% control of perennial weed species including *Convolvulus arvensis* L. and *C. rotundus* in soybean cropping systems. Similarly, Ameena et al. (2006) concluded stale seedbeds along with glyphosate applications were effective against *C. rotundus* in okra production systems. While there are disadvantages associated with stale seedbed, it is a viable component of integrated weed management programs in select cropping systems and can effectively increase weed control, while reducing herbicide inputs and overall production costs.

Chemical control

Herbicides are an integral component of *Cyperus* species management and should be used in combination with mechanical, cultural, and preventative tactics depending on cropping system. Preemergent- or preplant-incorporated herbicides are commonly used in combination with postemergent herbicides to optimally control *Cyperus* species. Common preemergent- or preplant-incorporated herbicides for *Cyperus* species control in agronomic crops include metolachlor, dimethenamid, acetochlor, and alachlor, which are all members of the chloroacetamide family and are commonly mixed with atrazine. Norflurazon and pendimethalin are also registered in select agronomic crops. S-ethyl dipropylthiocarbamate (EPTC) and butylate are members of the thiocarbamate family and are also registered in select agronomic crops, offering control of *C. esculentus*, but must be incorporated to obtain acceptable control. Common postemergent herbicides for *Cyperus* species include atrazine, bentazon, chlorimuron, glyphosate, halosulfuron, and pyridate in agronomic crops.

It is also important to note, other preemergent and postemergent herbicides, soil sterilants, and fumigants provide *Cyperus* species control in specialty cropping systems. Specifically, Gilreath and Santos (2004) concluded 90% *C. rotundus* reduction with 1,3-dichloropropene plus chloropicrin, whereas McAvoy and Freeman (2013) reported complete short-term *C. esculentus* control with dimethyl disulfide. Similarly, Monday et al. (2015) concluded ethalfluralin applied preemergent followed by halosulfuron postemergent as well as glufosinate applied postemergent reduced *C. esculentus* density. In summary, although *C. esculentus* and *C. rotundus* are difficult to control, various herbicides are efficacious against nutsedges with glyphosate being one of the most efficacious in tolerant cropping systems (Edenfield et al., 2005; Kumar et al., 2012; Webster et al., 2008).

Preemergent versus postemergent herbicides

The efficacy of various herbicides varies widely and depends on herbicide rate, the application timing, number of applications, growth stage, and tuber bank density, among other factors. Postemergent herbicides can provide acceptable control of many *Cyperus* species; however, in tolerant cropping systems, preemergent- or preplant-incorporated herbicides may still be used in combination with postemergent herbicide applications. To optimize efficacy, preplant or preemergent herbicides must be placed within the upper 5 cm of the soil surface to optimize absorption by emerging shoots and roots of germinating seeds or tubers. Depending on the herbicide, incorporation via mechanical or irrigation may be necessary to prevent off-target movement, thereby optimizing efficacy. Although preplant incorporated treatments may enhance *Cyperus* control, additional costs from equipment, time, and labor must be considered.

Postemergent herbicides including glyphosate, halosulfuron, imazaquin, MSMA, and trifloxysulfuron, among other herbicides, are also effective for *Cyperus* species control in select cropping systems. Edenfield et al. (2005) concluded sequential applications of glyphosate provided acceptable control of *C. rotundus*. In scenarios with high tuber bank density, sequential applications of preemergent and postemergent herbicides are recommended for acceptable *C. rotundus* control (Singh et al., 2016; Tuti and Das, 2011) although may not be necessary in all cropping systems.

Application timing

The effect of herbicide application timing is significant to efficacy for *C. esculentus* and *C. rotundus* control with active growth stages generally being most susceptible. Burke et al. (2008) concluded optimal *C. rotundus* control when herbicides were applied at earlier growth stages. Not only does optimal timing improve efficacy, previous research has concluded lower rates may provide the same level of control. Specifically, Lati et al. (2012) concluded *C. rotundus* control with trifloxysulfuron was higher when applied at earlier physiological timings. Furthermore, they reported that tuber density affected trifloxysulfuron efficacy, which concurs with reported trends on other species (Bussan et al., 2001; Dieleman et al., 1999). In specialty cropping systems, preemergent herbicides such as EPTC and fomesafen have been reported to provide inconsistent *C. esculentus* and *C. rotundus* control. Reed et al. (2016) concluded EPTC and fomesafen applications at early sprouting are most effective on *C. rotundus* as control decreased with nutsedge age.

Single versus sequential applications

Cyperus species control is enhanced with multiple applications of several postemergent herbicides. Hubbard et al. (2007) concluded *C. rotundus* control increased from 45% with a single trifloxysulfuron application to 90% with a sequential application, whereas Brecke et al. (2007) reported similar results with sulfosulfuron. Conversely, Edenfield et al. (2005) concluded >90% *C. rotundus* control was obtained with single or sequential glyphosate applications in soybean or cotton, which is likely due to differences in canopy architecture. Furthermore, Akin and Shaw (2001) and Edenfield et al. (2005) concluded glyphosate and select conventional herbicides reduced viable and total tuber density in glyphosate-tolerant cropping systems, which has implications for long-term control and management.

Optimizing herbicide efficacy

To ensure optimal herbicide efficacy, always select the appropriate herbicide based on proper weed identification, cropping system, and product label. Additionally, managers should consider stage of *C. esculentus* and *C. rotundus* growth, environmental conditions at application, temperature, soil moisture, soil type, organic matter content, application rate, carrier, carrier volume, and adjuvant selection, among other factors. Adjuvants used in combination with herbicides for nutsedge control should be selected based on the herbicide as well as spray solution quality. Paraquat efficacy on *C. rotundus* has been shown to increase when N was used as the carrier (Thangaraj and Rao, 1973), whereas many have reported ammonium sulfate significantly increased *C. rotundus* control with glyphosate (Suwunnamek and Parker, 1975).

Reasons for poor *Cyperus* species control with herbicides may be due to a number of factors including marginal herbicide absorption and/or translocation. As previously mentioned, proper herbicide timing that coincides with maximum absorption and translocation may increase herbicide efficacy. Additionally, properly selecting the correct spray additive as well as understanding the physiological growth stage is imperative for optimizing control with herbicides. Plant age impacts the amount absorbed and translocated of many herbicides including glyphosate (Pereira and Crabtree, 1986). Although the amount of glyphosate absorbed was higher in younger plants, total amount translocated to underground plant parts was higher in older plants (Pereira and Crabtree, 1986), which equates to increased efficacy and long-term management. Furthermore, with select herbicides and cropping systems, control may be optimized by herbicide placement or employing practices which mitigate off-target movement. Specifically, practices that increase EPTC and fomesafen contact with *C. esculentus* and *C. rotundus* may increase herbicide efficacy and consistency and may reduce tuber production, although other researchers concluded the herbicides did not affect tuber production (Felix and Newberry, 2012; Pereira et al., 1987).

Conclusion

Cyperus esculentus L. and *C. rotundus* L. are two of the most troublesome weeds in the world. These species are problematic due to prolific underground vegetative growth that consists of a network of rhizomes, tubers, and basal bulbs. *Cyperus* species are favored by high light intensity and tolerate a wide range of soil types; however, temperature, soil moisture, and photoperiod are the most influential factors that affect their prevalence. High growth rates and quick emergence of these species create competition during early crop development resulting in yield reductions that may lead to significant economic losses. Crop quality, weed control strategies, and harvest operations are often compromised. No single control tactic provides season-long control of *Cyperus* species; therefore integrated approaches that involve biological, cultural, and chemical options are most effective. Several pathogens and insects have shown potential for use as bioherbicides on *Cyperus* species. However, associated costs, application frequency, application technique, and possible antagonism from crop management practices limit their efficacy and use as bioherbicides for *Cyperus* species control. Cultural control methods consisting of mulches, tilling, crop rotation, cover crops, and stale seedbed techniques help reduce *Cyperus* species survival and competition enough to favor chemical control options. Efficient chemical control is best achieved through the use of sequential postemergent herbicide applications made to *Cyperus* species at early growth stages. Long-term control of *Cyperus* species can only be achieved through the persistent use of an integrated management approach.

References

Akbar, N., Jabran, K., Ali, M.A., 2011. Weed management improves yield and quality of direct seeded rice. Australian J. Crop. Sci. 5, 688.

Akin, D.S., Shaw, D.R., 2001. Purple nutsedge (*Cyperus rotundus*) and yellow nutsedge (*Cyperus esculentus*) control in glyphosate-tolerant soybean (*Glycine max*). Weed Technol. 15, 564–570.

Ameena, M., Kumari, V.L., George, S., 2006. Integrated management of purple nutsedge (*Cyperus rotundus* L.) in okra. Indian J. Weed Sci. 38, 81–85.

Andrews, F., 1940. A study of nut grass (*Cyperus rotundus* L.) in the cotton soil of Gezira. 1. The maintenance of life in the tuber. Ann. Botany 4, 177–193.

Auld, B.A., Hetherington, S.D., Smith, H.E., 2003. Advances in bioherbicide formulation. Weed Biol. Manag. 3, 61–67.

Bachie, O., McGiffen, M.E., 2013. Effects of summer cover cropping on weed population density and biomass in a subsequent broccoli crop. Int. Conf. Sustain. Environ. Agric. 57, 31–36.

Bailey, K.L., 2014. In: Abrol, D.P. (Ed.), The Bioherbicide Approach to Weed Control using Plant Pathogens, Integrated Pest Management: Current Concepts and Ecological Perspective. Elsevier, Academic Press, pp. 245–266.

Bangarwa, S.K., Norsworthy, J.K., Jha, P., Malik, M., 2008. Purple nutsedge (*Cyperus rotundus*) management in an organic production system. Weed Sci. 56, 606–613.

Barnes, J.P., Putnam, A.R., Burke, B.A., Aasen, A.J., 1987. Isolation and characterization of allelochemicals in rye herbage. Phytochemistry 26, 1385–1390.

Bell, R.S., Lachman, W.H., Rahn, E.M., Sweet, R.D., 1962. Life history studies as related to weed control in the Northeast. 1. Nutgrass. Rhode Isl. Agric. Exp. Stn. Bull. 364, 33.

Bendixen, L.E., 1973. Anatomy and sprouting of yellow nutsedge tubers. Weed Sci. 21, 501–503.

Bendixen, L.E., Nandihalli, U.B., 1987. Worldwide distribution of purple and yellow nutsedge (*Cyperus rotundus* and *C. esculentus*). Weed Technol. 1, 61–65.
Black, C.C., Chen, T.M., Brown, R.H., 1969. Biochemical basis for plant competition. Weed Sci. 17, 338–344.
Boardman, N.K., 1977. Comparative photosynthesis of sun and shade plants. Annu. Rev. Plant. Physiol. 28, 355–377.
Brecke, B.J., Hutto, K.C., Unruh, J.B., 2007. Sulfosulfuron: weed management and turfgrass tolerance. Proc. South. Weed Sci. Soc. 60, 114.
Bruckart, W.L., Johnson, D.R., Frank, J.R., 1988. Bentazon reduces rust-induced disease in yellow nutsedge, *Cyperus esculentus*. Weed Technol. 2, 299–303.
Bryson, C.T., Reddy, K.N., Molin, W.T., 2003. Purple nutsedge (*Cyperus rotundus*) population dynamics in narrow row transgenic cotton (*Gossypium hirsutum*) and soybean (*Glycine max*) rotation. Weed Technol. 17, 805–810.
Buker, R.S., Stall, W.M., Olson, S.M., Schilling, D.G., 2003. Season-long interference of yellow nutsedge (*Cyperus esculentus*) with direct-seeded and transplanted watermelon (*Citrullus lanatus*) 1. Weed Technol. 17, 751–754.
Burke, I.C., Troxler, S.C., Wilcut, J.W., Smith, W.D., 2008. Purple and yellow nutsedge (*Cyperus rotundus* and *C. esculentus*) response to postemergence herbicides in cotton. Weed Technol. 22, 615–621.
Bussan, A.J., Boerboom, C.M., Stoltenberg, D.E., 2001. Response of velvetleaf demographic processes to herbicide rate. Weed Sci. 49, 22–30.
Charudattan, R., 2001. Biological control of weeds by means of plant pathogens: significance for integrated weed management in modern agro-ecology. BioControl 46, 229–260.
Charudattan, R., Dinoor, A., 2000. Biological control of weeds using plant pathogens: accomplishments and limitations. Crop. Prot. 19, 691–695.
Cherry, M., 1973. Problems and progress in controlling nutgrass. SPAN. Agric. Rev. 16, 77–79.
Daramola, O.S., Adeyemi, O.R., Adigun, J.A., Adejuyigbe, C.O., 2020. Influence of row spacing and weed control methods on weed population dynamics in soybean (*Glycine max* L.). Int. J. Pest. Manag. 1–16.
Dev, I., Yaduraju, N.T., Ahuja, K.N., 1997. Efficacy of glyphosate with or without additives/herbicides for the control of purple nutsedge (*Cyperus rotundus* L.) in soybean. Indian J. Weed Sci. 29, 155–158.
Dieleman, J.A., Mortensen, D.A., Martin, A.R., 1999. Influence of velvetleaf (*Abutilon theophrasti*) and common sunflower (*Helianthus annuus*) density variation on weed management outcomes. Weed Sci. 81–89.
DiTomaso, J.M., 1995. Approaches for improving crop competitiveness through the manipulation of fertilization strategies. Weed Sci. 43, 491–497.
Dodet, M., Petit, R.J., Gasquez, J., 2008. Local spread of the invasive *Cyperus esculentus* (*Cyperaceae*) inferred using molecular genetic markers. Weed Res. 48, 19–27.
Drost, D.C., Doll, J.D., 1980. The allelopathic effect of yellow nutsedge (*Cyperus esculentus*) on corn (*Zea mays*) and soybeans (*Glycine max*). Weed Sci. 28, 229–233.
Durigan, J.C., 2005. Effects of plant densities and management of purple nutsedge on sugarcane yield and effect of growth stages and main way of herbicides contact and absorption on the control of tubers. J. Environ. Sci. Health, Part. B 40, 111–117.
Edenfield, M.W., Brecke, B.J., Colvin, D.L., Dusky, J.A., Shilling, D.G., 2005. Purple nutsedge (*Cyperus rotundus*) control with glyphosate in soybean and cotton. Weed Technol. 19, 947–953.
Eilenberg, J., Hajek, A., Lomer, C., 2001. Suggestions for unifying the terminology in biological control. BioControl 46, 387–400.

Felix, J., Newberry, G., 2012. Yellow nutsedge control and reduced tuber production with S-metolachlor, halosulfuron plus dicamba, and glyphosate in furrow-irrigated corn. Weed Technol. 26, 213–219.

Felix, J., Dauer, J.T., Hulting, A.G., Mallory-Smith, C., 2012. Yellow nutsedge (*Cyperus esculentus*) growth and tuber production in response to increasing glyphosate rates and selected adjuvants. Weed Technol. 26, 95–101.

Frick, K.E., Chandler, J.M., 1978. Augmenting the moth (*Bactra verutana*) in field plots for early-season suppression of purple nutsedge (*Cyperus rotundus*). Weed Sci. 26, 703–710.

Frick, K.E., Williams, R.D., Quimby Jr, P.C., Wilson, R.F., 1979. Comparative biocontrol of purple nutsedge (*Cyperus rotundus*) and yellow nutsedge (*C. esculentus*) with *Bactra verutana* under greenhouse conditions. Weed Sci. 27, 178–183.

Garg, D.K., Bendixen, L.E., Anderson, S.R., 1967. Rhizome differentiation in yellow nutsedge. Weeds 15, 124–128.

Gifford, E.M., Bayer, D.E., 1995. Developmental anatomy of *Cyperus esculentus* (yellow nutsedge). Int. J. Plant. Sci. 156, 622–629.

Gilreath, J.P., Santos, B.M., 2004. Efficacy of methyl bromide alternatives on purple nutsedge (*Cyperus rotundus*) control in tomato and pepper. Weed Technol. 18, 341–345.

Glaze, N.C., 1987. Cultural and mechanical manipulation of *Cyperus* spp. Weed Technol. 1, 82–83.

Grichar, W.J., Sestak, D.C., 2000. Effect of adjuvants on control of nutsedge (*Cyperus esculentus* and *C. rotundus*) by imazapic and imazethapyr. Crop. Prot. 19, 461–465.

Gupta, V.P., Kumar, V., Mishra, R.K., Thiagarajan, V., Datta, R.K., 2002. *Puccinia romagnoliana* Marie & Sacc.—a potential bioherbicide agent for biocontrol of purple nutsedge (*Cyperus rotundus* L.) in mulberry. J. Phytopathology 150, 263–270.

Hajjaj, B., El Oualkadi, A., 2019. Influence of tillage and straw mulching on purple nutsedge (*Cyperus rotundus* L.) in Olive Orchards. Int. J. Environment, Agric. Biotechnol. 4, 1732–1734.

Hauser, E.W., 1962. Development of purple nutsedge under field conditions. Weeds 10, 315.

Hershenhorn, J., Zion, B., Smirnov, E., Weissblum, A., Shamir, N., Dor, E., et al., 2015. *Cyperus rotundus* control using a mechanical digger and solar radiation. Weed Res. 55, 42–50.

Hill, E.R., Lachman, W.H., Maynard, D.N., 1963. Reproductive potential of yellow nutsedge by seed. Weeds 11, 160.

Holt, J.S., Orcutt, D.R., 1996. Temperature thresholds for bud sprouting in perennial weeds and seed germination in cotton. Weed Sci. 44, 523–533.

Horak, M.J., Holt, J.S., 1986. Isozyme variability and breeding systems in populations of yellow nutsedge. Weed Sci. 34, 538–543.

Horowitz, M., 1972. Growth, tuber formation, and spread of *Cyperus rotundus* L. from single tubers. Weed Res. 12, 348–363.

Horowitz, M., 1973. Competitive effects of *Cynodon dactylon*, *Sorghum halepense* and *Cyperus rotundus* on cotton and mustard. Exp. Agric. 9, 263–273.

Hubbard, L.R., Estes, A.G., McCarty, L.B., 2007. New options for purple nutsedge control in bermudagrass turf. Proc. South. Weed Sci. Soc. 60, 111.

Hutchinson, C.M., McGiffen, M.E., 2000. Cowpea cover crop mulch for weed control in desert pepper production. HortScience 35, 196–198.

Iqbal, J., Wright, D., 1997. Effects of nitrogen supply on competition between wheat and three annual weeds species. Weed Res. 37, 391–400.

Islam, M.N., Baltazar, A.M., De Datta, S.K., Karim, A.N.M.R., 2009. Management of purple nutsedge (*Cyperus rotundus* L.) tuber populations in rice-onion cropping systems. Philippine Agric. Scientist 92, 407–418.

Jansen, L.L., 1971. Morphology and photoperiodic responses of yellow nutsedge. Weed Sci. 19, 210–219.

Javaid, A., Bajwa, R., Rabbani, N., Anjum, T., 2007. Comparative tolerance of rice (*Oryza sativa* L.) genotypes to purple nutsedge (*Cyperus rotundus* L.) allelopathy. Allelopathy J. 20, 157–166.

Johnson III, W.C., Coble, H.D., 1986. Crop rotation and herbicide effects on the population dynamics of two annual grasses. Weed Sci. 34, 452–456.

Johnson III, W.C., Mullinix Jr., B.G., 2008. Potential weed management systems for organic peanut production. Peanut Sci. 35, 67–72.

Johnson III, W.C., Way, T.R., Beale, D.G., 2015. An undergraduate student project to improve mechanical control of perennial nutsedges with a peanut digger in organic crop production. Weed Technol. 29, 861–867.

Jordan-Molero, J.E., Stoller, E.W., 1978. Seasonal development of yellow and purple nutsedges (*Cyperus esculentus* and *C. rotundus*) in Illinois. Weed Sci. 26, 614–618.

Kadir, J., Charudattan, R., 2000. Dactylaria higginsii, a fungal bioherbicide agent for purple nutsedge (*Cyperus rotundus*). Biol. Control. 17, 113–124.

Kadir, J.B., Charudattan, R., Berger, R.D., 2000a. Effects of some epidemiological factors on levels of disease caused by *Dactylaria higginsii* on *Cyperus rotundus*. Weed Sci. 48, 61–68.

Kadir, J.B., Charudattan, R., Stall, W.M., Brecke, B.J., 2000b. Field efficacy of *Dactylaria higginsii* as a bioherbicide for the control of purple nutsedge (*Cyperus rotundus*) 1. Weed Technol. 14, 1–6.

Kanatas, P., Travlos, I., Papastylianou, P., Gazoulis, I., Kakabouki, I., Tsekoura, A., 2020. Yield, quality and weed control in soybean crop as affected by several cultural and weed management practices. Not. Botanicae Horti Agrobotanici Cluj-Napoca 48, 329–341.

Keeley, P.E., 1987. Interference and interaction of purple and yellow nutsedges (*Cyperus rotundus* and *C. esculentus*) with crops. Weed Technol. 1, 74–81.

Keeley, P.E., Thullen, R.J., 1975. Influence of yellow nutsedge competition on furrow-irrigated cotton. Weed Sci. 23, 171–175.

Keeley, P.E., Thullen, R.J., 1978. Light requirements of yellow nutsedge (*Cyperus esculentus*) and light interception by crops. Weed Sci. 26, 10–16.

Keeley, P.E., Thullen, R.J., Miller, J.H., 1970. Biological control studies on yellow nutsedge with *Bactra verutana* Zeller. Weed Sci. 18, 393–395.

Kumar, M., Das, T.K., Yaduraju, N.T., 2012. An integrated approach for management of *Cyperus rotundus* (purple nutsedge) in soybean-wheat cropping system. Crop. Prot. 33, 74–81.

Kunz, C., Sturm, D.J., Sökefeld, M., Gerhards, R., 2016. Weed suppression and early sugar beet development under different cover crop mulches. Plant. Prot. Sci. 52, 1–7.

Lapham, J., Drennan, D.S., 1990. The fate of yellow nutsedge (*Cyperus esculentus*) seed and seedlings in soil. Weed Sci. 38, 125–128.

Lati, R.N., Filin, S., Eizenberg, H., 2012. Effect of tuber density and trifloxysulfuron application timing on purple nutsedge (*Cyperus rotundus*) control. Weed Sci. 60, 494–500.

Leon, C.T., Shaw, D.R., Bruce, L.M., Watson, C., 2003. Effect of purple (*Cyperus rotundus*) and yellow nutsedge (*C. esculentus*) on growth and reflectance characteristics of cotton and soybean. Weed Sci. 51, 557–564.

Mahmood, A., Cheema, Z.A., 2004. Influence of sorghum mulch on purple nutsedge (*Cyperus rotundus* L.). Int. J. Agric. Biol. 6, 86–88.

Manley, B.S., Wilson, H.P., Hines, T.E., 2002. Management programs and crop rotations influence populations of annual grass weeds and yellow nutsedge. Weed Sci. 50, 112–119.

Marí, A.I., Pardo, G., Aibar, J., Cirujeda, A., 2020. Purple nutsedge (*Cyperus rotundus* L.) control with biodegradable mulches and its effect on fresh pepper production. Sci. Horticulturae 263, 109–111.

McAvoy, T., Freeman, J.H., 2013. Yellow nutsedge (*Cyperus esculentus*) control with reduced rates of dimethyl disulfide in combination with totally impermeable film. Weed Technol. 27, 515–519.

Mirsky, S.B., Ryan, M.R., Teasdale, J.R., Curran, W.S., Reberg-Horton, C.S., Spargo, J.T., et al., 2013. Overcoming weed management challenges in cover crop–based organic rotational no-till soybean production in the eastern United States. Weed Technol. 27, 193–203.

Moffett, J.E., McCloskey, W.B., 1998. Effects of soil moisture and yellow nutsedge (*Cyperus esculentus*) density on cotton (*Gossypium hirsutum*). Weed Sci. 46, 231–237.

Monday, T.A., Foshee, W.G., Blythe, E.K., Wehtje, G.R., Gilliam, C.H., 2015. Yellow nutsedge (*Cyperus esculentus*) control and tomato response to application methods of drip-applied herbicides in polyethylene-mulched tomato. Weed Technol. 29, 625–632.

Morales-Payan, J.P., Santos, B.M., Stall, W.M., Bewick, T.A., 1998. Interference of purple nutsedge (*Cyperus rotundus*) population densities on bell pepper (*Capsicum annuum*) yield as influenced by nitrogen. Weed Technol. 12, 230–234.

Morales-Payan, J.P., Santos, B.M., Stall, W.M., Bewick, T.A., 2000. Influence of nitrogen fertilization on the competitive interactions of cilantro (*Coriandrum sativum*) and purple nutsedge (*Cyperus rotundus*). J. Herbs, Spices Medicinal Plants 6, 59–66.

Muñiz, B.F., Tames, R.S., 1982. Seasonal patterns of growth regulators in tubers of yellow nutsedge (*Cyperus esculentus*). Weed Sci. 30, 83–86.

Neeser, C., Aguero, R., Swanton, C.J., 1997. Survival and dormancy of purple nutsedge (*Cyperus rotundus*) tubers. Weed Sci. 45, 784–790.

Nelson, K.A., Smoot, R.L., 2010. Yellow nutsedge (*Cyperus esculentus*) interference in soybean. Weed Technol. 24, 39–43.

Nishimoto, R.K., 2001. Purple nutsedge tuber sprouting. Weed Biol. Manag. 1, 203–208.

Nishimoto, R.K., Yip, C.P., Sweet, R.D., 1978. Some factors influencing atrazine activity on yellow nutsedge (*Cyperus esculentus*). Weed Sci. 26, 421–425.

Norsworthy, J.K., Malik, M.S., Jha, P., Riley, M.B., 2007. Suppression of *Digitaria sanguinalis* and *Amaranthus palmeri* using autumn-sown glucosinolate-producing cover crops in organically grown bell pepper. Weed Res. 47, 425–432.

Okoli, C.A.N., Shilling, D.G., Smith, R.L., Bewick, T.A., 1996. Distribution and genotypes of purple nutsedge (*Cyperus rotundus*) and yellow nutsedge (*C. esculentus*). Weed Sci. Soc. Am. 36, 47.

Okoli, C.A.N., Shilling, D.G., Smith, R.L., Bewick, T.A., 1997. Genetic diversity in purple nutsedge (*Cyperus rotundus* L.) and yellow nutsedge (*Cyperus esculentus* L.). Bio. Control 8, 111–118.

Osipitan, O.A., Dille, J.A., Assefa, Y., Knezevic, S.Z., 2018. Cover crop for early season weed suppression in crops: systematic review and meta-analysis. Agron. J. 110, 2211–2221.

Patterson, D.T., 1982. Shading responses of purple and yellow nutsedges (*Cyperus rotundus* and *C. esculentus*). Weed Sci. 30, 25–30.

Patterson, M.G., Buchanan, G.A., Street, J.E., Crowley, R.H., 1980. Yellow nutsedge (*Cyperus esculentus*) competition with cotton (*Gossypium hirsutum*). Weed Sci. 28, 327–329.

Peerzada, A.M., 2017. Biology, agricultural impact, and management of *Cyperus rotundus* L.: the world's most tenacious weed. Acta Physiolog. Plant. 39, 270.

Pereira, W., Crabtree, G., William, R.D., 1987. Herbicide action on purple and yellow nutsedge (*Cyperus rotundus* and *C. esculentus*). Weed Technol. 1, 92–98.

Pereira, W., Crabtree, G., 1986. Absorption, translocation, and toxicity of glyphosate and oxyfluorfen in yellow nutsedge (*Cyperus esculentus*). Weed Sci. 34, 923–929.

Phatak, S.C., Sumner, D.R., Wells, H.D., Bell, D.K., Glaze, N.C., 1983. Biological control of yellow nutsedge with the indigenous rust fungus *Puccinia canaliculata*. Science 219, 1446–1447.

Price, A.J., Williams, J.P., Duzy, L.A., McElroy, J.S., Guertal, E.A., Li, S., 2018. Effects of integrated polyethylene and cover crop mulch, conservation tillage, and herbicide application on weed control, yield, and economic returns in watermelon. Weed Technol. 32, 623–632.

Rao, J.S., Nagarajan, M., 1962. Relationship between moisture levels and viability of nutgrass tubers. Madras Agric. J. 49, 120–123.

Reed, T.V., Boyd, N.S., Dittmar, P.J., 2016. Application timing influences purple nutsedge (*Cyperus rotundus*) and yellow nutsedge (*Cyperus esculentus*) susceptibility to EPTC and fomesafen. Weed Technol. 30, 743–750.

Riemens, M.M., van der Weide, R.Y., Runia, W.T., 2008. Nutsedge, Biology and Control of *Cyperus rotundus* and *Cyperus esculentus*, Review of a Literature Survey (No. 3250100200). PPO AGV.

Rodrigues, A.C., Estelita, M.E.M., 2009. Morphoanatomy of the stem in *Cyperaceae*. Acta Botan. Brasil. 23, 889–901.

Roozkhosh, M., Eslami, S.V., Jami Al-Ahmadi, M., 2017. Effect of plastic mulch and burial depth on purple nutsedge (*Cyperus rotundus*) emergence and growth. Arch. Agron. Soil. Sci. 63, 1454–1464.

Santos, B.M., Morales-Payan, J.P., Stall, W.M., Bewick, T.A., 1998. Influence of purple nutsedge (*Cyperus rotundus*) density and nitrogen rate on radish (*Raphanus sativus*) yield. Weed Sci. 46, 661–664.

Schippers, P., Borg, S.J.T., Van Groenendael, J.M., Habekotte, B., 1993. What makes *Cyperus esculentus* (yellow nutsedge) an invasive species? A spatial model approach. Proceedings Brighton Crop Protection Conference-Weeds. British Crop Production Council, Hampshire, UK, pp. 495–504.

Shamsi, S.R.A., Al-Ali, F.A., Hussain, S.M., 1978. Temperature and light requirements for sprouting of chilled and unchilled tubers of purple nutsedge, Cyperus rotundus. Physiol. Plant. 44, 193–196.

Singh, V.P., Dhyani, V.C., Singh, S.P., Kumar, A., Manalil, S., Chauhan, B.S., 2016. Effect of herbicides on weed management in dry-seeded rice sown under different tillage systems. Crop. Prot. 80, 118–126.

Siriwardana, G., Nishimoto, R.K., 1987. Propagules of purple nutsedge (*Cyperus rotundus*) in soil. Weed Technol. 1, 217–220.

Smith, E.V., Fick, G.L., 1937. Nutgrass eradication studies: I. Relation of the life history of nutgrass, *Cyperus rotundus* L., to possible methods of control. J. Am. Soc. Agron. 29, 1007–1013.

Stoller, E.W., 1973. Effect of minimum soil temperature on differential distribution of *Cyperus rotundus* and *C. esculentus* in the United States. Weed Res. 13, 209–217.

Stoller, E.W., 1981. Yellow nutsedge: a menace in the Corn Belt. U.S. Dept. Agric. Technical Bull. 1642.

Stoller, E.W., Nema, D.P., Bhan, V.M., 1972. Yellow nutsedge tuber germination and seedling development. Weed Sci. 20, 93–97.

Stoller, E.W., Sweet, R.D., 1987. Biology and life cycle of purple and yellow nutsedges (*Cyperus rotundus* and *C. esculentus*). Weed Technol. 1, 66–73.

Stoller, E.W., Wax, L.M., 1973. Yellow nutsedge shoot emergence and tuber longevity. Weed Sci. 21, 76–81.

Stoller, E.W., Weber, E.J., 1975. Differential cold tolerance, starch, sugar, protein, and lipid of yellow and purple nutsedge tubers. Plant. Physiol. 55, 859–863.

Stoller, E.W., Woolley, J.T., 1983. The effects of light and temperature on yellow nutsedge (*Cyperus esculentus*) basal-bulb formation. Weed Sci. 31, 148–152.

Sun, W.H., Nishimoto, R.K., 1997. Dormancy release of purple nutsedge tuber buds by a single thermal pulse. J. Am. Soc. Horticultural Sci. 122, 306–309.

Sun, W.H., Nishimoto, R.K., 1999. Thermoperiodicity in shoot elongation of purple nutsedge. J. Am. Soc. Horticultural Sci. 124, 140–144.

Suwunnamek, U., Parker, C., 1975. Control of *Cyperus rotundus* with glyphosate: the influence of ammonium sulphate and other additives. Weed Res. 15, 13–19.

Teasdale, J.R., 1996. Contribution of cover crops to weed management in sustainable agricultural systems. J. Prod. Agric. 9, 475–479.

Teasdale, J.R., Mohler, C.L., 2000. The quantitative relationship between weed emergence and the physical properties of mulches. Weed Sci. 48, 385–392.

Tehranchian, P., Norsworthy, J.K., Nandula, V., McElroy, S., Chen, S., Scott, R.C., 2015. First report of resistance to acetolactate-synthase-inhibiting herbicides in yellow nutsedge (*Cyperus esculentus*): confirmation and characterization. Pest. Manag. Sci. 71, 1274–1280.

Thangaraj, M., Rao, T.S., 1973. Effect of Gramoxone and nitrogenous carriers on nitrogen metabolism of nutgrass tubers. Madras Agric. J 60, 1745–1750.

Thomas, P.E.L., 1969. Effects of desiccation and temperature on survival of *Cyperus esculentus* tubers and *Cynodon dactylon* rhizomes. Weed Res. 9, 1–8.

Thullen, R.J., Keeley, P.E., 1979. Seed production and germination in *Cyperus esculentus* and *C. rotundus*. Weed Sci 27, 502–505.

Thullen, R.J., Keeley, P.E., 1980. Competition between yellow nutsedge (*Cyperus esculentus*) and Japanese millet (*Echinochloa crus-galli* var. frumentacea). Weed Sci. 28, 24–26.

Tumbleson, M.E., Kommedahl, T., 1961. Reproductive potential of *Cyperus esculentus* by tubers. Weeds 9, 646–653.

Tumbleson, M.E., Kommedahl, T., 1962. Factors affecting dormancy in tubers of Cyperus esculentus. Bot. Gazette 123, 186–190.

Tuor, F.A., Froud-Williams, R.J., 2002. Influence of nitrogen on competition between purple nutsedge, maize and soybean. Int. J. Pest. Manag. 48, 73–79.

Tursun, N., Datta, A., Budak, S., Kantarci, Z., Knezevic, S.Z., 2015. Row spacing impacts the critical period for weed control. Phytoparasitica 44, 139–149.

Tuti, M.D., Das, T.K., 2011. Sequential application of metribuzin on weed control, growth and yield of soybean (*Glycine max*). Indian J. Agron. 56, 57–61.

Ueki, K. 1969. Studies on the control of nutsedge: on the germination of tubers. Proceedings 2nd Asian-Pacific Weed Control Inter-change. p. 355–369.

Van Wychen, L., 2019. 2019 Survey of the most common and troublesome weeds in broadleaf crops, fruits & vegetables in the United States and Canada. Weed Science Society of America National Weed Survey Dataset. Available. Available from: https://wssa.net/wp-content/uploads/2019-Weed-Survey_broadleaf-crops.xlsx.

Van Wychen, L., 2020. 2020 Survey of the most common and troublesome weeds in grass crops, pasture, and turf in the United States and Canada. Weed Science Society of America National Weed Survey Dataset. Available. Available from: https://wssa.net/wp-content/uploads/2020-Weed-Survey_grass-crops.xlsx.

Vencill, W.K., Banks, P.A., 1994. Effects of tillage systems and weed management on weed populations in grain sorghum (*Sorghum bicolor*). Weed Sci. 42, 541–547.

Wang, G., McGiffen Jr, M.E., Ogbuchiekwe, E.J., Butler, L., 2009. Economic return of purple and yellow nutsedge management in vegetable production of southern California. Crop. Prot. 28, 319–326.

Webster, T.M., 2005. Mulch type affects growth and tuber production of yellow nutsedge (*Cyperus esculentus*) and purple nutsedge (*Cyperus rotundus*). Weed Sci. 53, 834–838.

Webster, T.M., Grey, T.L., Davis, J.W., Culpepper, A.S., 2008. Glyphosate hinders purple nutsedge (*Cyperus rotundus*) and yellow nutsedge (*Cyperus esculentus*) tuber production. Weed Sci. 56, 735–742.

Wilcut, J.W., York, A.C., Wehtje, G.R., 1994. The control and interaction of weeds in peanut (*Arachis hypogaea*). Rev. Weed Sci. 6, 177–205.

Wilen, C.A., Holt, J., McCloskey, W.B., 1996. Predicting yellow nutsedge (*Cyperus esculentus*) emergence using degree-day models. Weed Sci. 44, 821–829.

William, R.D., Warren, G.F., 1975. Competition between purple nutsedge and vegetables. Weed Sci. 23, 317–323.

Williams, R.D., 1978. Photoperiod effects on the reproductive biology of purple nutsedge (*Cyperus rotundus*). Weed Sci. 26, 539–542.

Wills, G.D., 1975. Effect of light and temperature on growth of purple nutsedge. Weed Sci 23, 93–96.

Wills, G.D., 1987. Description of purple and yellow nutsedge (*Cyperus rotundus* and *C. esculentus*). Weed Technol. 1, 2–9.

Yandoc, C.B., Rosskopf, E.N., Pitelli, R.L.C.M., Charudattan, R., 2006. Effect of selected pesticides on conidial germination and mycelial growth of *Dactylaria higginsii*, a potential bioherbicide for purple nutsedge (*Cyperus rotundus*). Weed Technol. 20, 255–260.

Chapter 9

Digitaria ciliaris, *Digitaria ischaemum*, and *Digitaria sanguinalis*

Eric A.L. Jones, Diego J. Contreras and Wesley J. Everman
Department of Crop and Soil Sciences, North Carolina State University, Raleigh, NC, United States

Introduction

The *Digitaria* genus belongs to the Poaceae family. *Digitaria* is derived from the Latin word "Digitus" meaning finger-like, referring to the seed heads that resemble a hand with fingers (Bryson and DeFelice, 2010; Mitich, 1988). *Digitaria* spp. were cultivated as food and forage crops around the world before being abandoned in favor of maize (*Zea mays* L.) and wheat (*Triticum aestivum* L.) (Mitich, 1988). However, the characteristics of volunteer domesticated *Digitaria* species now considered the genus weedy in the early 20th century (Mitich, 1988). The common weedy *Digitaria* species include *D. ciliaris* (Retz.) Koeler, *D. ischaemum* (Schreb.) Schreb. ex Muhl., and *D. sanguinalis* (L.) Scop. Each species exhibits distinct morphological features. Despite the phenotypic differences, the *Digitaria* species are often considered together, and management guidelines for ''crabgrass'' sometimes do not distinguish between them (Derr, 2003; McCarty et al., 2001; Peters and Dunn, 1971; Watschke et al., 1995). Thus unless there are distinct differences, these three species will be referred to as "*Digitaria* spp." throughout the chapter.

Digitaria spp. are ubiquitous weeds across disturbed ecosystems globally (Holm et al., 1979a,b; Kim et al., 2002). *Digitaria* spp. are difficult to control due to similar morphology and physiology of desirable plants. Management options of *Digitaria* spp. for a respective disturbed ecosystem are essentially the same worldwide.

Chauhen-BMPCWS-1632754. DOI: https://doi.org/10.1016/B978-0-12-822917-0.00014-8

Biology

Vegetative structures

Stems

Stems exhibit erect or basally decumbent growth; ranging in height between 10 and 100 cm (Bryson and DeFelice, 2010). Stems of *D. ciliaris* and *D. sanguinalis* exhibit pubescence, whereas stems of *D. ischaemum* are glabrous (Bryson and DeFelice, 2010) (Fig. 9.1).

Leaves

Leaves of *Digitaria* spp. range from flat to involute (Fig. 9.2). *Digitaria ciliaris* leaves are prominently glabrous but can exhibit pubescence on the basal of leaves (Bryson and DeFelice, 2010). *Digitaria ischaemum* leaves are glabrous, whereas *D. sanguinalis* leaves are basally and dorsally pubescent (Bryson and DeFelice, 2010). *Digitaria ciliaris* exhibits a membranous ligule (1.0–3.0 mm) with a truncated, frayed margin (Bryson and DeFelice, 2010). *Digitaria ischaemum* exhibits a membranous ligule (1.0–2.0 mm) with even margins with sparse pubescence at the base of the leaf sheath (Bryson and DeFelice, 2010). *Digitaria sanguinalis* exhibits a membranous ligule (1.0–2.0 mm) with uneven teeth or margins (Bryson and DeFelice, 2010).

Roots

A fibrous root system is common to all *Digitaria* spp.

FIGURE 9.1 *Digitaria* spp. stems. Left to right: *D. ischaemum*, *D. sanguinalis*. *Photographs courtesy of Dr. Bridget Lassiter.*

FIGURE 9.2 *Digitaria* spp. leaves. Left to right: *D. ischaemum*, *D. sanguinalis*. *Photographs courtesy of Dr. Bridget Lassiter.*

Reproductive structures

Inflorescences

Digitaria spp. exhibit 2–9 finger-like inflorescence branches (Fig. 9.3). The length of inflorescence branches of *D. ciliaris* branches range from 3 to 21 cm, whereas *D. ischaemum* branches measure 10 cm and *D. sanguinalis* range from 4 to 18 cm, respectively (Bryson and DeFelice, 2010).

Seeds

Seeds exhibit similar morphology across *Digitaria* spp. Fruits are caryopsis and encompass the entire spikelet with glumes attached. At maturity, only the spikelet detaches from the inflorescence of *D. ciliaris and D. sanguinalis*,

FIGURE 9.3 *Digitaria sanguinalis* seedhead. The seedhead morphology is uniform across *Digitaria* spp. *Photograph courtesy of Dr. Bridget Lassiter.*

whereas bracts detach with the spikelet from the inflorescence of *D. ischaemum* (Bryson and DeFelice, 2010).

Germination and growth

The weedy *Digitaria* spp. are summer annuals; thus propagation occurs via seeds (Bryson and Defelice, 2010; Mitich, 1988). However, in the tropical regions where *Digitaria* spp. colonize, the annual species can begin to exhibit growth of a perennial species due to extended durations of favorable growing conditions (Mitich, 1988). *Digitaria* spp. germinate in the late spring to late summer when the air temperature is between 20°C and 30°C (Masin et al., 2006; Cardina et al., 2011; Oreja et al., 2019). Additionally, *Digitaria* spp. seed dormancy can be terminated by differential light absorption through a phytochrome-mediated system (Kobayashi et al., 2003; Oreja et al., 2017; Peters and Dunn, 1971). Seeds will exhibit extended germination throughout a growing season until freezing temperatures occur (Cardina et al., 2011; Gallart et al., 2010; Mohler and Callaway, 1995). Extant seeds on the soil surface will remain viable until conditions favor germination again (Kobayashi and Oyanagi, 2005). Nevertheless, the number of viable seeds carried over into the next year on the soil surface can be reduced by 10% of the initial number of seeds and viability significantly reduced (>50%) if the seeds remain in the soil for over a year (Kobayashi and

Oyanagi, 2005; Masin et al., 2006). The seed viability is virtually depleted if the seeds remain in the soil for 3 years (Masin et al., 2006). *Digitaria* spp. seedlings primarily emerge from depths of 3–4 cm; but emergence may be observed from depths as great as 6–8 cm (Chauhan and Johnson, 2008; Kobayashi and Oyanagi, 2005; Peters and Dunn, 1971). Seedlings that emerge later in season, even by a difference of several days, will result in less accumulated vegetative growth (Gallart et al., 2010).

Since *Digitaria* spp. are C_4 plants, vegetative growth will be dictated by light and temperature (Pearcy and Ehleringer, 1984). Thus plants emerging or growing in shade will result in smaller stature than plants going in full light (Shetty et al., 1982). Since *Digitaria* spp. exhibits decumbent/prostrate growth, additional vegetative size will occur close to the soil surface rather than vertically. *Digitaria* spp. vegetative size is plastic and remains the same when subjected to interspecific and/or intraspecific competition (Norsworthy et al., 2008; Souza et al., 2012). Once *Digitaria* spp. plants have grown to the fourth leaf stage, further vegetative growth is achieved by growing additional tillers (Peters and Dunn, 1971).

Reproduction

Digitaria spp. are determinate plants, meaning vegetative growth ceases when reproductive growth is initiated (Peters and Dunn, 1971). Flowering is day length-dependent and begins when the photoperiod is approximately 10–14 h (Peters and Dunn, 1971). *Digitaria* spp. are monoecious, thus plants primarily self-pollinate but cross-pollination from neighboring plants can occur. *Digitaria* spp. can also hybridize, which can result in trait introgression and incomplete dominance phenotypes (Ebinger, 1962). *Digitaria* spp. in lieu of interspecific and intraspecific competition can produce up to 150,000 seeds per plant (Mitich, 1988). When subjected to interspecific and intraspecific competition, *Digitaria* spp. can produce up to approximately 10,000 seeds per plant (Mohler and Callaway, 1995). *Digitaria* spp. can be more fecund if the plant is growing in a tropical region (Mitich, 1988). Similar to a reduction of vegetative growth in later emerging *Digitaria* spp., later emerging plants will incur a reduction of fecundity (Gallart et al., 2010; Mohler and Callaway, 1995).

Distribution

The *Digitaria* genus is native to Europe and is now distributed throughout all continents, excluding Antarctica (Bryson and DeFelice, 2010; Clayton et al., 2014; Holm et al., 1979b; Mitich, 1988). *Digitaria* spp. are ruderal and inhabit most disturbed areas. Such disturbed areas include: cultivated fields, no-till fields, pastures, lawns, turf, roadsides, railroads, and waste sites

(Bryson and DeFelice, 2010; Holm et al., 1977a,b; Lepschi and Macfarlane, 1997; Mitich, 1988; Peters and Dunn, 1971).

Management

Herbicide resistance

Digitaria spp. have evolved resistance to the acetyl-CoA-carboxylase-, acetolactate synthase-, 5-enolpyruvylshikimate-3-phosphate synthase- and photosystem II-inhibiting herbicides. Currently, 19 unique cases of herbicide-resistant *Digitaria* spp. have been reported in 12 countries (Heap, 2020). These herbicide-resistant *Digitaria* spp. populations inhabit environments of agronomic crops, horticultural crops, rangeland, and turf (Heap, 2020).

Agronomic crops

Digitaria spp. inhabit fields of prominent agronomic crops globally (Bryson and DeFelice, 2010; Chauhan and Johnson, 2008; Galon et al., 2018; Heap, 2020; Holm et al., 1979a; Mitich, 1988). *Digitaria* spp. not controlled can incur yield losses ranging from 5% to 100% in agronomic crops (Basinger et al., 2019a; Byrd and Coble, 1991; Galinato et al., 1999; Hugo et al., 2014; Smith et al., 1990). The critical period of *Digitaria* spp. control in agronomic crops ranges between 2 and 7 weeks after crop emergence before an unacceptable yield loss is incurred (Buchanan and Burns, 1970; Contreras, 2020; Everman et al., 2008; Eyherabide and Cendoya, 2002; Gantoli et al., 2013). *Digitaria* spp. is mainly controlled by herbicides despite the presence of herbicide-resistant populations in agronomic crops (Heap, 2020; Mahajan and Chauhan, 2013; Owen, 2016). *Digitaria* spp. can be controlled with herbicides at preplant burndown, preemergence, and postemergence application timings in most agronomic crops (Table 9.1).

Preplant burndown herbicide applications will usually consist of a nonselective herbicide, such as glyphosate or paraquat, to control the *Digitaria* spp. and other weed species present. Expected control of *Digitaria* spp. should be >85% with preplant burndown herbicide applications (Vollmer et al., 2019). Preplant herbicide applications are utilized where reduced- or no-till fields have been implemented to facilitate planting (Davis, 2010). Preplant herbicide applications can also terminate cover crops which will further facilitate planting (Davis, 2010). Herbicides with residual activity tank-mixed with nonselective herbicides can increase the duration of control of *Digitaria* spp. that have not germinated in case planting or subsequent herbicide applications are delayed (Jhala et al., 2013). Controlling *Digitaria* spp. prior to crop establishment or planting is crucial as plants growing without crop/desirable plant interference will grow more vigorously which could result in harder to control plants (Shetty et al., 1982).

TABLE 9.1 Herbicides that can be applied to agronomic crops to control *Digitaria* spp.

Site of action	WSSA group	Herbicide	Timing	Crop							
				Forage grass	Corn	Sorghum	Forage legumes	Soybeans	Rice	Cotton	Peanuts
ACCase	1	Clethodim	POST				x	x		x	x
	1	Fenoxaprop-P-ethyl	POST						x		
	1	Fluazifop-P-butyl	POST				x	x		x	x
	1	Quizalofop-P-ethyl	POST				x	x		x	x
	1	Sethoxydim	POST				x	x		x	x
ALS	2	Chlorimuron ethyl	POST				x	x			x
	2	Imazamox	POST				x	x			
	2	Imazapic	PRE								x
	2	Rimsulfuron	POST	x	x	x					
Microtubule	3	Pendimethalin	PRE				x	x		x	x
	3	Trifluralin	PRE				x	x		x	
Synthetic auxins	4	Quinclorac	POST	x		x			x		

(*Continued*)

TABLE 9.1 (Continued)

Site of action	WSSA group	Herbicide	Timing	Crop							
				Forage grass	Corn	Sorghum	Forage legumes	Soybeans	Rice	Cotton	Peanuts
PSII	5	Ametryn	POST-directed	x[b]	x[b]	x[b]					
	5	Atrazine	PRE/POST	x	x	x					
	5	Simazine	PRE	x	x	x					
	7	Diuron	PRE/POST-directed	x	x	x				x	
	7	Linuron	PRE/POST-directed	x[b]	x[b]	x[b]	x	x		x	
	7	Propanil	PRE						x		
EPSP	9	Glyphosate	POST	x[a]	x[a]	x[a]	x[a]	x[a]		x[a]	
Glutamine synthetase	10	Glufosinate	POST	x[a]	x[a]	x[a]	x[a]	x[a]		x[a]	
DBS	13	Clomazone	PRE				x	x	x	x	
VLCFA	15	Acetochlor	PRE	x	x		x	x		x	x
	15	Alachlor	PRE	x	x	x	x	x			

	15	Dimethenamid-P	PRE	x	x	x	x	x		x[a]	
	15	Flufenacet	PRE	x	x		x	x			
	15	Pyroxasulfone	PRE	x	x		x	x		x[a]	x[a]
	15	*S*-metolachlor	PRE	x	x	x	x	x		x	x
HPPD	27	Isoxaflutole	PRE/ POST	x	x						
	27	Mesotrione	PRE/ POST	x	x	x					
	27	Pyrasulfotole	PRE/ POST	x		x					
	27	Tembotrione	POST	x	x						
	27	Topramezone	POST	x	x						

WSSA, Weed Science Society of America; *ACCase*, acetyl-CoA-carboxylase; *ALS*, acetolactate synthase; *PSII*, photosystem II; *ESPS*, 5-enolpyruvyl-shikimate-3-phospate; *DBS*, diterpene biosynthesis; *PPO*, protoporphyrinogen oxidase; *VLCFA*, very long-chain fatty acid; *HPPD*, 4-hydroxyohenylpyruvate dioxygenase; *PRE*, preemergence; *POST*, postemergence.

[a]*Glyphosate- or glufosinate-resistant crop.*

[b]*POST-directed herbicide application.*

Preemergence herbicide applications control *Digitaria* spp. at the germination stage before the seedling emerges from the soil surface. Expected control of *Digitaria* spp. should be ≤85% with preemergence herbicide applications (Bhowmik and Bingham, 1996; Everman et al., 2009). Some farmers will recurrently apply preemergence herbicides during the growing season to control later germinating *Digitaria* spp. (Osterholt et al., 2017; Riar et al., 2013). While later germinating *Digitaria* spp. may not reduce crop yield, plants could produce seeds that will result in plants that have to be managed in following years (Everman et al., 2009).

Postemergence herbicide applications control *Digitaria* spp. when the plant has emerged from the soil surface. Expected control of *Digitaria* spp. should be ≥85% with postemergence herbicide applications (Burke et al., 2004; Hartzler and Foy, 1983; Parker et al., 2006). Postemergence herbicides applications should be executed before *Digitaria* spp. plants exceed 7 cm in height (2–5 leaf) to ensure control. Since herbicide-resistant *Digitaria* spp. inhabit agronomic crop fields, farmers should steward herbicides by means of rotating herbicide mode of action, tank-mixes, timely applications, and nonchemical control to increase the longevity of effective herbicides (Beckie, 2006).

Despite the intensive use of chemical control, biological, cultural, and mechanical management options are efficacious on *Digitaria* spp. and increased control can be realized if the different tactics are integrated (Liebman et al., 1997; Oreja et al., 2019; Sanyal et al., 2008). Granivores and seed decay comprise the primary means of biological control of *Digitaria* spp. (Lundgren, 2005; Mott and McKeon, 1997). Cultural practices will control and suppress infestations of *Digitaria* spp. (Masin et al., 2006). Such cultural practices include competitive crops/crop variety, crop rotation, fertilization, planting date, row spacing, and seeding rate (Bhowmik, 1997; Nalewaja, 1999; Sanyal et al., 2008). Mechanical practices that are utilized to manage *Digitaria* spp. include hand weeding, interrow cultivation, and preplant tillage. While biological, cultural, and mechanical practices will not be as efficacious as herbicides alone, integrating several practices together will result in high efficacy (Bhowmik, 1997; Liebman et al., 1997; Nalewaja, 1999).

Any input that increases crop vigor will contribute to cultural weed control. Crop rotation manages weeds by utilizing differences in the morphology, physiology, and production practices of the crop grown and to allow rotation of the herbicides used in the crop rotation sequence (Beckie, 2006). A simplistic example of crop rotation aiding in chemical weed control is *Digitaria* spp. will be less difficult to control in broad-leaf crops compared to monocot crops due to differential herbicide selectivity (Burke et al., 2004; Hartzler and Foy, 1983). In a more traditional sense, a crop with tall stature and faster canopy development will inhibit germination and growth of *Digitaria* spp. Early planting date, narrow row spacing, and increased crop seeding rate can result in faster canopy closure (Gill and Holmes, 1997; Nalewaja, 1999). Once canopy closure occurs less light will reach the soil surface, reducing *Digitaria* spp. germination, leading to less crop competition and seed inputs into the soil. Moreover, any delay in germination of *Digitaria* spp. will reduce vegetative growth and fecundity (Gallart et al., 2010).

Tillage and cultivation are efficacious when implemented when *Digitaria* spp. plants are small (less than 7 cm). While tillage is efficacious on *Digitaria* spp., reduced or no-till fields are beginning to become more common globally to reduce erosion and pollution (Gebhart et al., 1985; Holland, 2004; Teasdale et al., 1991). Reduced tillage may increase *Digitaria* spp. abundance/fecundity by leaving seeds on top of the soil, which could cause a reversion back to applying herbicides or implementing cultivation for control (Melander et al., 2013; Mohler and Callaway, 1995). Although no-till can increase the abundance and fecundity of *Digitaria* spp., the absence of tillage allows seeds to remain on the surface to be subjected to granivory (Franke et al., 2009; Lundgren, 2005). Cover crops, surface residue, and crop canopy can reduce and delay the germination of *Digitaria* spp.; the germination behavior of *Digitaria* spp. incurred by these treatments can be heritable to the next growing season's accession (Mohler and Callaway, 1995; Oreja et al., 2019). Cover crops will also promote the abundance of granivores that will consume *Digitaria* spp. seeds (Gallandt et al., 2005).

Horticultural crops

Digitaria spp. inhabit prominent horticulture crops globally (Bryson and DeFelice, 2010; Heap, 2020; Holm et al., 1979b; Mitich, 1988). *Digitaria* spp. not controlled can incur yield losses ranging from 5% to 100% in horticulture crops (Basinger et al., 2019b; Bertucci et al., 2018; Chaudhari et al., 2016; Monks and Schultheis, 1998). The critical period of *Digitaria* spp. control in horticultural crops ranges between 0 and 6 weeks after crop emergence or transplanting before an unacceptable yield loss is incurred (Bertucci et al., 2018; Chaudhari et al., 2016; Fu and Ashley, 2006; Monks and Schultheis, 1998).

Digitaria spp. can be controlled with herbicides at preplant, preemergence, and postemergence application timings in horticulture crops (Table 9.2). The efficacy of herbicide applications in horticultural crops will be similar to agronomic crops as previously described. While agronomic and horticultural crops share similar herbicide application timings, the number of registered herbicides in horticultural crops will be significantly less than the number of registered herbicides in agronomic crops (Fennimore and Doohan, 2008; Harr et al., 2002). Since there are limited herbicides registered in horticultural crops, herbicide-resistant *Digitaria* spp. is threatening for production (Wiederholt and Stoltenberg, 1996). While herbicides can be applied to control *Digitaria* spp. in horticultural crops, farmers may choose other means

TABLE 9.2 Herbicides that can be applied to horticulture crops to control *Digitaria* spp.

Site of action	WSSA group	Herbicide		Crop													
			Timing	Beans	Brassica	Cantaloupes	Carrots	Cucumbers	Garlic	Greens	Lettuce	Onions	Peppers	Potatoes	Sweet potatoes	Tomatoes	Watermelon
ACCase	1	Clethodim	POST	x	x	x	x	x	x	x	x	x	x	x	x	x	x
	1	Fluazifop	POST				x		x			x			x		
	1	Quizalofop−p−ethyl	POST	x													
	1	Sethoxydim	POST	x	x	x	x	x	x	x	x	x	x	x	x	x	x
ALS	2	Rimsulfuron	PRE/POST											x		x	
Microtubule	3	Benefin	PRE								x						
	3	DCPA	PRE		x	x		x[a]		x		x			x	x	x
	3	Ethalfluralin	PRE	x		x		x									x
	3	Pendimethalin	PRE				x		x			x	x	x		x	
	3	Pronamide	PRE								x						
	3	Trifluralin	PRE	x	x	x	x	x[a]		x			x			x	
PSII	5	Bensulfide	PRE		x	x	x	x	x	x	x	x	x			x	
	7	EPTC	PRE											x			x
	8	Linuron	PRE				x							x			
	8	Metribuzin	PRE				x							x		x	
DBS	13	Clomazone	PRE		x	x		x					x		x		
PPO	14	Oxyfluorfen	PRE										x			x	x
VLCFA	15	Dimethenamid-P	PRE	x					x			x					
	15	Napropamide	PRE		x								x		x	x	
	15	*S*-metolachlor	PRE	x										x		x	

WSSA, Weed Science Society of America; *ACCase*, Acetyl-CoA-carboxylase; *ALS*, acetolactate synthase; *PSII*, photosystem II; *DBS*, diterpene biosynthesis; *PPO*, protoporphyrinogen oxidase; *VLCFA*, very long-chain fatty acid; *PRE*, preemergence; *POST*, postemergence.
[a]*POST-directed herbicide application.*

of control as not to injure their crops which will result in reduced market value (Harr et al., 2002); therefore biological, cultural, and mechanical control tactics are more widely utilized to control *Digitaria* spp. infestations (Pannacci et al., 2017).

Similar to agronomic crops, the biological control of *Digitaria* spp. will be primarily by granivores and seed decay. Mechanical control of *Digitaria* spp. will largely be conducted by cultivation and hand weeding. While cultivation is efficacious on *Digitaria* spp., the cultivation must be implemented early as injury can be incurred by the crops exhibiting vineous and/or large vegetative growth (Monks and Schultheis, 1998; Pannacci et al., 2017). Subsequently, damage caused by cultivation can lead to reduced yield or reduced market value from cosmetic injury. Hand weeding is efficacious on *Digitaria* spp. but can be very laborious and expensive. Hand weeding can be conducted season long without any repercussions of crop damage or injury. While flaming is efficacious on many broad-leaved weeds, the tactic is not efficacious on *Digitaria* spp. as the apical meristem is protected within the leaf sheath (Cisneros and Zandstra, 2008).

Since horticultural crops do not generally produce a canopy closure due to morphology and wide row spacing; shading is artificially produced via plastic mulch to inhibit weed germination and growth. Plant residue is not favorable for inhibiting *Digitaria* spp. germination in horticultural crops as the residue can harbor insects and pathogens that can inflict plant injury, damage, or death (Alabouvette et al., 2006). Instead, plastic mulch can be laid down within the field rows to inhibit weed germination and interference with the crops. Plastic mulch will also heat the soil beneath which may reach temperatures high enough to inhibit seed germination (Horowitz et al., 1983). Crops can be planted/transplanted directly into the plastic mulch; however, *Digitaria* spp. can also utilize the openings in the plastic mulch to emerge and interfere with the crop (Norsworthy et al., 2008). Thus *Digitaria* spp. can be managed even where plastic mulch is utilized.

Rangeland and pastures

Digitaria spp., although historically grown in pastures as forage, are now considered weedy inhabitants of rangelands and pastures (Bryson and DeFelice, 2010; Heap, 2020; Holm et al., 1979b; Mitich, 1988). Control of *Digitaria* spp. in grass rangelands and pastures can be difficult as the methods to control the weedy grass may also impact the desirable grass species. Since *Digitaria* spp. can tolerate hot, dry conditions and possesses similar nutritional qualities as other forage grasses, some ranchers may allow their livestock to graze or include the weedy grasses in the feed bales (Mitich, 1998; Popay and Field, 1996). Furthermore, if the grazing is done selectively in areas of dense infestations of *Digitaria* spp., the population can be reduced significantly over several years (National Academy of Sciences,

1968; Popay and Field, 1996; Vitelli and Pitt, 2006). Overgrazing should be avoided as the bare, disturbed soil is favorable for *Digitaria* spp. and other weedy species colonization (National Academy of Sciences, 1968; Popay and Field, 1996; Vitelli and Pitt, 2006).

However, many ranchers prefer rangelands and pastures absent of weedy plants, such as the *Digitaria* spp., because weedy infestations can inhibit the germination and growth of desirable forage species. Continuous ground cover provided by seeding a perennial forage can inhibit the germination of *Digitaria* spp. (National Academy of Sciences, 1968; Vitelli and Pitt, 2006). Since the establishment of the perennial grass is a slow process, selective preemergence herbicides and fertilizers can be applied to the immature grass to promote growth (Choudhary et al., 2017; Gill and Holmes, 1997; Sheley et al., 1997; Sund, 1964; Vitelli and Pitt, 2006). Some pastures and rangelands with dense infestations of *Digitaria* spp. may require additional herbicide applications and reduced cuttings until the perennial grass reaches maturity. Grasses that require vegetative propagation, such as *Cynodon dactylon*, can be transplanted into pastures at a denser rate to inhibit germination and growth of *Digitaria* spp. facilitated by shading (Walker et al., 1998; Shetty et al., 1982). Mowing and cutting height of the grass can be increased in pastures and rangelands for dense canopy that does not allow *Digitaria* spp. to germinate or grow.

Pastures planted in legumes can simplify the control of *Digitaria* spp. compared to pastures planted in forage grasses. Select herbicides can be applied to legume pastures to control emerged *Digitaria* spp. without injury to the forage legume (National Academy of Sciences, 1968; Vitelli and Pitt, 2006) (Table 9.1). Similar cultural and mechanical control tactics in grass pastures can be implemented in legume pastures. Planting perennial legumes, increased seed rate, fertilization, and timely cuttings (Choudhary et al., 2017; Gill and Holmes, 1996; National Academy of Sciences, 1968; Sheley et al., 1996). If these practices are integrated and executed at the time of sowing, *Digitaria* spp. will likely be effectively controlled by the time the legume is established. Legume-grass pasture mixtures follow the same cultural and mechanical tactics previously described; however, chemical control will be confounded as there are minimal selective herbicides that can be applied to the legume-grass mixture (Table 9.1). Herbicides with preemergence-only activity can be applied to established stands of the legume-grass mixture to control germinating *Digitaria* spp.

Turf and ornamentals

Digitaria spp. are common weeds of disturbed ecosystems grown for asethetic appeal including ornamentals and turfgrass (Bryson and DeFelice, 2010; Heap, 2020; Holm et al., 1979b; Mitich, 1988; Neal and Williams, 1998). Turf and ornamentals can suffer reduced esthetics and recreational value due to infestations of *Digitaria* spp. (Dobbs and Potter, 2014; National academy of Sciences, 1968). While herbicides can be applied to directly control *Digitaria* spp. in turf and ornamentals, some sites (i.e., golf courses, gardens, nursery plants, etc.) and adjacent sites cannot tolerate injury incurred by herbicides, thus managers of these sites may decide not to apply herbicides (Busey et al., 2003). Moreover, *Digitaria* spp. can be difficult to control in turf and ornamentals with herbicides due to the lack of selectivity.

The most common and widely utilized method of weed control in turf is mowing (Busey, 2003; Dobbs and Potter, 2014; National Academy of Sciences, 1968). However, due to the decumbent and prostrate growth exhibited by *Digitaria* spp., plants can survive short mowing treatments (Busey, 2003; Mitich, 1988). Additionally, with a protected growing point within the leaf sheath, *Digitaria* spp. can initiate new meristem growth after mowings (Busey, 2003). Short mowing heights can also increase the density of *Digitaria* spp. in turf due to the ease of colonization (Busey, 2003). Increasing the mowing height can reduce the density of *Digitaria* spp. infestation. While increasing mowing height does not directly control *Digitaria* spp., taller turfgrass can create a denser canopy that inhibits germination/growth and increased granivore prevalence (Busey, 2003; Dobbs and Potter, 2014). Furthermore, higher mowing height can promote a deep, robust root system of the turfgrass that reduces need for water and chemical inputs that can additionally improve *Digitaria* spp. control (Dobbs and Potter, 2014). Thus mowing height should be set between 4 and 8 cm to suppress *Digitaria* spp. (Busey, 2003; Dobbs and Potter, 2014; DeBels et al., 2012). Mowing can directly control larger, erect-growing *Digitaria* spp. by excising the seedhead to reduce the amount of seeds returning to the soil seedbank.

Controlling *Digitaria* spp. prior to turf establishment or planting is crucial as plants growing without turfgrass interference will grow more vigorously which could result in harder to control plants (Busey, 2003; Shetty et al., 1982). Chemical control of *Digitaria* spp. can be a problem in turf due to the lack of selective herbicides. Herbicides are commonly applied at preemergence and postemergence timings in turf (Table 9.3). Nonselective herbicides may be applied preplant if the site is being renovated. Herbicides with only preemergence activity can be applied for *Digitaria* spp. control to established turfgrass without incurring injury (Bhowmik and Bingham, 1990; Gannon et al., 2015). Postemergence herbicides have been identified which provide selective control of *Digitaria* spp. in certain turfgrass species (Table 9.3). Cultural practices that increase turf density and vigor decrease the colonization and persistence of *Digitaria* spp. compared to unmaintained turf sites (Busey, 2003; Dobbs and Potter, 2014; National Academy of Sciences, 1968). Such cultural practices include aeration, fertilizer, and increased seeding/sprigging rate and fertilizer inputs (Turner and Van Acker, 2014). Reinfestations of *Digitaria* spp. in herbicide- and nonherbicide-treated lawns is

TABLE 9.3 Herbicides that can be applied to ornamentals and turfgrass to control *Digitaria* spp.

Site of action	WSSA herbicide group	Active ingredient	Application timing	Turfgrass		Ornamentals
				Cool season	Warm season	
ACCase	1	Clethodim	POST			x
	1	Fenoxaprop	POST	x	x	x
	1	Fluazifop	POST			x
	1	Sethoxydim	POST	x	x	x
ALS	2	Imazaquin	PRE	x	x	x
Microtubule synthesis	3	Benefin	PRE	x[a]	x[a]	x
	3	Dithiopyr	PRE	x	x	x
	3	Oryzalin	PRE	x[a]	x[a]	
	3	Pendimethalin	PRE	x[a]	x[a]	x
	3	Prodiamine	PRE	x[a]	x[a]	x
	3	Trifluralin	PRE	x[a]	x[a]	x
Synthetic auxins	4	Quinclorac	POST	x	x	
PSII	5	Metribuzin	PRE		x	

	5	Simazine	PRE			x
	7	Siduron	PRE	x		
	8	Bensulfide	PRE	x	x	
PPO	14	Oxadiazon	PRE			x
	14	Flumioxazin	PRE			x
	14	Oxyfluorfen	PRE			x
VLCFA synthesis	15	Dimethenamid-P	PRE		x[a]	x
	15	*S*-metolachlor	PRE		x[a]	x
	15	Napropamide	PRE	x[a]	x[a]	x
Nucleic acid	17	MSMA	PRE	x	x	
DHP	18	Asulam	POST			x
Cell wall synthesis	20	Dichlobenil	POST			
HPPD	27	Mesotrione	PRE/POST	x	x	
	27	Topramezone	POST	x	x	
Cellulose biosynthesis	29	Indaziflam	PRE	x[a]	x[a]	x

WSSA, Weed Science Society of America; *ACCase*, Acetyl-CoA-carboxylase; *ALS*, acetolactate synthase; *PSII*, photosystem II; *PPO*, protoporphyrinogen oxidase; *VLCFA*, very long-chain fatty acid; *MSMA*, monosodium acid methanearsonate; *DHP*, dihydropteroate; *HPPD*, 4-hydroxyohenylpyruvate dioxygenase; *PRE*, preemergence; *POST*, postemergence.

[a]*Herbicide application on established turfgrass.*

due to noncontrolled neighboring plant seed rain and emigration of seeds, thus calling for recurrent herbicide applications, continuous maintenance, and/or mowings each year (Busey, 2003; Peters and Dunn, 1971).

While ornamental plants are grown for esthetics, weed-interference in the container or at the field nurseries can impede the growth and development of ornamental plants (Altland et al., 2003; Berchielli-Robertson et al., 1990; Neal, 1999). Traditionally, weed control in nursery production was conducted by manual removal but has recently proved to be expensive (Fennimore and Doohan, 2008; Mathers, 2003). A typical regime for *Digitaria* spp. control with herbicides in nursery plant production would include preplant burndown, preemergence, and postemergence applications; which is efficacious (Altland et al., 2003; Neal, 1999) (Table 9.3). The herbicides applied for preplant burndown are typically nonselective with no residual activity so the ornamental plants can be planted directly into the container or field nurseries without incurring any injury. *Digitaria* spp. are common in container-grown ornamental plants and are controlled with preemergence herbicides applied three to five times a year to ensure no germination occurred (Mathers, 2003; Neal, 1999).

Conclusions

Digitaria spp. are pervasive in most disturbed ecosystems globally and if left unmanaged can be detrimental to esthetics and yield of desirable plants. Understanding the biology and ecology of *Digitaria* spp. can exploit new tactics and facilitate the management of these weeds in the plethora of disturbed ecosystems. While pervasive and ubiquitous in disturbed ecosystems, *Digitaria* spp. can be managed in each respective ecosystem with biological, chemical, cultural, and mechanical tactics effectively. While nonchemical and chemical management options are efficacious on *Digitaria* spp. inhabiting various ecosystems, no management option should be relied upon solely to reduce selection pressure and maintain efficacy longevity (Bhowmik, 1996; Liebman et al., 1997).

References

Alabouvette, C., Olivain, C., Steinberg, C., 2006. Biological control of plant diseases: the European situation. Eur. J. Plant. Pathol. 114, 329–341.

Altland, J.E., Gilliam, C.H., Wehtje, G., 2003. Weed control in field nurseries. Horttechnology 13, 9–14.

Basinger, N.T., Jennings, K.M., Monks, D.W., Jordan, D.L., Everman, W.J., Hestir, E.L., et al., 2019a. Interspecific and intraspecific interference of Palmer amaranth (*Amaranthus palmeri*) and large crabgrass (*Digitaria sanguinalis*) in soybean. Weed Sci. 67, 649–656.

Basinger, N.T., Jennings, K.M., Monks, D.W., Jordan, D.L., Everman, W.J., Hestir, E.L., et al., 2019b. Interspecific and intraspecific interference of Palmer amaranth (*Amaranthus palmeri*) and large crabgrass (*Digitaria sanguinalis*) in sweetpotato. Weed Sci. 67, 426–432.

Beckie, H.J., 2006. Herbicide-resistant weeds: management tactics and practices. Weed Technol. 20, 793–814.

Berchielli-Robertson, D.L., Gilliam, C.H., Fare, D.C., 1990. Competitive effects of weeds on the growth of container-grown plants. Hortscience 25, 77–79.

Bertucci, M.B., Jennings, K.M., Monks, D.W., Schulthesis, J.R., Louws, F.J., Jordan, D.L., et al., 2018. Critical period for weed control in grafted and nongrafted watermelon grown in plasticulture. Weed Sci. 67, 221–228.

Bhowmik, P.C., 1996. Weed biology: importance for weed management. Weed Sci. 45, 349–356.

Bhowmik, P.C., Bingham, S.W., 1990. Preemergence activity of dinitroaniline herbicides used for weed control in cool-season turfgrasses. Weed Technol. 4, 387–393.

Bryson, C.T., DeFelice, M.S., 2010. Weeds of the Midwestern United States and Central Canada. GA University of Georgia Press, Athens.

Buchanan, G.A., Burns, E.R., 1970. Influence of weed competition on cotton. Weed Sci. 18, 149–154.

Burke, I.C., Price, A.J., Wilcut, J.W., Jordan, D.L., Culpepper, A.S., Treadaway-Ducar, J., 2004. Annual grass control in peanut (*Arachis hypogaea*) with clethodim and imazapic. Weed Technol. 18, 88–92.

Busey, P., 2003. Cultural management of weeds in turfgrass: a review. Crop. Sci. 43, 1899–1911.

Busey, P., Broschat, T.K., Johnston, D.L., 2003. Injury to landscape and vegetable plants by volatile turf herbicides. Horttechnology 13, 650–653.

Byrd, J.D., Coble, H.D., 1991. Interference of selected weeds in cotton (*Gossypium hirsutum*). Weed Technol. 5, 263–269.

Cardina, J., Herms, C.P., Herms, D.A., 2011. Phenological indicators for emergence of large and smooth crabgrass (*Digitaria sanguinalis* and *D. ischaemum*). Weed Technol. 25, 141–150.

Chaudhari, S., Jennings, K.M., Monks, D.W., Jordan, D.L., Gunter, C.C., McGowen, S.J., et al., 2016. Critical period for weed control in grafted and nongrafted fresh market tomato. Weed Sci. 64, 523–530.

Chauhan, B.S., Johnson, D.E., 2008. Germination ecology of southern crabgrass (*Digitaria ciliaris*) and India crabgrass (*Digitaria longiflora*): two important weeds of rice in tropics. Weed Sci. 56, 722–728.

Choudhary, S., Chopral, N.K., Chopral, N.K., Singh, M., Kumar, R., Kushwaha, M., 2017. Influence of nitrogen levels and weed management practices on yield and quality of forage pearl millet (*Pennisetum glaucum* L.). Indian J. Anim. Nutr. 34, 64–69.

Cisneros, J.J., Zandstra, B.H., 2008. Flame weeding effects on several weed species. Weed Technol. 22, 290–295.

Clayton, W.D., Govaerts, R., Harman, K.T., Williamson, H., Vorontsova, M., 2014. World checklist of Poaceae. Royal Botanic Gardens, Richmond, UKKew. Available from: http://apps.kew.org/wcsp/.

Contreras D.J. 2020. Critical period of grass weed control in grain sorghum (Sorghum bicolor) and wheat (*Triticum aestivum*) (MS thesis). North Carolina State University.

Davis, A.S., 2010. Cover-crop roller-crimper contributes to weed management in no-till soybean. Weed Sci. 58, 300–309.

DeBels, B.T., Griffith, S.E., Kreuser, W.C., Melby, E.S., Soldat, D.J., 2012. Evaluation of mowing height and fertilizer application rate on quality and weed abundance of five home lawn grasses. Weed Technol. 26, 826–831.

Derr, J., 2003. Crabgrass control. Grounds Maint. 38, 28–32.

Dobbs, E.K., Potter, D.A., 2014. Conservation Biological Control and Pest Performance in Lawn Turf: Does Mowing Height Matter? Environmental Management 53, 648–659.

Ebinger, J.E., 1962. Validity of the grass species *Digitaria adscendens*. Brittonia 14, 248–253.

Everman, W.J., Clewis, S.B., Thomas, W.E., Burke, I.C., Wilcut, J.W., 2008. Critical period of grass vs. broadleaf weed interference in peanut. Weed Technol. 22, 68–73.

Everman, W.J., Clewis, S.B., York, A.C., Wilcut, J.W., 2009. Weed control and yield with flumioxazin, fomesafen, and *S*-metolachlor systems for glufosinate-resistant cotton residual weed management. Weed Technol. 23, 391–397.

Eyherabide, J.J., Cendoya, M.G., 2002. Critical periods of weed control in soybean for full field and in-furrow interference. Weed Sci. 50, 162–166.

Fennimore, S.A., Doohan, D.J., 2008. The challenges of specialty crop weed control, future directions. Weed Technol. 22, 364–372.

Franke, A.C., Lotz, L.A.P., Van Der Burg, W.J., Van Overbeek, L., 2009. The role of arable weed seeds for agroecosystem functioning. Weed Res. 49, 131–141.

Fu, R., Ashley, R.A., 2006. Interference of large crabgrass (*Digitaria saguinalis*), redroot pigweed (*Amaranthus retroflexus*), and hairy galinsoga (*Galinsoga ciliata*) with bell pepper. Weed Sci. 54, 364–372.

Galinato, M.I., Moody, K., Piggin, C.M., 1999. Upland Rice Weeds of South and Southeast Asia. International Rice Research Institute, Makati City (Philippines), p. 156.

Gallandt, E.R., Molloy, T., Lynch, R.P., Drummond, F.A., 2005. Effect of cover-cropping systems on invertebrate seed predation. Weed Sci. 53, 69–76.

Gallart, M., Mas, M.T., Verdú, A.M.C., 2010. Demography of *Digitaria sanguinalis*: effect of the emergence time on survival, reproduction, and biomass. Weed Biol. Manag. 10, 132–140.

Galon, L., Bagnara, M.A.M., Gabiatti, R.L., Júnior, F.W.R., Basso, F.J.M., Nonemacher, F., et al., 2018. Interference periods of weeds infesting maize crop. J. Agric. Sci. 10. Available from: https://doi.org/10.5539/jas.v10n10p197.

Gannon, T.W., Jeffries, M.D., Brosnan, J.T., Breeden, G.K., Tucker, K.A., Henry, G.M., 2015. Preemergence herbicide efficacy for crabgrass (*Digitaria* spp.) control in common bermudagrass managed under different mowing heights. HortSci 50, 546–550.

Gantoli, G., Ayala, V.R., Gerhards, R., 2013. Determination of the critical period for weed control in corn. Weed Technol. 27, 63–71.

Gebhart, M.R., Daniel, T.C., Schweizer, E.E., Allmaras, R.R., 1985. Conservation tillage. Science 230, 625–630.

Gill, G.S., Holmes, J.E., 1997. Efficacy of cultural control methods for combating herbicide-resistant *Lolium rigidum*. Pestic. Sci. 51, 352–358.

Harr, M.J., Fennimore, S.A., McGiffen, M.E., Lanini, W.T., Bell, C.E., 2002. Evaluation of preemergence herbicides in vegetable crops. Horttechnology 12, 95–99.

Hartzler, R.G., Foy, C.L., 1983. Efficacy of three postemergence grass herbicides for soybeans. Weed Sci. 31, 557–561.

Heap, I. 2020. International survey of herbicide resistant weeds. www.weedscience.org/in.asp (Accessed August 10, 2020).

Holland, J.M., 2004. The environmental consequences of adopting conservation tillage in Europe: reviewing the evidence. Agric. Ecosyst. Env. 103, 1–25.

Holm, L.G., Plucknett, D.L., Pancho, J.V., Herberger, J.P., 1977a. The World's Worst Weeds: Distribution and Biology. The University Press of Hawaii, Honolulu, Hawaii.

Holm, L., Pancho, J.V., Herberger, J.P., Plucknett, D.L., 1979b. A geographical atlas of world weeds. John Wiley and Sons, New York, New York.

Horowitz, M., Regev, Y., Herzlinger, G., 1983. Solarization for weed control. Weed Sci. 31, 170–179.

Hugo, E., Morey, L., Saayman-Du Toit, A.E.J., Reinhardt, C.F., 2014. Critical periods of weed control for naked crabgrass (*Digitaria nuda*), a grass weed in corn in South Africa. Weed Sci. 62, 647–656.

Jhala, A.J., Ramirez, A.H.M., Singh, M., 2013. Tank Mixing Saflufenacil, Glufosinate, and Indaziflam Improved Burndown and Residual Weed Control. Weed Technol. 27, 422–429.

Kim, T., Neal, J.C., Ditomaso, J.M., Rossi, F.S., 2002. A survey of weed scientists' perceptions on the significance of Crabgrasses (Digitaria spp.) in the United States. Weed Technol 16, 239–242.

Kobayashi, H., Miura, S., Oyanagi, A., 2003. Effect of winter cover crop on germination and fate of *Digitaria ciliaris* seed in no-tillage soybean. Weed Sci. Technol. 48, 234–235.

Kobayashi, H., Oyanagi, A., 2005. *Digitaria ciliaris* seed banks in untilled and tilled soybean fields. Weed Biol. Manag. 5, 53–61.

Lepschi, B.J., Macfarlane, T.D., 1997. *Digitaria aequiglumis* (Poaceae), a new weed for Western Australia. Nuytsia 11, 425–427.

Liebman, M., Gallandt, E.R., Jackson, L.E., 1997. Many little hammers: ecological management of crop-weed interactions. Ecol. Agric. 1, 291–343.

Lundgren, J.G., 2005. Ground beetles as weed control agents: effects of farm management on granivory. Am. Entomol. 51, 224–226.

Mahajan, G., Chauhan, B.S., 2013. Herbicide options for weed control in dry-seeded aromatic rice in India. Weed Technol. 27, 682–689.

Masin, R., Zuin, M.C., Otto, S., Zanin, G., 2006. Seed longevity and dormancy of four summer annual grass weeds in turf. Weed Res. 46, 362–370.

Mathers, H.M., 2003. Novel methods of weed control in containers. Horttechnology 13, 28–34.

McCarty, L.B., Everest, J.W., Hall, D.W., Murphy, T.R., Yelverton, F., 2001. Color Atlas of Turfgrass Weeds. Ann Arbor Press, Chelsea, MI, pp. 38–42.

Melander, B., Munier-Jolain, N., Charles, R., Wirth, J., Schwarz, J., van der Weide, R., et al., 2013. European perspectives on the adoption of nonchemical weed management in reduced-tillage systems for arable crops. Weed Technol. 27, 231–240.

Mitich, L.W., 1988. Intriguing world of weeds: crabgrass. Weed Technol. 2, 114–115.

Mohler, C.L., Callaway, M.B., 1995. Effects of tillage and mulch on weed seed production and seed banks in sweet corn. J. Appl. Ecol. 32, 627–639.

Monks, D.W., Schultheis, J.R., 1998. Critical weed-free period for large crabgrass (*Digitaria sanguinalis*) in transplanted watermelon (*Citrullus lanatus*). Weed Sci. 46, 53–532.

Mott, J.J., McKeon, G.M., 1977. A note on the selection of seed types by harvester ants in northern Australia. Aust. J. Ecol. 2, 231–235.

Nalewaja, J., 1999. Cultural practices for weed resistance management. Weed Technol. 13, 643–646.

National Academy of Sciences, 1968. Principles of plant and animal pest control, Weed Control. Chapter 2. The Ecology of Weeds, Vol 2. National Academy of Sciences, Publication 1975.

Neal, J.C., Williams, A., 1998. Eastern North Carolina nursery weed scouting program—1997. Proc. Northeastern Weed Sci. Soc. 52, 134.

Neal, J., 1999. Weeds and you. Nurs. Mgt. Prod. 15 (60–62), 64–65.

Norsworthy, J.K., Oliveira, M.J., Jha, P., Malik, M., Buckelew, J.K., Jennings, K.M., et al., 2008. Palmer amaranth and large crabgrass growth with plasticulture-grown bell pepper. Weed Technol. 22, 296–302.

Oreja, F.H., Batilla, D., De, I.F.E., 2017. Role of seed environment and covering structures on large crabgrass germination. South. Afr. J. Botany 111, 170–175.

Oreja, F.H., Batilla, D., De La Fuente, E.B., 2019. Digitaria sanguinalis seed dormancy release and seedling emergence are affected by crop canopy and stubble. Weed Res. 60, 111–120. Available from: https://doi.org/10.1111/wre.12392.

Osterholt, M.J., Webster, E.P., Blouin, D.C., McKnight, B.M., 2017. Overlay of residual herbicides in rice for improved weed management. Weed Technol. 33, 426–430.

Owen, M.D.K., 2016. Diverse approaches to herbicide-resistant weed management. Weed Sci. 64, 570–584.

Pannacci, E., Lattanzi, B., Tei, F., 2017. Non-chemical weed management strategies in minor crops: a review. Crop. Prot. 96, 44–58.

Parker, R.G., York, A.C., Jordan, D.L., 2006. Weed control in glyphosate-resistant corn as affected by preemergence herbicide and timing of postemergence herbicide application. Weed Technol. 20, 564–570.

Pearcy, R.W., Ehleringer, J., 1984. Comparative ecophysiology of C3 and C4 plants. Plant. Cell Env. 7, 1–13.

Peters, R.A., Dunn, S., 1971. Life History Studies as Related to Weed Control in the Northeast; 6: Large and Small Crabgrass. Northeast Regional Weed Control Technical Committee (NE-42). Connecticut Agricultural Research Station, Storrs, CT.

Popay, I., Field, R., 1996. Grazing animals as weed control agents. Weed Technol. 10, 217–231.

Riar, D.S., Norsworthy, J.K., Steckel, L.E., Stephenson, D.O., Eubank, T.W., Bond, J., et al., 2013. Adoption of best management practices for herbicide-resistant weeds in midsouthern United States cotton, rice, and soybean. Weed Technol. 27, 788–797.

Sanyal, D., Prasanta, C., Bhowmik, P.C., Randy, L., Anderson, R.L., Shrestha, A., 2008. Revisiting the perspective and progress of integrated weed management. Weed Sci. 56, 161–167.

Sheley, R.L., Svejcar, T.J., Maxwell, B.D., 1996. A theoretical framework for developing successional weed management strategies on rangeland. Weed Technol. 10, 766–773.

Shetty, S.V.R., Sivakumar, M.V.K., Ram, S.A., 1982. Effect of shading on the growth of some common weeds of the semi-arid tropics. Agron. J. 74, 1023–1029.

Smith, B.S., Murray, D.S., Green, J.D., Wanyahaya, W.M., Weeks, D.L., 1990. Interference of three annual grasses with grain sorghum (*Sorghum bicolor*). Weed Technol. 4, 245–249.

Souza, R.C., Dias, A.C., Figueiredo, M.R.A., Obara, F.E.B., Christoffoleti, P.J., 2012. Growth of the crabgrass species *Digitaria ciliaris* and *Digitaria nuda*. Planta Daninha 30, 317–325.

Sund, K.A., 1964. Evaluation of atrazine, simazine, monuron, and diuron on ten Hawaiian sugarcane plantations. Weeds 12, 215–219.

Teasdale, J.R., Beste, C.E., Potts, W.E., 1991. Response of weeds to tillage and cover crop residue. Weed Sci. 39, 195–199.

Turner, F.A., Van Acker, R.C., 2014. Fertilizer application has no effect on large (*Digitaria sanguinalis*) or smooth (*Digitaria ischaemum*) crabgrass germination and emergence in residential turfgrass in a Northern climate. Weed Sci. 62, 145–157.

Vollmer, K.M., VanGessel, M.J., Johnson, Q.R., Scott, B.A., 2019. Preplant and residual herbicide application timings for weed control in no-till soybean. Weed Technol. 33, 166–172.

Vitelli, J.S., Pitt, J.L., 2006. Assessment of current weed control methods relevant to the management of the biodiversity of Australian rangelands. Rangel. J. 28, 37–46.

Walker, R., Wehtje, G., Richburg, J.S., 1998. Interference and control of large crabgrass (*Digitaria sanguinalis*) and southern sandbur (*Cenchrus echinatus*) in forage bermudagrass (*Cynodon dactylon*). Weed Technol. 12, 707–711.

Watschke, T.L., Dernoeden, P.H., Shetlar, D.J., 1995. Managing Turfgrass Pests. Lewis Publishers, Boca Raton, FL, pp. 4–6.

Wiederholt, R.J., Stoltenberg, D.E., 1996. Similar fitness between large crabgrass (*Digitaria sanguinalis*) accessions resistant or susceptible to acetyl-coenzyme A carboxylase inhibitors. Weed Technol. 10, 42–49.

Chapter 10

Echinochloa colona and *Echinochloa crus-galli*

Adusumilli Narayana Rao*

Consultant Scientist (Weed Science) and General Secretary, Asian-Pacific Weed Science Society (APWSS), Hyderabad, India

**Corresponding author. e-mail address: anraojaya1@gmail.com*

Introduction

The genus *Echinochloa* (Poaceae), commonly referred to as barnyard grass, includes about 50 annual and perennial species that are widespread in both tropical and temperate regions of the world in dry or water-flooded soils (Michael, 2001, 2003). Among *Echinochloa* species, *Echinochloa crus-galli* (L.) Beauv. (barnyard grass) (annual, $2n = 42, 48, 54$) is the most prevalent species (Yabuno, 1983), followed by *Echinochloa colona* (L.) Link (junglerice) (annual, $2n = 36, 48, 54, 72$) and were ranked as the world's third and fourth worst weeds, respectively (Holm et al., 1991) (Photo 10.1).

Echinochloa colona is an annual C_4 grass, which is native to India, emerged as a serious and persistent threat in 35 cropping systems in more than 60 countries (Masood et al., 2016). It is widespread in the tropics and subtropics between 30°N and 30°S and is the most commonly reported weed of rice (Kraehmer et al., 2016) in South and Southeast Asia (Moody, 1989). It is often abundant on fertile soils, streamsides, rangelands and pastures, and on seasonally flooded landscapes and prefers low altitudes (CABI, 2019b). *Echinochloa colona* is a major weed in many crops, including rice (*Oryza sativa* L.), maize (*Zea mays* L.), cotton (*Gossypium hirsutum* L.), mungbean (*Vigna radiate* (L.) R. Wilczek), sorghum (*Sorghum bicolor* (L.) Moench), sugarcane (*Saccharum officinarum* L.), peanut (*Arachis hypogaea* L.), and cassava (*Manihot esculenta* Crantz) (Holm et al., 1991; Shabbir et al., 2019). *Echinochloa colona* was reported as an important associated weed species in 24 countries in dry-seeded rice, 12 countries in wet-seeded rice, and also in all countries growing transplanted rice (Rao et al., 2007). *Echinochloa colona* is considered as one of the most difficult grass weed to control in summer and in vegetable crops (Masood et al., 2016).

Biology and Management of Problematic Crop Weed Species. DOI: https://doi.org/10.1016/B978-0-12-822917-0.00013-6

PHOTO 10.1 Echinochloa colona and Echinochloa crus-galli in rice field. *Photo: A.N. Rao.*

Echinochloa crus-galli is originally from Europe and India and ranges from latitude 50°N–40°S, in both temperate and tropical habitats as a serious weed problem in 42 countries and found in another 27 more (Holm et al., 1991). It is widespread in Europe, Asia, and Australia, although is scarce in Africa. *Echinochloa crus-galli* is the most reported worst weed in rice fields (Rao et al., 2007) and was listed as a weed in 36 other crops (e.g., maize, mungbean, soya bean (*Glycine max* (L.) Merr.), lucerne (*Medicago sativa* L.), fibrous flax (*Linum usitatissimum* L.), vegetables, root crops, orchards and vineyards) in tropical and temperate parts of the world (Holm et al., 1991; Heller and Adamczewski, 2010; Von Redwitz et al., 2016). It is a principal weed in many other crops and countries (Holm et al., 1991).

Taxonomy

Both species are annuals with a C_4 photosynthetic pathway and belong to: Domain: Eukaryota; Kingdom: Plantae; Phylum: Spermatophyta; Subphylum: Angiospermae; Class: Monocotyledonae; Order: Cyperales; Family: Poaceae; Genus: *Echinochloa*; Kingdom: Plantae; Subkingdom: Tracheobionta; Super division: Spermatophyta; Division: Magnoliophyta; Class: Liliopsida; Subclass: Commelinidae; Order: Cyperales; Family: Poaceae; and Genus: *Echinochloa* P. Beauv. As per the phylogenetic analysis, *E. colona* diverged between 2.65 and 3.18 million years ago from *E. crus-galli* (Lee et al., 2017).

Echinochloa colona is a hexaploid or tetraploid species (Yabuno, 1962). *Echinochloa colona* is a tufted, erect to spreading and semiprostrate, grows up to 60 cm tall or more with culms freely branching at base and lower nodes and possesses fibrous roots. Sheath glabrous, ligule absent or occasionally represented by a rim of very short hairs, leaf blades glabrous, elongate, 5–15 mm wide, light green. The inflorescence is a terminal panicle erect or nodding, green or purple tinged, 10–20 cm long. The seed is caryopsis, ovate, obtuse, usually 2.5–3.5 mm long, white to hyaline.

Echinochloa crus-galli is robust, tufted annual grass, erect to decumbent, often branched at base, and rooting at the nodes, 20–150 cm tall. The culms stout, cylindrical, glabrous, filled with white spongy pith. The leaf sheaths glabrous, 9–13 cm long. Blades merging into the sheath, linear, with a broad, rounded base and acute top; without ligule, rough margined, glabrous or at the base with a few long hairs, smooth or the upper surface minutely bristly. The inflorescence is an apical panicle of 5–40 spikelike racemes, all turned to one side, with shortly ascending branches, the crowded spikelets at one side only, initially erect, later often bent down, 5–21 cm long. The seed is caryopsis, ovoid to obovoid, flat on one side (dorsally) and rounded on the other, abruptly pointed, 1.5–2 mm long. Vast phenotypic variation within *E. crus-galli* makes identification difficult and may have implications for farmers (Pratley et al., 2008).

A powerful tool for discriminating these weeds species is molecular markers (Tabacchi et al., 2006), which have advantages over morphological characterization in better identification of *Echinochloa* taxa and can help to plan the appropriate control strategies by the farmer. The complete chloroplast genomes of *E. crus-galli* (KT983256) and *E. colona* (KT983255) were sequenced and compared (Lee et al., 2017). Two different *E. crus-galli* varieties (*E. crus-galli* var. crus-galli, *E. crus-galli* var. praticola) and *E. colona* were successfully distinguished by chloroplast molecular marker #1 (psbA) (Zhang et al., 2017a). The molecular technique PCR-RFLP was used to distinguish at an early stage the species of *E. crus-galli* and *E. colona* (Romano et al., 2018).

Adverse impact or losses caused

Echinochloa crus-galli and *E. colona* infested crops production is frequently significantly reduced and necessitating effective control measures (CABI, 2019a,b). In Australia alone, the annual control and yield loss cost of *Echinochloa* species was estimated at AU$14.6 million (Llewellyn et al., 2016). *Echinochloa colona* is a less vigorous competitor than *E. crus-galli*, but *E. colona* poses serious problems in rice systems because of its occurrence in greater density (Zhang and Watson, 1997a). Normally these two species occur as a component of weed flora associated with the weed community in agroecosystems and other ecosystems. Hence the losses and adverse effects vary with the other weeds associated with them and several biotic and abiotic factors. These two weeds may be altering successional processes and outcompete native vegetation. The reduction in leaf photosynthetic rate, root biomass, grain filling stage's root oxidation activity, and biomass at maturity was caused by canopy light transmission interference by these two weeds which resulted in losses in rice grain yield and quality (Zhang et al., 2017b). In addition to losses of crop yield and quality, and additional weed control costs, both species are alternative hosts to a wide range of crop pests, namely, diseases, insects, and nematodes (Holms et al., 1991; CABI, 2019a,b).

Losses caused by *E. crus-galli*

The ecophysiology of this weed is well adapted to different crops. *Echinochloa crus-galli* reduces (6%–97%) crop yields (CABI, 2019a) and causes forage crops to fail by removing up to 80% of the available soil nitrogen (Holm et al., 1991). *Echinochloa crus-galli* has been proven to reduce the yields of rice (Rao and Matsumoto, 2017), potatoes (vanGessel and Renner, 1990), snap beans (Rahn et al., 1968), common beans (Saberali and Mohammadi, 2019), maize (Ionescu and şarpe, 2001; Ghanizadeh et al., 2014), grain sorghum (Smith et al., 1990), soybean (Ionescu and şarpe, 2001; Song et al., 2017); sunflower (Ionescu and şarpe, 2001), tomato (Bhowmik and Reddy, 1988), sugarbeets, green peas, and melons (Roche and Muzik, 1964). *Echinochloa crus-galli* is more competitive in direct-seeded than in transplanted rice in Asia (Rao et al., 2007). The crops yield losses due to *E. crus-galli* varies with it's emergence time (Cowan et al., 1998), crop establishment methods (Chin, 2001), crop variety grown (Mennan et al., 2012; Zhang et al., 2017a), crop density (Ottis and Talbert, 2005), intra/inter row occurrence of weed (vanGessel and Renner, 1990), cultural practices undertaken (Singh et al., 2018), and associated environmental factors (Rao et al., 2007). In rice, the economic thresholds of *E. crus-galli* was 2.3 plants/m^2 for DSR (Kim et al., 2012) and were between 0.298 and 1.078 plants/m^2 for transplanted rice (Moon et al., 2010). The rice yield was directly related to the efficacy of controlling *E. crus-galli* at flowering ($R^2 \approx 0.60$) (Marchesi and Chauhan, 2019). Yield losses were 3100, 1500, and 2000 kg/ha in maize, sunflower, and soybean, respectively (Ionescu and şarpe, 2001). *Echinochloa crus-galli* permitted to compete for 6, 9, 12, and 25 weeks reduced cotton yields an average of 21, 59, 90, and 97%, respectively, and a weed-free period of 9 weeks was required to prevent cotton yield reduction (Keeley and Thullen, 1991). The leaf area index has a critical role in weed (*E. crus-galli*)—crop (maize) competition (Ghanizadeh et al., 2014).

Many of the growth characteristics of *E. crus-galli* are similar to those of rice, making it a strong competitor to rice in fields across the world. Competition with *Echinochloa* species, especially *E. colona* and *E. crus-galli*, occurs from 15 days after rice emergence to maturity (Smith Jr., 1983), with yield reduction up to 70% (Ottis and Talbert, 2005) to 80% (Smith Jr., 1983; Ottis and Talbert, 2005). Worldwide losses of yield in rice due to *E. crus-galli* competition are estimated to be about 35% (Oerke and Dehne, 2004). The yield loss of rice varied with cultivation method with greater yield losses in direct-seeded rice than in transplanted rice (Chin, 2001) due to greater (2–3 times) weed biomass in direct-seeded rice than that in manually transplanted rice. Crop yield loss was higher in the wet season than in the dry season (Ni et al., 2004). Fertilizer applications favor the weed over crop (Holm et al., 1991; Saberali and Mohammadi, 2019). *Echinochloa crus-galli* has a negative effect on nodulation and N_2-fixation of soyabeans (You and Kahnt, 1990). The deeper (15–20 cm soil depth) placement of phosphorous reduced the competitive ability of *E. crus-galli* over rice in P-deficient soils, irrespective of the soil moisture regime (Weerarathne et al., 2015).

Echinochloa crus-galli is also a host to many viruses of rice and other grass crops (Holm et al., 1991). It is a host of *Striga asiatica*, which infests sorghum, maize, millet, sugarcane, rice, and tobacco in several countries (King, 1966). *Echinochloa crus-galli* interferes with harvesting of row crops and increases labor costs as the crop must be separated from the weed clumps (Maun and Barrett, 1986). The tough *E. crus-galli* plants, often still green and growing throughout crop plant senescence, can cause hindrance to machinery in crops harvests and drag clods into the equipment. The weed seeds also become a contaminant of crop seed (Photo 10.2) and thus reduce the crop grain quality while getting themselves disseminated through crop seed (Rao and Moody, 1990).

Losses caused by *E. colona*

Echinochloa colona is able to germinate throughout the growing season in several upland crops whenever favorable conditions are present and competes with crops causing severe yield losses. *Echinochloa colona* has

PHOTO 10.2 Harvesting and threshing of *Echinochloa crus-galli* with rice is resulting in it's seed contamination of rice seed and dissemination through rice seed in certain countries. *Photo: A.N. Rao.*

globally invaded rice fields causing severe grain yield loss of 27%–62% (Rabbani et al., 2011; Rao and Matsumoto, 2017). The critical period of *E. colona* interference with rice (Giza 178) lies between 20 and 40 days after transplanting (DAT) and a weed-free period of 20–60 DAT was suggested to minimize the regrowth of *E. colona* and prevent significant rice yield losses (Sheble, 2004). The increasing nitrogen rate has favored *E. colona* biomass production and has increased the competitive ability of the *E. colona* over rice regardless of planting density of dry-seeded rice (Awan et al., 2014). The management of the dominant *E. colona* has increased the aerobic rice yield by 3.92–7.84 fold (Gunawardane et al., 2013) compared to the unweeded control. Heavy infestations of *E. colona* may remove 60%–80% of the nitrogen from the soil (Holm et al., 1991), in addition to considerable amounts of macronutrients (Maun and Barrett, 1986).

Echinochloa colona root, shoot, and full competition has caused bush bean (*Phaseolus vulgaris* cv. top crop) yield reduction of 48%, 37%, and 53%, respectively (Premalal et al., 1998). The mungbean interference caused a reduction in height (17%–19%), tiller numbers (69%–82%), total shoot biomass (85%–91%), inflorescence numbers (74%–91%), and seed production (85%–95%) of both glyphosate-resistant and susceptible *E. colona* indicating no fitness penalty associated with resistance (Mutti et al., 2019). Thus mungbean interference can reduce *E. colona* growth and seed production (Mutti et al., 2019). The development of improved management strategies to minimize the abundance and further spread of *E. colona* was suggested to restore and conserve wetland biodiversity (Andrew, 2018).

Ecology

The success of these two problematic *Echinochloa* species as weeds may be attributed to their rapid growth, prolific seed production, seed dormancy, and adaptability to a wide range of environments. The cultural weed management strategies need to be developed and used along with herbicides in integrated management to maintain the sustainability of agricultural systems. However, the design of such strategies requires an understanding of the differential responses of weeds to microenvironmental factors in agroecosystems. The comparative ecological aspects of both the species are summarized in this review (Table 10.1).

Echinochloa colona cooccurs with *E. crus-galli* in rice agroecosystems but tends to dominate in rainfed agriculture. *Echinochloa colona* and *E. crus-galli* are examples of "crop mimicry" (Barrett, 1983) as their seedlings closely resemble the rice seedlings at the seedling stage and are sometimes unintentionally transplanted into fields together with rice (Rao and Moody, 1987, 1992). By the time the weed can be easily recognized by farmers and removed, crop yield losses may already be inevitable (Holm et al., 1991; Rao and Moody, 1987, 1992). Manual weeding is difficult as weeds (e.g., *E.*

TABLE 10.1 Comparative ecological characteristics of *Echinochloa colona* and *E. crus-galli*.

Characteristic	*E. colona*	*E. crus-galli*
The growth habit	Could be prostrate, decumbent, or open (Tahir, 2016)	Could be decumbent to open (Tahir, 2016)
The height	– Ranged from 65 to 94 cm (Tahir, 2016); – Varied with seeding date and was 45, 75, and 65 cm when sown on 1 June, 1 July and 15 July, respectively (Dhawan and Malik, 2003)	– Ranged from 87 to 126 cm (Tahir, 2016) – Height was 110 cm with 1 June and 1 July sowings and 75 cm when sown on 15 July (Dhawan and Malik, 2003).
Biomass accumulation	27 g and 29 g/three plants with sowing on 1 June and 1 July, respectively (Dhawan and Malik, 2003).	53.9 g and 39 g/three plants with sowing on 1 June and 1 July, respectively (Dhawan and Malik, 2003).
Photoperiodism	Photoperiod insensitive (Tahir, 2016)	Can grow and flower in photoperiods ranging from 8 to 16 h but prefers the latter (Holm et al., 1991)
Flowering commencement	3–4 weeks after emergence (CABI, 2019b)	About 40 days after emergence (Maun and Barrett, 1986)
Number of seeds produced	– Produced 9098–217,217 seeds per plant with an average of 539 seeds per panicle (Tahir, 2016) – 42,000 viable seeds in the life cycle of over 4 months (Kim and Moody, 1989).	– Produced 7186–71,494 seed/plant with an average of 345 seeds per panicle (Tahir, 2016). – In the absence of competition, produced 100,000–700,000 seeds per plant (Norris, 2003). When emerged with the crop, produced 16,500–35,500 seeds/plant in cotton and 2900–39,000 and seeds/plant in rice (Bagavathiannan et al., 2012a).

(*Continued*)

TABLE 10.1 (Continued)

Characteristic	*E. colona*	*E. crus-galli*
		– Mean seed production ranged from 20,000/plant (emerged with crop) to 7300/plant (emerged 35 days after crop emergence) (Bagavathiannan et al., 2011).
1000 seed weight	4.5 g (Dhawan, 2009).	9.5 g (Dhawan, 2009); 14–27 g (Norris, 2003).
Seed viability	Seeds viable for less than 2 years (Dhawan, 2009)	Seeds viable for: >2 years (Dhawan, 2009); 6–8 years under dry storage (Maun and Barrett, 1986); <15 years in irrigated sandy loam soil (Dawson and Bruns, 1975).
Seed dormancy	Seeds may have a short postharvest dormancy but this rarely lasts beyond 2 months in dry storage (Sen, 1981)	Seed dormancy ranges from 0 to 48 months (WSSA, 2020); 3–4 months (Soerjani et al., 1986)
Period of germination	Showed germination from April to September and did not germinate from October to March in Punjab, India (Dhawan, 2009).	Same as in *E. colona* (Dhawan, 2009).
Seed burial depth influence on seed germination and seedling emergence	– Greatest germination on the soil surface (0 cm) (photoblastic) (Chun and Moody, 1982). – The emergence declined exponentially with increasing seed burial depth, and no seedlings emerged from seeds buried at 6 cm (Chauhan and Johnson, 2009).	– Greatest seeds germinatiion when seeds were placed on the soil surface (92%) with declining emergence as burial depth in soil increased (0.4 cm burial depth caused 50% decline) and no seedlings emerged from the depth of 6 cm (Mohammad et al., 2010), 8 cm (Chauhan and Johnson, 2011), or 15 cm

(Continued)

TABLE 10.1 (Continued)

Characteristic	*E. colona*	*E. crus-galli*
		(Abouziena et al., 2015).
Light and germination	– Light is required for germination (Holm et al., 1991, Chun and Moody, 1986). – No germination occurred in the dark. But when the seed coat was removed, germination percentage in the dark increased markedly. – Less germination observed at lower levels of daylight intensity. The germination of seeds exposed to light for 1 h each day was significantly lower than that of seeds exposed to light for 2 h or more each day. Once light requirement was satisfied, additional illumination did not increase germination (Chun and Moody, 1987)	– Higher germination percentages under the higher light intensity than under the lower light intensity (Ikeda et al., 2003). – Light was more effective for germination under the lower temperature (15°C, 20°C, and 25°C) than under the optimum temperature (30°C) (Ikeda et al., 2003). – Some germination in complete darkness and increased light exposure promoted the germination (Abouziena et al., 2015; Mohammad et al., 2015). – Short photoperiods promoted germination more. Increasing temperature from 25°C to 30°C, caused partial reduction in light requirement for germination (Mohammad et al., 2015)
Shade effect on growth	Shade reduces growth and seed production (Chauhan, 2013)	Susceptible to shading (Maun and Barrett, 1986)
Soil pH	– Seeds germinated at pH range (6.6–7.9) (Dhawan, 2009). – Germination was not significantly affected by the pH range from 3 · to 9 and was decreased at pH 10 (Chun and	– Seeds germinated over a wider pH range (5.6–10.0) (Dhawan, 2009) – A high percentage of seed germination at pH 5 and decreased to 61.5% at pH 4 and to

(*Continued*)

TABLE 10.1 (Continued)

Characteristic	*E. colona*	*E. crus-galli*
	Moody, 1982; Chauhan and Johnson, 2009)	11% at pH 9 (Sadeghloo et al., 2013). – The germination was 38% at pH 4, optimum germination was at pH 6, and was 0% at pH 9, indicating that its germination was more tolerant of acidic than basic solutions (Abouziena et al., 2015).
Salinity and germination, seedling growth	– The germination was: unaffected at 0.1% or less salt concentration; reduced in 0.5% and 1.0% salt solutions and entirely inhibited at 5.0% (Chun and Moody, 1982) – The salinity (as measured by electrical conductivity [EC]) that inhibited 50% shoot biomass of was 13 dS/m (Chauhan et al., 2013)	– The salinity levels for 50% inhibition of maximum germination was 307 mM (Sadeghloo et al., 2013). Salinity stress up to 250 mM had no effect on seed germination (Sadeghloo et al., 2013). – Germination was 100% at 50 mM NaCl, some at 250 mM NaCl (33.3%), and none at 300 mM NaCl (Abouziena et al., 2015). – 50% shoot biomass inhibition was at 15 dS/m (Chauhan et al., 2013)
Moisture level for germination	– A relatively high moisture level is required for germination; the critical moisture level was −5.4 bars water potential (Chun and Moody, 1985) – Tolerated (no effect on germination at) −3.4 bars, delayed, and decreased germination at −4.6 bars and no germination at −9.8 bars (Chun and Moody, 1982)	– Drought levels for 50% inhibition of maximum germination was -0.5 MPa. – Seed germination was completely inhibited by a potential of −9 MPa (Abouziena et al., 2015) and −1.0 MPa (Sadeghloo et al., 2013).

(*Continued*)

TABLE 10.1 (Continued)

Characteristic	*E. colona*	*E. crus-galli*
Water depth and seed germination	No germination in flooded soil (under a water depth of at least 15 mm) (Chun and Moody, 1985).	– No germination in water deeper >12 cm (Soerjani et al., 1986). – Flooding with 10 cm deep water immediately after seeding was effective in suppressing germination (Estioko et al., 2014).
Oxygen and germination	Tolerant of anaerobiosis and germinated in the absence of oxygen (Mujer et al., 1993).	Can germinate and grow for extended periods of time in an anaerobic environment (Kennedy et al., 1980).
Seedling growth and moisture stress	Affected at a moisture stress of −0.8 to −1.0 MPa (Dhawan, 2009)	Was not affected at a moisture stress of −0.8 to −1.0 MPa (Dhawan, 2009)
Flooding after germination and establishment	– Once established it is unaffected by flooding to a depth of 5 cm. Survives in both flooded and nonflooded rice fields but will die upon submergence (Civico and Moody, 1979) – With leaves that extended above the water surface at the start of flooding, it matured under flooding depths of 5 and 10 cm (Namuco and Piggin, 1999) – Flooding with 100-mm-deep water immediately after seeding caused greater reductions in emergence, shoot and root growth than *E. crus-galli* (Estioko et al., 2014)	– The *E. crus-galli* seedlings decreased by 90% with flooding to 1.3 cm (Smith and Fox, 1973) – Flooding with 100-mm-deep water immediately after seeding was effective in suppressing growth. Delaying flooding for 2 or 4 days was less injurious (Estioko et al., 2014) – Early and deep flooding significantly suppressed growth of *E. crus-galli* seedlings. In flooded conditions, with increased water depth, the weed allocated more biomass to shoots at the expense of roots (Chauhan and Johnson, 2011)

(*Continued*)

TABLE 10.1 (Continued)

Characteristic	*E. colona*	*E. crus-galli*
Soil moisture content and the establishment and growth	Promoted in moist soils but that excessive water was detrimental to its growth (Kapoor and Ramakrishnan, 1974)	The growth and seed production were higher in flooded conditions than in aerobic conditions when grown alone but not when grown with rice interference (Chauhan and Abugho, 2013a)
Drought effect on growth	Growth reduced when subjected to drought stress (Chun and Moody, 1985).	Drought stress reduced growth to different degrees depending on the biotype (Wu et al., 2019).
Soil temperature and germination	The optimum germination conditions: 30°C day/ 20°C night temperature cycle (Chauhan and Johnson, 2009)	– The optimal soil temperatures for germination are: 20°C–30°C (Soerjani et al., 1986); between 27°C and 31°C, and the range of adequate temperatures increased with seed age (Martinková et al., 2006). – The minimum, optimum and maximum germination temperatures were: 9.6°C, 31°C, and 45.5°C (Mohammad et al., 2015); 5°C, 38°C, and 45°C, respectively (Sadeghloo et al., 2013). – Alternating temperatures in a day, such as 25°C (8 h)/10°C (16 h) and 25°C (8 h)/5°C (16 h), were effective in promoting the germination in darkness (Ikeda et al., 2003).
Crop Residues on emergence and growth	Seedling emergence and weed biomass decreased with increasing residue amounts and most substantial reductions in emergence occurred with 6 t/ha of residue (Chauhan and Abugho, 2013b).	Seedling emergence and seedling biomass were reduced by the addition of high level (6 t/ha) of rice residue to the soil surface (Chauhan and Johnson, 2011).

crus-galli) and rice seedlings look similar (Rao and Moody, 1987, 1988), and the labor needed for hand weeding would be too high for the farmer to bear the cost.

Echinochloa colona reproduces primarily through seeds, which have a short dormancy period compared with other species in the same genus (Chauhan and Johnson, 2009) and is able to reproduce through nodes too (Masood et al., 2016). *Echinochloa colona* is characterized by a high relative growth rate together with a high dry matter partitioning into leaves, during seedling and juvenile stages which enhances the species competitive ability and may ensure a resource turnover from vegetative to reproductive structures (Hegazy et al., 2005). *Echinochloa colona* begins to produce flower buds early in its lifespan ensuring limited seed production even in short growing season years (Hegazy et al., 2005). The seed production data is important to assess weed management methods in totality (Norris, 2007; Rao et al., 2017) and for long-term weed populations predictions using simulation models (von Redwitz et al., 2016). Seed production for *E. crus-galli* highly variable (Table 10.1). *Echinochloa crus-galli*'s high capacity for seed production allows large populations to rapidly establish. *Echinochloa crus-galli* growth and seed production increased with increases in N rate from 0 to 100 and 200 kg N ha^{-1} (Chauhan and Abugho, 2013a). Delaying *E. crus-galli* emergence in rice and/or adopting better competitive rice cultivars can reduce weed seed production substantially (Singh et al., 2017). *Echinochloa crus-galli* populations can also expand when lower nodes of stems develop roots (OLA and MAFF, 2002).

Mature *E. crus-galli* seeds exhibited a high degree of primary dormancy at the time of ripening (Benvenuti et al., 1997 which was partially and gradually lost after weeks of storage. However, dormancy-breaking most often occurred after seed piercing near the embryo (Benvenuti et al., 1997). There was a tendency for seeds with secondary dormancy such as *E. crus-galli* var. crus-galli to have less longevity (Suzuki, 2003). The soil-imposed germination inhibition was partially relieved by flushing the seed-soil matrix with an oxygen-free gas (N_2) (Benvenuti et al., 1997). Heat shock (Benvenuti et al., 1997), application of N-compounds in conjunction with red light (DongSheng et al., 2010), treatment with GA_3 or with H_2SO_4 (Wu et al., 2007) were found effective for *E. crus-galli* seed dormancy-breaking. *Echinochloa crus-galli* seed distribution was confined to only 15 cm depth with about 50% of seeds in the uppermost 5 cm of clay and sandy clay loam fields soil of irrigated and rainfed rice (Kim and Chun, 1996). Seeds buried over 8 cm remain viable for at least 3 years, and some seeds are viable even after 13 years (Maun and Barrett, 1986).

The plasticity in morphology and phenology may contribute to the success of *E. crus-galli* as a weed (Norris, 1996). *Echinochloa crus-galli* germinates over a wide temperature interval ranging from 13°C to 40°C (Shipley and Parent, 1991). The germination percentage was highest under natural light, filtrated light and red light, followed by green light, and far-red light and control (dark) (Ikeda et al., 2004). The enhancement of seed germination by red light was reduced in the presence of far-red light and vice versa indicating the strong possibility of phytochrome involvement in the regulation of seed germination in both species by light (Ikeda et al., 2004).

Echinochloa colona required a relatively high moisture level for germination but seeds do not germinate in flooded soil, under a water layer of at least 15 mm (Civico and Moody, 1979). Thus it can survive in both flooded and nonflooded rice fields. The growth of *E. colona* was reduced under drought stress resulting in significantly shorter plants with reduced number of spikes, seeds per panicle, and seed weight (Chun and Moody, 1985). *Echinochloa colona* emerges 2–3 days after seeding (DAS) and reaches the two-leaf stage at 8 DAS (Chun and Moody, 1986). Seedling emergence may vary according to seed population origin (Martinková et al., 2006). Its emergence was significantly greater in a single-cropping system of rice than in a double-cropping system of rice followed by barley and/or strawberry (Kim and Chun, 1996). Flooding depth inhibited seedling emergence where light availability was crucial for increasing the chances of emergence (Benvenuti et al., 1997). *Echinochloa colona* showed greater reductions in emergence, shoot, and root growth than *E. crus-galli* under water submergence (Estioko et al., 2014). The response to flooding in *E. crus-galli* is related to the formation of root aerenchyma which developed in a polar direction from the basal parts to the apical parts as the roots aged (Yoneyama and Takeuchi, 2000). The quantitative trait loci (QTLs) affecting flooding tolerance and other physiological and morphological traits in *E. crus-galli* were analyzed using a restriction fragment length polymorphism map, and two QTLs were detected for flood tolerance, and one was mapped on linkage group 24 (Fukao et al., 2004). Other traits, including plant morphology, heading date, number of leaves, and plant height, were highly correlated.

Seeds of both species can be dispersed by wind, water, or as a contaminant in soil, crops seed, waterways, insects, birds, the feet of larger animals, and anthropogenic causes (machinery, contaminated seed, etc.) (OLA and MAFF, 2002; Rao et al., 2017; CABI, 2019a, 2019b). *Echinochloa crus-galli* has potential endozoochory dispersal and recuperation of the species in the feces was 23.4% (Viero et al., 2018). The quarantine period for animals that ingest *E. crus-galli* seeds should be at least 7 days. A large number of seeds capable of germination were excreted along with cattle feces, especially dormant seeds (Ikeda et al., 2006). Increased time within the gut and digestive tract of birds reduces the probability of viability (Wongsriphuek et al., 2008). Rice weeds may spread through the manure of cattle feeding on forage rice silage that is contaminated by weed seeds and seeds of *E. crus-galli* (L.) Beauv. var. formosensis Ohwi, in forage rice died completely as a result of ensiling for approximately 90–180 days (Koarai et al., 2015). Burying manure by moldboard plowing decreased or even canceled the exogenous organic matter effect of weed seed contamination (Colbach et al., 2013). The red-eared sliders (*Trachemys scripta elegans*) and common snapping turtles (*Chelydra serpentina*) are capable of dispersing *E. crus-galli* seeds to areas near pond margins and to more distant locations during overland movements (Kimmons and Moll, 2010). The seeds of both these noxious weeds often enter rice fields with crop seeds (Rao and Moody, 1990; Holm et al., 1991). They may be transferred from neighboring fields through seed has the potential to spread through irrigation water and the movement of farm machinery through infested fields and on the feet, fur, feathers, and skin of rodents, birds, and larger animals, including human (Rao et al., 2017).

The crop yield loss caused by weeds capable of releasing allelochemicals in their surrounding environment is also attributed to the allelopathic influence of the weeds on the crop (Xuan et al., 2006) in addition to crop–weed competition. Extracts of *E. colona* and *E. crus-galli* could reduce the germination and seedling growth of rice (Sitthinoi et al., 2017), field bean, and sugar beet; maize; soybean; wheat (Sitthinoi et al., 2017). The root exudates of *E. crus-galli* suppressed the growth of rice, lettuce (*Lactuca sativa* L.), and *Monochoria vaginalis* (Burm.f.) C.Presl., during the early growth stages (Xuan et al., 2006). *Echinochloa crus-galli* inhibited the growth of cockscomb (*Celosia cristata* L. var. kunze), timothy (*Phleum pratense* L.), cress (*Lepidium sativum* L.), amaranth (*Amaranthus viridis* L.), rice, lettuce, and *E. crus-galli* itself, at the germination and in the early growth period (Yamamoto et al., 1999; Son et al., 2010). A larger number of identified allelochemicals belong to phenolics, sequiterpenes, flavonoids, terpenes, alkaloids, and their derivatives and exhibit strong phytotoxic activity (Yamamoto et al., 1999; Jabran et al., 2015). In the *E. crus-galli* genome, Guo et al. (2017) have identified gene cluster involved in the biosynthesis of an allelochemical 2,4-dihydroxy-7-methoxy-1,4-benzoxazin-3-one which plays a critical role in *E. crus-galli* competition with rice and its invasiveness in rice fields. Allelochemicals were also suggested for their potential usefulness as bioherbicide for weed control in agriculture (Goma and AbdElgawad, 2012). However, the strong phytotoxic action caused by these weeds infested soil against weeds and crops can be attributed to the interaction of many chemicals released by them to the soils (Khanh et al., 2008). Hence, greater research should concentrate on identifying the phytotoxins released by these species roots into the arable land, as well as their transformation in nature, in order to understand the chemical interaction of these noxious rice weeds against crops and their surrounding other weeds.

Integrated management

Various management strategies, including cultural techniques, hand weeding, mechanical control, and chemical herbicides, are available to manage *E. crus-galli* and *E. colona* in different ecosystems. The response of these weed populations to chemical or nonchemical or biological control may be affected by phenotypic and genetic variability among their populations caused by agricultural practices, crop characteristics, geographic location, and herbicide pressure (Kapoor and Ramakrishnan, 1974; Norris, 1996; Altop and Mennan, 2011; Nguyen, 2015). Thus the best long-term management strategy for both the species is to integrate both chemical and nonchemical options to manage survivors and prevent the replenishment of new seeds into the soil seedbank (Wu et al., 2004; Rao et al., 2019). The management components that were found effective in managing both the species and which can be used in integrated weed management (IWM) are discussed.

Crop competitiveness

Cultural weed management strategies are needed for the sustainability of cropping systems. Weed-competitive crop cultivars (Caton et al., 2003), narrow row spacing (Wu et al., 2004; Fahad et al., 2014), and high seeding rates (Chauhan and Abugho, 2013a; Dhillon and Bhullar, 2016) are some of the nonchemical crop competitiveness enhancing strategies that could help manage these weeds.

The crop cultivar selection

Crop cultivars with faster early growth and seedling vigor with rapid canopy closure have a greater ability to suppress weeds through resource competition (Ramesh et al., 2017). Rice cultivars HKR 1002 and HKR 126 have a suppressive effect on *E. colona*, whereas cultivars HKR 46, HKR 126 and DRR H 1 were found to have a suppressive effect on *E. crus-galli* (Dhawan and Malik, 2003). Delaying *E. crus-galli* emergence in rice and/or adopting better competitive rice cultivars can reduce the weed seed production substantially (Singh et al., 2017). Growing weed-suppressive Asian rice cultivars (PI 312777, Guichao, and Teqing) in conjunction with reduced herbicide rates could be an effective and economical *E. crus-galli* management strategy for rice (Gealy et al., 2003). The root–root competition may play a significant role in the natural suppression of *E. crus-galli* by PI 312777 (Gealy, 2001). In Brazil, IRGA 421, IRGA 416, and IRGA 417 were observed as the most competitive rice cultivars (Galon et al., 2007). In China, nontransgenic cultivar Yuefu was superior to transgenic cultivar 99-1 in competing with *E. crus-galli* (Zhang et al., 2005). Increasing the crop density of the low tillering new plant type reduced yield loss due to weeds in the Philippines (Ni et al., 2004). The screening of lines in standard nurseries for competitiveness against weeds was found to be possible using nondestructive scoring of plant cover, very early heights (i.e., vigor), tillering ability, and early height growth rates (m/day) (Caton et al., 2003).

Crop seeding rate and crop density

In rice seedling nurseries, significantly lowered weed density as compared to control (unweeded) was observed with the increased seed rate of 100 g/m^2 as practiced by farmers but the weed control efficiency was quite low against *E. crus-galli* (36.8%) (Dhillon and Bhullar, 2016). Irrespective of water regime and N rate, the growth and seed production of *E. crus-galli* declined with increases in rice density, suggesting that growth and seed production of *E. crus-galli* can be greatly reduced by increasing the rice seeding rate (Chauhan and Abugho, 2013a). Narrow crop row spacing and increased crop density, measures to delay in *E. colona* emergence and early weed control could be important components of IWM strategies of *E. colona* in aerobic rice (Chauhan and Johnson, 2010), sorghum (Wu et al., 2004), and maize (Fahad et al., 2014).

Tillage and mulching

Echinochloa crus-galli was the dominant species when tillage was done twice (Shahgholi et al., 2015), probably due to bringing up of the seeds to the soil surface and their exposure to light. The cultivation at night-time had inhibited the germination of *E. crus-galli* by 77.7% (Magyar and Kónya, 2008). Soil inversion by tillage to bury weed seeds below their maximum depth of emergence, use of crop residue as mulch and early flooding of the crop could serve as important tools for managing *E. crus-galli* and *E. colona*. The seed burial in saturated soil conditions prohibited *E. colona* germination (Pons, 1982). The rice bran (powder and pellet formulations) application inhibited *E. crus-galli* var. formosensis seedling growth due to the decrease in oxidation–reduction potential and dissolved oxygen content in rice field water with the rice barn treatment (Muroi et al., 2005). These management options, however, would need to be compatible with other crop management requirements.

Allelopathy

Allelopathy plays an important role in natural and agricultural ecosystems, particularly in crop ecosystems (Khanh et al., 2007; Jabran et al., 2015), and it could potentially serve as an important IWM component weed management strategy in crop production.

The residues of crops sorghum + sunflower + brassica (each incorporated at 2 g/kg soil) inhibited *E. colona* germination by 22% and those of sorghum + brassica (each incorporated at 3 g/kg soil) while increasing the time of *E. colona* germination initiation by 17% (Khaliq et al., 2011). The use of sunflower leaf extract (10%–15% (wt./vol) as a preemergence bioherbicide for inhibiting seedling growth of *E. crus-galli* without injuring the rice seedlings, was suggested (Dilipkumar and Chuah, 2013). The growth of *E. crus-galli* was severely suppressed in the presence of 0.9% rice straw in hydroponic culture due to shoot's high Fe content, and it was suggested that the addition of organic materials into the soil may help to selectively suppress *E. crus-galli* (Nozoe et al., 2010). The crop residues utilization in rice fields has long been recognized as an important source to improve the soil organic matter status and to suppress the growth of weeds.

Allelopathy of crop cultivars as a means *Echinochloa* species management was studied mostly in rice. The allelopathic potential of rice cultivars on *E. crus-galli* was demonstrated (Gealy et al., 2003, 2005; Khanh et al., 2007). The rice varieties Taichung native 1 (Lee et al., 2003); BR17 (Noguchi et al., 2009); Duchungjong (Jung et al., 2004); Asian cultivars (PI 312777, Guichao and Teqing) (Gealy et al., 2003, 2019); Buldo and Agudo (Ahn et al., 2005); Herathbanda, Gonabaru, Godaheenati, Batapola el, and Sula (Wathugala and Ranagalage, 2014); Ld 365, Masuran, and Kaluheenati (Ranagalage and Wathugala, 2016); and Aeron 1(Sunyob et al., 2015) were found to have an allelopathic effect on *E. crus-galli*. Allelopathic rice cv. PI 312777 had greater control ability for broadleaf weeds in summer rice fields and *E. crus-galli* under water deficiency stress (Xia et al., 2009). In rice, two gene clusters involved in the biosynthesis of the allelopathic diterpene phytoalexins momilactone A and phytocassane were identified on Chromosomes 4 and 2, respectively (Shimura et al., 2009; Swaminathan et al., 2009). These diterpene phytoalexins can significantly inhibit the growth of *E. crus-galli*. Nine QTLs controlling allelopathic effects of rice on *E. crus-galli* were identified on chromosome no. 1, 2, 3, 4, 5, 8, 9, and 12. QTLs on chromosome no. 1 and 5 showed the largest effect on the allelopathic effect and explain 36.5% of total phenotypic variation (Lee et al., 2005)

Echinochloa crus-galli biomass was reduced in mixed-culture with rice cultivars, in particular, allelopathic cultivars. The allantoin (5-ureidohydantoin) concentration in rice cultivars was positively related to *E. crus-galli* biomass (Sun et al., 2012). The allantoin stimulated the growth of *E. crus-galli* once released from rice or added to the soil. Hence, the allelopathic rice cultivars respond by decreasing production of allantoin on detecting *E. crus-galli* and thus regulating the growth of *E. crus-galli* by competition with *E. crus-galli* (Sun et al., 2012).

In addition to the above nonchemical measures, the competition for light and nutrients (Graf and Hill, 1992), shade (Kobayashi, 2006), early flooding to increase seedling mortality (Sahid and Hossain, 1995); subsurface steam injection to prevent the seeds germination in the soil (Vidotto et al., 2009); repeated clipping reduced weed total biomass (Butler et al., 2013) were also tested successfully to manage *E. crus-galli*.

Biocontrol

The possibility of using indigenous plant pathogensas biological control agents to manage these two *Echinochloa* spp. in rice was investigated

(Zhang et al., 1996). A fungus, *Exserohilum monoceras* (Drechs.) K. J. Leonard & E. G. Suggs, isolated from diseased *Echinochloa* spp. in the Philippines, caused blight symptoms on *E. crus-galli* and *E. colona*, and it was nonpathogenic to rice (Zhang et al., 1996; Zhang and Watson, 1997b). Drop inoculation gave better weed control than spray inoculation of *E. monoceras* which exhibited high herbicidal activity against *E. crus-galli* var. crus-galli (long and short awn biotypes), E. crus-galli var. formosensis, and *E. colona* in Japan (Tsukamoto et al., 1997). An inoculum dose of 2.5 to 5.0 × 107 conidia/m^2 killed seedlings of *Echinochloa* spp. at the 1.5-leaf stage in the greenhouse (Zhang et al., 1996). The minimum dew period to achieve 100% mortality was 16 h for *E. colona*, 12 h for *E. crus-galli* (Zhang and Watson, 1997a). When the inoculum was formulated as an oil emulsion or when applied as a dry powder, the level of primary infection increased significantly and 90% mortality was achieved at 14–20 days after inoculation (DAI) (Zhang, 1996). Complete mortality of *E. crus-galli* plants at 10 DAI was observed when it was subjected to postinoculation high levels of humidity (80%–90%) periods of 12 or 24 h (Tuat et al., 1999). Seedlings at the 3.0–3.5-leaf stages were the most susceptible to *E. monoceras*, with 99.6% disease incidence and more than 80% mortality (Huang et al., 2001).

Drechslera monoceras has been tested on *E. crus-galli* in rice; better control being observed under low (25/15 day/night) temperatures (Hirase et al., 2004). *Fusarium equiseti* was suggested for consideration as a probable bioherbicide for controlling *E. crus-galli* at its 2–3-leaf growth stage (Motlagh, 2011). The 1.5–2.0-leaf stage was the most susceptible to *Drechslera monoceras* (*Setosphaeria monoceras*), exhibiting 100% disease incidence and 70.7%–75.2% mortality (Huang et al., 2001). *Echinochloa crus-galli* was completely infested with BC-1 21 DAI, and fresh weight inhibition was 93.7% and has potential to be developed into an effective microbial herbicide with economic and social benefits (Li et al., 2016). A fungal pathogen, *Helminthosporium gramineum* Rabehn f.sp. echinochloae, has been developed as a mycoherbicidal agent for the control of *E. crus-galli* in China (Geng et al., 2009). One particular virulent strain each of *Cochliobolus lunatus* (Li et al., 2013) and *Bipolaris eleusines* (Zhang et al., 2014) were found as highly pathogenic to *E. crus-galli*. Hence, it was suggested that this strain could be a potential mycoherbicide for *E. crus-galli* control in rice in future as it was safe to rice.

Use of goats to graze on weeds including *E. colona* during the off-season prior to planting of maize, resulted in significant control and weed seed reservoir depletion (Jebarathnam and Kathiresan, 2007).

Harvest weed seed control

Harvest weed seed control (HWSC) is an alternative nonchemical approach to weed management that targets escaped weed seeds at the time of crop harvest. There would be some benefit to using HWSC for *E. crus-galli*, but the use of this practice on mitigating herbicide resistance would be less pronounced because of the lesser seed retention (41% of its seed) or early seed shatter (32% of its seed) (Schwartz-Lazaro et al., 2017).

Seed predation

Postdispersal seed predation can cause severe seed losses in plant populations. However, it is unclear whether such losses can help to lower weed densities, as an aid in weed control in arable fields (Westerman et al., 2014). The activity density of all cricket species in the field was 2.4–6.0 individuals/trap/day. The cricket species differed in their spatial distribution. The activity density of *Teleogryllus emma* peaked at 5 m from the field edge, whereas that of *Velarifictorus micado* peaked 15 m from the edge. In contrast, the activity density of *Loxoblemmus* spp., including *L. campestris*, was higher on the paddy levee than in the field. The proportion of invertebrate seed predation on *E. crus-galli* var. crus-galli (23.9%–40.9%/day) did not differ significantly between the examined distances from the field edge. The crickets can help deplete weed seed banks over a large portion of paddy fields (Ichihara et al., 2014). Postdispersal seed predators contribute to biological weed control by removing newly produced weed seeds before they can enter the seedbank and the effect of *E. crus-galli* seed predation on weed population dynamics was observed to differ between years and fields (Pannwitt et al., 2017).

Usage

Both the species were reported to be used in different ways, which are listed below.

Forage

E. crus-galli was tested as forage in the United States. In Australia as a part of drought mitigation programs, it has been used for nourishing sheep. Toxic quantities of nitrates have been measured in *E. crus-galli*, although these are not usually of concern. *Echinochloa colona* is also much valued as a fodder. *Echinochloa colona* was identified as the salt-tolerant plants useful as a novel feed resource for livestock under saline conditions and could be used for poverty alleviation by improving the income of poor small farmers on saline lands (Khanum et al., 2011).

Edible seed

The grain of *E. colona* has occasionally been cultivated as a grain and used as food by a few people. It produces edible seeds that can be ground into meal or flour and made into mush or bread. *Echinochloa colona* has high fiber content preferred in diets of patients suffering with diabetes, cardio vascular-related disorders, and constipation. It shows high antimicrobial and antioxidant and plays a significant role in phyto-remediation. *Echinochloa crus-galli* is extensively cultivated in Japan, Korea, and northern China (De Wet et al., 1983).

Crop improvement

An efficient protocol was developed for plant regeneration through somatic embryogenesis using cultured leaf base and leaf tip segments of *E. colona* (Samantaray et al., 1997). This will help into utilize emerging biotechnologies such as somatic hybridization and/or recombinant DNA technology to accomplish transfer to crops the traits like tolerance to heavy metals and other biotic and abiotic stresses of *E. colona*.

Medicinal use

Echinochloa colona has wound healing, antioxidant and antimicrobial properties and its medicinal use literature was reviewed (Singh and Sharma, 2018; Ray and Chatterjee, 2017). *Echinochloa colona* has high fiber content preferred in diets of patients suffering with diabetes, cardio vascular-related disorders, and constipation.

Reclamation

Echinochloa crus-galli has the greatest potential for use in vegetative restoration of problematic soils due to its high salt (Yuan et al., 2014) and metal (Samantaray, 1991) tolerance. In Egypt, *E. colona* is used for the reclamation of saline soils and is recommended for cultivation as a fodder crop in salinity affected areas (Sherif, 2007). The aqueous extract of aerial plant parts of *E. colona* was used as a reducing and stabilizing agent for the synthesis of the silver nano particles (AgNPs) by ecofriendly bio reduction method (Kumar et al., 2016).

Herbicide use

Effective herbicides

Echinochloa colona and *E. crus-galli* control may include several nonchemical management options as discussed above. However, herbicides, either preemergent or postemergent, are widely used for their control (Bajwa et al., 2015; Masood et al., 2016; Shabbir et al., 2019) as they tend to be economical due to shortage and increased labor wages for hand weeding (Rao and Matsumoto, 2017).

A few of the several herbicides that were found effective in managing *E. colona* include propanil, azimsulfuron, bensulfuron-methyl, molinate, oxadiazon, penoxsulam, profoxydim (Salas et al., 2009); bispyribac-sodium (Gunawardane et al., 2013; Guo et al., 2016), cyhalofop-butyl (Esquivel and Valle, 2009; Salas et al., 2009), fenoxaprop-P-ethyl (Guo et al., 2016), quinclorac (Martinez et al., 1999); cyhalofop-butyl mixtures with pendimethalin and clomazone (Esquivel and Valle, 2009); penoxsulam, metamifop (Guo et al., 2016) in rice; the mixture of trifluralin and pendimethalin in combination with hoeing in peanut (Punia and Malik 1999); preplant application of trifluralin in soybean (Balyan and Malik, 1998); fenoxaprop-P-ethyl, clodinafop-propargyl, and cyhalofop-butyl in blackgram (Rao, 2008); soil incorporation of atrazine mixed with metolachlor at sorghum planting (Wu et al., 2004). The sequential application treatments using glyphosate and paraquat gave 100% control of glyphosate-susceptible and glyphosate-resistant *E. colona* as it can emerge in multiple flushes (Widderick et al., 2013).

Bensulfuron-methyl + quinclorac; benthiocarb + peopanil; benthiocarb + pyrazosulfuron-ethyl; bispyribac-sodium; cinmethylin and fenoxapropethyl; cyclosulfamuron + pendimethalin; cyhalofop-butyl + bensulfuron fb bentazon/MCPA; fenoxapropethy; fentrazamide + propanil; florpyrauxifen−benzyl; glufosinate ammonium (prior to tillage); molinate; molinate + 2,4-D; molinate + 2,4-D IBE; molinate + 2,4-D isobutyl ester; molinate + bensulfuron; penoxsulam + cyhalofop-butyl; pretilachlor + safener; pyrazosulfuron-ethyl; quinclorac; quinclorac + bensulfuron; quinclorac + cinosulfuron + pretilachlor are a few of several herbicides that were found to be effective in managing *E. crus-galli* in rice (Rao et al., 2019), the crop in which it predominates across the world. The delayed premergence herbicide application of (clomazone plus quinclorac plus pendimethalin plus thiobencarb or the premergence herbicide (clomazone) application of followed by postemergence herbicide (quinclorac or quinclorac plus imazethapyr) were found to be effective to control of herbicide-resistant *E. crus-galli* biotypes (Wilson et al., 2014). The herbicide efficacy depends on environmental conditions and under unfavorable weather conditions delayed application (Chauhan and Abugho, 2012). The most common herbicides for control of *E. crus-galli*, both in cereals, vegetables, oil seed, and potato, belong to the group acetyl-CoA carboxylase inhibitors (Vencill et al., 2014) and herbicides in this group have developed resistant weeds in cereals making weed control complicated and expensive. Thus the importance of early-season control of *E. crus-galli* and an integrated approach to weed management with residual herbicides to control later-emerging weeds in glyphosate-resistant cotton was highlighted (Werth et al., 2008). Crop rotations are highly recommended to manage herbicide-resistant *E. crus-galli* populations as rotations allow for the utilization of different herbicides with alternative modes of action that reduce the *E. crus-galli* infestation, delaying the evolution of resistance (Talbert and Burgos, 2007).

Herbicides continue to play a major role in managing both species. However, herbicides must be used in rotations and integration with other nonchemical methods like flooding (Dorji et al., 2013); high seeding rate

(Tehranchian et al., 2014); competitive cultivar (Ramesh et al., 2017), narrow row spacing (15 cm) and delaying *E. colona* emergence (Fahad et al., 2014); strategic tillage (Walker et al., 2013); conservation practices such as zero tillage (Singh et al., 2015); targeting the peak emergence periods (Bagavathiannan et al., 2011) to avoid the herbicide resistance occurrence and increase weed management efficacy. The long-term IWM should focus on employing diversified chemical and nonchemical management methods at different stages of these weeds life cycle to deplete their weed seed bank while preventing seed production of survived weeds to minimize replenishment of the weed seedbank.

Herbicide resistance and management

The heavy reliance upon and repeated use of these herbicides in crops have led to the evolution of herbicide resistance, in both the *Echinochloa* species, to different herbicides (Table 10.2). *Echinochloa crus-galli* acquired or evolved target-site and nontarget-site herbicide resistance making it among the most pernicious herbicide-resistant weeds of the world (Norsworthy et al., 2014). From a global standpoint, there is considerable evidence that both the species have the propensity to develop resistance to most of the herbicide groups (Table 10.2). The spread of herbicide resistance in these two species poses a serious threat to crop production.

Resistant populations showed a higher degree of resistance at 30°C than at 20°C temperature while there was no variation in response of susceptible populations at both the temperatures (Nguyen, 2015). The glyphosate-resistant *E. colona* plants were more moisture and salinity stress tolerant at germination, more competitive and produced more aboveground dry biomass and seeds than the glyphosate-susceptible plants when grown together or alone (Shrestha et al., 2018).

However, there was a negative correlation between the adaptability of resistant population to environment and the level of resistance, and it showed a certain degree of fitness cost at the stage of germination (Zhang et al., 2018).

The studies on the role of pollen-mediated gene flow in the spread of herbicide resistance in *E. crus-galli* have revealed gene flow within a short distance of up to 10 m and in one event at 20 m (Bagavathiannan et al., 2012a). Gene flow declined with distance from the source with an average frequency of 3% at 0.5 m, although frequencies of up to about 10% were observed at this distance. Wind speed exhibited a positive influence on gene flow, whereas air temperature and relative humidity influenced negatively on gene flow (Bagavathiannan et al., 2012b). Therefore efforts to manage and monitor for herbicide resistance in these species need to be continued.

To mitigate rapid resistance evolution under a changing climate, measures to reduce the herbicide selection pressure are to be implemented. The measures including reduction of weed population size through reduction of

TABLE 10.2 Herbicide reistance of *E. colona* and *E. crus-galli* (Original information source: Heap, 2020).

	E. colona	*E. crus-galli*	Both *E. colona* and *E. crus-galli*
Herbicides	Only *E. colona*: Ametryn; fluazifop-P-butyl; metribuzin	Only *E. crus-galli*: bensulfuron-methyl; butachlor; clomazone; cyanazine; flucetosulfuron; foramsulfuron; imazamox; imazosulfuron; metamifop; molinate; nicosulfuron; pendimethalin; penoxsulam; profoxydim; pyrazosulfuron-ethyl; pyribenzoxim; pyriminobac-methyl; simazine; quizalofop-P-ethyl; thiobencarb; quizalofop-P-tefuryl	Atrazine, Azimsulfuron, bispyribac-sodium, cyhalofop-butyl, fenoxaprop-P-ethyl, glyphosate; haloxyfop-P-methyl; imazapic; imazapyr; imazethapyr; propanil; quinclorac
Multiple resistance: four sites of action		Fenoxaprop-P-ethyl, imazamox, imazethapyr, propanil, and quinclorac (MOA in monocots) [ACCase inhibitors (A/1); ALS inhibitors (B/2); PSII inhibitor (Ureas and amides) (C2/7); Cellulose inhibitors (L/26) [United States (Mississippi) Rice]	
Multiple resistance: three sites of action	1. Azimsulfuron, fenoxaprop-P-ethyl, and propanil [ACCase inhibitors (A/1); ALS inhibitors (B/2); PSII inhibitor (Ureas and amides) (C2/7)] (Costa Rica, Rice)	1. Cyhalofop-butyl, penoxsulam, and quinclorac (MOA in monocots); [ACCase inhibitors (A/1); ALS inhibitors (B/2); Cellulose inhibitors (L/26)]; (Brazil, Rice)	

(*Continued*)

TABLE 10.2 (Continued)

	E. colona	*E. crus-galli*	Both *E. colona* and *E. crus-galli*
	2. Bispyribac-sodium, cyhalofop-butyl, and quinclorac (MOA in monocots) [ACCase inhibitors (A/1); ALS inhibitors (B/2); Cellulose inhibitors (L/26)] (Colombia, Rice) 3. Bispyribac-sodium, cyhalofop-butyl, and propanil [ACCase inhibitors (A/1); ALS inhibitors (B/2); PSII inhibitor (Ureas and amides)] (C2/7) (Venezuela, Rice) 4. Imazethapyr, propanil, and quinclorac (MOA in monocots) [ALS inhibitors (B/2); PSII inhibitor (Ureas and amides) (C2/7); Cellulose inhibitors (L/26)] (USA, Rice)		
Multiple resistance: two sites of action	1. Bispyribac-sodium, cyhalofop-butyl, imazapic, and imazapy [ACCase inhibitors (A/1); ALS inhibitors (B/2)] (Costa Rica, Rice)	1. Cyhalofop-butyl, fenoxaprop-P-ethyl, molinate, and thiobencarb = benthiocarb [ACCase inhibitors (A/1); Lipid Inhibitors (N/8)]; 'United	

(*Continued*)

TABLE 10.2 (Continued)

	E. colona	*E. crus-galli*	Both *E. colona* and *E. crus-galli*
	2. fluazifop-P-butyl, and glyphosate [ACCase inhibitors (A/1); EPSP synthase inhibitors (G/9)] (Venezuela, Rice)	States (California); Rice] 2. Propanil, and quinclorac [PSII inhibitor (Ureas and amides) (C2/7); Synthetic Auxins (O/4)]; (United States (Arkansas); Rice) 3. Bispyribac-sodium, cyhalofop-butyl, and penoxsulam [ACCase inhibitors (A/1); ALS inhibitors (B/2) }; (Turkey, Rice); 4. Butachlor, and propanil [PSII inhibitor (Ureas and amides) (C2/7); Long chain fatty acid inhibitors (K3/15)]; (Thailand, Rice) 5. Azimsulfuron, bensulfuron-methyl, bispyribac-sodium, cyhalofop-butyl, fenoxaprop-P-ethyl, flucetosulfuron, halosulfuron-methyl, imazosulfuron, metamifop, pyrazosulfuron-ethyl, pyribenzoxim, and pyriminobac-methyl [ACCase	

(*Continued*)

TABLE 10.2 (Continued)

	E. colona	*E. crus-galli*	Both *E. colona* and *E. crus-galli*
		inhibitors (A/1); ALS inhibitors (B/2)]; (South Korea, Rice) 6. Butachlor, and propanil; [PSII inhibitor (Ureas and amides) (C2/7); Long chain fatty acid inhibitors (K3/15)]; (Philippines, Rice) 7. azimsulfuron, bispyribac-sodium, cyhalofop-butyl, imazamox, penoxsulam, and profoxydim; [ACCase inhibitors (A/1); ALS inhibitors (B/2)]; (Italy, Rice) 8. bispyribac-sodium, imazethapyr, penoxsulam, and quinclorac; [ALS inhibitors (B/2); Synthetic Auxins (O/4)]; (Brazil, Rice}	
Countries [herbicide, crops]	Iran [Ametryn and metribuzin; Sugarcane]; Bolivia [fluazifop-P-butyl; rice and soybean]; Venizuela [fluazifop-P-butyl, Rice]	South Korea [bensulfuron-methyl; Rice]; China and Philippines [butachlor, Rice]; United States (Arkansas) [clomazone, rice]; United States (Maryland) [cyanazine, maize, and Cropland]; South Korea	Canada (Ontario), Czech Republic, France, Poland, Spain, United States (Maryland) [Atrazine, maize, and Cropland]; Italy, South Korea, [Azimsulfuron, maize, and Rice]; Brazil, Italy [bispyribac-sodium, rice]; Brazil, Italy,

(*Continued*)

TABLE 10.2 (Continued)

	E. colona	*E. crus-galli*	Both *E. colona* and *E. crus-galli*
		[flucetosulfuron, rice]; France [foramsulfuron, maize and rice]; Italy, Ukraine, United States (Mississippi) [imazamox, maize, and Rice]; South Korea [imazosulfuron, rice]; South Korea [metamfop, rice]; United States (California) [molinate, rice]; Germany, Italy, Spain, Ukraine [nicosulfuron, maize, rice]; Bulgaria [pendimethalin, orchards]; Brazil, China, France, Italy, Turkey, Ukraine [penoxsulam, rice, maize]; Italy [profoxydim, rice]; South Korea [pyrazosulfuron-ethyl, rice]; South Korea [pyribenzoxim, rice]; South Korea [pyriminobac-methyl, rice]; simazine; China and Thailand [quizalofop-P-ethyl, rice and soybean]; China and United States (California) [thiobencarb, rice]; Thailand [quizalofop-P-tefuryl, rice]	Japan, South Korea, Thailand, Turkey, United States (California) [cyhalofop-butyl rice]; China, Egypt, South Korea, Thailand, United States (California and Mississippi) [fenoxaprop-P-ethyl; rice, soybean]; Argentina [glyphosate, maize]; Uruguay [imazapic, rice]; Ukraine and Uruguay [imazapyr; rice]; Brazil, United States (Mississippi and (Louisiana) [imazethapyr, rice]; Greece, Italy, Japan, sri Lanka, Thailand, United States (Arkansas, Texas, Missouri, Mississippi and Louisiana) [propanil, rice]; Brazil, China, United States (Arkansas, Mississippi and Louisiana); Uruguay [quinclorac, rice]

the soil seedbank, ensuring complete control of current infestations with multiple herbicide modes of action in mixture and in sequence, augmenting herbicides with mechanical control where possible, rotation with weed-competitive crops, use of weed-competitive cultivars and weed-suppressive cover crops, and other practices recommended for IWM of these two weed species.

Climate change impact

Changes in the environment, specifically rising temperature and increasing atmospheric carbon dioxide concentration [CO_2], can alter the growth and physiology of weedy plants. These changes could alter herbicide efficacy, crop–weed interaction, and weed management. The knowledge of the weeds' biological responses to changing conditions helps to reduce their long-term population development by targeting management measures at specific growth phases of the weeds. As the atmospheric CO_2 concentration increases, it is anticipated that the competitive ability of C_3 crops could be enhanced relative to C_4 weeds in agricultural systems. The competitiveness of C_3 crop rice could be enhanced relative to C_4 weeds in response to rising [CO_2] in situ but such a response could be contingent on the supply of N (Wu et al., 2006). However, the response of *E. crus-galli* to CO_2 was not limited by N stress, showing increase in growth. Thus for rice cropping systems with N limited supply, rising atmospheric [CO_2] could still exacerbate competitive losses, even from C_4 weeds (Wu et al., 2008). Climate change could promote the altitudinal spread of these C_4 weed species since they can benefit from a warmer climate and the impact of climate conditions vary between species (Satrapová et al., 2013).

Methane emission by *Echinochloa* species

The comparison of methane emissions by rice and *E. crus-galli* plants grown on paddy soil in cement pools revealed that *E. crus-galli* emitted 11 times and rice 8 times as much methane as *C. difformis* (Min et al., 1994). The diurnal changes of methane emission peaks during 8.00–11.00 h and 14.00–20.00 h for rice and *E. crus-galli*, respectively (Min et al., 1994). Methanogenic activity in rhizospheric soils was higher than that in nonrhizospheric soils, and the methane was released from the junction of the nodal plate and leaf sheath of *E. crus-galli* and rice (Min et al., 1994).

Impact of climate change on growth of two *Echinochloa* species and herbicide efficacy

Very limited work was done on the impact of climate change on growth of two *Echinochloa* species and the herbicide efficacy in managing them.

Echinochloa colona plants grown under high [CO_2] or high temperature were taller than those in ambient conditions (Refatti et al., 2019). The greater height, tillers number, leaves and biomass of both glyphosate-resistant, and susceptible biotypes of *E. colona* occurred, with no effect on seed production or the amount of photosynthesis pigments, at elevated CO_2 and well-watered conditions (Mollaee et al., 2020). The reaction to changing climate conditions is species-specific and for *E. crus-galli* neutral (Burger et al., 2014). *Echinochloa crus-galli* occurs abundantly in European maize fields, and its predominance is likely to further increase (Peters and Gerowitt, 2014) as *E. crus-galli* was observed profited from the climate change conditions. The germination of buried seeds was stimulated by exposure to CO_2 at 30 dm^3/m^3 and concluded that soil CO_2 is responsible for causing intermittent flushes of germination (Yoshioka et al., 1998).

The CO_2 enrichment generated an increased root biomass and induced an earlier development of *E. crus-galli* inflorescences (Potvin and Strain, 1985). The rising atmospheric [CO_2] could alter the competition between rice and *E. crus-galli* in rice fields in favor of rice (Qing et al., 2011). Atmospheric CO_2 enrichment may impact arbuscular mycorrhizae development and function, which could have subsequent effects on host plant species interactions by differentially affecting plant nutrient acquisition. Jun et al. (2009) found that elevated CO_2 also increased mycorrhizal colonization of C_4 *E. crus-galli* but did not affect mycorrhizal colonization of C_3 upland rice. Mycorrhizal inoculation increased the shoot biomass ratio of C_4 *E. crus-galli* to C_3 upland rice under both CO_2 concentrations but had a greater impact under the elevated than ambient CO_2 level. Jun et al. (2009) concluded that CO_2 stimulation of mycorrhizae and their nutrient acquisition may impact the competitive interaction of C_4 *E. crus-galli* and C_3 upland rice under future CO_2 scenarios.

The impact of climate change on the efficacy of herbicides has revealed that the growth temperature has no influence on glyphosate susceptibility of the sensitive *E. colona* population but the level of glyphosate resistance in resistant populations of *E. colona* populations showed a significant increase at 30°C compared with 20°C indicating the greater difficulty in managing these resistant populations in summer (Nguyen et al., 2016). Similarly, the high [CO_2] or high temperatures have increased the resistance level of multiple-resistant *E. colona* to cyhalofop-butyl, whereas the herbicide efficacy on susceptible *E. colona* population was not reduced by these factors (Refatti et al., 2019).

Conclusion

To ensure control of these two weed species under climate change conditions various weed management strategies are necessary and a greater focus needs to be made on an IWM approach. Deeper understanding of multiple resistance mechanisms is needed to manage *E. crus-galli* and *E. colona*, which have evolved multiple resistance to herbicides. Use of herbicide-resistant crops provides effective control of these two *Echinochloa* species but should include measures to avoid evolution of herbicide-resistant populations to herbicides used in those crops. To mitigate rapid resistance evolution under changing climate and to reduce the herbicide selection pressure, the measures that need to be implemented include reduction of weed population size by soil seedbank reduction, use of herbicides that have multiple modes of action in the mixture and sequence and adoption of IWM strategies including mechanical control, use of rotations with crops and crop cultivars that are weed-competitive, use of weed-suppressive cover crops, preventive measures, and other practices and their integration with herbicides. The integrated management approach, rather than using any of the weed management methods alone, helps in managing these weeds.

References

Abouziena, H.F., El-Awadi, M.E.S., El-Saeid, H.M., El-Metwally, I.M., El-Desouki, I.R., Singh, M., 2015. Influence of environmental factors, flooding period and seeding depth on germination and emergence of barnyardgrass [*Echinochloa crus-galli* (L.) Beauv.]. Int. J. Chemtech Res. 8, 459–467.

Ahn, J.K., Hahn, S.J., Kim, J.T., Khanh, T.D., Chung, I.M., 2005. Evaluation of allelopathic potential among rice (*Oryza sativa* L.) germplasm for control of *Echinochloa crus-galli* P. Beauv in the field. Crop. Prot. 24, 413–419.

Altop, E.K., Mennan, H., 2011. Genetic and morphologic diversity of *Echinochloa crus-galli* populations from different origins. Phytoparasitica. 39, 93–102.

Andrew, M., 2018. Invasive species *Echinochloa colona* reduces abundance and diversity of resident plant communities in tropical wetland. Open J. Ecol. 8, 483–494.

Awan, T.H., Chauhan, B.S., Sta. Cruz, P.C., 2014. Growth plasticity of Junglerice (*Echinochloa colona*) for resource use when grown with different rice (*Oryza sativa*) planting densities and nitrogen rates in dry-seeded conditions. Weed Sci. 62, 571–587.

Bagavathiannan, M.V., Norsworthy, J.K., Smith, K.L., Burgos, N., 2011. Seedbank size and emergence pattern of barnyardgrass (*Echinochloa crus-galli*) in Arkansas. Weed Sci. 59, 359–365.

Bagavathiannan, M.V., Norsworthy, J.K., Smith, K.L., Neve, P., 2012a. Seed production of barnyardgrass (*Echinochloa crus-galli*) in response to time of emergence in cotton and rice. J. Agric. Sci. 150, 717–724.

Bagavathiannan, M.V., Norsworthy, J.K., Smith, K.L., Neve, P., 2012b. Pollen-mediated movement of herbicide resistance in barnyardgrass. B.R. Wells Rice Res. Stud.—Ark. Agric. Exp. Station, Univ. Ark. Syst. 600, 142–148.

Bajwa, A.A., Jabran, K., Shahid, M., Ali, H.H., Chauhan, B.S., Ehsanullah, 2015. Eco-biology and management of *Echinochloa crus-galli*. Crop. Prot. 75, 151–162.

Balyan, R.S., Malik, R.K., 1998. Carpetweed (*Trianthema portulacastrum* L.) and barnyardgrass (*Echinochloa colona* L.) control in soybean (*Glycine max*). Indian J. Weed Sci. 30, 153–157.

Barrett, S.C.H., 1983. Crop mimicry in weeds. Econ. Bot. 37, 255–282.

Benvenuti, S., Macchia, M., Bonari, E., 1997. Ecophysiology of germination and emergence of *Echinochloa crus-galli* L. seeds. Riv. Agron. 31, 925–933.

Bhowmik, P.C., Reddy, K.N., 1988. Effects of Barnyardgrass (*Echinochloa crus-galli*) on growth, yield, and nutrient status of transplanted tomato (*Lycopersicon esculentum*). Weed Sci. 36, 775–778.

Burger, J., Edler, B., Gerowitt, B., Steinmann, H.H., 2014. Predicting weed problems in maize cropping by species distribution modelling. Julius-Kühn-Archiv 443, 379–386.

Butler, R.A., Brouder, S.M., Johnson, W.G., Gibson, K.D., 2013. Response of four summer annual weed species to mowing frequency and height. Weed Technol. 27, 798–802.

CABI, 2019a. *Echinochloa crus-galli* (barnyard grass). Invasive species compendium. http://www.cabi.org/isc/datasheet/20367 (accessed 19 September 2020).

CABI, 2019b. *Echinochloa colona* (junglerice). Invasive species compendium. https://www.cabi.org/isc/datasheet/20368 (accessed 19 September 2020).

Caton, B.P., Cope, A.E., Mortimer, M., 2003. Growth traits of diverse rice cultivars under severe competition: implications for screening for competitiveness. Field Crop. Res. 83, 157–172.

Chauhan, B.S., 2013. Shade reduces growth and seed production of *Echinochloa colona*, *Echinochloa crus-galli*, and *Echinochloa glabrescens*. Crop. Prot. 43, 241–245.

Chauhan, B.S., Abugho, S.B., 2012. Effect of growth stage on the efficacy of postemergence herbicides on four weed species of direct-seeded rice. Sci. World J. 7, Article ID 123071. 7 p.

Chauhan, B.S., Abugho, S.B., 2013a. Effects of water regime, nitrogen fertilization, and rice plant density on growth and reproduction of lowland weed *Echinochloa crus-galli*. Crop. Prot. 54, 142–147.

Chauhan, B.S., Abugho, S.B., 2013b. Effect of crop residue on seedling emergence and growth of selected weed species in a sprinkler-irrigated zero-till dry-seeded rice system. Weed Sci. 61, 403–409.

Chauhan, B.S., Johnson, D.E., 2010. Implications of narrow crop row spacing and delayed *Echinochloa co*lona and *Echinochloa crus-galli* emergence for weed growth and crop yield loss in aerobic rice. Field Crop. Res. 117, 177–182.

Chauhan, B.S., Johnson, D.E., 2009. Seed germination ecology of Junglerice (*Echinochloa colona*): a major weed of rice. Weed Sci. 57, 235–240.

Chauhan, B.S., Johnson, D.E., 2011. Ecological studies on *Echinochloa crus-galli* and the implications for weed management in direct-seeded rice. Crop. Prot. 30, 1385–1391.

Chauhan, B.S., Abugho, S.B., Amas, J.C., Gregorio, G.B., 2013. Effect of salinity on growth of barnyardgrass (*Echinochloa crus-galli*), horse purslane (*Trianthema portulacastrum*), junglerice (*Echinochloa colona*), and rice. Weed Sci. 61, 244–248.

Chin, D.V., 2001. Biology and management of barnyardgrass, red sprangletop and weedy rice. Weed Biol. Manage 1, 37–41.

Chun, J.C., Moody K., 1982. Effect of some factors on germination and growth of *Echinochloa colona* (L.) Link. Cropping Systems Tuesday Seminar, May 25, 1982, The International Rice Research Institute, Los Banos, Laguna, Philippines.

Chun, J.C., Moody, K., 1985. Some factors affecting germination and growth of *Echinochloa colona*. Korean J. Weed Sci. 5, 103–108.

Chun, J.C., Moody, K., 1986. Growth, development, and morphological characteristics of *Echinochloa colona*. Korean J. Weed Sci. 6, 1–6.

Chun, J.C., Moody, K., 1987. Seed dormancy and germination behavior of *Echinochloa colon*a. Korean J. Weed Sci. 7, 12–18.

Civico, R.S.A., Moody, K., 1979. The effect of time and depth of submergence on growth and development of some weed species. Philippine J. Weed Sci. 6, 41–49.

Colbach, N., Tschudy, C., Meunier, D., Houot, S., Nicolardot, B., 2013. Weed seeds in exogenous organic matter and their contribution to weed dynamics in cropping systems. A simulation approach. Eur. J. Agron. 45, 7–19.

Cowan, P., Weaver, S.E., Swanton, C.J., 1998. Interference between pigweed (*Amaranthus* spp.), barnyardgrass (*Echinochloa crus-galli*), and soybean (*Glycine max*). Weed Sci. 46, 533–539.

Dawson, J.H., Bruns, V.F., 1975. Longevity of barnyardgrass, green foxtail and yellow foxtail seeds in soil. Weed Sci. 23, 437–440.

De Wet, J., Rao, K.E.P., Mengesha, M., Brink, D., 1983. Domestication of Sawa Millet (*Echinochloa colona*). Econ. Bot. 37, 283–291.

Dhawan, R.S., 2009. Factors effecting germination, emergence and establishment of two *Echinochloa* species. Ann. Plant. Prot. Sci. 17, 220–224.

Dhawan, R.S., Malik, R.K., 2003. Biology of *Echinochloa* species under different dates of sowing and in association with different rice (*Oryza sativa*) accessions. Indian J. Agric. Sci. 73, 594–597.

Dhillon, B.S., Bhullar, M.S., 2016. Effective post-emergence herbicides for weed control in rice nurseries. Indian J. Weed Sci. 48, 272–274.

Dilipkumar, M., Chuah, T.S., 2013. Allelopathic effects of sunflower leaf extract and selected pre-emergence herbicides on barnyardgrass. J. Trop. Agric. Food Sci. 41, 309–318.

DongSheng, T., Hamayun, M., Khan, A.L., Shinwari, Z.K., Kim, Y.H., Kang, S.M., et al., 2010. Germination of some important weeds influenced by red light and nitrogenous compounds. Pak. J. Bot. 42, 3739–3745.

Dorji, S., Chauhan, B.S., Baltazar, A.M., Johnson, D., 2013. Effect of flooding depth and pretilachlor rate on emergence and growth of three rice weeds: junglerice (*Echinochloa colona*), smallflower umbrella sedge (*Cyperus difformis*), and ludwigia (*Ludwigia hyssopifolia*). Can. J. Plant. Prot. 1, 43–48.

Esquivel, V.A.E., Valle, O.H.T., 2009. Alternatives to propanil for controlling *Echinochloa colona* (L.) Link in rain-fed rice. Agron. Mesoam. 20, 111–119.

Estioko, L.P., Miro, B., Baltazar, A.M., Merca, F.E., Ismail, A.M., Johnson, D.E., 2014. Differences in responses to flooding by germinating seeds of two contrasting rice cultivars and two species of economically important grass weeds. AoB Plants 2014, plu064.

Fahad, S., Hussain, S., Saud, S., Hassan, S., Muhammad, H., Shan, D., et al., 2014. Consequences of narrow crop row spacing and delayed *Echinochloa colona* and *Trianthema portulacastrum* emergence for weed growth and crop yield loss in maize. Weed Res. 54, 475–483.

Fukao, T., Paterson, A.H., Hussey, M.A., Yamasue, Y., Kennedy, R.A., Rumpho, M.E., 2004. Construction of a comparative RFLP map of *Echinochloa crus-galli* toward QTL analysis of flooding tolerance. Theor. Appl. Genet. 108, 993–1001.

Galon, L., Agostinetto, D., Moraes, P.V.D., Tironi, S.P., Magro, T.D., 2007. Estimation of grain yield loss in rice (*Oryza sativa*) cultivars due to interference by barnyardgrass (*Echinochloa* spp.). Planta Daninha 25, 697–707.

Gealy, D.R., 2001. Determination of root distribution of weed-suppressive rice (*Oryza sativa*) and barnyardgrass (*Echinochloa crus-galli*) in field soil using ^{13}C isotope analysis. Res. Ser.—Ark. Agric. Exp. Sta. 485, 69–74.

Gealy, D.R., Wailes, E.J., Estorninos Jr., L.E., Chavez, R.S.C., 2003. Rice cultivar differences in suppression of barnyardgrass (*Echinochloa crus-galli*) and economics of reduced propanil rates. Weed Sci. 51, 601–609.

Gealy, D., Ottis, B., Talbert, R., Moldenhauer, K., Yan, W., 2005. Evaluation and improvement of allelopathic rice germplasm at Stuttgart, Arkansas, USA. In: Proc. Fourth World Congress on Allelopathy, August 21–26, 2005, Charles Sturt University, Wgga Wagga, Australia.

Gealy, D.R., Rohila, J.S., Boykin, D.L., 2019. Genetic potential of rice under alternate-wetting-and-drying irrigation management for barnyardgrass (*Echinochloa crus-galli*) suppression and grain yield production. Weed Sci. 67, 453–462.

Geng, R.M., Zhang, J.P., Yu, L.,Q., 2009. *Helminthosporium gramineum* Rabehn f.sp. echinochloae conidia for biological control of barnyardgrass. Weed Sci. 57, 554–561.

Ghanizadeh, H., Lorzadeh, S., Aryannia, N., 2014. Effect of weed interference on Zea mays: growth analysis. Weed Biol. Manage. 14, 133–137.

Goma, N.H., AbdElgawad, H.R., 2012. Phytotoxic effects of *Echinochloa colona* (L.) Link. (Poaceae) extracts on the germination and seedling growth of weeds. Span. J. Agric. Res. 10, 492–501.

Graf, B., Hill, J.E., 1992. Modelling the competition for light and nitrogen between rice and *Echinochloa crus-galli*. Agric. Syst. 40, 345–359.

Gunawardane, W.G.N., Ariyaratne, M., Bandaranayake, P., et al., 2013. Control of Echinochloa colona in aerobic rice: effect of different rates of seed paddy and post-plant herbicides in the dry zone of Sri Lanka. The role of weed science in supporting food security by 2020. In: Proc. 24th Asian-Pacific Weed Science Society Conf., Bandung, Indonesia, October 22–25, 2013. pp. 431–437.

Guo, C.L., Qin, J.L., Wang, Y.H., Ma, Y.F., Huang, H.Y., Liang, H.B., 2016. Biological activity of five herbicides to *Echinochloa colonum* (L.) Link and *Digitaria sanguinalis* (L.) Scop. and their secutity for upland rice. J. South. Agric. 47, 389–394.

Guo, L., Qiu, J., Ye, C., Jin, G., Mao, L., Zhang, H., et al., 2017. *Echinochloa crus-galli* genome analysis provides insight into its adaptation and invasiveness as a weed. Nat. Commun. 8, 1031. Available from: https://doi.org/10.1038/s41467-017-01067-5.

Heap, I., 2020. The International Survey of Herbicide Resistant Weeds. Online. Internet. Monday, November 2, 2020. Available www.weedscience.org

Hegazy, A.K., Fahmy, G.M., Ali, M.I., Gomaa, N.H., 2005. Growth and phenology of eight common weed species. J. Arid. Environ. 61, 171–183.

Heller, K., Adamczewski, K., 2010. The effect of weather conditions on the status and degree of weed population in fibrous flax. Fragm. Agron. 27, 63–69.

Hirase, K., Yoshigai, S., Nishida, M., Shinmi, T., 2004. Influence of water management, application timing and temperature on eYcacy of MTB-951, a mycoherbicide using Drechslera monoceras to control *Echinochloa crus-galli* L. Weed Biol. Manage 4, 72–74.

Holm, L.G., Plucknett, D.L., Pancho, J.V., Herberger, J.P., 1991. The World's Worst Weeds: Distribution and Biology. The University Press of Hawaii, Malabar, Florida, p. 609.

Huang, S.W., Watson, A.K., Duan, G.F., Yu, L.Q., 2001. Preliminary evaluation of potential pathogenic fungi as bioherbicides of barnyardgrass (*Echinochloa crus-galli*) in China. Int. Rice Res. Notes 26, 35–36.

Ichihara, M., Inagaki, H., Matsuno, K., Saiki, C., Mizumoto, S., Yamaguchi, S., et al., 2014. Postdispersal weed seed predation by crickets in a rice paddy field after irrigation water recedes. JARQ, Jpn. Agric. Res. Q. 48, 63–69.

Ikeda, K., Gotoh, T., Tobisa, M., Shimojo, M., Masuda, Y., 2003. Influence of light and temperature on germination of *Echinochloa crus-galli* (L.) Beauv. and *Digitaria adscendens* (H.B. K.) Henr. seeds. J. Japanese Grassl. Soc. 49, 28–32.

Ikeda, K., Hayashi, K., Eto, T., Gotoh, T., Tobisa, M., Shimojo, M., et al., 2004. Effect of different light quality on the germination of *Echinochloa crus-galli* (L.) Beauv. seeds and Digitaria adscendens (H.B.K.) Henr. seeds. J. Japanese Grassl. Soc. 49, 640–644.

Ikeda, K., Hayashi, K., Gotoh, T., Etou, T., Tobisa, M., Shimojo, M., et al., 2006. Viability of *Echinochloa crus-galli* (L.) Beauv. *and Digitaria adscendens* (H.B.K.) Henr. seeds excreted in feces from cattle. J. Japanese Grassl. Soc. 51, 171–178.

Ionescu, N.E., şarpe, N., 2001. The competition intensity between row-crops and Echinochloa crus-galli, 68. Analele Institutului de Cercetări pentru Cereale şi Plante Tehnice, Fundulea, pp. 327–339.

Jabran, K., Mahajan, G., Sardana, V., Chauhan, B.S., 2015. Allelopathy for weed control in agricultural systems. Crop. Prot. 72, 57–65.

Jebarathnam, T.G., Kathiresan, R., 2007. Biological control of weeds using goats and its' impact on weed seed bank in maize. In: Proc. 21st Asian-Pacific Weed Sci. Soc. (APWSS) Conference, 2–6 October 2007, Colombo, Sri Lanka. pp. 196–199.

Jun, T.J., Xu, L.M., Chen, X., Hu, S.J., 2009. Interaction between C_4 barnyard grass and C_3 upland rice under elevated CO_2: impact of mycorrhizae. Acta Oecol. 35, 227–235.

Jung, W.S., Kim, K.H., Ahn, J.K., Hahn, S.J., Chung, I.M., 2004. Allelopathic potential of rice (*Oryza sativa* L.) residues against *Echinochloa crus-galli*. Crop. Prot. 23, 211–218.

Kapoor, P., Ramakrishnan, P.S., 1974. Soil factors influencing the distribution of ecotype populations in *Echinochloa colonum* (L.) Link. (Gramineae). Bot. J. Linn. Soc. 69, 65–78.

Keeley, P.E., Thullen, R.J., 1991. Growth and interaction of barnyardgrass (*Echinochloa crus-galli*) with cotton (*Gossypium hirsutum*). Weed Sci. 39, 369–375.

Kennedy, R.A., Barrett, S.C.H., VanderZee, D.V., Rumpho, M.E., 1980. Germination and seedling growth under anaerobic conditions in *Echinochloa crus-galli* (Barnyard grass). Plant. Cell Env. 3, 243–248.

Khaliq, A., Matloob, A., Cheema, Z.A., Farooq, M., 2011. Allelopathic activity of crop residue incorporation alone or mixed against rice and its associated grass weed jungle rice (*Echinochloa colona* [L.] Link). Chil. J. Agric. Res. 3, 418–421.

Khanh, T., Xuan, T., Chung, I., 2007. Rice allelopathy and the possibility for weed management. Ann. Appl. Biol. 151, 325–339.

Khanh, T.D., Xuan, T.D., Chung, I.M., Tawata, S., 2008. Allelochemicals of barnyardgrass-infested soil and their activities on crops and weeds. Weed Biol. Manage 8, 267–275.

Khanum, S.A., Hussain, M., Hussain, H.N., Awan, A.R., Mehmood, K., 2011. Economic utilisation of saline marginal lands for animal production. FAO Anim. Prod. Health Proc. 11, 107–110.

Kim, J.S., Chun, J.C., 1996. Effect of rice cultural practices on emergence of *Echinochloa crus-galli* and *Monochoria vaginalis*. Korean J. Weed Sci. 16, 171–175.

Kim, S.C., Moody, K., 1989. Adaptation strategy in dry matter and seed production of rice and weed species. Korean J. Weed Sci. 9, 183–200.

Kim, S.K., Kim, S.Y., Won, J.G., Shin, J.H., Kim, H.Y., 2012. Effect of densities of *Echinochloa crus-galli* and *Cyperus serotinus* in direct-seeding flooded rice on rice yield and quality, and economic threshold level of the weeds. Korean J. Weed Sci. 32, 44–51.

Kimmons, J.B., Moll, D., 2010. Seed dispersal by red-eared sliders (*Trachemys scripta elegans*) and common snapping turtles (*Chelydra serpentina*). Chelonian Conserv. Biol. 9, 289–294.

King, L.J., 1966. Weeds of the World. Interscience Publisher:, Inc., New York.

Koarai, A., Hattori, I., Suzuki, T., Sumiyoshi, T., Ohdan, H., Sato, K., et al., 2015. Seed viability of paddy weeds ensiled by forage rice. J. Weed Sci. Technol. 60, 93–100.

Kobayashi, H., 2006. Weed ecology and cultural control in no-tillage soybean. Bull. Natl Agric. Res. Cent. Tohoku Reg. 105, 97–154.

Kraehmer, H., Jabran, K., Mennan, H., Chauhan, B.S., 2016. Global distribution of rice weeds: a review. Crop. Prot. 80, 73–86.

Kumar, L.D., Sankar, S.S., Dr, Venkatesh, P., Kalarani, D.H., 2016. Green synthesis of silver nanoparticles using aerial parts extract of *Echinochloa colona* and their characterization. Eur. J. Pharm. Med. Res. 3, 325–328.

Lee, S.B., Kim, K.H., Hahn, S.J., Chung, I.M., 2003. Evaluation of screening methods to determine the allelopathic potential of rice varieties against *Echinochloa crus-galli* Beauv. var. oryzicola ohwi. Allelopathy J. 12, 37–52.

Lee S.B., Seo K.I., Koo J.H., Hur H.S., Shin J.C., 2005. QTLs and molecular markers associated with rice allelopathy. Proc. 4th World Congress on Allelopathy, "Establishing the Scientific Base", Wagga Wagga, New South Wales, Australia, 21–26 August 2005. pp. 505–507. http://www.regional.org.au/au/allelopathy/2005/2/4/2615_leesb.htm

Lee, J.R., Kim, J.W., Lee, I.Y., Ahn, J.H., 2017. Comparison of the complete genomes of two *Echinochloa* species: barnyard grass and jungle rice. Mitochondrial DNA. Part. B 2, 512–513.

Li, J., Wei, T., Sun, A.R., Ni, H.W., 2013. Evaluation of *Curvularia lunata* strain B6 as a potential mycoherbicide to control barnydrgrass (*Echinochloa crus-galli*). J. Integr. Agric. 12, 1201–1207.

Li, J., Li, M., Gao, X., Fang, F., Dong, L.H., 2016. Isolation and biological characteristics of the biological control fungi BC-1 for *Echinochloa crusgalli*. ACTA Pratacult. Sin. 25, 164–171.

Llewellyn, R., Ronning, D., Ouzman, J., Walker, S., Mayfield, A., Clarke, M., 2016. Impact of weeds on Australian grain production: the cost of weeds to Australian grain growers and the adoption of weed management and tillage practices. Report for GRDC, CSIRO Australia.

Magyar, L., Kónya, Z., 2008. Hungarian results of the photocontrol of weeds. Magy. Gyomkutatás és Technológia 9, 41–52.

Marchesi, C., Chauhan, B.S., 2019. The efficacy of chemical options to control *Echinochloa crus-galli* in dry-seeded rice under alternative irrigation management and field layout. Crop. Prot. 118, 72–78.

Martinez, N.L., Salva, A.P., Finch, R.P., Prado, R.D., 1999. Molecular markers indicate intraspecific variation in die control of *Echinochloa* species with quinclorac. Weed Sci. 47, 310–315.

Martinková, Z., Honěk, A., Lukas, J., 2006. Seed age and storage conditions influence germination of barnyardgrass (*Echinochloa crus-galli*). Weed Sci. 54, 298–304.

Masood, A.P., Bajwa, A.A., Ali, H.H., Chauhan, B.S., 2016. Biology, impact, and management of *Echinochloa colona* (L.) Link. Crop. Prot. 83, 56–66.

Maun, M.A., Barrett, S.C.H., 1986. The biology of Canadian weeds. 77. Echinochloa crus-galli (L.) Beauv. Can. J. Plant. Sci. 66, 739–759.

Mennan, H., Ngouajio, M., Sahin, M., Isik, D., Altop, E.K., 2012. Competitiveness of rice (*Oryza sativa* L.) cultivars against *Echinochloa crus-galli* (L.) Beauv. in waterseeded production systems. Crop. Prot. 41, 1–9.

Michael, P.W., 2001. The taxonomy and distribution of Echinochloa species (barnyardgrasses) in the Asian-Pacific region, with a review of pertinent biological studies. In: Proc. 18th Asian-Pacific Weed Sci. Soc. Conf. Manila, Philippines. pp. 57–66.

Michael, P.W., 2003. Echinochloa. In: Barkworth, M.E., Long, S., Piep, M. (Eds.), Flora of North America. Oxford University Press, New York, p. 390. , 392-403.

Min, H., Chen, M.C., Qian, Z.S., 1994. Biological characteristics of methane emission of *Oryza sativa*, *Panicum crus-galli* and *Cyperus difformis* grown on paddy soil. Pedosphere 4, 165–172.

Mohammad, E., Koocheki, A., Mahallati, M.N., Kumleh, A.,S., 2015. Investigating the germination response of two *Echinochloa* species to temperature and photoperiod with emphasis on invasiveness. Iran. J. Field Crop. Sci. 45, 639–648.

Mohammad, E., Koocheki, A. Shahdi, A., Yaghoubi, B., 2010. The effects of seed burial, and flooding depth on the emergence and growth of *Echinochloa crus-galli*. Proc. 3rd Iranian Weed Sci. Cong. Volume 1: Weed biology and ecophysiology, Babolsar, Iran, 17–18 February 2010. pp. 60–63.

Mollaee, M., Mobli, A., Chauhan, B.S., 2020. The response of glyphosate-resistant and glyphosate-susceptible biotypes of *Echinochloa colona* to carbon dioxide, soil moisture and glyphosate. Sci. Rep. 10, 329. Available from: https://doi.org/10.1038/s41598-019-57307-9.

Moody, K., 1989. Weeds reported in rice in South and Southeast Asia. International Rice Research Institute, Philippines.

Moon, B.C., Cho, S.H., Kwon, O.D., Lee, S.G., Lee, B.W., Kim, D.S., 2010. Modelling rice competition with *Echinochloa crus-galli* and *Eleocharis kuroguwai* in transplanted rice cultivation. J. Crop. Sci. Biotech. 13, 121–126.

Motlagh, M.R.S., 2011. *Fusarium equiseti* (Corda) Saccardo as biological control agent of barnyardgrass (*Echinochloa crus-galli* L.) in rice fields. J. Food Agric. Env. 9, 310–313.

Mujer, C.V., Rumpho, M.E., Lin, J.J., Kennedy, R.A., 1993. Constitutive and inducible aerobic and anaerobic stress proteins in the Echinochloa complex and rice. Plant. Physiol. 101, 217–226.

Muroi, Y., Kobayashi, K., Takai, Y., 2005. Effect of powder and pellet formulation of rice bran on growth of *Echinochloa crus-galli* (L.) Beauv. var. formosensis Ohwi. J. Weed Sci. Technol. 50, 169–175.

Mutti, N.K., Mahajan, G., Jha, P., Chauhan, B.S., 2019. The response of glyphosate-resistant and glyphosate-susceptible biotypes of junglerice (*Echinochloa colona*) to mungbean interference. Weed Sci. 67, 419–425.

Namuco, O.S., Piggin, C., 1999. Weed seedling survival under flooded conditions. In: Proc. 17th Asian-Pacific Weed Control Conf., Bangkok, Thailand. pp. 645–649.

Nguyen, T.H., 2015. Evolution and spread of glyphosate resistant barnyard grass (*Echinochloa colona* (L.) Link) from Australia (PhD Thesis). The University of Adelaide, SA, Australia.

Nguyen, T.H., Malone, J.M., Boutsalis, P., Shirley, N., Preston, C., 2016. Temperature influences the level of glyphosate resistance in barnyardgrass (*Echinochloa colona*). Pest. Manag. Sci. 72, 1031–1039.

Ni, H.W., Moody, K., Robles, R.P., 2004. Analysis of competition between wet-seeded rice and barnyardgrass (*Echinochloa crus-galli*) using a response-surface model. Weed Sci. 52, 142–146.

Noguchi, H.K., Salam, M.A., Kobayashi, T., 2009. A quick seeding test for allelopathic potential of bangladesh rice cultivars. Plant. Prod. Sci. 12, 47–49.

Norris, R.F., 1996. Morphological and phenological variation in barnyardgrass (Echinochloa crus-galli) in California. Weed Sci. 44, 804–814.

Norris, R.F., 2003. *Echinochloa crus-galli* (barnyardgrass) seed rain under irrigated conditions. Asp. Appl. Biol. 69, 163–170.

Norris, R.F., 2007. Weed fecundity: current status and future needs. Crop. Prot. 26, 182–188.

Norsworthy, J.K., Wilson, M.J., Scott, R.C., Gbur, E.E., 2014. Herbicidal activity on acetolactate synthase-resistant barnyardgrass (*Echinochloa crus-galli*) in Arkansas, USA. Weed Biol. Manage. 14, 50–58.

Nozoe, T., Shinano, T., Tachibana, M., Uchino, A., 2010. Tolerance of rice (*Oryza sativa* L.) and *Echinochloa* weeds to growth suppression by rice straw added to paddy soil in relation to iron toxicity. Plant. Prod. Sci. 13, 314–318.

Oerke, E.C., Dehne, H.W., 2004. Safeguarding production-losses in major crops and the role of crop protection. Crop. Prot. 23, 275–285.

OLA and MAFF, 2002. Open Learning Agency (OLA) and British Columbia Ministry of Agriculture, Food, and Fisheries (MAFF). 2002. Guide to weeds in British Columbia. Open Learning Agency. Burnaby, British Columbia, Canada. p. 195.

Ottis, B.V., Talbert, R.E., 2005. Rice yield components as affected by cultivarand seeding rate. Agron. J. 97, 1622–1625.

Pannwitt, H., Westerman, P.R., Mol, F.D., Selig, C., Gerowitt, B., 2017. Biological control of weed patches by seed predators; responses to seed density and exposure time. Biol. Control. 108, 1–8.

Peters, K., Gerowitt, B., 2014. Important maize weeds profit in growth and reproduction from climate change conditions represented by higher temperatures and reduced humidity. J. Appl. Bot. Food Qual. 87, 234–242.

Pons, T., 1982. Factors affecting weed seed germination and seedling growth in lowland rice in Indonesia. Weed Res. 22, 155–161.

Potvin, C., B.R. Strain., 1985. Effects of CO2 enrichment and temperature on growth in two C4 weeds, Echinochloa crus-galli and Eleusine indica. Can. J. Botany 63, 1495–1499.

Pratley, J.E., Broster, J.C., Michael, P., 2008. *Echinochloa* spp. in Australian rice fields—species distribution and resistance status. Aust. J. Agric. Res. 59, 639–645.

Premalal, K.P.S.B., Sangakkara, U.R., Damme, P.V., Bulcke, R., 1998. Effect of root and shoot competition of two weeds on Phaseolus vulgaris beans. In: Proc. 50th International Symposium on Crop Protection, Gent, Belgium. pp. 727–734.

Punia, S.S., Malik, R.K., 1999. Effect of dinitroaniline herbicides on the control of weeds in groundnut (*Arachis hypogaea* L.). In: Proc. 17th Asian-Pacific Weed Sci. Soc. Conf., Thailand. pp. 180–185.

Qing, Z., Liu, B., Gilna, B., Zhang, Y.L., Zhu, C.W., Ma, H.L., et al., 2011. Elevated CO_2 effects on nutrient competition between a C_3 crop (*Oryza sativa* l.) and a C_4 weed (*Echinochloa crus-galli* L.). Nutr. Cycl. Agroecosyst. 89, 93–10493-104.

Rabbani, N., Bajwa, R., Javaid, A., 2011. Interference of five problematic weed species with rice growth and yield. Afr. J. Biotechn 10, 1854–1862.

Rahn, E.M., Sweet, R.D., Vengris, J., Dunn, S., 1968. Life history studies as related to weed control in the Northeast. 5. Barnyardgrass. Agric. Exp. Stn. Univ. Del. Bull. 368, 1–46.

Ramesh, K., Rao, A.N., Chauhan, B.S., 2017. Role of crop competition in managing weeds in rice, wheat, and maize in India: a review. Crop. Prot. 95, 14–21.

Ranagalage, A.S., Wathugala, D.L., 2016. Allelopathic potential of some Sri Lankan traditional rice (*Oryza sativa* L.) cultivars against *Echinochloa crus-galli* L. Tropical Agric. Res. Ext. 19, 190–200.

Rao, A.S., 2008. Effect of time and dose of post-emergence herbicides on *Echinochloa colona* (L.) Link. in blackgram grown as relay crop. Indian J. Weed Sci. 40, 165–168.

Rao, A.N., Matsumoto, H. (Eds.), 2017. Weed Management in Rice in the Asian-Pacific Region. Asian-Pacific Weed Science Society (APWSS). The Weed Science Society of Japan, Japan and Indian Society of Weed Science, India.

Rao, A.N., Moody, K., 1987. Rice yield losses caused by transplanted Echinochloa glabrescens and possible control methods. In: Proc. 11th Asian Pacific Weed Sci. Soc. Conf. 29th Nov-5th Dec 1987, Taipei, Taiwan. pp. 203–210.

Rao, A.N., Moody, K., 1988. Weed control in rice seedling nurseries. Crop. Prot. 7, 202–206.

Rao, A.N., Moody, K., 1990. Weed seed contamination in rice seed. Seed Sci. Technol. 18, 139–146.

Rao, A.N., Moody, K., 1992. Competition between Echinochloa glabrescens and rice (*Oryza sativa*). Trop. Pest. Manage 38, 25–29.

Rao, A.N., Johnson, D.E., Sivaprasad, B., Ladha, J.K., Mortimer, A.M., 2007. Weed management in direct-seeded rice. Adv. Agron. 93, 153–255.

Rao, A.N., Brainard, D.C., Kumar, V., Ladha, J.K., Johnson, D.E., 2017. Preventive weed management in direct-seeded rice: targeting the weed seedbank. Adv. Agron. 144, 45–142.

Rao, A.N., Dixit, S., Juraimi, A.S., 2019. *Echinochloa crus-galli* (L.) P. Beauv (English Name: Barnyardgrass; Malay Name: Rumput sambau). In: Karim, S.M.R., Juraimi, A.S., Rao, A.N., Mispan, M.S., Masilamany, D., Norhafizah, M.Z., Nahar, L. (Eds.), 2019. Invasive Weeds of Malaysia and Their Sustainable Management. UPM Publisher., Universiti Putra Malaysia, Malaysia, pp. 118–138.

Ray, S., Chatterjee, A., 2017. Nutritional and biological importance of the weed *Echinochloa colona*: a review. Int. J. Food Sci. Biotechnol. 2, 31–37.

Refatti, J.P., Avila, L.A.D., Camargo, E.R., Ziska, L.H., Oliveira, C., Salas-Perez, R., et al., 2019. High [CO2] and temperature increase resistance to cyhalofop-butyl in multiple-resistant *Echinochloa colona*. Front. Plant. Sci. 10, 529.

Roche Jr., B.F., Muzik, T.J., 1964. Physiological study of *Echinochloa crus-galli* (L.) Beauv. and the response of its biotypes to sodium 2,2-dichloropropionate. Agron. J. 56, 155–160.

Romano, Y., Mendoza, F., Palmerín, J.A., Quiles, J.M., Amaro, I., Osuna, M.D., 2018. Use of molecular markers for the characterization of rice crop weeds in Extremadura. XVI Congreso de la Sociedad Española de Malherbología, SEMh 2017, Pamplona-Iruña, España, 25–27 octubre, 2017. pp. 337–341.

Saberali, S.F., Mohammadi, K., 2019. The above-ground competition between common bean (Phaseolus vulgaris L.) and barnyardgrass (*Echinochloa crus-galli* L.) affected by nitrogen application. Phytoparasitica. 47, 451–460.

Sadeghloo, A., Asghari, J., Ghaderi-Far, F., 2013. Seed germination and seedling emergence of velvetleaf (*Abutilon theophrasti*) and Barnyardgrass (*Echinochloa crus-galli*). Planta Daninha 31, 259–266.

Sahid, I.B., Hossain, M.S., 1995. The effects of flooding and sowing depth on the survival and growth of five rice-weed species. Plant. Prot. Q. 10, 139–142.

Salas, I., Cirujeda, A., Cavada, F.S., Aibar, J., Zaragoza, C., 2009. The genus Echinochloa in the rice fields of Northeastern Spain: importance and control. XII Congresso da SEMh/XIX Congresso da ALAM/II Congresso da IBCM, iSA Press, Lisbao, Portugal. Spanish Weed Science Society, Madrid, Spain. pp. 95–98.

Samantaray, S., 1991. Studies on the Effect of Minewastes on the Vegetation of the Adjoining Regions of Sukinda Chromite Mine (PhD thesis). Utkal University, Orissa.

Samantaray, S., Rout, G.R., Das, R., 1997. Regeneration of plants via somatic embryogenesis from leaf base and leaf tip segments of *Echinochloa colona*. Plant. Cell Tissue Organ. Cult. 47, 119–125.

Satrapová, J., Hyvönen, T., Venclová, V., Soukup, J., 2013. Growth and reproductive characteristics of C4 weeds under climatic conditions of the Czech Republic. Plant. Soil. Env. 59, 309–315.

Schwartz-Lazaro, L.M., Green, J.K., Norsworthy, J.K., 2017. Seed retention of Palmer amaranth (*Amaranthus palmeri*) and barnyardgrass (*Echinochloa crus-galli*) in soybean. Weed Technol. 31, 617–622.

Sen, D.N., 1981. Ecological approaches to Indian weeds. Ecological approaches to Indian weeds. Geobios International. P.O. Box 14, Jodhpur 342 001 India, viii + 301 pp.

Shabbir, A., Chauhan, B.S., Walsh, M.J., 2019. Biology and management of *Echinochloa colona* and *E. crus-galli* in the northern grain regions of Australia. Crop. Pasture Sci. 70, 917–925.

Shahgholi, H., Makarian, H., Shokati, B., Talaei, G.H., Asgharipour, M.R., 2015. Tillage methods affect germination and species similarity of soil weed seeds bank? Acta Technol. Agric. 18, 97–101.

Sheble, S.M., 2004. Junglerice (*Echinochloa colona*) interference and critical period of control in drill-seeded rice (*Oryza sativa* L.). Egypt. J. Agric. Res. 82, 319–329.

Sherif, E.A.A., 2007. *Echinochloa colona* (L.) Link., a promising species to cultivate salt affected soils in arid lands. American-EuroAsian J. Agric. Environ. Sci. 2, 767–774.

Shimura, K., Okada, A., Okada, K., Jikumaru, Y., Ko, K.W., Toyomasu, T., et al., 2009. Identification of a biosynthetic gene cluster in rice for momilactones. J. Biol. Chem. 282, 34013–34018.

Shipley, B., Parent, M., 1991. Germination responses of 64 wetland species in relation to seed size, minimum time to reproduction and seedling relative growth rate. Funct. Ecol. 5, 111.

Shrestha, A., DeSouza, L.L., Yang, P., Sosnoskie, L., Hanson, B.D., 2018. Differential tolerance of glyphosate-susceptible and glyphosate-resistant biotypes of junglerice (*Echinochloa colona*) to environments during germination, growth, and intraspecific competition. Weed Sci. 66, 340–346.

Singh, S., Sharma, N.P., 2018. Medicinal potential of weed *Echinochloa colona* (l.) Link: a review. Int. J. Pharmacognosy Phytochem. Res. 10, 98–102.

Singh, S., Bhullar, M.S., Chauhan, B.S., 2015. Seed bank dynamics and emergence pattern of weeds as affected by tillage systems in dry direct-seeded rice. Crop. Prot. 67, 168–177.

Singh, M., Bhullar, M.S., Chauhan, B.S., 2017. Relative time of weed and crop emergence is crucial for managing weed seed production: a study under an aerobic rice system. Crop. Prot. 99, 33–38.

Singh, M., Bhullar, M.S., Gill, G., 2018. Integrated weed management in dry-seeded rice using stale seedbeds and post sowing herbicides. Field Crop. Res. 224, 182–191.

Sitthinoi, P., Lertmongkol, S., Chanprasert, W., Vajrodaya, S., 2017. Allelopathic effects of jungle rice [*Echinochloa colona* (L.) Link] extract on seed germination and seedling growth of rice. Agric. Nat. Resour. 51, 74–78.

Smith, R.J. Jr., 1983. Weed of major economic importance in rice and yield losses due to weed competition. In: Proc. Conference on Weed Control in Rice. International Rice Research Institute, Los Banos, Philippines. pp. 19–36.

Smith Jr., R.J., Fox, W.T., 1973. Soil water and growth of rice and weeds. Weed Sci. 21, 61–63.

Smith, B.S., Murray, D.S., Green, J.D., Wanyahaya, W.M., Weeks, D., 1990. Interference of three annual grasses with grain sorghum (*Sorghum bicolor*). Weed Technol. 4, 245–249.

Soerjani, M., Kostermans, A.J.G.H., Tjitrosoepomo, G., 1986. Weeds of Indonesia. Balai Pustaka, Jakarta, Indonesia, p. 716.

Son, D.H., Thil, H.L., Kato-Noguchi, H., 2010. Allelopathic potential and isolation process of allelopathic substances in barnyardgrass (*Echinochloa crus-galli*). Omonrice 17, 143–146.

Song, J.S., Kim, J.W., Im, J.H., Lee, K.J., Lee, B.W., Kim, D.S., 2017. The effects of single- and multiple-weed interference on soybean yield in the Far-Eastern region of Russia. Weed Sci. 65, 371–380.

Sun, B., Kong, C.H., Wang, P., Qu, R., 2012. Response and relation of allantoin production in different rice cultivars to competing barnyardgrass. Plant. Ecol. 213, 1917–1926.

Sunyob, N.B., Juraimi, A.S., Hakim, M.A., Man, A., Selamat, A., Alam, M.A., 2015. Competitive ability of some selected rice varieties against weed under aerobic condition. Int. J.Agric. Biol. 17, 61–70.

Suzuki, M., 2003. Dormancy and germination characteristics of buried main upland weeds in Akita Prefecture. J. Weed Sci. Technol. 48, 130–139.

Swaminathan, S., Morrone, D., Wang, Q., Fulton, D.B., Peters, R.J., 2009. CYP76M7 is an ent-cassadiene C11α-hydroxylase defining a second multifunctional diterpenoid biosynthetic gene cluster in rice. Plant. Cell. 21, 3315–3325.

Tabacchi, M., Tabacchi, R., Spada, A., Ferrero, A., 2006. Morphological traits and molecular markers for classification of *Echinochloa* species from Italian rice fields. Weed Sci. 54, 1086–1093.

Tahir, H., 2016. Characterization of *Echinochloa* spp. in Arkansas (MSc thesis). Master of Science in Crop, Soil and Environmental Sciences, University of Arkansas, Arkansas, USA. p. 163.

Talbert, R.E., Burgos, N.R., 2007. History and management of herbicide-resistant barnyardgrass (*Echinochloa crus-galli*) in Arkansas rice. Weed Technol. 21, 324–331.

Tehranchian, P., Norsworthy, J.K., Johnson, D.B., Schrage, B.W., Bell, H.D., Bararpour, M.T., et al., 2014. Evaluation of conventional and hybrid rice, seeding rate, and herbicide program on barnyardgrass control in Clearfield rice. B.R. Wells Rice Research Studies—Arkansas Agricultural Experiment Station, University of Arkansas System. 617, pp. 234–238.

Tsukamoto, H., Gohbara, M., Tsuda, M., Fujimori, T., 1997. Evaluation of fungal pathogens as biological control agents for the paddy weed, *Echinochloa* species by drop inoculation. Ann. Phytopathol. Soc. Japan. 63, 366–372.

Tuat, N.V., Hanh, T.H., Hetherington, S.D., Chin, D.V., Auld, B.A., 1999. Potential use of *Exserohilum monoceras* as a biological control agent for *Echinochloa* spp. (barnyard grass) in Vietnam. Plant. Prot. Q. 14, 136–138.

vanGessel, M.J., Renner, K.A., 1990. Redroot pigweed (Amaranthus retroflexus) and barnyardgrass (*Echinochloa crus-galli*) interference in potatoes (*Solanum tuberosum*). Weed Sci. 33, 338–343.

Vencill, W., Nichols, R., Webster, T., Moss, S., 2014. Framework for an expert evaluation for the evolution of weed resistance. Julius-Kühn-Archiv . Available from: https://doi.org/10.5073/jka.2014.443.004.

Vidotto, F., Letey, M., Follis, F., Saglia, A.A., 2009. Use of steam for the devitalization of weed seeds in the soil. Protezione delle Colture 2, 127–128.

Viero, J.L.C., Schaedler, C.E., Azevedo, E.B.D., Santos, J.V.A.D., Scalcon, R.D.M., et al., 2018. Endozoochorous dispersal of seeds of weedy rice (*Oryza sativa* L.) and barnyardgrass (*Echinochloa crus-galli* L.) by cattle. Ciência Rural. 48, e20170650.

Von Redwitz, C., Pannwitt, H., Gerowitt, B., 2016. About the interplay of sensitive and resistant biotypes in weed populations—simulation exercises for *Echinochloa crus-galli* in maize crops. Julius-Kuhn- Arch. 452, 93–99.

Walker, S., Widderick, M.J., Cook, T., Rice, L., 2013. Innovative approaches to manage glyphosate resistant weeds in sub-tropical grain region of Australia. In: Proc. 24th Asian Pacific Weed Science Society Conference, Indonesia. pp. 284–288.

Wathugala, D.L., Ranagalage, A.S., 2014. Allelopathic potential of plant residues of thirty Sri lankan tra-ditional rice (*Oryza sativa* L.) varieties on barnyardgrass (*Echinochloa crus-galli*). Trop. Agric. Res. Extn 17, 115–123.

Weerarathne, L.V.Y., Suriyagoda, L.D.B., Marambe, B., 2015. Barnyard grass (*Echinochloa crus-galli* (L.) P.Beauv) is less competitive on rice (*Oryza sativa* L.) when phosphorus (P) is applied to deeper layers in P-deficient and moisture-limited soils. Plant. Soil. 391, 1–7.

Werth, J.A., Preston, C., Roberts, G.N., Taylor, I.N., 2008. Weed management impacts on the population dynamics of barnyardgrass (*Echinochloa crus-galli*) in glyphosate-resistant cotton in Australia. Weed Technol. 22, 190–194.

Westerman, P.R., Pannwitt, H., Selig, C., Mol, F.D., Gerowitt, B., 2014. Testing for the weed control potential of seed predators in agroecosystems. IOBC/WPRS Bull. 100, 143–146.

Widderick, M.J., Bell, K.L., Boucher, L.R., Walker, S.R., 2013. Control by glyphosate and its alternatives of glyphosate-susceptible and glyphosate-resistant *Echinochloa colona* in the fallow phase of crop rotations in subtropical Australia. Weed Biol. Manage. 13, 89–97.

Wilson, M.J., Norsworthy, J.K., Scott, R.C., Gbur, E.E., 2014. Program approaches to control herbicide-resistant barnyardgrass (*Echinochloa crus-galli*) in Midsouthern United States rice. Weed Technol. 28, 39e46.

Wongsriphuek, C., Dugger, B.D., Bartuszevige, A.M., 2008. Dispersal of wetland plant seeds by mallards: influence of gut passage on recovery, retention, and germination. Wetlands. 28, 290–299.

WSSA, 2020. http://wssa.net/wp-content/themes/WSSA/WorldOfWeeds/barnyardgrass.html. Weed Science Society of America (WSSA), USA.

Wu, H., Walker, S., Osten,V., Taylor, I., Sindel, B., 2004. Emergence and persistence of barnyard grass (*Echinochloa colona* (L.) Link) and its management options in sorghum. Fourteenth Australian Weeds Conference, Wagga Wagga, NS, Australia. pp. 538–541.

Wu, Z.C., Qing, Z., Ziska, L.H., Guo, Z.J., Bing, Z.Z., Zhao, H.W., et al., 2006. Effect of free-air CO2 enrichment (FACE) on root growth in C3 crop (rice, *Oryza sativa*) and C4 weed (barnyardgrass, *Echinochloa crus-galli*). J. Ecol. Nat. Environ. 22, 1–4.

Wu, S.G., Wang, Q., Zhao, X.P., Wu, C.X., Chen, L.P., Shen, J.L., 2007. Dormancy and dormancy breaking in barnyard grass (*Echinochloa crus-galli*). Acta Agric. Zhejiangensis 19, 225–228.

Wu, Z.C., Qing, Z., Ziska, L.H., Guo, Z.J., Bing, Z.Z., Gang, L., 2008. Effect of nitrogen supply on carbon dioxide-induced changes in competition between rice and barnyardgrass (*Echinochloa crus-galli*). Weed Sci. 56, 66–71.

Wu, L.M., Fang, Y., Yang, H.N., Bai, L.Y., 2019. Effects of drought-stress on seed germination and growth physiology of quinclorac-resistant *Echinochloa crus-galli*. PLoS One 14, e0214480.

Xia, S.X., Wang, H.B., Lin, H.F., He, H.B., Lu, J.C., Zeng, C.M., et al., 2009. Effects of weed suppression by different allelopathic rice varieties under dry-raising condition. Chin. J. Eco-Agric 17, 842–846.

Xuan, T.D., Chung, I.M., Khanh, T.D., Tawata, S., 2006. Identification of phytotoxic substances from early growth of barnyardgrass (*Echinochloa crus-galli*) root exudates. J. Chem. Ecol. 32, 895–906.

Yabuno, T., 1962. Cytotaxonomic studies on the two cultivated species and the wild relatives in the genus *Echinochloa*. Cytologia. 27, 296–305.

Yabuno, T., 1983. Biology of *Echinochloa* species. In: Weed control in rice. International Rice Research Institute and International Weed Science Society, Manila, Philippines. pp. 307–318.

Yamamoto, T., Tomita, K.H., Kosemura, S., Yamamura, S., Yamada, K., Hasegawa, K., 1999. Allelopathic substance exuded from a serious weed, germinating barnyarngrass (*Echinochloa crus-galli* L.) roots. J. Plant. Growth Reg. 18, 65–66.

Yoneyama, K., Takeuchi, Y., 2000. Responses of barnyardgrass (*Echinochloa crus-galli* (L) Beauv. var. crus-galli) to water drained, saturated and flooded soils in root aerenchyma formation. J. Weed Sci. Technol. 45, 21–25.

Yoshioka, T., Satoh, S., Yamasue, Y., 1998. Effect of increased concentration of soil CO_2 on intermittent flushes of seed germination in *Echinochloa crus-galli* var. crus-galli. Plant, Cell Environ. 21, 1301–1306.

You, Z.G., Kahnt, G., 1990. Influence of barnyardgrass (*Echinochloa crus-galli*) on the nodulation and N_2-fixation of soyabeans. Z. Pflanzenkrankh. Pflanzenschutz 12, 285–291.

Yuan, M.H., Sheng, L.B., Wei, L.X., Wei, L.Z., 2014. Germination response to differing salinity levels for 18 grass species from the saline-alkaline grasslands of the Songnen Plain, China. Pak. J. Bot. 46, 1147–1152.

Zhang, W.M., 1996. Biological control of Echinochloa species with pathogenic fungi (PhD thesis). McGill University, Montréal, Québec, Canada.

Zhang, W.M., Watson, A.K., 1997a. Effect of dew period and temperature on the ability of Exserohilum monoceras to cause seedling mortality of *Echinochloa* species. Plant. Dis. 81, 629–634.

Zhang, W.M., Watson, A.K., 1997b. Host range of Exserohilum monoceras, a potential bioherbicide for the control of *Echinochloa* species. Can.J. Bot. 75, 685–692.

Zhang, W.M., Moody, K., Watson, A.K., 1996. Responses of *Echinochloa* species and rice (Oryza sativa) to indigenous pathogenic fungi. Plant. Dis. 80, 1053–1058.

Zhang, H.J., Jiang, S.R., Liu, X., Liu, T.G., Li, X.J., Ni, H.W., 2005. A comparison of competitiveness between glufosinate-tolerant transgenic rice cultivar 99-1 and non-transgenic cultivar Yuefu against *Echinochloa crus-galli*. Plant. Prot. 31, 59–63.

Zhang, J.P., Peng, G., Duan, G.F., Zhou, Y.J., Yang, S.A., Yu, L.Q., 2014. *Bipolaris eleusines*, a potential mycoherbicide candidate for control of barnyardgrass (*Echinochloa crus-galli*). Biocontrol Sci. Technol. 24, 839–846.

Zhang, J.P., Fu, F., Liu, C., Lin, Z.X., Wang, Y., Ye, C.Y., et al., 2017a. Chloroplast DNA markers for *Echinochloa* taxa. Weed Res. 57, 355–360.

Zhang, Z., Gu, T., Zhao, B., Yang, X., Peng, Q., Li, Y., et al., 2017b. Effects of common Echinochloa varieties on grain yield and grainquality of rice. Field Crop. Res. 203, 163–172.

Zhang, W., Zhang, L., Li, R., Chen, G.Q., Dong, L.Y., 2018. The response of seed germination of penoxsulam-resistant Echinochloa crusgalli (L.) P. Beauv from different genetic backgrounds to adverse conditions. J. Nanjing Univ. Nat. Sci. 41, 1045–1053.

Chapter 11

Ipomoea hederacea, *Ipomoea lacunosa*, and *Ipomoea purpurea*

Eric A.L. Jones, Diego J. Contreras and Wesley J. Everman
Department of Crop and Soil Sciences, North Carolina State University, Raleigh, NC, United States

Introduction

The *Ipomoea* genus belongs to the Convolvulaceae family. The genus name *Ipomoea* is derived from the Greek words *ips* which means "a worm" and *homoios* which means "resembling," referring to the wormlike twining growth (DeFelice, 2001). The weedy species of the *Ipomoea* genus are commonly referred to as "morningglory" as an entity. The *Ipomoea* spp. has been referred to as both a beautiful ornamental flowering plant and a prolific weed depending on circumstances (DeFelice, 2001). *Ipomoea* spp. originated in the tropical Americas and were collected for food and ornamental purposes in Europe and North America (Bryson and DeFelice, 2010; DeFelice, 2001). By the early 18th century, selected *Ipomoea* spp. escaped cultivation to become a weed (Holm et al., 1977; Pursh, 1814). The common weedy *Ipomoea* species include *Ipomoea hederacea* (L.) Jacq, *Ipomoea lacunosa* (L.), and *Ipomoea purpurea* (L.) Roth. These species can be distinguished apart phenotypically, but the methodology of control is equal for the species.

Ipomoea spp. are ubiquitous weeds across disturbed ecosystems globally (DeFelice, 2001; Holm et al., 1979). *Ipomoea* spp. are difficult to control due to differential morphology of desirable plants. Management options of *Ipomoea* spp. for a respective disturbed ecosystem are similar worldwide. While herbicides are commonly used to control *Ipomoea* spp. in field and vegetable crops, biological, cultural, and mechanical tactics can also be utilized to effectively control plants inhabiting disturbed areas.

Chauhen-BMPCWS-1632754. DOI: https://doi.org/10.1016/B978-0-12-822917-0.00008-2

Biology

Vegetative structures

Cotyledons

Cotyledons are unique to each *Ipomoea* spp. (Fig. 11.1). *Ipomoea hederacea* cotyledons are deeply indented with rounded points (Bryson and DeFelice, 2010). *Ipomoea lacunosa* cotyledons are deeply indented with long, pointed lobes (Bryson and DeFelice, 2010). *Ipomoea purpurea* cotyledons are indented with rounded lobes (Bryson and DeFelice, 2010).

Stems

Highly branched; vining to climbing growth reaching up to 2.0 m or more. Stems of *I. hederacea* are densely pubescent (Bryson and DeFelice, 2010). Stems of *I. lacunosa* are glabrous to sparsely pubescent (Bryson and DeFelice, 2010). Stems of *I. purpurea* are tomentose (Bryson and DeFelice, 2010).

Leaves

Leaves of *Ipomoea* spp. are highly plastic (Fig. 11.2). *Ipomoea hederacea* has alternate leaves (5.0–12.0 cm long; 4.0–10.0 cm wide); either heart-shaped (entireleaf) or with three lobes (ivyleaf) (Bryson and DeFelice, 2010). *Ipomoea lacunosa* has alternate leaves (2.0–10.0 cm long; and 1.5–10.0 cm wide); variably heart-shaped or 3–5 lobes, with purple

FIGURE 11.1 *Ipomoea* spp. cotyledons. Left to right: *I. hederacea*, *I. lacunosa*, and *I. purpurea*.

FIGURE 11.2 *Ipomoea* spp. leaves. Left to right: *I. hederacea*, *I. lacunosa*, and *I. purpurea*.

margins. Petioles are 1.0–3.0 cm long (Bryson and DeFelice, 2010). *Ipomoea purpurea* has loosely pubescent alternate leaves (1.0–12.0 cm long; 1.0–12.0 cm wide); heart-shaped with margins entire or rarely five-lobed. Petioles are 1.0–14.0 cm long (Bryson and DeFelice, 2010).

Roots

Roots are fibrous with a distinct taproot. *Ipomoea lacunosa* possesses a more slender taproot when compared to other *Ipomoea* spp. (Bryson and DeFelice, 2010).

Reproductive structures

Inflorescences

Ipomoea spp. inflorescences consist of axillary clusters that are either solitary to a few or also groups of 3–5 (Fig. 11.3). Flowers are funnelform with corollas measuring between 2.0 and 6.0 mm long and differ in color by species (Bryson and DeFelice, 2010). The flowers of *I. hederacea* are blue with a white center, and the inflorescence turns purple with senescence (Bryson and DeFelice, 2010). The flowers of *I. lacunosa* are entirely white (Bryson and DeFelice, 2010). The flowers of *I. purpurea* have a purple inflorescence which can possess a blue to white hue with a white center (Bryson and DeFelice, 2010).

Seeds

Seeds of *Ipomoea* spp. exhibit similar phenotypes. Seeds of *I. hederacea* are wedge-shaped ranging from dull gray to black, pubescent, 4.0–5.0 mm long, prominent hilum scar (Bryson and DeFelice, 2010). Seeds of *I. lacunosa* are wedge-shaped, large (5.0–6.0 mm long; 4.5–5.0 mm wide); glossy black, nonpubescent, with a smooth hilum scar (Bryson and DeFelice, 2010). Seed shape and size of *I. purpurea* is similar to *I. hederacea* but are nonpubescent, brown to black, with a smooth hilum scar (Bryson and DeFelice, 2010).

FIGURE 11.3 *Ipomoea* spp. flowers. Left to right: *I. purpurea, I. hederacea*, and *I. lacunose*.

Germination and growth

Ipomoea spp. can germinate over a wide range of temperatures. High germination, ranging from 80% to 100%, occurred at day/night temperatures of 25/15°C to 30/20°C (Cole and Coats, 1973; Oliveira and Norsworthy, 2006; Singh et al., 2012). Photoperiod is negligible for *Ipomoea* spp. germination. Under continuous light and dark conditions, *Ipomoea* spp. still germinated >80% (Cole and Coats, 1973; Norsworthy, 2004; Oliveira and Norsworthy, 2006; Singh et al., 2012). *Ipomoea* spp. are capable of emerging in saturated soils with delayed emergence and reduced germination (Oliveira and Norsworthy, 2006). The physical barrier to seed dormancy in *Ipomoea* spp. is an impermeable seed coat (Chandler et al., 1977). In turn, the impermeable seed coat facilitates the prolonged persistence in the soil, up to 50 years (Elmore et al., 1990; Toole and Brown, 1946). Maximum germination of *Ipomoea* spp. occurs at depths ranging from 1 to 4 cm, but seedlings can emerge from depths as deep as 10 cm with a significantly reduced germination rate (Gomes et al., 1978; Oliveira and Norsworthy, 2006; Singh et al., 2012). *Ipomea* spp. are C_3 plants; thus vegetative growth is limited by CO_2 (Pearcy and Ehleringer, 1984). Since *Ipomoea* spp. are not light limited, plants may not realize growth reductions in shaded environments (Murdock et al., 1986; Pearcy and Ehleringer, 1984). Plants that emerge during the hotter months can accumulate more vegetative growth compared to plants that emerge earlier in the growing season (Thullen and Keely, 1983). *Ipomoea* spp. grow prostrate but have the innate ability to detect neighboring plants/structures and begin to exhibit vining growth upon the neighboring plants/structures (Cordes and Bauman, 1984; Price and Wilcut, 2007).

Reproduction

The weedy *Ipomoea* spp. are summer annuals, thus propagate via seeds (DeFelice, 2001). *Ipomoea* spp. are monoecious and self-pollinate, but cross-pollination and hybridization can occur via insect pollinators or wind deposition (Bryson et al., 2008; DeFelice, 2001; Hull-Sanders et al., 2005). Photoperiod dictates flower initiation in *Ipomoea* spp. and occurs when the day/night photoperiod is approximately 13/11 h (Klingman and Oliver, 1996; Senseman and Oliver, 1993; Simonsen and Stinchcombe, 2010). *Ipomoea* spp. can produce 5000–26,000 seeds per plant in lieu of interspecific and intraspecific competition (Crowley and Buchanan, 1982; Gomes et al., 1978; Senseman and Oliver, 1993). *Ipomoea* spp. seed production can be reduced (>85%) with interspecific and intraspecific competition (Keeley et al., 1986; Senseman and Oliver, 1993). Plants that emerge during hotter months will produce more flowers and be more fecund (Thullen and Keely, 1983).

Distribution

The *Ipomoea* genus originates from tropical America (*I. hederacea* and *I. purpurea*) or subtropical North America (*I. lacunosa*) and is distributed across all continents, excluding Antarctica (Bryson and DeFelice, 2010; DeFelice, 2001; Holm et al., 1979). *Ipomoea* spp. are ruderals and can be found in cultivated areas, fields, pastures, fencerows, roadsides and disturbed and waste sites (Bryson and DeFelice, 2010; DeFelice, 2001; Holm et al., 1977).

Management

Herbicide resistance

There have been no confirmed herbicide-resistant *Ipomoea* spp. populations globally (Heap, 2020). However, *Ipomoea* spp. are tolerant to many herbicides making control difficult (Baucom and Mauricio, 2008; Higgins et al., 1988; Sunderland and Coble, 1994).

Agronomic crops

Ipomoea spp. inhabit the fields of prominent agronomic crops globally (Bryson and DeFelice, 2010; DeFelice, 2001; Holm et al., 1977). *Ipomoea* spp. not controlled can reduce yield ranging up to 90% in agronomic crops (Higgins et al., 1988; Norsworthy and Oliver, 2002; Norsworthy and Oliveria, 2004; Rogers et al., 1995). The critical period of *Ipomoea* spp. control in agronomic crops ranges between 3 and 8 weeks after crop emergence before an unacceptable yield loss is incurred (Everman et al., 2008; Keeley et al., 1986; Norsworthy and Oliveria, 2004; Oliver et al., 1976; Rogers et al., 1995; Wilson and Cole, 1966). *Ipomoea* spp. are primarily controlled with herbicides in agronomic crops, despite the genus being tolerant to some commonly used herbicides (Baucom and Mauricio, 2008; Higgins et al., 1988; Sunderland and Coble, 1994). *Ipomoea* spp. can be controlled with herbicides at preplant burndown, preemergence and postemergence, and harvest aid application timings in agronomic crops (Table 11.1).

Preplant burndown herbicide applications will usually consist of a nonselective herbicide, such as glyphosate or paraquat, to control the *Ipomoea* and other weed species present. Expected control of *Ipomoea* spp. should be $\geq 85\%$ with effective preplant burndown herbicide applications (Vollmer et al., 2019). Preplant burndown herbicide applications are most commonly utilized in reduced- or no-till fields to facilitate planting (Davis, 2010). Preplant herbicide applications can also terminate cover crops to further aid in achieving a timely planting date (Davis, 2010). Herbicides with residual activity tank-mixed with nonselective herbicides can increase the duration of control of *Ipomoea* spp. that have not germinated in case planting or

TABLE 11.1 Herbicides that can be applied to agronomic crops to control *Ipomoea* spp.

Site of action	WSSA group	Active ingredient	Application Timing	Crop							
				Monocot forage	Corn	Sorghum	Legume forage	Soybean	Rice	Cotton	Peanuts
ALS	2	Chloransulam	POST				x	x			
	2	Diclosulam	PRE		x						x
	2	Flumetsulam	PRE				x	x			
	2	Imazamox	POST				x	x			
	2	Imazaquin	PRE/POST				x	x			
	2	Imazethapyr	POST						x		
	2	Nicosulfuron	POST	x	x	x					
	2	Prosulfuron	POST	x	x	x					
	2	Pyrithiobac	POST							x	
	2	Rimsulfuron	POST	x	x	x					
	2	Thifensulfuron	POST	x	x	x					
	2	Thifensulfuron	POST	x	x	x	x	x			
	2	Trifloxysulfuron	POST							x	
Synthetic Auxins	4	2,4-D	POST	x	x	x	x	x[a]	x		
	4	2,4-DB	POST				x	x			x

	4	Dicamba	POST	x	x	x	x	x[a]			
	4	Florpyrauxifen	POST						x		
	4	Quinclorac	POST	x	x	x			x		
	4	Triclopyr	POST						x		
PSII	5	Atrazine	PRE/POST	x	x	x					
	6	Bromoxynil	POST	x	x	x					
	7	Diuron	PRE							x	
	7	Flumeturon	PRE							x	
EPSPS	9	Glyphosate	POST	x[a]	x[a]		x[a]	x[a]	x[a]		
Glutamine synthase	10	Glufosinate	POST	x[a]	x[a]		x[a]	x[a]	x[a]		
PBS	13	Clomazone	PRE				x	x		x	
PPO	14	Acifluorfen	POST								x
	14	Carfentrazone	POST	x	x	x			x		x
	14	Fomesafen	PRE/POST				x	x		x	
	14	Lactofen	POST				x	x		x	x
	14	Saflufenacil	PRE/POST	x	x	x	x	x	x	x	
VLCFA	15	Dimethinamid-p	PRE	x	x	x	x	x		x	x
	15	Pyroxasulfone	PRE				x	x			
	15	*S*-metalochlor	PRE	x	x	x	x	x			x

(*Continued*)

TABLE 11.1 (Continued)

Site of action	WSSA group	Active ingredient	Application Timing	Crop							
				Monocot forage	Corn	Sorghum	Legume forage	Soybean	Rice	Cotton	Peanuts
HPPD	27	Mesotrione	PRE/POST	x	x	x					
	27	Pyrasulfotole	POST	x	x	x					
	27	Tembtrione	POST	x	x						
	27	Tolpyralate	POST	x	x	x					
	27	Topramezone	PRE/POST	x	x	x					

WSSA, Weed Science Society of America; *ALS*, acetolactate synthase; *PSII*, photosystem II; *ESPS*, 5-enolpyruvyl-shikimate-3-phospate; *DBS*, diterpene biosynthesis; *PPO*, protoporphyrinogen oxidase; *VLCFA*, very long-chain fatty acid; *HPPD*, 4-hydroxyohenylpyruvate dioxygenase; *PRE*, preemergence, *POST*, postemergence.
[a] *2,4-D, Dicamba, Glyphosate- or glufosinate-resistant crop.*

subsequent herbicide applications are delayed (Jhala et al., 2013). Controlling *Ipomoea* spp. prior to crop establishment or planting is crucial as plants growing without crop/desirable plant interference will grow more vigorously which could result in harder to control plants (Murdock et al., 1986; Shetty et al., 1982).

Preemergence herbicide applications control *Ipomoea* spp. at the germination stage before the seedling emerges from the soil surface. Expected control of *Ipomoea* spp. should be $\leq$85% with preemergence herbicide applications (Crowley et al., 1979; Everman et al., 2009). Some growers will recurrently apply preemergence herbicides during the growing season to provide residual control of later germinating *Ipomoea* spp. (Osterholt et al., 2017; Riar et al., 2013). While later germinating *Ipomoea* spp. may not reduce crop yield, plants may produce seed that will result in plants that have to be managed in following years (Everman et al., 2009; Gallandt, 2006).

Postemergence herbicide applications control *Ipomoea* spp. when the plant has emerged from the soil surface. Expected control of *Ipomoea* spp. should be $\geq$85% with postemergence herbicide applications for acceptable control (Everman et al., 2009; Palmer et al., 2000; Parker et al., 2006). Postemergence herbicide applications should be executed before *Ipomoea* spp. plants exceed 7 cm in size (cotelydon to four-leaf) and begin vining, or running, onto other plants to ensure control. Desiccants can be used to reduce the negative impact that *Ipomoea* spp. cause during crop harvest. Desiccants consist of postemergence herbicide applications that cause premature senescence of crops and weeds which facilitates the harvesting process. Harvest aid desiccants applied include paraquat and glyphosate. Caution should be exercised when recurrently applying glyphosate in glyphosate-tolerant crops as *I. lacunosa* and *purpurea* could become more prevalent and reduce harvest efficiency as these species exhibit tolerance to this herbicide (Burke et al., 2009; Hilgenfeld et al., 2004; Owen, 2008). Since herbicide-resistant *Ipomoea* spp. have not evolved, farmers should steward herbicides by means of rotating herbicide mode of action, tankmixes, timely applications, and nonchemical control to increase the longevity of effective herbicides (Beckie, 2006).

Despite the intensive use of chemical control, biological, cultural, and mechanical management options are efficacious on *Ipomoea* spp. and increased control can be realized if the different tactics are integrated (Bhowmik, 1996; Liebman et al., 1997; Sanyal et al., 2008). Granivores and seed decay comprise the primary means of biological control of *Ipomoea* spp. (Lundgren, 2005, Mott and McKeon, 1977). Cultural practices will control and suppress infestations of *Ipomoea* spp. Such cultural practices include competitive crops/crop variety, crop rotation, fertilization, planting date, row spacing, and increased crop seeding rate (Bhowmik, 1996; Nalewaja, 1999; Sanyal et al., 2008; Place et al., 2010). Mechanical practices that are utilized to manage *Ipomoea* spp. include hand weeding, interrow cultivation, and preplant tillage. While biological, cultural, and mechanical practices will not be as efficacious as herbicides alone, integrating several practices together will result in greater efficacy (Bhowmik, 1996; Liebman et al., 1997; Nalewaja, 1999).

Any input that increases crop vigor will contribute to cultural weed control. Crop rotation aids in weed management by utilizing differences in morphology, physiology, and production practices of the crop grown and to allow rotation of the herbicides used in the crop rotation sequence (Ball, 1992; Harper, 1956; Place et al., 2010). A crop with tall stature and faster canopy development will suppress germination and growth of *Ipomoea* spp. Early planting date, narrow row spacing, and increased crop seeding rate can result in faster canopy closure (Gill and Holmes, 1997; Nalewaja, 1999: Place et al., 2010). Once canopy closure occurs less light will reach the soil surface, reducing *Ipomoea* spp. germination, leading to less crop competition and seed inputs into the soil.

Tillage burying seeds will not significantly control *Ipomoea* spp., as the large seeds can emerge from depths up to 10 cm (Gomes et al., 1978; Kapusta, 1979; Oliveira and Norsworthy, 2006; Singh et al., 2012; Wilcut et al., 1987). However, since reduced or no-till fields are becoming more common globally to reduce erosion and pollution, small-seeded weeds may be selected in no-till fields over *Ipomoea* spp. (Buhler et al., 1997; Gebhart et al., 1985; Holland, 2004). Cultivation is efficacious on *Ipomoea* spp. as the emerged plant (along with apical meristem) is excised from the soil surface. Caution should be exercised when trying to control large *Ipomoea* spp. plants with interrow cultivation as plants can regrow from excised branches.

While the *Ipomoea* spp. can directly reduce crop yield via competition and interference, the reduction of crop yields can be incurred indirectly. *Ipomoea* spp. can grow and reproduce on the ground surface; however, plants appear to grow toward erect plants or structures that provide support for a climbing growth habit and light interception (Price and Wilcut, 2007). *Ipomoea* spp. interfere with harvest efficiency by tangling and wrapping around crops that result in crop lodging and obstructions of harvest machinery (Elmore et al., 1990; Johnson, 1971; Wood et al., 1999).

Horticultural crops

Ipomoea spp. inhabits the fields of prominent horticultural crops, globally (Bryson and DeFelice, 2010; DeFelice, 2001; Heap, 2020; Schutte, 2014). *Ipomea* spp. not controlled can reduce yield ranging from 25% to 60% (Monaco et al., 1981; Seems et al., 2003). The critical period of *Ipomea* spp. control in horticultural crops ranges between 2 and 6 weeks after crop emergence before an unacceptable yield loss is incurred (Seems et al., 2003). The critical period of *Ipomoea* spp. control in horticultural crops may be negligible as weed control is conducted season long.

While agronomic and horticultural crops share similar herbicide application timings, the number of registered herbicides in horticultural crops will be significantly less than the number of registered herbicides in agronomic crops (Fennimore and Doohan, 2008; Glaze and Hall, 1990; Harr et al., 2002) (Table 11.2). While herbicides can be applied to control *Ipomoea* spp. in horticultural crops, farmers may choose other means of control as not to injure their crops which will result in reduced market value (Harr et al., 2002). *Ipomoea* spp. can be difficult to control in some horticulture crops with herbicides, such as sweet potatoes [*Ipomoea batatas* (L.) Lam.] as both weed and crop share physiological similarities (Glaze and Hall, 1990). Since chemical control is not as intensive in horticultural crops as agronomic crops, biological, cultural, and mechanical control tactics are utilized to control *Ipomoea* spp. infestations.

Since horticultural crops do not generally produce a canopy closure due to morphology and wide row spacing, shading is artificially produced to inhibit weed germination and growth. Plant residue is not favorable in horticultural crops as the residue can harbor insects and pathogens that can inflict plant injury, damage, or death (Alabouvette et al., 2006). Instead, plastic mulch can be laid down in the field rows to inhibit weed germination and interference with the crops (Horowitz et al., 1983; Price et al., 2018). Plastic mulch will also heat the soil beneath which could reach temperatures that could kill *Ipomoea* spp. seeds (Horowitz et al., 1983). Crops will be planted/transplanted directly into the plastic mulch; *Ipomoea* spp. can also utilize the openings in the plastic mulch to emerge and interfere with the crop (Norsworthy et al., 2008). Thus *Ipomoea* spp. will need to be managed if plastic mulch is laid down.

Similar to agronomic crops, the biological control of *Ipomoea* spp. will be primarily by granivores and seed decay. Cultivation can provide control on *Ipomoea* spp. at the beginning of the development of horticultural crops (Price et al., 2018). Mechanical control of *Digitaria* spp. will largely be conducted by cultivation and hand weeding. While cultivation is efficacious on *Digitaria* spp., the cultivation must be implemented early as injury can be incurred by the crops exhibiting vineous and/or large vegetative growth (Monks and Schultheis, 1998; Pannacci et al., 2017). *Ipomoea* spp. can be a serious problem in horticultural crops as both can exhibit vining growth, become entangled, and compromise select control measures (Glaze and Hall, 1990; Price et al., 2018). An example of a compromised control measure would be trying to interrow cultivate *Ipomoea* spp. in sweet potato or watermelon [*Citrullus lanatus* (Thumb.) Matsum. and Nakai] as the cultivation tines can remove and/or injure the crop's vining growth which would result in a lower crop yield. Similar to agronomic crops, late infestations of *Ipomoea* spp. that entangle horticultural crops causing lodging and reduced harvest efficiency. Hand weeding is efficacious on *Digitaria* spp. but can be very laborious and expensive. Hand weeding can be conducted season long without any repercussions of crop damage or

TABLE 11.2 Herbicides that can be applied to horticulture crops to control *Ipomoea* spp.

Site of action	WSSA group	Herbicide	Timing	Crops													
				Beans	Brassica	Cantaloupes	Carrots	Cucumbers	Garlic	Greens	Lettuce	Onions	Peppers	Potatoes	Sweet potato	Tomatoes	Watermelon
ALS	2	Imazethapyr	PRE/POST	x													
	2	Halosulfuron	POST	x													
	2	Rimsulfuron	PRE/POST											x		x	
	2	Trifloxysulfuron	POST													x	
PSII	5	Metribuzin	PRE													x	
	6	Bentazon	POST	x													
	8	EPTC	PRE											x			
DBS	13	Clomazone	PRE			x		x					x		x		x
PPO	14	Carfentrazone	POST-directed		x	x	x	x	x	x	x	x	x	x	x	x	x
	14	Flumioxazin	PRE						x					x			
	14	Fomesafen	PRE/POST	x													
	14	Oxyfluorfen	PRE		x				x			x	x			x	
VLCFA	15	Dimethinamid-P	PRE						x			x		x			
	15	*S*-metalochlor	PRE											x	x		

WSSA, Weed Science Society of America, *ALS*, acetolactate synthase; *PSII*, photosystem II; *DBS*, diterpene biosynthesis; *PPO*, protoporphyrinogen oxidase; *VLCFA*, very long-chain fatty acid; *PRE*, preemergence, *POST*, postemergence.

injury. While flaming is efficacious on many broad-leaved weeds, the tactic is not efficacious on *Digitaria* spp. as the apical meristem is protected within the leaf sheath (Cisneros and Zandstra, 2008).

Rangeland and pastures

Ipomoea spp. are not common weeds of rangeland and pastures; plants inhabiting adjacent cultivated fields or gardens can encroach and colonize in rangeland and pastures (DeFelice, 2001). Control of *Ipomoea* spp. in grass rangelands and pastures can be relatively simple with applying a synthetic auxin herbicide (i.e., 2,4-D or picloram) to the area infested (Vitelli and Pitt, 2006). Some ranchers may allow their livestock to graze or include the weedy *Ipomoea* spp. plants in the feed bales (National Academy of Sciences, 1968; Popay and Field, 1996). Furthermore, if the grazing is done selectively in areas of dense infestations of *Ipomoea* spp., the population can be reduced significantly over several years (National Academy of Sciences, 1968; Popay and Field, 1996; Vitelli and Pitt, 2006). Overgrazing should be avoided as the bare, disturbed soil is favorable for *Ipomoea* spp. and other weedy species colonization (National Academy of Sciences, 1968; Popay and Field, 1996; Vitelli and Pitt, 2006). Ipomoea spp. can be controlled further with controlled field burnoff and regular forage cuttings (Vitelli and Pitt, 2006).

Pastures planted in legumes can complicate the chemical control of *Ipomoea* spp. compared to pastures planted in forage grasses due to selectivity. However, herbicides can be applied to legume pastures to control germinating or emerged *Ipomoea* spp. without injury to the forage legume (National Academy of Sciences, 1968; Vitelli and Pitt, 2006) (Table 11.1). Similar cultural and mechanical control tactics in grass pastures can be implemented in legume pastures. Planting perennial legumes, increased seed rate, fertilization, and timely cuttings (Choudhary et al., 2017; National Academy of Sciences, 1968; Sheley et al., 1996; Vitelli and Pitt, 2006). If these practices are integrated and executed at the time of sowing, *Ipomoea* spp. will likely be effectively controlled by the time the legume is established. Legume-grass pasture mixtures follow the same cultural and mechanical tactics previously described; however, chemical control will be confounded as there are minimal selective herbicides that can be applied to the legume-grass mixture (Table 11.1). Herbicides with preemergence-only activity can be applied to established stands of the legume-grass mixture to control germinating *Ipomoea* spp.

Turf and ornamentals

Globally, *Ipomoea* spp. often inhabit turf and ornamental environments both as a desirable plant and as a weed (Bryson and DeFelice, 2010; DeFelice, 2001; Holm et al., 1977; Holm et al., 1979). Turf and ornamentals can suffer reduced esthetic value and recreational uses due to infestations of *Ipomoea* spp. (Dobbs and Potter, 2014; National academy of Sciences, 1968). Since *Ipomoea* spp., specifically *I. purpurea*, are often grown as ornamentals, the twining vine growth can be a problem if the plant begins to wrap around neighboring shrubbery, flowers, and trees (DeFelice, 2001). Once they entwine a desirable plant, it is often best to cut them off at the base and let the vine dry up and decay (DeFelice, 2001). *Ipomoea* spp. grown as ornamentals can also become rogue and encroach areas where the plant is not desired (i.e., turfgrass, gardens, adjacent properties, etc.) (DeFelice, 2001).

The most common and widely utilized method of weed control in turf is mowing (Busey, 2003; Dobbs and Potter, 2014; National Academy of Sciences, 1968). However, due to the prostrate growth and vining growth on neighboring plants/structures exhibited by *Ipomoea* spp., plants can survive short mowing treatments (Busey, 2003; Price and Wilcut, 2007). Mowing can directly control larger, erect-growing *Ipomoea* spp. by excising the flowers to reduce the amount of seed returning to the soil seedbank. Increasing the mowing height can reduce the density of *Ipomoea* spp. infestation; while not directly controlling *Ipomoea* spp., taller turfgrass can create a denser canopy that inhibits germination/growth (Busey, 2003; Dobbs and Potter, 2014). Furthermore, increasing the mowing height can promote a deep, robust root system in turfgrass which reduces the need for water and chemical inputs that can additionally improve *Ipomoea* spp. control (Dobbs and Potter, 2014). Therefore mowing height should be set between 4 and 8 cm to suppress *Ipomoea* spp. (Busey, 2003; Dobbs and Potter, 2014; DeBels et al., 2012).

While ornamental plants are grown for esthetics, weed-interference in the container or at field nurseries can impede the growth and development of the desirable ornamental species (Altland et al., 2003; Berchielli-Robertson et al., 1990; Neal, 1999). While *Ipomoea* spp. are uncommon in container sites, *Ipomoea* species are pervasive infield nursery sites (Neal and Williams, 1998). Traditionally, weed control in nursery containers or fields is conducted by manual removal, but has recently proved to be expensive (Fennimore and Doohan, 2008; Mathers, 2003). A typical regime for *Ipomoea* spp. control with herbicides in field nurseries is to include preplant burndown, preemergence, and postemergence herbicide applications (Altland et al., 2003). While *Ipomoea* spp. are not common in container-grown ornamental plants, containers are typically treated with granular preemergence herbicides three to five times a year to ensure no weed germination occurred (Mathers, 2003). If *Ipomoea* spp. germinate through the recurrent herbicide applications, plants can be removed from the containers by hand.

Conclusion

Ipomoea spp. are pervasive in most disturbed ecosystems globally and if left unmanaged can be detrimental to esthetics and yield of desirable plants.

Understanding the biology and ecology of *Ipomoea* spp. can exploit new tactics and facilitate the management of these weeds in the plethora of disturbed ecosystems. While pervasive and ubiquitous in disturbed ecosystems, *Ipomoea* spp. can be managed in each respective ecosystem with biological, chemical, cultural, and mechanical tactics effectively. While nonchemical and chemical management options are efficacious on *Ipomoea* spp. inhabiting various ecosystems, no management option should be relied upon solely to reduce selection pressure and maintain efficacy longevity (Bhowmik, 1996; Liebman et al., 1997).

References

Alabouvette, C., Olivain, C., Steinberg, C., 2006. Biological control of plant diseases: the European situation. Eur. J. Plant. Pathol. 114, 329–341.

Altland, J.E., Gilliam, C.H., Wehtje, G., 2003. Weed control in field nurseries. Horttechnology 13, 9–14.

Ball, D.A., 1992. Weed seedbank response to tillage, herbicides, and crop rotation sequence. Weed Sci. 40, 654–659.

Baucom, R.S., Mauricio, R., 2008. Constraints on the evolution of tolerance to herbicide in the common morning glory: resistance and tolerance are mutually exclusive. Evolution 62, 2842–2854.

Beckie, H.J., 2006. Herbicide-resistant weeds: management tactics and practices. Weed Technol. 20, 793–814.

Berchielli-Robertson, D.L., Gilliam, C.H., Fare, D.C., 1990. Competitive effects of weeds on the growth of container-grown plants. Hortacience 25, 77–79.

Bhowmik, P.C., 1996. Weed biology: importance for weed management. Weed Sci. 45, 349–356.

Bryson, C.T., DeFelice, M.S., 2010. Weeds of the Midwestern United States and Central Canada. GA University of Georgia Press, Athens.

Bryson, C.T., Reddy, K.N., Burke, I.C., 2008. Morphological comparison of morningglory (*Ipomoea* and *Jacquemontia* spp.) populations from the Southeastern United States. Weed Sci. 56, 692–698.

Buhler, D.D., Hartzler, R.G., Forcella, F., 1997. Implications of weed seedbank dynamics to weed management. Weed Sci. 45, 329–336.

Burke, I.C., Reddy, K.N., Bryson, C.T., 2009. Pitted and hybrid morningglory accessions have variable tolerance to glyphosate. Weed Technol. 23, 592–598.

Busey, P., 2003. Cultural management of weeds in turfgrass: a review. Crop. Sci. 43, 1899–1911.

Chandler, J.M., Munson, R.L., Vaughan, C.E., 1977. Purple moonflower: emergence, growth, reproduction. Weed Sci. 25, 163–167.

Choudhary, S., Chopral, N.K., Chopral, N.K., Singh, M., Kumar, R., Kushwaha, M., 2017. Influence of nitrogen levels and weed management practices on yield and quality of forage pearl millet (*Pennisetum glaucum* L.). Indian J. Anim. Nutr. 34, 64–69.

Cisneros, J.J., Zandstra, B.H., 2008. Flame weeding effects on several weed species. Weed Technol. 22, 290–295.

Cole, A.W., Coats, G.E., 1973. Tall morningglory germination response to herbicides and temperature. Weed Sci. 21, 443–446.

Cordes, R.C., Bauman, T.T., 1984. Field competition between ivyleaf morningglory (*Ipomoea hederacea*) and soybeans (*Glycine max*). Weed Sci. 32, 364–370.

Crowley, R.H., Buchanan, G.A., 1982. Variations in seed production and the response to pests of morningglory (*Ipomoea*) species and smallflower morningglory (*Jacquemontia tamnifolia*). Weed Sci. 30, 187–190.

Crowley, R.H., Teem, D.H., Buchanan, G.A., Hoveland, C.S., 1979. Responses of *Ipomoea* spp. and *Cassia* spp. to preemergence applied herbicides. Weed Sci. 27, 531–535.

Davis, A.S., 2010. Cover-crop roller-crimper contributes to weed management in no-till soybean. Weed Sci. 58, 300–309.

DeBels, B.T., Griffith, S.E., Kreuser, W.C., Melby, E.S., Soldat, D.J., 2012. Evaluation of mowing height and fertilizer application rate on quality and weed abundance of five home lawn grasses. Weed Technol. 26, 826–831.

DeFelice, M.S., 2001. Intriguing World of weeds: tall morningglory, *Ipomoea purpurea* (L.) Roth—flower or foe? Weed Technol. 15, 601–606.

Dobbs, E.K., Potter, D.A., 2014. Conservation biological control and pest performance in lawn turf: does mowing height matter? Env. Manage. 53, 648–659.

Elmore, C.D., Hurst, H.R., Austin, D.F., 1990. Biology and control of morningglories (*Ipomoea* spp.). Rev. Weed Sci. 5, 83–114.

Everman, W.J., Clewis, S.B., Thomas, W.E., Burke, I.C., Wilcut, J.W., 2008. Critical period of grass vs. broadleaf weed interference in peanut. Weed Technol. 22, 68–73.

Everman, W.J., Clewis, S.B., York, A.C., Wilcut, J.W., 2009. Weed control and yield with flumioxazin, fomesafen, and *S*-metolachlor systems for glufosinate-resistant cotton residual weed management. Weed Technol. 23, 391–397.

Fennimore, S.A., Doohan, D.J., 2008. The challenges of specialty crop weed control, future directions. Weed Technol. 22, 364–372.

Gallandt, E.R., 2006. How can we target the weed seedbank? Weed Sci. 54, 588–596.

Gebhart, M.R., Daniel, T.C., Schweizer, E.E., Allmaras, R.R., 1985. Conservation tillage. Science 230, 625–630.

Glaze, N.C., Hall, M.R., 1990. Cultivation and herbicides for weed control in sweet potato (*Ipomoea batatas*). Weed Technol. 4, 518–523.

Gomes, L.F., Chandler, J.M., Vaughan, C.E., 1978. Aspects of germination, emergence, and seed production of three Ipomoea taxa. Weed Sci. 26, 245–248.

Harper, J.L., 1956. The evolution of weeds in relation to resistance to herbicides. In: Proceedings of the 3rd British Weed Control Conference, 5–8 November, British Weed Control Council, Farnham, UK, pp. 179–188.

Harr, M.J., Fennimore, S.A., McGiffen, M.E., Lanini, W.T., Bell, C.E., 2002. Evaluation of preemergence herbicides in vegetable crops. Horttechnology 12, 95–99.

Heap, I. 2020. International survey of herbicide resistant weeds. www.weedscience.org/in.asp (Accessed August 10, 2020).

Higgins, J.M., Whitwell, T., Murdock, E.C., Toler, J.E., 1988. Recovery of pitted morningglory (*Ipomoea lacunosa*) and ivyleaf morningglory (*Ipomoea hederacea*) following applications of acifluorfen, fomesafen, and lactofen. Weed Sci. 36, 345–353.

Hilgenfeld, K.L., Martin, A.R., Mortensen, D.A., Mason, S.C., 2004. Weed management in a glyphosate resistant soybean system: weed species shifts. Weed Technol. 18, 284–291.

Holland, J.M., 2004. The environmental consequences of adopting conservation tillage in Europe: reviewing the evidence. Agric. Ecosyst. Env. 103, 1–25.

Holm, L.G., Plucknett, D.L., Pancho, J.V., Herberger, J.P., 1977. The World's Worst Weeds: Distribution and Biology. The University Press of Hawaii, Honolulu, Hawaii.

Holm, L., Pancho, J.V., Herberger, J.P., Plucknett, D.L., 1979. A Geographical Atlas of World Weeds. John Wiley and Sons, New York, New York.

Horowitz, M., Regev, Y., Herzlinger, G., 1983. Solarization for weed control. Weed Sci. 31, 170–179.

Hull-Sanders, H.M., Eubanks, M.D., Carr, D.E., 2005. Inbreeding depression and selfing rate of *Ipomoea hederacea* var. integriuscula (Convolvulaceae). Am. J. Bot. 92, 1871–1877.

Jhala, A.J., Ramirez, A.H.M., Singh, M., 2013. Tank mixing saflufenacil, glufosinate, and indaziflam improved burndown and residual weed control. Source: Weed Technol. 27, 422–429.

Kapusta, G., 1979. Seedbed tillage and herbicide influence on soybean (*Glycine max*) weed control and yield. Weed Sci. 5, 520–526.

Keeley, P.E., Thullen, R.J., Carter, C.H., 1986. Influence of planting date on growth of ivyleaf morningglory (*Ipomoea hederacea*) in cotton (*Gossypium hirsutum*). Weed Sci. 34, 906–910.

Klingman, T.E., Oliver, L.R., 1996. Existence of ecotypes among populations of entireleaf morningglory (*Ipomoea hederacea var. Integriuscula*). Weed Sci. 44, 540–544.

Liebman, M., Gallandt, E.R., Jackson, L.E., 1997. Many little hammers: ecological management of crop-weed interactions. Ecol. agriculture 1, 291–343.

Lundgren, J.G., 2005. Ground beetles as weed control agents: effects of farm management on granivory. Am. Entomol. 51, 224–226.

Mathers, H.M., 2003. Novel methods of weed control in containers. Horttechnology 13, 28–34.

Monaco, T.J., Grayson, A.S., Sanders, D.C., 1981. Influence of four weed species on the growth, yield, and quality of direct-seeded tomatoes (*Lycopersicon esculentum*). Weed Sci. 29, 394–397.

Monks, D.W., Schultheis, J.R., 1998. Critical weed-free period for large crabgrass (*Digitaria sanguinalis*) in transplanted watermelon (*Citrullus lanatus*). Weed Sci. 46, 53–532.

Mott, J.J., McKeon, G.M., 1977. A note on the selection of seed types by harvester ants in northern Australia. Aust. J. Ecol. 2, 231–235.

Murdock, E.C., Banks, P.A., Toler, J.E., 1986. Shade development effects on pitted morningglory (*Ipomoea lacunosa*) interference with soybeans (*Glycine max*). Weed Sci. 34, 711–717.

Nalewaja, J., 1999. Cultural practices for weed resistance management. Weed Technol. 13, 643–646.

National Academy of Sciences, 1968. Principles of plant and animal pest control, Weed Control. Chapter 2. The ecology of weeds, Vol 2. National Academy of Sciences, Publication 1975.

Neal, J.C., Williams, A., 1998. Eastern North Carolina nursery weed scouting program—1997. Proc. Northeastern Weed Sci. Soc. 52, 134.

Neal, J., 1999. Weeds and you. Nurs. Mgt. Prod. 15, 60–62. 64–65.

Norsworthy, J.K., 2004. Soybean canopy formation effects on pitted morningglory (*Ipomoea lacunosa*), common cocklebur (*Xanthium strumarium*), and sicklepod (*Senna obtusifolia*) emergence. Weed Sci. 52, 954–960.

Norsworthy, J.K., Oliver, L.D., 2002. Pitted morningglory interference in drill-seeded glyphosate- resistant soybean. Weed Sci. 50, 26–33.

Norsworthy, J.K., Oliveria, M.J., 2004. Comparison of the critical period for weed control in wide- and narrow-row corn. Weed Sci. 52, 802–807.

Norsworthy, J.K., Oliveira, M.J., Jha, P., Malik, M., Buckelew, J.K., Jennings, K.M., et al., 2008. Palmer amaranth and large crabgrass growth with plasticulture-grown bell pepper. Weed Technol. 22, 296–302.

Oliveira, M.J., Norsworthy, J.K., 2006. Pitted morningglory (*Ipomoea lacunosa*) germination and emergence as affected by environmental factors and seeding depth. Weed Sci. 54, 910–916.

Oliver, L.R., Frans, R.E., Talbert, R.E., 1976. Field competition between tall morningglory and soybeans. I. Growth analysis. Weed Sci. 24, 482–488.

Osterholt, M.J., Webster, E.P., Blouin, D.C., McKnight, B.M., 2017. Overlay of residual herbicides in rice for improved weed management. Weed Technol. 33, 426–430.

Owen, M.D.K., 2008. Weed species shifts in glyphosate-resistant crops. Pest. Manag. Sci. 64, 377–387.

Palmer, E.W., Shaw, D.R., Holloway, J.C., 2000. Broadleaf weed control in soybean (*Glycine max*) with CGA-277476 and four postemergence herbicides. Weed Technol. 14, 617–623.

Pannacci, E., Lattanzi, B., Tei, F., 2017. Non-chemical weed management strategies in minor crops: a review. Crop. Prot. 96, 44–58.

Parker, R.G., York, A.C., Jordan, D.L., 2006. Weed control in glyphosate-resistant corn as affected by preemergence herbicide and timing of postemergence herbicide application. Weed Technol. 20, 564–570.

Pearcy, R.W., Ehleringer, J., 1984. Comparative ecophysiology of C_3 and C_4 plants. Plant. Cell Env. 7, 1–13.

Place, G.T., Reberg-Horton, S.C., Jordan, D.L., 2010. Interaction of cultivar, planting pattern, and weed management tactics in peanut. Weed Sci. 58, 442–448.

Popay, I., Field, R., 1996. Grazing animals as weed control agents. Weed Technol. 10, 217–231.

Price, A.J., Wilcut, J.W., 2007. Response of ivyleaf orningglory (*Ipomoea hederacea*) to neighboring plants and objects. Weed Technol. 21, 922–927.

Price, A.J., Williams, P.J., Duzy, L.A., McElroy, J.S., Guertal, E.A., Li, S., 2018. Effects of integrated polyethylene and cover crop mulch, conservation tillage, and herbicide application on weed control, yield, and economic returns in watermelon. Weed Technol. 32, 623–632.

Pursh, F., 1814. Flora Americae Septentrionalis, Volume 1. Cochrane, London, p. 358.

Riar, D.S., Norsworthy, J.K., Steckel, L.E., Stephenson, D.O., Eubank, T.W., Bond, J., et al., 2013. Adoption of best management practices for herbicide-resistant weeds in midsouthern United States cotton, rice, and soybean. Weed Technol. 27, 788–797.

Rogers, J.B., Murray, D.S., Verhalen, L.M., Claypool, P.L., 1995. Ivyleaf morningglory (*Ipomoea hederacea*) interference with cotton (*Gossypium hirsutum*). Weed Technol. 10, 107–114.

Schutte, B.J., 2014. Measuring interference from midseason tall morningglory (*Ipomoea purpurea*) to develop a model for teaching weed weedbank effects on chile pepper. Weed Technol. 31, 155–164.

Sanyal, D., Prasanta, C., Bhowmik, P.C., Randy, L., Anderson, R.L., Shrestha, A., 2008. Revisiting the perspective and progress of integrated weed management. Weed Sci. 56, 161–167.

Seems, J.E., Creamer, N.G., Monks, D.W., 2003. Critical weed-free period for 'Beauregard' sweetpotato (*Ipomoea batatas*). Weed Technol. 17, 686–695.

Senseman, S.A., Oliver, L.R., 1993. Flowering patterns, seed production, and somatic polymorphism of three weed species. Weed Sci. 41, 418–425.

Sheley, R.L., Svejcar, T.J., Maxwell, B.D., 1996. A theoretical framework for developing successional weed management strategies on rangeland. Weed Technol. 10, 766–773.

Shetty, S.V.R., Sivakumar, M.V.K., Ram, S.A., 1982. Effect of shading on the growth of some common weeds of the semi-arid tropics. Agron. J. 74, 1023–1029.

Simonsen, A.K., Stinchcombe, J.R., 2010. Quantifying evolutionary genetic constraints in the ivyleaf morningglory, *Ipomea hederacea*. Int. J. Plant. Sci. 171, 972–986.

Singh, M., Ramirez, A.H.M., Sharma, S.D., Jhala, A.J., 2012. Factors affecting the germination of tall morningglory (*Ipomoea purpurea*). Weed Sci. 60, 64–68.

Sunderland, S.L., Coble, H.D., 1994. Differential tolerance of morningglory species (*Ipomoea* spp.) to DPX-PE350. Weed Sci. 42, 227–232.

Thullen, R.J., Keely, P.E., 1983. Germination, growth, and seed production of *Ipomoea hederacea* when planted at monthly intervals. Weed Sci. 31, 837–840.

Vitelli, J.S., Pitt, J.L., 2006. Assessment of current weed control methods relevant to the management of the biodiversity of Australian rangelands. Rangel. J. 28, 37–46.

Vollmer, K.M., VanGessel, M.J., Johnson, Q.R., Scott, B.A., 2019. Preplant and residual herbicide application timings for weed control in no-till soybean. Weed Technol. 33, 166–172.

Wilcut, J.W., Wehtje, G.R., Walker, R.H., 1987. Economics of weed control in peanuts (*Arachis hypogaea*) with herbicides and cultivation. Weed Sci. 35, 711–715.

Wilson, H.P., Cole, R.H., 1966. Morningglory competition in soybeans. Weeds 14, 49–51.

Wood, M.L., Murray, D.S., Westerman, B.R., Verhalen, L.M., Claywood, P.L., 1999. Full-season interference of *Ipomoea hederacea* with *Gossypium hirsutum*. Weed Sci. 47, 693–696.

Chapter 12

Lolium rigidum and *Lolium multiflorum*

Hugh J. Beckie[1] **and Marie Jasieniuk**[2]
[1]School of Agriculture and Environment, The University of Western Australia, Perth, WA, Australia, [2]Department of Plant Sciences, University of California, Davis, Davis, CA, United States

Introduction

The genus *Lolium* probably evolved from within the genus *Festuca* (Catalán et al., 2004; Torrecilla et al., 2003) in the Middle East. Subsequent migration of agriculturalists introduced the genus worldwide (Balfourier et al., 2000). *Lolium rigidum* Gaudin and *L. multiflorum* Lam. belong to the Poaceae/Gramineae—grass family. Synonyms of *L. rigidum* include *Lolium perenne* L. subsp. *rigidum* (Gaudich.) A. Love & D. Love, and *Lolium multiflorum* Lam. var. *rigidum* (Gaud.) Trab. Common names are rigid ryegrass, annual ryegrass, Wimmera ryegrass, stiff darnel, and Swiss ryegrass (Anonymous, 2019b). Synonyms of *L. multiflorum* include *L. perenne* L. subsp. *multiflorum* (Lam.) Husnot, *L. perenne* L. var. *multiflorum* (Lam.) Parnell, *L. multiflorum* subsp. *italicum* (Braun) Schinz & Keller, *Lolium italicum* A. Braun, *Lolium aristatum* Lagasca, *Lolium strictum* C. Presl, and *Festuca perennis* (L.) Columbus & J.P. Sm. (Baldwin et al., 2012; Beddows, 1973). The species is highly variable in several morphological traits, for example, awn length, spikelet, and floret number, such that Terrell (1968) listed 121 synonyms. Common names include Italian ryegrass, annual ryegrass, and westerworld ryegrass (Anonymous, 2019a; Beddows, 1973).

The botanical description of *L. rigidum* outlined below is derived from Anonymous (2019b) and Monaghan (1980). It is an annual grass, 0.3–1 m high with a fibrous root system and having a variable habit from dense prostrate to more erect types. Stems are geniculate and reddish-purple at the base that extends upward as the plant matures. Leaf blades are dark green, hairless, flat, with the upper surface evenly ribbed, the lower surface smooth and shiny, 50–250 mm long and 3–5 mm wide. The young leaves are rolled when in bud; the auricles are small and narrow. The ligule is white,

Biology and Management of Problematic Crop Weed Species. DOI: https://doi.org/10.1016/B978-0-12-822917-0.00017-3

translucent, and wider than it is long. Leaf sheath is hairless, with fine longitudinal ribs as in the leaf blades, reddish-purple at the base. The inflorescence is a spike up to 30 cm long, with the spikelets on alternating sides, and embedded in the rachis. Spikelets comprise 3–10 fertile and 0–2 rudimentary florets, laterally flattened, green, 10–25 mm long, and 2–5 mm wide. Only the terminal spikelet has two approximately equal glumes; otherwise only one glume subtends each spikelet, lanceolate, 10–15 mm long, three-quarters as long, or as long as the spikelet with the outer surface finely veined and ribbed like the upper surface of the leaf blade. Lemma is lanceolate, about 5 mm long, five-nerved, usually with no awn. Palea is similar to lemma in shape and size, and two nerved with tiny hairs. The three anthers are yellow. The seeds are relatively flat, 4–6 mm long, 1 mm wide, and straw colored. *L. rigidum* is diploid with a chromosome count of $2n = 14$ (Taylor and Evans, 1976).

Lolium rigidum may be mistaken for *L. perenne* or *Phalaris* spp. The glume is half the length of the spikelet in *L. perenne*, that is, shorter than for *L. rigidum* (Anonymous, 2019b). *Phalaris* spp. have dull, silver-green leaves and red sap, whereas *L. rigidum* has shiny, bright green leaves and clear sap (Anonymous, 2008). Although electrophoretic and morphological studies can differentiate among *Lolium* spp., visual identification in the field can be difficult when considerable hybridization and introgression has occurred where different species cohabitate, especially between *L. rigidum* and *L. multiflorum* (Bennett et al., 2000) (Fig. 12.1).

Lolium rigidum is native to the Mediterranean region (southern Europe, northern Africa), eastern Europe, southwestern Asia, the Middle East, and Indian subcontinent (Anonymous, 2019b; Bennett et al., 2000). It was introduced or spread to southern Africa, central Europe, North and South America, and Oceania (Australia); it is well adapted to Mediterranean or subtropical, and to a lesser extent, temperate climatic conditions across variable soil substrates. The species has a high degree of genetic variability and phenotypic plasticity, enabling it to rapidly adapt to a range of climatic or edaphic conditions (Balfourier et al, 1998; Gill, 1996b; Kloot, 1983). Highly productive varieties of *L. rigidum* were selected and introduced into southern Australia in the late 1800s as a pasture species, and quickly became well adapted to the region (Goggin et al., 2012). With a shift to continuous cropping in the 1980s, it became a major crop weed and also occurs in ruderal (noncrop disturbed) areas (Fig. 12.2).

Morphological characteristics used to identify *L. multiflorum* have been described by many authors. The following species description is derived from Anonymous (2019a), Beddows (1973), and Clayton et al. (2020). *L. multiflorum* is a bunchgrass. Plants produce a few to moderate numbers of erect or decumbent stems that are 30–90 cm long and slightly flattened in cross-section. Mature leaf blades are flat, green to dark green, hairless, up to 400 mm long, and 5–10 mm wide with upper surfaces that are evenly

FIGURE 12.1 *Lolium rigidum*: seedling (juvenile) stage (A) and reproductive stage (B; http://www.plantillustrations.org); *Lolium multiflorum*: mature plant (C; http://www.plantillustrations.org).

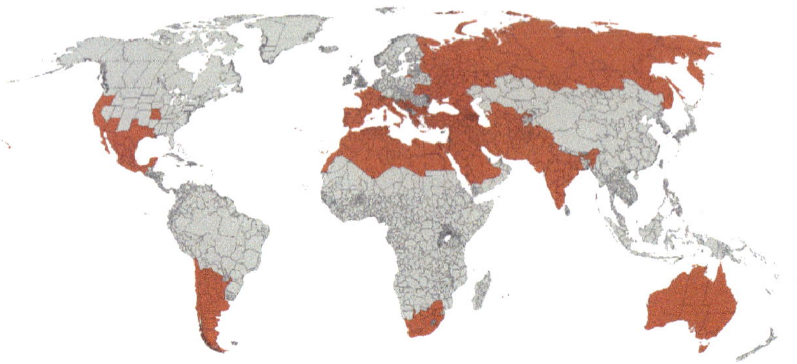

FIGURE 12.2 Global distribution of *Lolium rigidum. Dataset from: Anonymous, 2019b. Lolium rigidum (Rigid Ryegrass). Invasive Species Compendium, CABI, Switzerland. https://www.cabi.org/isc/datasheet/31167#:~:text=Top%20of%20page%20L.,up%20to%201%20m%20high.*

ribbed and lower surfaces that are smooth and shiny. Auricles are small and narrow; ligules are short and membranous. Young leaf blades are rolled in the bud.

The inflorescence is composed of racemes 100–300 mm long and comprise 8–25 mm long spikelets that break up at maturity by disarticulating below each fertile floret (Fig. 12.1C). Spikelets alternate with one another along the rachis, which is recessed opposite each spikelet, and comprise 5–20 laterally flattened florets with smaller sterile florets at the apex. Each floret has a terminal awn up to 10 mm long. The terminal spikelet has two approximately equal glumes; remaining spikelets have only one glume up to half the length of the spikelet. Lemma is lanceolate and five nerved; palea is similar to lemma in shape and size and two nerved with tiny hairs. Each floret has three yellow, occasionally tinged red, anthers and two feathery stigmas, which protrude slightly before or at the same time as the anthers when the floret opens. Each floret produces one caryopsis (single-seeded fruit).

The species *L. multiflorum* is thought to be native to central and southern Europe, northwest Africa, and southwest Asia (Hubbard, 1968) but is now found in temperate regions of all continents (Balfourier et al., 2000; Charmet et al., 1996; Clayton et al., 2020). According to Lacefield et al. (2003), it was first reported to be cultivated in northern Italy followed by cultivation in France, Switzerland, and England. It was first introduced into North America in the early colonial days and rapidly became an important forage grass, especially in the southern United States. Today, the species is cultivated as a forage crop throughout much of its worldwide range (Anonymous, 2019a).

As a weed, *L. multiflorum* establishes, grows, and reproduces best in disturbed habitats, such as cereal grain fields, orchards, vineyards, and roadsides, although it can spread into relatively undisturbed grasslands

(Anonymous, 2019a; Beddows, 1973). It prefers fertile, mesic, and well-drained soils in areas with mild climates (Anonymous, 2019a; Lacefield et al., 2003).

Biology and ecology

Seeds of *L. rigidum* germinate from 5°C to 35°C, with an optimum temperature of 11°C in darkness and 27°C in light (Cocks and Donald, 1973; Gramshaw, 1976). However, another study found that germination rate increased as temperature increased from 5°C to 29°C in the dark (McWilliam et al., 1970). Under field conditions, the optimum temperature for seed germination is about 10°C for buried seeds and 25°C for seeds on the soil surface (Cook et al., 2005). The seed germination rate declines from 2- to 10-cm depth in the soil profile, with generally no germination and seedling emergence beyond 10-cm depth (Gill, 1996b; Gramshaw and Stern, 1977; Smith, 1968).

The majority (60%–80%) of *L. rigidum* propagules in the seed bank germinate and emerge as a single cohort early in the following growing season soon after significant rainfall, although subsequent flushes can occur over the growing season depending on the timing of precipitation events (Anonymous, 2019b; Fernandez-Quintanilla et al., 2000; Monaghan, 1980). Flowering is photoperiod (8–24 h) and temperature insensitive (Cooper, 1956). *L. rigidum* is an obligate outcrossing, wind-pollinated C_3 species with copious pollen production (Gill, 1996b; McCraw and Spoor, 1983). Because *L. rigidum* populations are widespread and abundant across the southern cropping regions of Australia, pollen dispersal occurs within fields and among adjacent fields. In a study conducted in Western Australia, interpopulation pollen-mediated flow of herbicide-resistant (HR) genes was documented at a distance of 3000 m (Busi et al., 2008). In another study in Western Australia, Busi et al. (2011) confirmed that gene flow could cause a substantial increase in the frequency of HR genes of *L. rigidum* plants infesting commercial organic fields located at considerable distance (up to 2.5 km) from neighboring conventional fields. In both studies, an extensive source of HR alleles (i.e., multiplicative effect of frequency of HR alleles and weed population abundance) promoted long-distance gene flow over time. In Europe, within-population genetic and phenotypic variation in *L. rigidum* are high, with pollen-mediated gene flow measured between populations being over twice that of seed-mediated gene flow (Balfourier et al., 2000).

Lolium rigidum can readily hybridize with *L. perenne* L. and *L. multiflorum* and can form intergeneric hybrids with a number of Festuca species (Beadle et al., 1972; Terrell, 1968). Where *L. rigidum* and *L. multiflorum* co-occur, substantial introgression between the species has occurred. In a survey in Italy, 40%–60% of sampled plants were hybrids of these two species (Dinelli et al., 2002).

Lolium rigidum is a prolific seed producer, and crop field infestations of the species can produce up to 45,000 seeds per m^2 or 1000 seeds per plant (Gill, 1996a; Rerkasem et al., 1980; Stanton et al., 2002). The lightweight seeds are readily transported by natural (wind, water, animals) and anthropogenic (e.g., farm machinery) vectors (Anonymous, 2008). Regional seed movement of *L. rigidum* via contaminated crop seed or livestock feed is a recurring problem (Michael et al., 2010), and grain contaminated with HR *L. rigidum* even impacts internationally traded wheat (Shimono et al., 2010).

Harvested seeds have a short period (about 8–9 weeks) of primary or innate dormancy (Steadman, 2004). The rate of dormancy release during the after-ripening period is positively correlated with temperature (base 5°C) and seed moisture content (6%–18%) (Gallagher et al., 2004). The seed bank of the species is relatively short lived. On average, 15%–30% of the seed bank can persist into the following year (Chauhan et al., 2006; Fernandez-Quintanilla et al., 2000), and studies estimate seed bank longevity to a maximum of 4 years (Narwal et al., 2008).

Lolium rigidum inflorescences (spikes) infested with the pathogen *Corynebacterium* or *Clavibacter* sp. residing in galls caused by nematodes (*Anguina* sp.) can cause livestock poisoning (annual ryegrass toxicity) due to the ingestion of corynetoxins (Lanigan et al., 1976; Stynes et al., 1979). *L. rigidum* is also susceptible to ergot fungus (*Claviceps purpurea*); the contaminated grain is downgraded and can result in toxicity to mammals when ingested. As a host to the cereal disease, take-all, caused by Rhizoctonia fungus (*Gaeumannomyces graminis*), *L. rigidum* may perpetuate soil inoculum levels (Cotterill and Sivasithamparam, 1988).

Lolium multiflorum is also a C_3 grass, with a summer annual, winter annual, or biennial life cycle (Anonymous, 2019a; Beddows, 1973) depending on the climatic conditions of the geographical region it inhabits. Similar to *L. rigidum*, it is a diploid ($2n = 14$) with an obligate outcrossing mating system due to self-incompatibility (Beddows, 1973; Guan et al., 2017). Self-incompatibility in Lolium species is determined by two multiallelic loci whose effect in the pollen is gametophytic (Fearon et al, 1983). The self-incompatibility system limits self-pollination to extremely low levels (~1% in *L. multiflorum*; Fearon et al., 1983). Self-incompatibility and obligate outcrossing, together with wind-mediated pollen movement, results in high genetic variation within populations and low genetic differentiation among populations as observed in California, USA (Karn and Jasieniuk, 2017). High genetic variation within populations has been suggested to account for the propensity of annual *Lolium* spp. to rapidly evolve resistance to many different herbicide modes of action (Powles and Yu, 2010).

As described previously, *L. multiflorum* readily hybridizes with the other self-incompatible Lolium species, *L. rigidum* and *L. perenne* (Bulinska-Radomska and Lester, 1985; Dinelli et al., 2002; Terrell, 1968), which would further increase the amount of genetic variation within populations available for herbicide selection. Intergeneric hybridizations between *L. multiflorum* and *Festuca* spp. have also been detected (Dinelli et al., 2002, 2004). However, the extent to which plant fitness is reduced in hybrid offspring due to outbreeding depression or increased due to the introduction of novel adaptive traits is unknown for crosses with the more genetically distant group.

Lolium multiflorum reproduces solely by seeds. Individual plants are prolific seed producers. In Arkansas, USA, a single plant was capable of producing up to 45,000 seeds (Bararpour et al., 2017). Seed retention on a plant at harvest is reasonably high; approximately 60% of seed had not shattered at the time of crop harvest in Washington State, USA (D.J. Lyon, unpublished data, cited in Walsh et al., 2018).

Persistence of *L. multiflorum* seeds in the soil seed bank varies with tillage system. Based on the size and shape of seeds, Ghersa and Martinez-Ghersa (2000) predicted, using models of Bekker et al. (1998), that *L. multiflorum* would not form a persistent soil seed bank in no-tillage crop production systems since roughly 90% of seeds would be found in the top layer of soil (0–5 cm) and be exposed to several mortality factors. To provide insight into the influence of no-tillage versus tillage on seed bank dynamics, Ichihara et al. (2009) compared the fate of seeds after-ripened in the soil with that of after-ripened seeds on the soil surface. The study showed that germination and seedling emergence were earlier, higher, and increased with increasing temperature during the summer for seeds on the soil surface compared with seeds buried in the soil. The authors suggested that no-tillage systems may promote the rapid depletion of *L. multiflorum* seeds in wheat (*Triticum aestivum* L.) fields as a result of seed predation together with seed germination and herbicidal control before the crop is sown.

Like *L. rigidum*, seeds of *L. multiflorum* exhibit primary dormancy at maturity. Considerable research has focused on determining the processes and environmental conditions that break dormancy. Rodriguez et al. (1998) showed that for seeds stored dry under ambient environmental conditions, after-ripening was required for approximately 4 months to break dormancy. Stanisavljevic et al. (2011) found that seeds exhibited the highest germination percentage and seedling vigor (shoot length, root length, and fresh weight) following dry storage for 9–11 months after harvest. In a parallel study of seeds collected 5 and 6 years later from the same location, maximum germination was observed following dry storage for 6 months (Stanisavljevic et al., 2020). For hydrated (imbibed) seeds, Rodriguez et al. (1998) showed that dormancy could be overcome by exposing hydrated seeds to mild cool temperatures (15°C or below) for short periods (24–96 h).

After dormancy is broken, germination of *L. multiflorum* seeds has been shown to be enhanced by exposure to high red (R)/far-red (FR) light ratios; the response is presumed to be phytochrome-mediated (Deregibus et al., 1994). High R/FR light ratios increased the germination of seeds stored dry at a constant temperature as well as seeds exposed to field conditions and the seasonal temperature changes typical of the field. In contrast, *L. multiflorum* plants treated with a low R/FR light ratio throughout the photoperiod or with FR light at the end of the day developed longer main shoots, tillers, and remaining shoots (Casal et al., 1987). Results indicate that vegetative grass shoots respond to light quality similarly to the internodes of dicotyledons.

Herbicidal treatment of *L. multiflorum* plants in crop fields has been hypothesized to select for seed traits (Mortimer, 1999) but research results have been mixed. Gundel et al. (2008) conducted a long-term field experiment that evaluated the relationship between treatment with diclofop-methyl at different doses for 5 years and changes in seed germination rate, seed aging, and resistance level of subsequent generations of plants. The study concluded that there was a general positive relationship between time to 50% seed germination and diclofop-methyl dose, but resistance level and seed aging rate were not related to herbicide dose. However, Ghersa et al. (1994) found in an earlier study that seedling emergence in the field of a diclofop-methyl-susceptible biotype was earlier than that of the resistant biotype. The study also noted that seeds in the soil seed bank with a greater degree of dormancy were more likely to be resistant.

Effects of salinity stress and alkalinity stress on the percentage and rate of seed germination in *L. multiflorum* were investigated by Lin et al. (2018). Germination percentage and germination rate were both inhibited by increasing salinity and alkalinity. The inhibitory effect of high salinity on germination was greatest at 10°C–20°C, whereas the inhibitory effect of high alkalinity was greatest at 25°C–35°C. Results indicated that salinity stress and alkalinity stress differ in their effects on the germination of *L. multiflorum*. Moreover, the tolerance of seeds to the stresses is influenced by the interactive effects of temperature and salinity or alkalinity.

Neotyphodium spp. are fungal endophytes that infect and live asymptomatically, and presumably mutualistically, with many C_3 grass species (Gundel et al., 2006). In temperate grasslands of Argentina, *L. multiflorum* plants infected with the seed-transmitted Neotyphodium endophyte produced significantly more vegetative tillers and allocated more biomass to roots and seeds than endophyte-free plants (Vila-Aiub et al., 2005a). However, endophytes did not appear to have a strong influence on the germination behavior of *L. multiflorum* seeds (Gundel et al., 2006). Rather, the effect of the endophytes on the germination behavior of *L. multiflorum* seeds depended on environmental conditions. No effects on dormancy were observed when seeds were incubated at 10°C, but the endophyte increased seed dormancy levels at higher incubation temperatures.

Impacts and management

L. rigidum is an economically important weed in cereal and other annual crops as well as orchards and vineyards in Mediterranean to temperate

TABLE 12.1 Wheat yield loss due to *Lolium multiflorum* interference and yield loss of various field crops due to *Lolium rigidum* interference.

Crop	Weed density (plants per m^2)	Yield loss (%)	Reference
Wheat			
L. multiflorum			
	10–100	4–42	Liebl and Worsham (1987)
	93	61	Appleby et al. (1976)
	20–200	21–59	Appleby and Brewster (1992)
	25–400	42–92	Hashem et al. (1998)
L. rigidum	287–328	22–40	Lemerle et al. (1995)
	300	45–55	Lemerle et al. (2013)
	100–700	10–50	Reeves (1976)
	50–300	5–70	Medd et al. (1985)
	450	42	Smith and Levick (1974)
Barley			
	90–179	18–23	Paynter and Hills (2009)
	287–328	10–55	Lemerle et al. (1995)
Oat			
	287–328	2–14	Lemerle et al. (1995)
Rye			
	287–328	14–20	Lemerle et al. (1995)
Triticale			
	287–328	5–24	Lemerle et al. (1995)
Canola			
	287–328	9–30	Lemerle et al. (1995)
	300	32	Reeves (1976)
Lupin			
	40	34	Arnold et al. (1985)
	90	45	Allen (1977)
	287–328	100	Lemerle et al. (1995)
Field pea			
	287–328	100	Lemerle et al. (1995)

climates (Izquierdo et al., 2003) and represents the most serious case of herbicide resistance globally (Heap, 2020). In the last 40 years, it has been the top weed in field crops in Australia causing annual revenue losses of nearly AUD 100 million (Llewellyn et al., 2016). Weed densities up to 300 plants per m^2 can cause cereal crop yield losses of up to 75% (Table 12.1).

Its biological characteristics, a propensity to evolve herbicide resistance, widespread use of no-tillage, and an almost exclusive reliance on herbicidal control, have collectively resulted in rapid evolution of herbicide resistance in the species. *L. rigidum* in Australia is often present in crop fields, and herbicide resistance that was initially reported in 1982 (Heap and Knight 1982) has evolved to herbicides in many different groups and is widespread across the Australian grain-growing region (Boutsalis et al., 2012; Owen et al., 2014a,b). In Western Australia, the overreliance on postemergence herbicides has led to the rapid selection of high levels of multiple resistance in populations ($>95\%$ frequency), particularly to acetyl-CoA carboxylase (ACCase) and acetolactate synthase (ALS) inhibitors (Llewellyn and Powles, 2001; Owen et al., 2014a,b) (Table 12.2).

Because of the rapid evolution of multiple herbicide resistance in *L. rigidum* and the significant cost associated with herbicide resistance management, alternative chemical and nonchemical means to control the weed have been developed and deployed in fields infested with HR plants. In Australia, growers have responded to weed resistance to postemergence herbicides by relying upon soil-applied preemergence herbicides, such as trifluralin, propyzamide, prosulfocarb, and pyroxasulfone. The incidence of resistance to these preemergence herbicides is currently low relative to postemergence herbicides (Boutsalis et al., 2012; Broster et al., 2011; Owen et al., 2014b). However, some *L. rigidum* populations in South Australia exhibit resistance to most soil-applied preemergence herbicides (Brunton et al., 2019). Additionally, there is increasing incidence of resistance to glyphosate in *L. rigidum* populations in Australia, Europe (Spain, France), and the America (Fernández-Moreno et al., 2017; Heap, 2020; Simarmata and Penner, 2008).

Populations of *L. rigidum* in Australia are adapting to increasing use of preemergence herbicides (including burndown treatments involving glyphosate or paraquat) by enhanced seed dormancy and delayed seedling emergence in-crop (Owen et al., 2011). In that study, seed dormancy was positively correlated with herbicide resistance levels in *L. rigidum* populations; the most fit individuals that emerge later after the reduction in phytotoxicity of preemergence herbicide residues in soil are HR biotypes that then survive the application of commonly used in-crop herbicides (ACCase or ALS inhibitors).

Similar to other troublesome weeds, there has been increasing focus on depletion of the seed bank as the best strategy to manage *L. rigidum*. Mitigation of *L. rigidum* population abundance requires diverse herbicidal and nonherbicidal tools or practices, such as effective preemergence

TABLE 12.2 Reported herbicide-resistant biotypes of *Lolium rigidum* and *Lolium multiflorum.*

Biotype	Group	Country	Situation	Year
L. rigidum				
ACCase inhibitor	1	Spain	Wheat	1992
		Saudi Arabia	Cereals	1992
		France	Wheat	1993
		Tunisia	Cereals	1996
		Chile	Cereals, canola, lupin	1997
		Greece	Wheat	1997
		Israel	Wheat, alfalfa	1998
		Iran	Wheat	2007
ALS inhibitor	2	Australia	Wheat, barley	1984
		France	Wheat	2006
		Greece	Wheat	2009
Microtubule inhibitor	3	Australia	Cropland, pasture	1984
Photosystem-II inhibitor-triazine	5	Israel	Orchard, ruderal	1979
		Australia	Cropland, pasture, ruderal	1988
		Spain	Orchard	1992
Photosystem-II inhibitor-urea/amide	7			
Lipid inhibitor	8	Australia	Wheat	2015
EPSP synthase inhibitor	9	Australia	Cereals, canola, ruderal	1996
		United States	Orchard	1998
		South Africa	Vineyard	2001
		France	Vineyard, orchard	2005

(*Continued*)

TABLE 12.2 (Continued)

Biotype	Group	Country	Situation	Year
		Spain	Orchard	2006
		Italy	Vineyard, orchard	2007
		Greece	Orchard	2016
Glutamine synthase inhibitor	10	Greece	Vineyard, orchard	2017
Carotenoid biosynthesis	11	Australia	Ruderal	1988
DOXP inhibitor	13			
PPO inhibitor	14			
VLCFA inhibitor	15	Australia	Cropland	1984
Photosystem-I electron diverter	22	South Africa	Vineyard, orchard	2002
Mitosis inhibitor	23			
Multiple	1,2,3	Australia	Wheat, barley, canola	1982
	1,2,3,8,13,15,23	Australia	Wheat, barley	1982
	1,7	Spain	Wheat	1992
	1,2	South Africa	Wheat	1993
		Chile	Wheat	2003
		Israel	Wheat, horticulture	2013
	1,2,3,9	Australia	Wheat	1999
	1,9,22	South Africa	Vineyard	2003
	1,2,9	Israel	Wheat	2007
	9,11	Australia	Vineyard	2008
	1,22	Australia	Pasture	2010
	1,2,5,9,22	Australia	Pasture	2010
	9,22	Australia	Vineyard	2013
	9,14	Spain	Orchard	2016

(*Continued*)

TABLE 12.2 (Continued)

Biotype	Group	Country	Situation	Year
L. multiflorum				
ACCase inhibitor	1	United States	Wheat	1987
		United Kingdom	Cereals, canola	1990
		France	Wheat	1993
		Italy	Wheat, alfalfa	1995
		Chile	Wheat, canola, lupin	1998
		Argentina	Cereals	2009
ALS inhibitor	2	United States	Ruderal	1995
		Italy	Wheat	2005
		Brazil	Wheat	2010
		United Kingdom	Cereals	2012
Photosystem-II inhibitor-urea/amide	7	United Kingdom	Wheat	1990
EPSP synthase inhibitor	9	Chile	Orchard	2001
		Brazil	Soybean, orchard	2003
		United States	Orchard, vineyard, wheat	2004
		Spain	Orchard	2006
		Argentina	Cereals	2007
		Japan	Rice	2011
		Switzerland	Orchard	2011
		New Zealand	Vineyard	2012
Glutamine synthase inhibitor	10	United States	Vineyard, orchard	2015
VLCFA inhibitor	15	France	Wheat	2018
		United Kingdom	Wheat	2018

(Continued)

TABLE 12.2 (Continued)

Biotype	Group	Country	Situation	Year
		United States	Wheat	2018
Multiple	1,2	United States	Wheat	1995
		Italy	Wheat	2002
		France	Wheat	2003
		Chile	Wheat, canola, lupin	2005
		Denmark	Wheat	2010
		Brazil	Wheat	2016
	2,9	Chile	Wheat	2002
		Argentina	Wheat	2010
		Italy	Wheat	2012
		Brazil	Corn, soybean, wheat	2017
	1,9	Chile	Lupin	2006
		Italy	Wheat	2008
		Brazil	Corn, soybean, wheat	2010
	1,2,15	United States	Wheat, canola, pulses	2005
	1,2,9	Chile	Barley	2007
	9,10	United States	Orchard	2010
	9,10,11	New Zealand	Vineyard	2015
	1,9,22	United States	Alfalfa, vineyard, orchard	2015
	1,2,9,22	United States	Alfalfa, orchard	2016

ACCase, Acetyl-CoA carboxylase; ALS, acetolactate synthase; *DOXP*, 4-deoxy-D-xylulose-5-phosphate synthase; *EPSP*, 5-enolpyruvylshikimate-3-phosphate; *PPO*, protoporphyrinogen oxidase; *VLCFA*, very long-chain fatty acid.
Adapted from Heap, I.M., 2020. International Herbicide-Resistant Weed Database. http://www.weedscience.org.

herbicides, competitive crops and cultivars (Lemerle et al., 1995), high crop seeding rates (Lemerle et al., 2013; Medd et al., 1985; Paynter and Hills, 2009), glyphosate or paraquat application during weed seed development (crop/spray topping) (Steadman et al., 2006), and harvest weed seed control (HWSC). Numerous studies conducted in Australia have consistently shown that maintaining high crop seeding rates, narrow (<30 cm) crop row spacing, seeding early, and seeding in an east-west orientation promotes *L. rigidum* growth and fecundity suppression through enhanced crop competition. These important cultural practices are widely promoted through the WeedSmart extension platform (WeedSmart, 2020). Because seed retention at harvest time is high, *L. rigidum* is well suited for management by HWSC practices. On average, recruitment the following growing season is reduced by 60% by all HWSC practices (Walsh et al., 2017). Annual usage of HWSC in Western Australian farms from 2002 to 2011 resulted in *L. rigidum* plant densities declining from an average of 35 plants per m^2 to 0.5 plants per m^2, a level that was consistently maintained (Walsh et al., 2013, 2018). Periodic moldboard plowing to bury *L. rigidum* seeds deep in the profile (> 10 cm) can reduce seedling recruitment by 98% (Vila-Aiub et al., 2005b). Populations in ruderal areas also need to be controlled before flowering to mitigate seed bank return and HR gene flow across the landscape.

L. multiflorum is a major agricultural weed worldwide in annual crops, such as cereal grains, canola (*Brassica napus* L.), soybean [*Glycine max* (L.) Merr.], cotton (*Gossypium hirsutum* L.), and pulses, and in perennial crops, such as alfalfa (*Medicago sativa* L.), vine, and fruit and nut tree crops (Anonymous, 2019a). However, studies of the economic losses and crop yield reductions caused by *L. multiflorum* infestations have focused on wheat (Table 12.1). The species is the most troublesome weed of wheat in the southern United States (Webster and Nichols, 2012). In the western United States, winter wheat yields were found to be reduced by 61% when the density of *L. multiflorum* was 93 plants per m^2 (Appleby et al., 1976) and were reduced from 42% to 92% as densities increased from 25 to 400 plants per m^2 in another study (Hashem et al., 1998). Even 20 plants per m^2 reduced wheat yield by an average of 38% in Oregon, USA in 1990 (Appleby and Brewster, 1992). Yield losses were also carried over to the following year due to higher weed densities in plots where the heaviest infestations occurred the previous year (Appleby et al., 1976). In the southern United States, winter wheat yields were reduced an average of 4.2% for every 10 plants per m^2 increase from 0 to 100 *L. multiflorum* plants per m^2 (Liebl and Worsham, 1987). In Argentina, wheat aboveground dry matter and yield were reduced at densities of 100 to 150 plants per m^2 of *L. multiflorum* (Acciaresi et al., 2001a,b). In Italy, the economic threshold for the control of *L. multiflorum* with herbicides was estimated at between 25 and 35 plants per m^2 (Zanin et al., 1993).

L. multiflorum is very competitive with wheat for mineral nutrients, water, space, and light (Acciaresis et al., 2003; Carson et al., 1999; Forcella,

1984; Hashem et al., 1998) and causes reduced photosynthesis, growth, and tiller production in wheat (Carson et al., 1999; Stone et al., 1998). It also contributes to crop lodging and grain quality reductions and dockage at harvest (Hulting et al., 2012). Several studies have manipulated different aspects of wheat agronomy in an attempt to increase crop competitiveness. Appleby and Brewster (1992) reported that increasing the wheat seeding rate from 50 to 150 kg/ha^1 reduced *L. multiflorum* biomass by an average of 26% across a range of weed densities (20–200 plants per m^2). In addition, the spatial arrangement of wheat plants, varying from 5 to 40 cm apart resulting in differing levels of rectangularity, significantly affected the competitiveness of *L. multiflorum* (Hashem et al., 1998). Moreover, wheat cultivars varied considerably in their ability to suppress *L. multiflorum* (Appleby and Brewster, 1992; Acciaresi et al., 2001b).

The first case of herbicide resistance in *L. multiflorum* worldwide was reported in 1987 in wheat in the United States (ACCase inhibitor; Table 12.2). Resistance to another important mode-of-action herbicide to control the weed (ALS inhibitor) was also first documented in the United States in 1995. That year, the first multiple-HR biotype (ACCase + ALS inhibitor) was reported in wheat in the United States. Where resistance to these modes of action are widespread, pyroxasulfone applied preemergence is commonly used (Hulting et al., 2012). Glyphosate and paraquat are routinely used as a burndown herbicide prior to crop planting.

Resistance to glyphosate in *L. multiflorum* was first reported in a Chilean orchard in 2001 but has since been reported in many other countries (Heap, 2020). Resistance to glufosinate, a glutamine synthase inhibitor, was first reported in 2015 in the United States (Avila-Garcia and Mallory-Smith, 2011). Resistance to this herbicide mode of action remains relatively rare globally, although its incidence is increasing as it has now also been confirmed in Japan (Kurata et al., 2017), New Zealand (Ghanizadeh et al., 2015), and Spain (Fernández et al., 2017). Similar to *L. rigidum*, the number of multiple-HR biotypes continues to increase in *L. multiflorum* (Heap 2020) and resistance to up to four modes of action has recently been reported (Tehranchian et al., 2018). Tillage and HWSC are the key large-scale nonherbicide tools to control HR *L. multiflorum* and prevent seed set (Beam et al., 2019).

Conclusion

Both *L. rigidum* and *L. multiflorum* will continue to rank high in economic importance globally because of their strong competitive ability, high fecundity, readily dispersed seed, high genetic diversity, and propensity for evolving herbicide resistance. The ability to rapidly adapt to different climatic and edaphic conditions will further challenge management, both herbicidal and nonherbicidal. Adoption of effective integrated cultural practices, including competitive crops or cultivars and HWSC, in conjunction with newer pre-emergence herbicides have proven effective in the management of *L. rigidum* and to a lesser extent, *L. multiflorum*. More nonherbicidal management studies are particularly needed on the latter species.

Unraveling the genomes of both species will greatly facilitate a better understanding of the genetic regulation of their growth and reproduction as well as the mechanistic basis of evolved herbicide resistance that now impacts the majority of global populations. Availability of this critical weed genomic information combined with rapidly developing and cost-effective molecular and genetic tools heralds an exciting new era in the potential molecular control of pernicious and troublesome weedy species as exemplified in the *Lolium* genus.

References

Acciaresi, H.A., Chidichimo, H.O., Sarandón, S.J., 2001a. Wheat-*Lolium multiflorum* competence: effect of nitrogen application onto Argentinean varieties aggressivity. Cereal Res. Commun. 29, 451–458.

Acciaresi, H.A., Chidichimo, H.O., Sarandón, S.J., 2001b. Traits related to competitive ability of wheat (*Triticum aestivum*) varieties against Italian ryegrass (*Lolium multiflorum*). Biol. Agric. Hort 19, 275–286.

Acciaresi, H.A., Chidichimo, H.O., Sarandón, S.J., 2003. Shoot and root competition in a *Lolium multiflorum*-wheat association. Biol. Agric. Hort. 21, 15–33.

Allen, J.M., 1977. Weeds in grain lupins. 1. The effect of weeds on grain lupin yields. Aust. J. Exp. Agric. Anim. Husb. 17, 112–116.

Anonymous, 2008. Annual Ryegrass (*Lolium rigidum*). Murrumbidgee Catchment Management Authority, New South Wales Department of Primary Industries, Sydney, NSW, Australia. http://archive.lls.nsw.gov.au/__data/assets/pdf_file/0008/495332/archive-annual-ryegrass.pdf.

Anonymous 2019a. *Lolium multiflorum* (Italian Ryegrass). Invasive Species Compendium, CABI, Switzerland. https://www.cabi.org/isc/datasheet/31165#: ~ :text = L.,mean%20that%20is%20highly%20adaptable.

Anonymous 2019b. *Lolium rigidum* (Rigid Ryegrass). Invasive Species Compendium, CABI, Switzerland. https://www.cabi.org/isc/datasheet/31167#: ~ :text = Top%20of%20page%20L.,up%20to%201%20m%20high.

Appleby, A.P., Brewster, B.D., 1992. Seeding arrangement on winter wheat (*Triticum aestivum*) grain yield and interaction with Italian ryegrass (*Lolium multiflorum*). Weed Technol 6, 820–823.

Appleby, A.P., Olson, P.D., Colbert, D.R., 1976. Winter wheat yield reduction from interference by Italian ryegrass. Agron. J. 68, 463–466.

Arnold, G.W., Weeldenberg, J., Grassia, A., 1985. Competition between Wimmera ryegrass and narrow-leafed lupins. Aust. J. Exp. Agric. 25, 824–831.

Avila-Garcia, W.V., Mallory-Smith, C., 2011. Glyphosate-resistant Italian ryegrass (*Lolium perenne*) populations also exhibit resistance to glufosinate. Weed Sci 59, 305–309.

Baldwin, B.G., Goldman, D., Keil, D.J., Patterson, R., Rosatti, T.J., Wilken, D. (Eds.), 2012. The Jepson Manual Vascular Plants of California, Thoroughly Revised and Expanded. second ed. University of California Press.

Balfourier, F., Charmet, G., Rvel, C., 1998. Genetic differentiation within and between natural populations of perennial and annual ryegrass (*Lolium perenne* and *L. rigidum*). Heredity 81, 100–110.

Balfourier, F., Imbert, C., Charmet, G., 2000. Evidence for phylogeographic structure in *Lolium* species related to the spread of agriculture in Europe. A cpDNA study. Theor. Appl. Genet. 101, 131–138.

Bararpour, M.T., Norsworthy, J.K., Burgos, N.R., Korres, N.E., Gbur, E.E., 2017. Identification and biological characteristics of ryegrass (*Lolium* spp.) accessions in Arkansas. Weed Sci 65, 350–360.

Beadle, N.C.W., Evans, O.D., Carolin, R.C., 1972. Flora of the Sydney Region. A.H. & A.W. Read, Sydney, Australia. p. 639.

Beam, S.C., Mirsky, S., Cahoon, C., Haak, D., Flessner, M., 2019. Harvest weed seed control of Italian ryegrass [*Lolium perenne* L. ssp. *multiflorum* (Lam.) Husnot], common ragweed (*Ambrosia artemisiifolia* L.) and Palmer amaranth (*Amaranthus palmeri* S. Watson). Weed Technol 33, 627–632.

Beddows, A.R., 1973. Biological flora of the British Isles. *Lolium multiflorum* Lam. J. Ecol 61, 587–600.

Bekker, R.M., Bakker, J.P., Grandin, U., Kalamees, R., Milberg, P., Poschlod, P., et al., 1998. Seed size, shape and vertical distribution in the soil: indicators of seed longevity. Funct. Ecol. 12, 834–842.

Bennett, S.J., Hayward, M.D., Marshall, D.F., 2000. Morphological differentiation in four species of the genus *Lolium*. Genet. Resourc. Crop Evol 47, 247–255.

Boutsalis, P., Gill, G., Preston, C., 2012. Incidence of herbicide resistance in rigid ryegrass (*Lolium rigidum*) accross southeastern Australia. Weed Technol 26, 391–398.

Broster, J., Koetz, E., Wu, H., 2011. Herbicide resistance levels in annual ryegrass (*Lolium rigidum* Gaud.) in southern New South Wales. Plant Prot. Q 26, 22–28.

Brunton, D.J., Boutsalis, P., Gill, G., Preston, C., 2019. Resistance to multiple PRE herbicides in a field-evolved rigid ryegrass (*Lolium rigidum*) population. Weed Sci 66, 581–585.

Bulinska-Radomska, Z., Lester, R.N., 1985. Relationships between five species of *Lolium* (Poaceae). Plant Syst. Evol. 148, 169–175.

Busi, R., Yu, Q., Barrett-Lennard, R., Powles, S., 2008. Long distance pollen-mediated flow of herbicide resistance genes in *Lolium rigidum*. Theor. Appl. Genet. 117, 1281–1290.

Busi, R., Michel, S., Powles, S.B., Délye, C., 2011. Gene flow increases the initial frequency of herbicide resistance alleles in unselected *Lolium rigidum* populations. Agric. Ecosyst. Environ. 142, 403–409.

Carson, K.H., Cralle, N.T., Chandler, J.M., Miller, T.D., Bovey, R.W., Senseman, S.A., et al., 1999. *Triticum aestivum* and *Lolium multiflorum* interaction during drought. Weed Sci 47, 440–445.

Casal, J.J., Sánchez, R.A., Deregibus, V.A., 1987. The effect of light quality on shoot extension growth in three species of grasses. Ann. Bot. 59, 1–7.

Catalán, P., Torrecilla, P., Rodríguez, J.A., Olmstead, R.G., 2004. Phylogeny of the festucoid grasses of subtribe Loliinae and allies (Poeae, Pooideae) inferred from ITS and trnL-F sequences. Mol. Phylogenet. Evol. 31, 517–541.

Charmet, G., Balfourier, F., Chatard, V., 1996. Taxonomic relationships and interspecific hybridization in the genus *Lolium* (grasses). Genet. Resour. Crop Evol 43, 319–327.

Chauhan, B.S., Gill, G., Preston, C., 2006. Influence of tillage systems on vertical distribution, seedling recruitment and persistence of rigid ryegrass (*Lolium rigidum*) seed bank. Weed Sci 54, 669–676.

Clayton, W.D., Vorontsova, M.S., Harman, K.T., Williamson, H., 2020. GrassBase - The Online World Grass Flora. http://www.kew.org/data/grasses-db.html.

Cocks, P.S., Donald, C.M., 1973. The germination and establishment of two annual pasture grasses (*Hordeum leporinum* Link and *Lolium rigidum* Gaud.). Aust. J. Agric. Res. 24, 1–10.

Cook, T., Moore, J., Peltzer, S., 2005. Integrated Weed management in Australian Cropping Systems. Cooperative Research Centre for Australian Weed management, Adelaide, Australia. pp. 149–153.

Cooper, J.P., 1956. A rapid technique for strain identification in the ryegrasses. J. Br. Grassl. Soc 11, 42–47.

Cotterill, P.J., Sivasithamparam, K., 1988. Importance of the proportion of grassy weeds within legume crops in the perpetuation of *Gaeumannomyces graminis* var. *tritici*. Plant Pathol 37, 337–343.

Deregibus, V.A., Casal, J.J., Jacobo, E.J., Gibson, D., Kauffman, M., Rodriguez, A.M., 1994. Evidence that heavy grazing may promote the germination of *Lolium multiflorum* seeds via phytochrome-mediated perception of high red/far-red ratios. Func. Ecol 8, 536–542.

Dinelli, G., Bonetti, A., Lucchese, C., Catizone, P., Bravin, F., Zanin, G., 2002. Taxonomic evaluation of Italian populations of *Lolium* spp. resistant and susceptible to diclofop- methyl. Weed Res 42, 156–165.

Dinelli, G., Bonetti, A., Marotti, I., Minelli, M., Catizone, P., 2004. Characterization of Italian populations of *Lolium* spp. resistant and susceptible to diclofop by inter simple sequence repeat. Weed Sci 52, 554–563.

Fearon, C.H., Hayward, M.D., Lawrence, M.J., 1983. Self-incompatibility in ryegrass. V. Genetic control, linkage and seed-set in diploid *Lolium multiflorum* Lam. Heredity 50, 35–45.

Fernández, P., Alcántara, R., Osuna, M.D., Vila-Aiub, M.M., De Prado, R., 2017. Forward selection for multiple resistance across the non-selective glyphosate, glufosinate and oxyfluorfen herbicides in *Lolium* weed species. Pest Manag. Sci. 73, 936–944.

Fernández-Moreno, P.T., Travlos, I., Brants, I., De Prado, R., 2017. Different levels of glyphosate-resistant *Lolium rigidum* L. among major crops in southern Spain and France. Sci. Rep. 7, e13116.

Fernandez-Quintanilla, C., Barroso, J., Recasens, J., Sans, X., Torner, C., Sanchez del Arco, M. J., 2000. Demography of *Lolium rigidum* in winter barley crops: analysis of recruitment, survival and reproduction. Weed Res 40, 281–291.

Forcella, F., 1984. Wheat and ryegrass competition for pulses of mineral nitrogen. Aust. J. Exp. Agric. Anim. Husb. 24, 421–425.

Gallagher, R.S., Steadman, K.J., Crawford, A.D., 2004. Alleviation of dormancy in annual ryegrass (*Lolium rigidum*) seeds by hydration and after-ripening. Weed Sci 52, 968–975.

Ghanizadeh, H., Harrington, K.C., James, T.K., 2015. Glyphosate resistant *Lolium multiflorum* and *Lolium perenne* populations from New Zealand are also resistant to glufosinate and amitrole. Crop Prot 78, 1–4.

Ghersa, C.M., Martínez-Ghersa, M.A., Brewer, T.J., Roush, M.L., 1994. Selection pressures for diclofop-methyl resistance and germination time of Italian ryegrass. Agron. J. 86, 823–828.

Ghersa, C.M., Martinez-Ghersa, M.A., 2000. Ecological correlates of weed seed size and persistence in the soil under different tilling systems: implications for weed management. Field Crops Res 67, 141–148.

Gill, G.S., 1996a. Why annual ryegrass is a problem in Australian agriculture. Plant Prot. Q 11 (Suppl. 1), 193–194.

Gill, G.S., 1996b. Ecology of annual ryegrass. Plant Prot. Q 11 (suppl. 1), 195–198.

Goggin, D.E., Powles, S.B., Steadman, K.J., 2012. Understanding *Lolium rigidum* seeds: the key to managing a problem weed? Agronomy 2, 222–239.

Gramshaw, D., 1976. Temperature/light interactions and the effect of seed source on germination of annual ryegrass (*Lolium rigidum* Gaud.) seeds. Aust. J. Agric. Res. 27, 779–786.

Gramshaw, D., Stern, W.R., 1977. Survival of annual ryegrass (*Lolium rigidum* Gaud.) in a Mediterranean type environment. II. Effects of short-term burial on persistence of viable seed. Aust. J. Agric. Res. 28, 93–101.

Guan, X., Yuyama, N., Stewart, A., Ding, C., Xu, N., Kiyoshi, T., et al., 2017. Genetic diversity and structure of *Lolium* species surveyed on nuclear simple sequence repeat and cytoplasmic marker. Front. Plant Sci. 8, 584.

Gundel, P.E., Maseda, P.H., Ghersa, C.M., Benech-Arnold, R.L., 2006. Effects of the *Neotyphodium* endophyte fungus on dormancy and germination rate of *Lolium multiflorum* seeds. Aust. Ecol. 31, 767–775.

Gundel, P.E., Martínez-Ghersa, M.A., Ghersa, C.M., 2008. Dormancy, germination and ageing of *Lolium multiflorum* seeds following contrasting herbicide selection regimes. Eur. J. Agron. 28, 606–613.

Hashem, A., Radosevich, S.R., Roush, M.L., 1998. Effect of proximity factors on competition between winter wheat (*Triticum aestivum*) and Italian ryegrass (*Lolium multiflorum*). Weed Sci 46, 181–190.

Heap, I.M., 2020. International Herbicide-Resistant Weed Database. http://www.weedscience.org.

Heap, I., Knight, R., 1982. A population of ryegrass tolerant to the herbicide diclofop-methyl. J. Aust. Inst. Agric. Sci. 48, 156–157.

Hubbard, C.E., 1968. Grasses, second ed. Penguin Books, Harmondsworth.

Hulting, A.G., Dauer, J.T., Hinds-Cook, B., Curtis, D., Koepke-Hill, R.M., Mallory-Smith, C., 2012. Management of Italian ryegrass (*Lolium perenne* ssp. *multiflorum*) in western Oregon with preemergence applications of pyroxasulfone in winter wheat. Weed Technol 26, 230–235.

Ichihara, M., Yamashita, M., Sawada, H., Kida, Y., Asai, M., 2009. Influence of after-ripening environments on the germination characteristics and seed fate of Italian ryegrass (*Lolium multiflorum*). Weed Biol. Manag. 9, 217–224.

Izquierdo, J., Recasens, J., Fernández-Quintanilla, C., Gill, G., 2003. Effects of crop and weed densities on the interactions between barley and *Lolium rigidum* in several Mediterranean locations. Agronomie 23, 529–536.

Karn, E., Jasieniuk, M., 2017. Genetic diversity and structure of *Lolium perenne* ssp *multiflorum* in California vineyards and orchards indicate potential for spread of herbicide resistance via gene flow. Evol. Appl. 10, 616–629.

Kloot, P.M., 1983. The genus *Lolium* in Australia. Aust. J. Bot 31, 421–435.

Kurata, K., Ichihara, M., Ishida, Y., Shimono, Y., Tominaga, T., 2017. Glufosinate-resistant Italian ryegrass populations emerge from glyphosate-resistant populations in Japan. Int. Inv. J. Agric. Soil Sci 5, 21–25.

Lacefield, G., Collins, M., Henning, J., Phillips, T., Rasnake, M., Spitaleri, R., et al., 2003. Annual Ryegrass. Kentucky Cooperative Extension Service AGR-179, Lexington, KY.

Lanigan, G.W., Payne, A.L., Frahn, J.L., 1976. Origin of toxicity in parasitised annual ryegrass (*Lolium rigidum*). Aust. Vet. J 52, 244–246.

Lemerle, D., Verbeek, B., Coombes, N., 1995. Losses in grain yield of winter crops from *Lolium rigidum* competition depend on crop species, cultivar and season. Weed Res 35, 503–509.

Lemerle, D., Lockley, P., Koetz, E., Diffey, S., 2013. Herbicide efficacy for control of annual ryegrass (*Lolium rigidum* Gaud.) is influenced more by wheat seeding rate than row spacing. Crop Pasture Sci 64, 708–715.

Liebl, R., Worsham, A.D., 1987. Interference of Italian ryegrass (*Lolium multiflorum*) in wheat (*Triticum aestivum*). Weed Sci 35, 819–823.

Lin, J., Hua, X., Peng, X., Dong, B., Yan, X., 2018. Germination responses of ryegrass (annual vs. perennial) seed to the interactive effects of temperature and salt-alkali stress. Front. Plant Sci. 9, 1458.

Llewellyn, R.S., Powles, S.B., 2001. High levels of herbicide resistance in rigid ryegrass (*Lolium rigidum*) in the wheat belt of Western Australia. Weed Technol 15, 242–248.

Llewellyn, R.S., Ronning, D., Ouzman, J., Walker, S., Mayfield, A., Clarke, M., 2016. Impact of Weeds on Australian Grain Production: the Cost of Weeds to Australian Grain Growers and the Adoption of Weed Management and Tillage Practices. Report for GRDC, CSIRO, Australia. 112 p.

McCraw, L.M., Spoor, W., 1983. Self-incompatibility in *Lolium* species. 1. *Lolium rigidum* Gaud. and *L. multiflorum* L. Heredity 1, 21–27.

McWilliam, J.R., Clements, R.J., Dowling, P.M., 1970. Some factors influencing the germination and early seedling development of pasture plants. Aust. J. Agric. Res. 21, 19–32.

Medd, R.W., Auld, B.A., Kemp, D.R., Murison, R.D., 1985. The influence of wheat density and spatial arrangement on annual ryegrass, *Lolium rigidum* Gaudin, competition. Aust. J. Agric. Res. 36, 361–371.

Michael, P.J., Owen, M.J., Powles, S.B., 2010. Herbicide-resistant weed seeds contaminate grain sown in the Western Australian grainbelt. Weed Sci 58, 466–472.

Monaghan, N.M., 1980. The biology and control of *Lolium rigidum* as a weed of wheat. Weed Res 20, 117–121.

Mortimer, A.M., 1999. Phenological adaptation in weeds – an evolutionary response to the use of herbicides? Pestic. Sci. 51, 299–304.

Narwal, S., Sindel, B.M., Jessop, R.S., 2008. Dormancy and longevity of annual ryegrass (*Lolium rigidum*) as affected by soil type, depth, rainfall, and duration of burial. Plant Soil 310, 225–234.

Owen, M.J., Michael, P.J., Renton, M., Steadman, K.J., Powles, S.B., 2011. Towards large- scale prediction of *Lolium rigidum* emergence: II. Correlation between dormancy and herbicide resistance levels suggests an impact of cropping systems. Weed Res 51, 133–141.

Owen, M.J., Martinez, N.J., Powles, S.B., 2014a. Multiple herbicide-resistant *Lolium rigidum* (annual ryegrass) now dominates across the Western Australian grain belt. Weed Res 54, 314–324.

Owen, M.J., Martinez, N.J., Powles, S.B., 2014b. Widespread occurrence of multiple herbicide resistance in Western Australian annual ryegrass (*Lolium rigidum*) populations. Aust. J. Agric. Res. 58, 711–718.

Paynter, B.H., Hills, A.L., 2009. Barley and rigid ryegrass (*Lolium rigidum*) competition is influenced by crop cultivar and density. Weed Technol 23, 40–48.

Powles, S.B., Yu, Q., 2010. Evolution in action: plants resistant to herbicides. Annu. Rev. Plant Biol. 61, 317–347.

Reeves, T.G., 1976. Effect of annual ryegrass (*Lolium rigidum* Gaud.) on yield of wheat. Weed Res 16, 57–63.

Rerkasem, K., Stern, W.R., Goodchild, N.A., 1980. Associated growth of wheat and annual ryegrass. I. Effect of varying total density and proportion in mixtures of wheat and annual ryegrass. Aust. J. Agric. Res. 31, 649–658.

Rodriguez, A.M., Jacobo, E.J., Deregibus, V.A., 1998. Germination behaviour of Italian ryegrass in flooding Pampa rangelands. Seed Sci. Res. 8, 521–528.

Shimono, Y., Takiguchi, Y., Konuma, A., 2010. Contamination of internationally traded wheat by herbicide-resistant *Lolium rigidum*. Weed Biol. Manag. 10, 219–228.

Simarmata, M., Penner, D., 2008. The basis for glyphosate resistance in rigid ryegrass (*Lolium rigidum*) from California. Weed Sci 56, 181–188.

Smith, D.F., 1968. The growth of barley grass (*Hordeum leporinum*) in annual pasture. 1. Germination and establishment in comparison with other annual pasture species. Aust. J. Exp. Agric. Anim. Husb. 8, 478–483.

Smith, D.F., Levick, G.R.T., 1974. The effect of infestation by *Lolium rigidum* Gaud. (annual ryegrass) on yield of wheat. Aust. J. Agric. Res. 25, 381–393.

Stanisavljevic, R., Đjokic, D., Milenkovic, J., Đukanovic, L., Stevovic, V., Simic, A., et al., 2011. Seed germination and seedling vigour of Italian ryegrass, cocksfoot, and timothy following harvest and storage. Ciênc. Agrotec., Lavras 35, 1141–1148.

Stanisavljevic, R., Postic, D., Štrbanovic, R., Tabakovic, M., Jovanovic, S., Milenkovic, J., et al., 2020. Effect of seed storage on seed germination and seedling quality of *Festulolium* in comparison with related forage grasses. Tropical Grasslands-Forrajes Tropicales 8, 125–132.

Stanton, R., Piltz, J., Pratley, J., Kaiser, A., Hudson, D., Gill, G., 2002. Annual ryegrass (*Lolium rigidum*) seed survival and digestibility in cattle and sheep. Aust. J. Exp. Agric. 42, 111–115.

Steadman, K.J., 2004. Dormancy release during hydrated storage in *Lolium rigidum* seeds is dependent on temperature, light quality, and hydration status. J. Exp. Bot. 55, 929–937.

Steadman, K.J., Eaton, D.M., Plummer, J.A., Ferris, D.G., Powles, S.B., 2006. Late-season non-selective herbicide application reduces *Lolium rigidum* seed numbers, seed viability, and seedling fitness. Aust. J. Agric. Res. 57, 133–141.

Stone, M.J., Cralle, H.T., Chandler, J.M., Miller, T.D., Carson, K.H., 1998. Above and below-ground interference of wheat (*Triticum aestivum*) by Italian ryegrass (*Lolium multiflorum*). Weed Sci 46, 438–441.

Stynes, B.A., Petterson, D.S., Lloyd, J., Payne, A.L., Lanigan, G.W., 1979. The production of toxin in annual ryegrass, *Lolium rigidum*, infested with a nematode, *Anguina* sp., and *Corynebacterium rathayi*. Aust. J. Agric. Res. 30, 201–209.

Taylor, I.B., Evans, G.M., 1976. The effect of B chromosomes on homologuegous pairing in species hybrids. Chromosoma 57, 25–32.

Tehranchian, P., Nandula, V., Jugulam, M., Puttac, K., Jasieniuk, M., 2018. Multiple resistance to glyphosate, paraquat and ACCase-inhibiting herbicides in Italian ryegrass populations from California: confirmation and mechanisms of resistance. Pest Manag. Sci. 74, 868–877.

Terrell, E., 1968. A taxonomic revision of the genus *Lolium*. USDA Tech. Bull 1392, 1–65.

Torrecilla, P., Rodríguez, J.A., Stancik, D., Catalán, P., 2003. Systematics of *Festuca* L. Sects. Eskia Willk., Pseudatropis Kriv., Amphigenes (Janka) Tzvel., Pseudoscariosa Kriev. and Scariosae Hack. based on analysis of morphological characters and DNA sequences. Plant Syst. Evol. 239, 113–139.

Vila-Aiub, M.M., Gundel, P.E., Ghersa, C.M., 2005a. Fungal endophyte infection changes growth attributes in *Lolium multiflorum* Lam. Aust. Ecol 30, 49–57.

Vila-Aiub, M.M., Neve, P., Steadman, K.J., Powles, S.B., 2005b. Ecological fitness of a multiple herbicide-resistant *Lolium rigidum* population: dynamics of seed germination and seedling emergence of resistant and susceptible phenotypes. J. Appl. Ecol. 42, 288–298.

Walsh, M.J., Newman, P., Powles, S., 2013. Targeting weed seeds in-crop: a new weed control paradigm for global agriculture. Weed Technol 27, 431–436.

Walsh, M.J., Aves, C., Powles, S.B., 2017. Harvest weed seed control systems are similarly effective on rigid ryegrass. Weed Technol 31, 178–183.

Walsh, M.J., Broster, J.C., Schwartz-Lazaro, L.M., Norsworthy, J.K., Davis, A.S., Tidemann, B. D., et al., 2018. Opportunities and challenges for harvest weed seed control in global cropping systems. Pest Manag. Sci. 74, 2235–2245.

Webster, T.M., Nichols, R.L., 2012. Changes in the prevalence of weed species in the major agronomic crops of the southern United States: 1994/1995 to 2008/2009. Weed Sci 60, 145–157.

WeedSmart, 2020. The Big 6: Crop Competition. https://www.weedsmart.org.au/big-6/crop-competition.

Zanin, G., Berti, A., Toniolo, L., 1993. Estimation of economic thresholds for weed control in winter wheat. Weed Res 33, 459–467.

Chapter 13

Weedy rice (*Oryza* spp.)

Masilamany Dilipkumar[1], Virender Kumar[2], Beng-Kah Song[3], Kenneth M. Olsen[4], Tse-Seng Chuah[5], Sharif Ahmed[6] and Sheng Qiang[7]

[1]Rice Research Center, Malaysian Agricultural Research and Development Institute (MARDI), MARDI Seberang Perai, Kepala Batas, Malaysia, [2]International Rice Research Institute, Los Baños, Philippines, [3]School of Science, Monash University Malaysia, Bandar Sunway, Malaysia, [4]Department of Biology, Washington University in St Louis, St. Louis, MO, United States, [5]Faculty of Plantation and Agrotechnology, University of Teknologi Mara, Arau, Malaysia, [6]International Rice Research Institute (IRRI) Bangladesh Office, Dhaka, Bangladesh, [7]Weed Research Laboratory, Nanjing Agricultural University, Nanjing, China

Introduction

Rice is a crop of paramount importance for global food and nutritional security as rice is a staple cereal for more than half of the world population, the majority of whom are in rapidly growing low-income countries (Maclean et al., 2013). Worldwide, rice is annually grown on 167 million hectares with a production of 782 million tons (MT) of paddy, out of which 90% of rice is grown (146 million hectares) and produced (705 MT paddy) in Asia (FAOSTAT, 2018; Table 13.1). Rice contributes to a major share of total calorie and protein intake (Table 13.1). It is projected that an additional 63 MT of milled rice would be needed by 2030 to meet the global rice demand of the rising world population which is expected to be ~8.5 billion (RICE, 2016). In the future, this additional rice has to be produced with lower water, labor, and agrochemicals while buffering the risks of climate change to ensure global food security and environmental sustainability. Closing the existing rice yield gaps caused by weed competition is one approach to achieve this target as actual global yield losses in rice due to weeds is about 10% (Oerke, 2005).

Worldwide, rice is cultivated primarily by two methods—puddled transplanted rice (PTR) and direct-seeded rice (DSR). In PTR, rice seedlings are first grown in a nursery, and then seedlings are uprooted and transplanted into puddled soil (wet tillage) in the main field. In DSR, rice seeds are sown directly in the main field and it eliminates the process of nursery raising, seedling uprooting, and transplanting. DSR is further divided into three

Biology and Management of Problematic Crop Weed Species. DOI: https://doi.org/10.1016/B978-0-12-822917-0.00007-0

TABLE 13.1 Area, production, productivity, share of rice in total calorie and protein intake worldwide and Asia.

Region	Area (million ha)	Production (million tons)	Productivity (t/ha)	Rice calorie intake (% of total/ day)[a]	Rice protein intake (% of total/ day)[a]
World	167	782	4.68	19.0	12.7
Asia	146	705	4.83	28.6	19.3
South Asia	62	251	4.02	28.8	21.9
Southeast Asia	50	221	4.42	46.5	35.8
Eastern Asia	33	231	6.97	26.3	15.5

FAOSTAT. 2018. FAOSTAT Database. Rome, Italy: Food and Agriculture Organization of the United Nations (FAO). Retrieved July 30, 2020, from http://www.fao.org/faostat/en/#home.
[a]*Based on 2011 data.*

types: (1) dry-DSR (dry ungerminated seeds are sown in nonpuddled soil), (2) wet-DSR (pregerminated seeds are sown on puddled soil), and (3) water-seeding [pregerminated seeds are sown in standing water on either puddled (in Asia) or nonpuddled soil (in the United States and Australia)] (Kumar and Ladha, 2011; Rao et al., 2017). PTR is a predominant method in Asian and African countries, whereas rice is grown by DSR methods in the United States, Australia, and Europe.

Recently, farmers in Asia are shifting from PTR to DSR in response to the major drivers of rural change in the region, especially rising labor, water scarcity, and cost of cultivation as DSR saves on labor, water, and cost of cultivation and hence increases farmers' net income (Kumar and Ladha, 2011; Rao et al., 2017). DSR has also emerged as an option to reduce greenhouse gas emissions (Chakraborty et al., 2017; Kumar and Ladha, 2011). Because of these drivers of agricultural change and benefits, DSR is now practiced widely in many Asian countries including Malaysia, Sri Lanka, Vietnam, Thailand, Cambodia, and the Philippines, and many other countries including China and India are going through this transition from PTR to DSR. In the future, more area is expected to shift from PTR to DSR as these drivers become stronger.

Weedy rice (*Oryza* spp.), also known as red rice in the United States, has emerged as one of the major global threats to rice production in DSR-based systems (Ziska et al., 2015). In countries where DSR is the dominant crop establishment method (e.g., the United States, Brazil) or where transplanting is widely replaced with DSR (e.g., Malaysia, Sri Lanka, Thailand, Vietnam), weedy rice has emerged as dominant weed species and a threat to the sustainability of rice production (Ziska et al., 2015). With the increasing shift from PTR to DSR in several other countries in Asia and Africa, the global problem of weedy rice is expected to be exacerbated further in the future.

Weedy rice grows unintentionally and competes vigorously with cultivated rice and is similar (morphologically, genetically, and phenologically) to cultivated rice and often (not always) a red pericarp. These factors make weedy rice highly competitive and difficult-to-control weed in rice. It has been reported that weedy rice can cause severe yield losses ranging from 15%–100% (Kumar and Ladha, 2011). In addition to yield losses, weedy rice infestation reduces the grain quality leading to a reduced value of rice at harvest, and therefore, ultimately reduces farmers' net income (Ottis et al., 2005).

There are limited in-crop control options for weedy rice because of its close similarity to cultivated rice. Even hand-weeding is challenging at an early stage because it is difficult to differentiate weedy rice seedlings from rice seedlings. Also, success in selective control (in-crop) of weedy rice with herbicides has not been achieved because rice and weedy rice respond similarly to herbicides (Chauhan, 2012a). So far, selective in-crop weedy rice control has been achieved with herbicide-tolerant (HT) rice varieties only (Burgos et al., 2008). It is, therefore, important to develop and deploy integrated options to manage weedy rice which are based on a strong foundation of knowledge of biology and ecology of weedy rice.

The purpose of this chapter is to synthesize up-to-date knowledge of weedy rice on (1) its origin and dispersal; (2) biology and ecology; and (3) management so that researchers, extensionists, and farmers can utilize this information to minimize the risks associated with weedy rice in their region.

Weedy rice: its origin and dispersal

Origin of weedy rice

Hypotheses for the origins of weedy rice have traditionally considered four potential models: (1) descent from wild rice populations that have invaded agricultural habitats; (2) emergence from crop-wild hybrids; (3) emergence from intervarietal hybridization within domesticated rice; and (4) direct descent from rice varieties or landraces through dedomestication (feralization). In the last two decades, the availability of high-quality rice reference genomes, together with increasingly cost-effective sequencing and genotyping technologies, has enabled multiple genetic studies of weedy rice populations around the world (Li and Olsen, 2020; Qiu et al., 2020). This accumulated research has made it clear that with few exceptions (described below), the vast majority of weedy rice strains are descended primarily from domesticated ancestors (model #4). These studies have also definitively established that weedy rice has evolved multiple times independently from different crop varieties and that strains that have convergently evolved shared weediness traits such as seed shattering and dormancy.

Another important insight from these recent population genetic studies is that the dynamics of weedy rice evolution differ markedly between world regions within and outside the distribution range of common wild rice (*Oryza rufipogon* Griffiths). In regions of tropical Asia where the wild progenitor species occurs, gene flow between wild, cultivated, and weedy rice appears to add a level of genetic complexity that is absent from weedy rice outside the geographical range of the wild species (Neik et al., 2019). For this reason, following sections first examine weedy rice origins and evolution in areas outside the range of common wild rice, and then consider weedy rice evolution within areas of overlap (Fig. 13.1).

Regions without O. rufipogon

Most weedy rice population genetic and evolutionary studies have been conducted in regions where there is no common wild rice present. These include the United States, Brazil, Colombia, Italy, Spain, northern China, Japan, Taiwan, and South Korea. Among these countries, the occurrence and spread of weedy rice in the United States has been particularly well documented. Both neutral marker [e.g., simple sequence repeat (SSR), single nucleotide polymorphism (SNP); Li et al., 2017; Reagon et al., 2010] and candidate

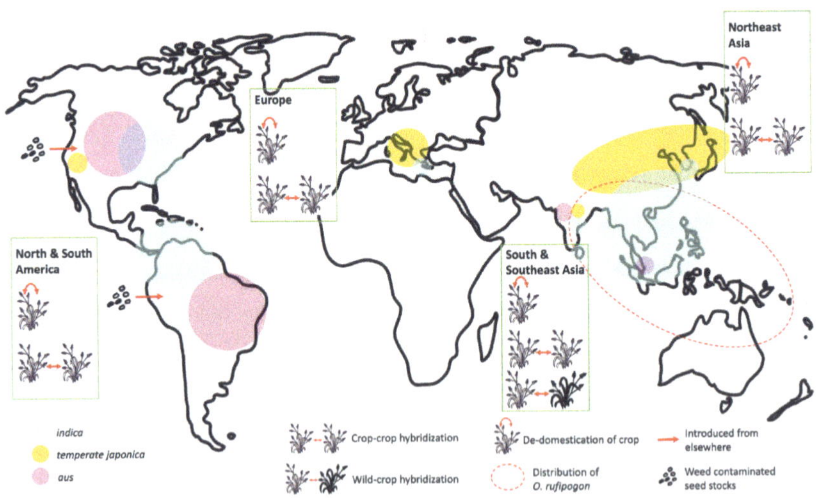

FIGURE 13.1 Global occurrence and ancestry of weedy rice.

gene analyses [e.g., *sh4*, conferring shattering (Thurber et al., 2010); *Bh4*, controlling hull color (Vigueira et al., 2013); *Rc*, conferring pericarp color and seed dormancy, (Gross et al., 2010)] revealed that the US weedy rice strains are genetically closest to cultivated rice grown in Asia. Specifically, the two major southern US weedy rice morphotypes, referred to as strawhull awnless (SH) and blackhull awned (BHA) weeds, are genetically most similar to domesticated *indica* and *aus* varieties, respectively. Given that neither *indica* nor *aus* rice has been commercially cultivated in the United States, it is likely that these weeds were accidentally introduced through contaminated seed stocks (Reagon et al., 2010). Studies using SSR and genome-wide SNP datasets further substantiate the dedomestication model for the US weedy rice. A recent study using SNPs derived from whole-genome resequencing indicated that SH weeds diverged from *indica* rice much later than BHA strains from *aus* varieties (Li et al., 2017). Similar to weedy rice morphotypes in the southern United States, Californian strawhull awned (SHA) weedy rice, which is genetically related to temperate *japonica* rice planted in California, appears to have been introduced from outside the United States (Kanapeckas et al., 2016).

Using morphological and genetic markers including SSRs and genome-wide SNPs, similar dedomestication results have been inferred for weedy rice populations studied in Latin America (Federici et al., 2001; Hoyos et al., 2020; Qiu et al., 2020), northern China (Cao et al., 2007; Qiu et al., 2017; Qiu et al., 2020), Italy (Grimm et al., 2013), South Korea (Kim et al., 2016; Vigueira et al., 2019), and Japan (Akasaka et al., 2009; Kawasaki et al., 2009; Qiu et al., 2020). In Central and South America, the presence of the reproductively compatible (AA genome) wild rice species, *O. glumaepatula* Steud., has led to speculation that this native wild species might be contributing to the evolution of local weedy strains (Hoyos et al., 2020; Qiu et al., 2020). However, morphological, genetic, and genomic studies (Arrieta-Espinoza et al., 2005; Qiu et al., 2020) do not support this hypothesis.

In China, population genetic analyses have revealed that most crop- and wild-like weedy rice strains are genetically closer to local *japonica* cultivars than to *indica* or common wild rice, suggesting a role for *japonica* dedomestication in their origin (Cao et al., 2007; Sun et al., 2013; Qiu et al., 2017). Using whole-genome resequencing, similar observations of dedomestication in China were recently reported (Qiu et al., 2020), with weedy rice accessions showing the highest kinship with co-occurring cultivars or varieties from nearby regions. Genome scans further indicate that selection on genomic regions that were not targets of selection during domestication may have facilitated feralization of cultivated rice and conferred weedy adaptations in present-day weedy rice strains.

Whole-genome sequence analysis of South Korean weedy rice has suggested a crop–weed hybridization mechanism where local landraces were found to have crossed with co-occurring modern rice cultivars (He et al., 2017).

In contrast, weedy rice strains in Bhutan were found to have originated from natural hybridization between local *indica* and *japonica* cultivars (Ishikawa et al., 2005). A recent population genomic study of Latin American weedy populations revealed that approximately 54% of the 95 weedy rice genomes examined comprise *aus–indica* admixed genotypes, suggesting that many Latin American weedy rice strains originated through recent hybridization of local weedy rice populations. Similarly, an *indica–japonica* hybridization origin has been reported for weedy rice in eastern China, where *indica* cultivars were recently replaced by *japonica* rice (Qiu et al., 2014).

Regions with O. rufipogon

As the wild progenitor of Asian rice, common wild rice has played an important role in contributing to modern rice breeding (Morishima et al., 1984; Oka, 1988). It is distributed over tropical and subtropical countries from western India through Southeast Asia and southern China. In South and Southeast Asia, where co-occurrence of cultivated and common wild rice is common, genetic diversity of weedy rice populations is influenced by contemporary introgression from proximal wild populations (Pusadee et al., 2012; Song et al., 2014; Wedger et al., 2019).

In Malaysia, for example, SSRs and genome-wide SNPs were used to demonstrate hybridization between the common wild rice and cultivated or weedy rice (Song et al., 2014; Neik et al., 2019; Vigueira et al., 2019). Specifically, Malaysian common wild rice is most likely the major contributor to the evolution of brown-striped-like and BHA-like weedy rice that are abundant in parts of Peninsular Malaysia. As a complement to the neutral marker studies, findings of heterozygous copies of candidate genes (e.g., *sh4*, *Bh4*, *Rc*, *An-1*) in Malaysian weedy rice samples suggest recent introgression into weed strains from both common wild rice and modern elite cultivars (Song et al., 2014; Cui et al., 2016; Neik et al., 2019).

Such intercrossing and hybridization between rice crops and local *O. rufipogon* were also found in Thailand (Pusadee et al., 2012). Natural gene flow occurring in fields has contributed to the wild-like adaptive traits of weedy rice types (e.g., easy seed shattering, presence of awns, open panicles), which promote their invasiveness in rice planting areas (Niruntrayakul et al., 2009; Wedger et al., 2019). Intercrosses among wild, weedy, and cultivated rice, although occurring at low rates (<1%), would assist the migration of alleles from wild rice to cultivars, or vice versa (Cao et al., 2006; Chen et al., 2004; Gealy et al., 2003).

In South Asia, wild-like weedy rice genomic analyses based on genome-wide SNP data demonstrate a major wild rice ancestry contribution (Huang et al., 2017). Although this weedy rice evolution suggests adaptive introgression from local common wild rice populations, weedy rice samples collected from Bangladesh, Myanmar, India, Nepal, Pakistan, and Sri Lanka show a heterogeneous genetic makeup, suggesting that cultivated varieties (*aus* and *indica*) also play a major role in the evolution of the weedy rice (Qiu et al., 2020). Such crop-to-crop and crop-to-weed allelic introgression in weedy rice populations was also found in Malaysia (Neik et al., 2019; Song et al., 2014), Thailand (Pusadee et al., 2012), southern China (Qiu et al., 2020), and Taiwan (Wu et al., 2020), indicating that continuous introgression from rice cultivars can contribute to the adaptive and crop mimicry nature of weedy rice.

Dispersal of weedy rice

Knowledge of weedy rice dispersal mechanisms in DSR systems can help to prevent its spread and in identifying management strategies based on dispersal prevention. The use of contaminated rice seeds is one of the most important dispersal mechanisms for weedy rice (Singh et al., 2013; Rao et al., 2017). This mechanism plays an even more vital role in regions where farmers use their saved seeds instead of certified seeds, and where farmers use high seeding rates in DSR systems (Rao et al., 2017). Another important mechanism of weedy rice dispersal reported is through farm machinery/equipment movement within a field and from one field to another without cleaning (Singh et al., 2013; Gao et al., 2018). Widespread occurrence of weedy rice in several countries (e.g., Malaysia, Vietnam, China, and Cambodia) in Southeast Asia is linked with the increased use of combine harvester on a rental basis (Vaughan et al., 1995). Other possible dispersal mechanisms of weedy rice reported include irrigation channels or irrigation water, heavy winds or storms, and flooding.

Biology and ecology of weedy rice

Knowledge of ecology and biology of weedy rice can help in designing weedy rice control tactics. Understanding of factors affecting seed dormancy, seed shattering, seed longevity, seed mortality, and emergence and growth of seedlings would help in identifying weak points for weedy rice management. These have been discussed below:

Seed dormancy

Seed dormancy is an adaptive trait contributing to its success to adapt and survive in rice ecosystems. A large variation in seed dormancy of weedy has been reported with no dormancy (Xia et al., 2011) to few days to several years depending on the biotype and seed storage conditions after its harvest (Vidotto and Ferrero, 2000; Tseng et al., 2013). This variation in dormancy in weedy rice is attributed to genetic factors [>10 interacting quantitative trait loci (QTL) reported to regulate dormancy] and environmental factors [weather conditions during seed development and storage conditions (moisture and temperature)] (Ziska et al., 2015; Leopold et al., 1988). Leopold et al. (1988) reported that weedy rice lost dormancy faster at 6%–14% moisture content but dormancy loss was very low when the moisture content was < 5% or >18% (Leopold et al., 1988). This information can be exploited for weedy rice management by using tillage intelligently to bring grain moisture in the range of 6%–14% and then using stale seedbed. Dormancy-breaking chemicals have been found effective in breaking weedy rice dormancy in laboratory experiments (Cohn, 2002); however, factors that contribute to the release of weedy rice dormancy under field conditions are not yet fully characterized (Ziska et al., 2015). Therefore there is a need for more research to identify dormancy-breaking factors under field conditions, so that weedy rice seedbank can be minimized using techniques like stale seedbed. Generally, light is not required for the germination of weedy rice but some photoblastic weedy rice biotypes from Korea have been reported, whose germination will be increased with exposure to light (Chung and Paek, 2003)

Seed longevity

In weedy rice, seed longevity is associated with seed dormancy and presence of red pericarp (Xia et al., 2011). The longevity of weedy rice in the soil is influenced by dormancy, seed burial depth, soil moisture, and varied with biotypes (Noldin, 2000). It has been reported that the longevity of weedy rice increased with an increase in seed burial depth (Noldin et al., 2006). For example, at Beaumont, Texas, when weedy rice biotypes were buried at 5-cm depth, all biotypes lost viability within 5 months except three biotypes whose viability was <1%. However, when weedy rice seeds were buried at 25 cm, seeds of nine biotypes remained viable (ranged from <1% to 12%) even after 2 years. Seeds of cultivated rice did not survive longer than 5 months irrespective of burial depth. In Sri Lanka also, it was found that weedy rice seeds can remain viable for up to 2 years and higher viability when buried deeper (30 cm) than at shallow depth (15 cm), whereas cultivated rice lost viability within 16 weeks of burial (Ratnasekera, 2015). A study conducted in Vietnam found that weedy rice seed viability declined faster under moist conditions than under submerged conditions (Chin et al., 2000). They found 52% and 27% of seeds remained viable after 4 months under submerged and moist conditions, respectively. Under room conditions, *indica*-type weedy rice from Korea had longer seed viability than *japonica*-type weedy rice. However, overwinter survival of weedy rice seeds in the field was observed higher in *japonica*-type weedy rice than *indica*-type weedy rice (50% germination vs 10%) (Kyoung et al., 1999). The country-specific knowledge and factors influencing weedy rice seed longevity would help in the development of ecological-based weed management including designing crop rotation duration.

Seed decay/mortality

Understanding factors influencing weedy rice seed decay in the soil can have important implications for management targeting weed seedbank of weedy rice. However, information on factors affecting the decay of weedy rice in DSR systems is meager. Soil moisture has been probably best documented for its influence on weedy rice seed decay. For example, winter flooding in Italy between rice crops reduced the viability of weedy rice seeds on the soil surface by >95% compared to a reduction by 26%–76% when the field is left dry (Fogliatto et al., 2010). The study concluded that the reduction in viability was partly due to seed decay of nongerminated seeds under low-temperature conditions and flooding. However, another study conducted in Korea by Baek and Chung (2012) also observed that winter flooding reduced the germination of weedy rice but the effect was not as dramatic as reported by Fogliatto et al. (2010). They observed that >60% weedy rice seeds could overwinter under flooded conditions, whereas in dry conditions, about 90% weedy rice seeds could overwinter.

Seed shattering

Seed shattering trait is one of the seed dispersal mechanisms for weedy rice (Ziska et al., 2015). Early shattering in weedy rice may also prevent drying of seeds and loss of seed dormancy as high moisture content confers greater levels of dormancy (Leopold et al., 1988). A large variation in the time and degree of seed shattering has been reported in weedy rice biotypes. Across the US rice belt, on average, seed shattering in weedy rice starts 15 days after anthesis (Noldin et al., 1999). In Vietnam, the onset of the shattering starts from 7 to 15 days after flowering, and the degree of shattering ranged from 20% to 94% (Chin et al., 2000). In Sri Lanka also, the extent of seed shattering is quite variable among biotypes (Ratnasekera, 2015). These results suggest that roughing should not be delayed after anthesis and should be done maximum at anthesis to avoid seed shattering.

Emergence and early growth

Weedy rice emergence generally declines with an increase in seed burial depth under both moist as well as flooded conditions. Weedy rice could sprout from soil up to 8 cm of flooding depth (Chauhan, 2012b) and had 50% or even higher emergence rate at 10 cm of flooding depth (Ghosh et al., 2017a,b). The emergence rate of weedy rice seedlings was 92%–45% in 0–5 cm saturated soil layers (Vidotto and Ferrero, 2000). Although the emergence rate gradually decreased with the increase in soil depth (Zhang et al., 2019), it could still emerge when the soil depth reached 10 cm under saturated conditions (Vidotto and Ferrero, 2000). Weedy rice has early growth survival strategies for persisting in the field. The seedling growth vigor is attributed to stronger photosynthetic ability in leaves than cultivated rice at early seedling stages (Dai et al., 2017). Early tillering and early maturity are the two important traits associated with weedy rice competitiveness against the commercial variety (Estorninos et al., 2002). Shivrain et al. (2009) reported 12 biotypes of weedy rice produced 226–462 tillers/m^2, whereas the cultivar of inbred rice produced 250 tillers/m^2 at 1:19 (weedy rice:cultivated rice) ratio. Early maturity is a key survival strategy that allows weedy rice to escape from harvesting, which has been reported in many countries including India, China, Costa Rica, and Malaysia (Zhao et al., 2018). The early maturity in weedy rice is attributed to the combination of early flowering and shorter grain-filling stages due to faster grain filling (7–21 days) than cultivated rice (Zhao et al., 2018).

Diversity in weedy rice

Weedy rice is rich in morphological polymorphism and phenotypic plasticity (Andres et al., 2015). In Malaysia, different weedy rice biotypes were reported: (1) taller, shorter, and similar height as the cultivated rice (Fig. 13.2); (2) erect, horizontal, and descending flag leaves (Fig. 13.3); (3) open and closed panicles (Fig. 13.3); (4) easy shattering and nonshattering seeds (Fig. 13.4); and (5) awned and awnless seeds (Fig. 13.4) (Dilipkumar et al., 2020). The morphology, karyotype, and genetic diversity of weedy rice in different regions of the world have been extensively studied (Suh et al., 1997). Tang and Morishima (1997) classified 24 weedy rice populations from China, Japan, and South Korea into wild-like rice type and cultivated-like rice type according to the seed shattering, dormancy, and 1000 grain weight. The classification criteria of phenol reaction, potassium chlorate resistance, lemma hair length, and the isozyme analysis of 11 loci further demonstrates various classification of *indica*-like, *indica*-like autotype, and *japonica*-like autotype (Tang and Morishima, 1997). According to

FIGURE 13.2 Weedy rice height taller (A), shorter (B), and similar (C) as cultivated rice (Dilipkumar et al., 2020).

FIGURE 13.3 Weedy rice with erect (A), horizontal (B), and descending flag leaves (C), open (D), and closed (E) panicles (Dilipkumar et al., 2020).

FIGURE 13.4 Weedy rice seeds with easy shattering (A), nonshattering (B) characteristics, awned (C), and awnless (D) seeds (Dilipkumar et al., 2020).

the color of awns and glumes, weedy rice in the United States can be divided into two categories: yellow awn rice husk and awn black husk (Noldin et al., 1999), with the former accounting for the majority (Shivrain et al., 2010a,b). Their morphological differences may originate from different Asian cultivated rice varieties (Londo and Schaal, 2007). In conclusion, global weedy rice may be roughly divided into four types (Suh et al., 1997): (1) wild-type weedy rice: similar to wild rice, such as easy shattering, long awn, black or purple glume, and red pericarp; (2) cultivated-type weedy rice: similar characteristics with cultivated rice, such as weak shattering, awnless or short awn, yellow or brown glume, and white pericarp; (3) *japonica* weedy rice: shorter grain length, less tillers, shorter plant height, and seed setting; and (4) *indica* weedy rice: longer grain length, more tillers, higher plant height, larger seed setting, but slightly smaller 1000 grain weight. Phenotypic diversity in weedy rice is reflection of underlying genetic variability and just phenotypic diversity information is not enough but intervention at the molecular level to understand genetic diversity is critical in devising management strategy (Rathore et al., 2016)

Management of weedy rice

Greater morphological and genetic diversity and close similarity of weedy rice to cultivated rice make it difficult to obtain the best control of weedy rice with any single method. Therefore to manage weedy rice in DSR systems effectively and sustainably, an integrated approach based on the combination of preventive, cultural, mechanical, chemical, and biotechnological methods is required (Fig. 13.5).

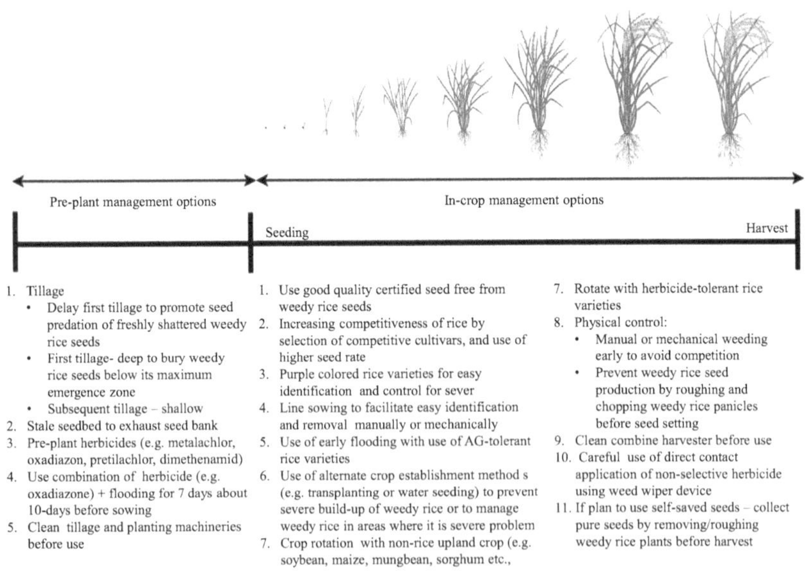

FIGURE 13.5 Integrated weed management options before sowing (preplant period) and after sowing (in-crop period) for weedy rice control.

Preventive method

As part of integrated weed management strategy, early detection and prevention are the first steps to minimize the infestation and spread of weedy rice in DSR systems. Increasing awareness of farmers, extension personnel, and researchers about the potential risks and problems imposed by weedy rice and the importance of its prevention is the first important step to develop successful strategies to combat the weed. A survey conducted by Mishra et al. (2017) revealed that about 58% of farmers in tribal areas of Madhya Pradesh, India do not know that weedy rice is a type of weed in the rice field. Monitoring for the emergence of weedy rice in a new area is important to avoid its invasion and for making timely management decisions. In DSR, the use of rice seeds contaminated with weedy rice is considered one of the main factors contributing to introduction and spread of weedy rice to new areas (Kumar and Ladha, 2011; Ziska et al., 2015). In many Asian countries where DSR is widely practiced, farmers use their own-saved seeds and the quality of farmers' own-saved seeds is often poor with weed seed contaminants including weedy rice. Noldin (2000) estimated that if paddy seed contaminated with two weedy rice seeds per kg of paddy seed is planted can produce 100 kg/ha weedy rice seeds within three seasons. The contamination level generally far exceeds this level in most of the Asian countries. For example, in Cambodia, 92% of surveyed paddy seed samples were contaminated with 524 weedy rice seeds per kg of paddy seed (Chhun et al., 2020). In Vietnam, more than one-third of surveyed paddy seed samples were found contaminated with weedy rice, and 314 weedy rice seeds were found per kg of paddy seed (Mai et al., 2000). In Thailand, up to 4000 weedy rice seeds in 1 kg of paddy seed were found (Maneechote et al., 2004). In the United States, Norsworthy et al. (2007) reported that rice seed contamination is one of the important factors of weedy rice spread in Arkansas as weedy rice seeds have been observed even in certified seeds.

Therefore it is widely accepted that the use of certified seeds free from weedy rice seeds is one of the effective strategies to prevent the introduction and spread of weedy rice in DSR. Chauhan et al. (2014) in a field study from Sri Lanka found that the use of clean rice seeds in the broadcast DSR method reduced weedy rice seed production by 29%–41% and increased rice yield compared to farmers' practice of broadcast DSR with their own-saved seeds. Besides, weedy rice also spreads rapidly within and between fields through agricultural implements (Saha et al., 2014). Machinery should be cleaned before moving it to new areas to prevent contamination. For example, combine harvesting is considered a major cause for the spread of weedy rice in several Southeast Asian countries (Vaughan et al., 1995).

Cultural and mechanical methods

Stale seedbed technique and tillage

Many studies reported that weed establishment in the forthcoming crop can be dramatically decreased if using the stale seedbed technique (Singh et al.,

2018; Kumar et al., 2013; John and Mathew, 2001). This method involves wetting the soil to stimulate weedy rice to establish well before sowing the rice and then emerged weedy rice seedlings are killed either by a nonselective herbicide (paraquat, glyphosate, or glufosinate) or by shallow tillage (Chauhan, 2012a). In Europe and South America, stale seedbed is practiced to keep weedy rice under check (Ferrero, 2003; Foloni, 1999). In Brazil, under a no-till system, stale seedbed with two cycles of irrigations with killing emerged weedy rice 15 days after irrigation was effective in reducing the weedy rice seedbank (Foloni, 1999). A successful stale seedbed method interconnected with soil tillage systems. Zheng et al. (2019) proposed to detect the distribution of weedy rice in soil seedbank before the stale seedbed method is practiced. If the distribution of seeds is mainly concentrated in the topsoil layer, induce germination and eradication methods could be applied as stated above. However, if the distribution of seeds is high in the deep soil layer, deep tillage could be performed first to bring back the seeds to the shallow soil depth before applying the stale seedbed technique.

Tillage also plays an important role in weedy rice control. Since weedy rice emergence declines with an increase in seed burial depth, higher emergence from top 0–5 cm and emergence ceases at 8–10 cm depth (Vidotto and Ferrero, 2000; Chauhan, 2012b). Therefore if the stale seedbed technique is not applied, then deep tillage can be used to bury weedy rice seeds deeper (below the maximum depth of their emergence; > 10 cm) to reduce weedy rice recruitment in the DSR crop, especially if the weed seedbank is concentrated in the topsoil layer prior to tillage (Massoni et al., 2013). Exposing newly shattered weedy rice seeds for predation by delaying first tillage after rice harvest is found an effective strategy to reduce its seedbank in Italy (Fogliatto et al., 2010).

Early flooding and role of anaerobic germination tolerant rice cultivars

Early flooding has been an effective tool for weed control in PTR as the emergence and growth of the majority of rice weeds including weedy rice is highly suppressed by flooding (Rao et al., 2017; Chauhan, 2012b). But early flooding for weed suppression in DSR has limited scope as rice is also sensitive to early flooding (anaerobic conditions) at the early emergence/crop establishment stage, and hence, can affect crop establishment negatively. Currently, in DSR systems, flooding is established only after rice emergence/establishment, and by then most of the weeds including weedy rice also emerge when flooding is not very effective. Water seeding with a higher seed rate is practiced in some areas where weedy rice is becoming a severe problem (Azmi and Johnson, 2009). In Malaysia, a 30 to 50% higher seed rate is used in the water-seeding system compared to the seed rate used in the wet-DSR system (Dilipkumar et al., 2017).

The availability of anaerobic germination (AG) tolerant rice cultivars—tolerant to anaerobic conditions/flooding during early emergence/crop establishment—would enable the use of early flooding for weed control including weedy rice without any adverse effect on crop establishment. International Rice Research Institute has identified AG-tolerant QTLs (AG1 and AG2) and introgressed these OTLs into high-yielding rice cultivars such as IR64, PSB Rc 82, and Ciherang-Sub 1 (Mondal et al., 2020). In a greenhouse study, a combination of early flooding (24–72 h after seeding) of 5-cm depth and pretilachlor + safener with AG-tolerant rice varieties suppressed 80%–90% emergence of weedy rice without affecting crop establishment (Kumar et al., 2017).

Improving rice competitiveness and purple-colored cultivars

Rice cultivar with strong weed competitiveness is another important tool for weedy rice management. In Malaysia, taller rice cultivars with higher tillering ability suppressed the weedy rice population (Azmi and Abdullah, 1998). The hybrid variety was found higher-yielding and suppressed the growth and seed production of weedy rice than inbred varieties in Arkansas, United States (Ottis et al., 2005). Similarly, Shivrain et al. (2009) also reported that hybrid varieties reduced the tillering and competitive effect of weedy rice biotypes. Practices such as the use of high seeding rates, which improve early canopy, cover shift competition in favor of crop than weeds. For example, in Malaysia, a four times higher seeding rate increased the yield of the DSR crop infested with weedy rice (Azmi et al., 2000). There is a need for more research in identifying high-yielding and weedy rice-suppressive rice cultivars. Growing cultivars with purple-colored leaves or purple-stemmed base may help to reduce the weedy rice seedbanks, where weedy rice emergence can be easily differentiated and removed from the field. Farmers in Himachal Pradesh, India have adapted to cultivate purple leaves rice cultivar "R585" in weedy rice infested fields (Kaushik et al., 2011).

Rice establishment methods

Line sowing instead of broadcasting facilitates the identification of weedy rice at an early stage that emerges between rows and can be easily removed manually or using mechanical tools like finger weeder or cono weeder (Chauhan, 2012a). A field study conducted in Sri Lanka indicated that row seeding reduced weedy rice panicle density from 39–49 panicles/m^2 to 15–30 panicles/m^2 and increased rice grain yield (Chauhan et al., 2014). Alternate rice establishment methods such as transplanting (manually or mechanically) and water seeding are proven effective establishment methods for weedy rice control. Hence, the introduction of these methods for one or two seasons if weedy rice has become a serious problem in DSR can help in keeping the weedy rice problem under check. In Malaysia, weedy rice panicle density declined by 66% with the use of mechanical transplanted rice for two consecutive seasons compared to continuous DSR (Azmi et al., 2005). In Korea, the reduction in weedy rice panicle density was 90% with three consecutive seasons of mechanical transplanted rice (Kim et al., 2000). In Sri Lanka, rice seedling broadcast and manual transplant reduced weedy rice seed production by 71%–87% and 95%–98% as compared to broadcast DSR (Chauhan et al., 2014).

Physical methods

Manual weeding or roguing or chopping of weedy rice panicles before seed production is another effective option for minimizing weedy rice infestation in DSR. However, the chopping or roughing technique is not feasible if the field is infested with weedy rice accession with the same or shorter plant height compared to the cultivated rice. Mechanical weeding can reduce the labor requirement, but line sowing is a prerequisite for mechanical weeding.

Crop rotation

Crop rotation helps to break the weedy rice life cycle, enables the use of other herbicides to control weedy rice, and also makes it easy to identify weedy rice—all leading to better control of weedy rice. In the southern United States, the rice–soybean rotation has been the most effective and widely adopted rotation to minimize weedy rice infestation (Burgos et al., 2008). In Uruguay, Scavino et al. (2013) observed a low problem of weedy rice in DSR because of the adoption of crop rotations with pastures (Scavino et al., 2013). Andres et al. (2012) in Brazil found a 90% reduction in weedy rice when rice was rotated with 2 years of cultivation of soybean (*Glycine max* (L.) Merr.), corn (*Zea mays* L.), or sorghum (*Sorghum bicolor* (L.) Moench). Similarly, in Asia, the weedy rice problem can be drastically reduced if one rice crop, preferably dry season, in a double or triple rice cropping system is rotated with non-rice upland crops such as soybean, corn, or mungbean (Chauhan, 2013). Alternate herbicides that are effective for weedy rice control but could not be used in rice can be used in these crops. In the Philippines, glyphosate-tolerant corn is commercialized and hence by using glyphosate-tolerant corn in rotation with rice enables farmers to use glyphosate to control weeds including weedy rice.

In the dry season, farmers in Central Vietnam rotate rice with mungbean (*Vigna radiata* (L.) R. Wilczek) to manage weedy rice because voluntary and weedy rice seedlings struggle to thrive in mungbean due to inadequate soil moisture (Watanabe et al., 1998). Ulguim et al. (2018) recorded high weed density, mostly weedy rice with continuous rice cultivation whereas systems with crop rotation and crop diversification reduced the weedy rice infestation level with a trade-off higher infestation of perennial species.

Chemical method including herbicide-tolerant rice technology

Weedy rice management using herbicide can only feasible before rice sowing. Any chemical control activity made after sowing the rice may give adverse effects on rice growth and yield. Metolachlor and oxadiazon, for example, effectively inhibited germination of weedy rice without affecting rice if sprayed at least 15 days before rice planting (Estorninos et al., 2005). In European rice conditions, pretilachlor or dimethenamid sprayed alone or in combination at 25 days before rice planting could obtain more than 75% of weedy rice control (Ferrero et al., 1999). In Peninsular Malaysia, application of oxadiazon at 500 g ai/ha in 5-cm standing water just after the second tillage and retained the water in the field up to 7 days before rice sowing suppressed weedy rice (Dilipkumar et al., 2020). This technique could significantly destroy the superficial weed seed bank, where the density of weedy rice could be reduced to 30%–45% in the first season and the reduction rate could be increased up to 75% in the following season. There are limited in-crop herbicide options for selective control of weedy rice other than HT-rice varieties. However, weedy rice plants which are taller than cultivated rice can be selectively controlled by carefully treating these plants with nonselective herbicides such as glyphosate/glufosinate/paraquat or cycloxidim at 15%–20% and 5% concentration, respectively using a weed wiper/wick applicator (Ferrero et al., 1999; Jose et al., 2020). Jose et al. (2020) recently developed a hand-held weed wiper device to minimize risks of herbicide dripping which can be used at 60–65 days after sowing (DAS) to selectively kill weedy rice panicles.

The introduction of HT-rice variety made selective control of weedy rice possible with the use of the herbicide. One of the successful HT-rice technologies that have been commercialized globally is Clearfield rice. Clearfield rice also known as imidazolinone (IMI)-tolerant rice was developed through mutagenesis of the acetolactate synthase locus which offers an opportunity to selectively control weedy rice with IMI herbicides (Croughan, 2003). At the beginning of introducing this technology, it provided 95%–100% weedy rice control in most cases (Burgos et al., 2008). In Malaysia, fields planted with Clearfield rice reduced the weedy rice population effectively, hence doubled the rice yield from 3.5 to 7.0 t/ha (Sudianto et al., 2013). Despite the benefit of the technology, the evolution of IMI-resistant weedy rice has been reported in some countries, including the United States, Brazil, Colombia, Costa Rica, Italy, and Malaysia (Sudianto et al., 2013; Dilipkumar et al., 2018a,b). Population genetic studies indicate that the IMI-resistant weedy rice was mainly originated from the gene from IMI-tolerant cultivars and distributed by seed migration (Song et al., 2009; Shivrain et al., 2010a,b).

Based on the survey conducted in Malaysia, most farmers cultivated Clearfield rice for more than two consecutive seasons, and some extended the planting to seven seasons. Ignorance of the stewardship guidelines such as purchasing noncertified seeds, cultivating Clearfield rice without using

IMI herbicide, or spraying the herbicide at the wrong time and at lower than recommended rate may have resulted in the escape of the resistance trait from Clearfield rice to weedy rice or wild rice (Dilipkumar et al., 2017). To help combat IMI-resistant weedy rice and complement the Clearfield technology, Badische Anilin- und SodaFabrik (BASF) recently developed another new HT-rice technology namely, Provisia rice that confers resistance to the acetyl coenzyme-A carboxylase inhibitor herbicide (Mankin et al., 2014). Considering all the experiences gained from the Clearfield rice technology, the new Provisia rice should be used wisely and carefully. The evolution of multiple-resistant weedy rice populations is possible if ignoring past experiences gained from the Clearfield rice technology. For the longevity of HR-rice technology, it is important that farmers should not continuously use these varieties and should be rotated with other HR-rice varieties with different modes of action or with conventional non-HR-rice varieties along with compliance with the stewardship program.

Conclusion

Weedy rice is the most troublesome weed of cultivated rice and it has emerged as a threat to DSR-based rice production systems. Its close similarity to the cultivated rice has complicated the implementation of the systematic weedy rice control strategy. In the absence of selective herbicides for weedy rice control in conventional rice cultivars, diversified weed management with multiple tactics prior to sowing and after sowing should be integrated for effective control such as optimization of tillage schedule, stale seedbed technique, preplant application of preemergence herbicides (e.g., pretilachor or oxadiazon) with flooding before rice sowing, use of clean seeds and machinery, a high seeding rate of rice, rotation of rice establishment methods, line sowing, roguing, and crop rotation. HT-rice could be another effective tool for weedy rice control. However, prior to the commercialization of HT-rice, it is important to assess potential risks associated with the technology, and stewardship guidelines should be developed to minimize the risk and for the long-term sustainability of the technology.

Acknowledgments

This review is the output of the Direct-Seeded Rice Consortium—a public–private multi-stakeholder research for development platform on direct-seeded rice, convened by the International Rice Research Institute.

References

Akasaka, M., Ushiki, J., Iwata, H., Ishikawa, R., Ishii, T., 2009. Genetic relationships and diversity of weedy rice (*Oryza sativa L.*) and cultivated rice varieties in Okayama Prefecture, Japan. Breed. Sci. 59, 401–409.

Andres, A., Concenco, G., Theisen, G., Galon, L., Tesio, F., 2012. Management of red rice (*Oryza sativa*) and barnyardgrass (*Echinochloa crus-galli*) grown with sorghum with reduced rate of atrazine and mechanical methods. Exp. Agric. 48, 587–596.

Andres, A., Fogliatto, S., Ferrero, A., Viddoto, F., 2015. Growth variability of Italian weedy rice populations grown with or without cultivated rice. J. Crop. Sci. 55 (1), 394–402.

Azmi, M., Abdullah, M.Z., 1998. A Manual for the Identification and Control of Padi Angin (Weedy Rice). MARDI Publication, Kuala Lampur, Malaysia.

Azmi, M., Johnson, D.E., 2009. Crop establishment options for lowland irrigated rice in relation to weed infestation and grain yield. J. Trop. Agric. Fd. Sci. 37 (1), 111–117.

Azmi, M., Abdullah, M.Z., Mislamah, B., Baki, B.B., 2000. Management of weedy rice (*Oryza sativa L.*): the Malaysian experience. In: Baki, B.B., et al., (Eds.), Wild and Weedy Rice in Rice Ecosystems in Asia e a Review. Limited Proceedings No. 2. International Rice Research Institute, Los Baños (Philippines), pp. 91–96. , Asian Pacific Weed Science Society.

Azmi, M., Muhamad, H., Johnson, D.E., 2005. Impact of weedy rice infestation on rice yield and influence of crop establishment technique. In: Proceedings of the 20th Asian-Pacific Weed Science Society Conference, Ho Chi Minh City, Vietnam.

Arrieta-Espinoza, G., Sánchez, E., Vargas, S., Lobo, J., Quesada, T., Espinoza, A.M., 2005. The weedy rice complex in Costa Rica. I. Morphological study of relationships between commercial rice varieties, wild *Oryza* relatives and weedy types. Genet. Resour. Crop. Evol. 52, 575–587.

Baek, J., Chung, N., 2012. Seed wintering and deterioration characteristics between weedy and cultivated rice. Rice. Available from: https://doi.org/10.1186/1939-8433-5-21.

Burgos, N.R., Norsworthy, J.K., Scott, R.C., Smith, K.L., 2008. Red rice status after five years of Clearfield rice technology in Arkansas. Weed Technol. 22, 200–208.

Cao, Q., Lu, B.R., Xia, H., Rong, J., Sala, F., Spada, A., et al., 2006. Genetic diversity and origin of weedy rice (*Oryza sativa f. spontanea*) populations found in North-eastern China revealed by simple sequence repeat (SSR) markers. Ann. Bot. 98 (6), 1241–1252.

Cao, Q.J., Li, B., Song, Z.P., Caim, X.X., Lum, B.R., 2007. Impact of weedy rice populations on the growth and yield of direct-seeded and transplanted rice. Weed Biol. Manag 7 (2), 97–104.

Chakraborty, D., Ladha, J.K., Rana, D.S., Jat, M.L., Gathala, M.K., Yadav, S., et al., 2017. A global analysis of alternative tillage and crop establishment practices for economically and environmentally efficient rice production. Sci. Rep. 7 (1). Available from: https://doi.org/10.1038/s41598-017-09742-9.

Chauhan, B.S., 2012a. Weed ecology and weed management strategies for dry seeded rice in Asia. Weed Technol. 26, 1–13.

Chauhan, B.S., 2012b. Weedy rice (*Oryza sativa*) II. Response of weedy rice to seed burial and flooding depth. Weed Sci. 60 (3), 385–388.

Chauhan, B.S., 2013. Strategies to manage weedy rice in Asia. Crop. Prot. 48, 51–56.

Chauhan, B.S., Abeysekera, A.S.K., Wickramarathe, M.S., Kulatunga, S.D., Wickrama, U.B., 2014. Effect of rice establishment methods on weedy rice (*Oryza sativa L.*) infestation and grain yield of cultivated rice (*O. sativa L.*) in Sri Lanka. Crop. Prot. 55, 42–49.

Chen, L.J., Lee, D.S., Song, Z.P., Suh, H.S., Lu, B., 2004. Gene flow from cultivated rice (*Oryza sativa*) to its weedy and wild relatives. Ann. Bot. 93, 67–73.

Chhun, S., Kumar, V., Martin, R.J., Srean, P., Hadi, B.A.R., 2020. Weed management practices of smallholder rice farmers in Northwest Cambodia. Crop. Prot. 135, 104793. Available from: https://doi.org/10.1016/j.cropro.2019.04.017.

Chin, D.V., Hien, T.V., Thiet, L.V., 2000. Weedy Rice in Vietnam, in Wild and Weedy Rice in Rice Ecosystems in Asia-A Review. International Rice Research Institute, Los Baños (Philippines), pp. 45–50.

Chung, N.J., Paek, N.C., 2003. Photoblastism and ecophysiology of seed germination in weedy rice. Agron. J. 95, 184–190.

Cohn, M.A., 2002. Reflections on investigating dormancy-breaking chemicals – a balance of logic and luck. Weed Sci. 50 (2), 261–266.

Croughan, T.P., 2003. Clearfield rice: it's not a GMO. La. Agric. 46, 24–26.

Cui, Y., Song, B.K., Li, L., Li, Y., Huang, Z., Caicedo, A.L., et al., 2016. Little white lies: pericarp color provides insights into the origins and evolution of southeast Asian weedy rice. G3 6 (12), 4105–4114.

Dai, L., Song, X., He, B., Valverde, B.E., Qiang, S., 2017. Enhanced photosynthesis endows seedling growth vigour contributing to the competitive dominance of weedy rice over cultivated rice. Pest. Manag. Sci. 73 (7), 1410–1420.

Dilipkumar, M., Chuah, T.S., Goh, S.S., Sahid, I., 2017. Weed management issues, challenges, and opportunities in Malaysia. Crop. Prot. Available from: https://doi.org/10.1016/j.cropro.2017.08.027.

Dilipkumar, M., Burgos, N.R., Chuah, T.S., Ismail, S., 2018a. Cross-resistance to imazapic and imazapyr in a weedy rice (*Oryza sativa*) biotype found in Malaysia. Planta Daninha . Available from: https://doi.org/10.1590/S0100-83582018360100058.

Dilipkumar, M., Lee, L.C., Chuah, T.S., 2018b. Effective method to control weedy rice in direct-seeded rice culture. In: Proceedings of the 5th International Rice Congress., Marina Bay Sands, Singapore, pp. 145.

Dilipkumar, M., Badrulhadza, A., Mohd Khusairy, K., Mohd Shahril, F.A.R., Chong, T.V., Chuah, T.S., 2020. Manual of Weedy Rice Management Technology (in Malay Language). Malaysian Agricultural Research and Development Institute, Serdang, Selangor, Malaysia, pp. 4–7.

Estorninos, L.E., Gealy, D.R., Talbert, R.E., 2002. Growth response of rice (*Oryza sativa* L.) in replacement series study. Weed Technol. 16, 401–406.

Estorninos, L.E., Gealy, D.R., Talbert, R.E., Gbur, E.E., 2005. Rice and red rice interference. I. Response of red rice (*Oryza sativa*) to sowing rates of tropical japonica and indica rice cultivars. Weed Sci. 53, 676–682.

FAOSTAT. 2018. FAOSTAT Database. Rome, Italy: Food and Agriculture Organization of the United Nations (FAO). Retrieved July 30, 2020, from http://www.fao.org/faostat/en/#home

Federici, M.T., Vaughan, D., Norihiko, T., Kaga, A., Xin, W.W., Koji, D., et al., 2001. Analysis of Uruguayan weedy rice genetic diversity using AFLP molecular markers. Electron. J. Biotechnol. 4 (3), 5–6.

Ferrero, A., 2003. Weedy rice, biological features and control. In: Weed Management for Developing Countries: Addendum 1, Labrada, R. (Ed.), FAO Plant Production and Protection Paper 120, Add. 1. FAO, Rome. pp. 89–107.

Ferrero, A., Vidotto, F., Balsari, P., Airoldi, G., 1999. Mechanical and chemical control of red rice (*Oryza sativa L. var. sylvatica*) in rice (*Oryza sativa L.*) preplanting. Crop. Prot. 18 (4), 245–251.

Fogliatto, S., Vidotto, F., Ferrero, A., 2010. Effects of winter flooding on weedy rice (*Oryza sativa L.*). Crop. Prot. 29, 1232–1240.

Foloni, L.L., 1999. New weed management system in no till irrigated rice aiming to improve red rice control. In: Vol. 1. Proceedings of the Brighton Crop Protection Conference: Weeds, Brighton, UK. pp. 369–372.

Gao, P., Zhang, Z., Sun, G., Yu, H., Qiang, S., 2018. The within-field and between-field dispersal of weedy rice by combine harvesters. Agron. Sustain. Dev. Available from: https://doi.org/10.1007/s13593-018-0518-2.

Gealy, D.R., Mitten, D.H., Rutger, J.N., 2003. Gene flow between red rice (*Oryza sativa*) and herbicide-resistant rice (*O. sativa*): implications for weed management. Weed Technol. 17 (3), 627–645.

Ghosh, D., Rathore, M., Brahmachari, K., Singh, R., Kumar, B., 2017a. Impact of burial and flooding depths on Indian weedy rice. Crop. Prot. 100, 106–110.

Ghosh, D., Singh, U.P., Brahmachari, K., Singh, N.K., Das, A., 2017b. An integrated approach to weed management practices in direct-seeded rice under zero-tilled rice–wheat cropping system. Int. J. Pest. Manag. 63, 37–46.

Grimm, A., Fogliatto, S., Nick, P., Ferrero, A., Vidotto, F., 2013. Microsatellite markers reveal multiple origins for Italian weedy rice. Ecol. Evol. 3, 4786–4798.

Gross, B.L., Reagon, M., Hsu, S., Caicedo, A.L., Jia, Y., Olsen, K.M., 2010. Seeing red: the origin of grain pigmentation in US weedy rice. Mol. Ecol. 19 (16), 3380–3393.

He, Q., Kim, K., Park, Y., 2017. Population genomics identifies the origin and signatures of selection of Korean weedy rice. Plant. Biotechnol. J. 15 (3), 357–366.

Hoyos, V., Plaza, G., Li, X., Caicedo, A.L., 2020. Something old, something new: evolution of Colombian weedy rice (*Oryza* spp.) through de novo de-domestication, exotic gene flow, and hybridization. Evol. Appl. Available from: https://doi.org/10.1111/eva.12955.

Huang, Z., Young, N.D., Reagon, M., Hyma, K.E., Olsen, K.M., Jia, Y., et al., 2017. All roads lead to weediness: patterns of genomic divergence reveal extensive recurrent weedy rice origins from South Asian *Oryza*. Mol. Ecol. 26 (12), 3151–3167.

Ishikawa, R., Toki, N., Imai, K., Sato, Y., Yamagishi, H., Shimamoto, Y., et al., 2005. Origin of weedy rice grown in Bhutan and the force of genetic diversity. Genet. Resour. Crop. Evol. 52 (4), 395–403.

John, P.S., Mathew, R., 2001. Stale seedbed-an alternate technology for preplanting to achieve total weed control in direct-seeded lowland rice. Int. Rice Res. Newsl. 26, 67–68.

Jose, N., Abraaham, C.T., Mathew, R., Kumari, L.S., 2020. Novel wiper device for the management of weedy rice. Indian. J. Weed Sci. 52, 27–31.

Kanapeckas, K.L., Vigueira, C.C., Ortiz, A., Gettler, K.A., Burgos, N.R., Fischer, A.J., et al., 2016. Escape to ferality: the endoferal origin of weedy rice from crop rice through de-domestication. PLoS One 11 (9), e0162676.

Kaushik, R.P., Shekhar, J., Srivastava, A., 2011. Status paper on rice in Himachal Pradesh. In Rice Knowledge Management Portal. Directorate of Rice Research: Rajendranagar, Hyderabad. pp. 1–34.

Kawasaki, A., Imai, K., Ushiki, J., Ishii, T., Ishikawa, R., 2009. Molecular constitution of weedy rice (*Oryza sativa L.*) found in Okayama prefecture, Japan. Breed. Sci. 59 (3), 229–236.

Kim, S.Y., Son, Y., Park, S.T., Kim, H.Y., 2000. Cultural management to control weedy rice in paddy field. Korean J. Crop. Sci. 45, 232–236.

Kim, T.S., He, Q., Kim, K., Yoon, M., Ra, W., Li, F.P., et al., 2016. Genome-wide resequencing of KRICE_CORE reveals their potential for future breeding, as well as functional and evolutionary studies in the post-genomic era. BMC Genom. 17, 408.

Kumar, V., Ladha, J.K., 2011. Direct seeding of rice: recent developments and future research needs. Adv. Agron. 111, 297–413.

Kumar, V., Singh, S., Chhokar, R.S., Malik, R.K., Brainard, D.C., Ladha, J.K., 2013. Weed management strategies to reduce herbicide use in zero-till rice wheat cropping systems of the Indo-Gangetic Plains. Weed Technol. 27, 241–254.

Kumar, V., J. Opeña, K. Valencia, O. Namuco, T. Migo, S. Dixit et al., 2017. Innovative solution for the management of weedy rice and other weeds in wet direct-seeded rice systems. In: Proceedings of the 26th Asian Pacific Weed Science Society Conference, 19–22 September 2017, Kyoto Japan.

Kyoung, E.S., Kim, J.K., Kim, J.S., Ma, S.Y., Ryang, W.S., 1999. Germination characteristics of some red rice accessions. Korean J. Crop. Sci. 44, 316–320.

Leopold, A.C., Glenister, R., Cohn, M.A., 1988. Relationship between water content and after ripening in red rice. Physiol. Plant. 74, 659–662.

Li, L.F., Olsen, K.M., 2020. Population genomics of weedy crop relatives: Insights from weedy rice. In: Proceedings of the Population Genomics. Crop Plants. https://doi.org/10.1007/13836_2020_77.

Li, L., Li, Y., Jia, Y., Caicedo, A.L., Olsen, K.M., 2017. Signatures of adaptation in the weedy rice genome. Nat. Genet. 49, 811–814.

Londo, J.P., Schaal, B.A., 2007. Origins and population genetics of weedy red rice in the USA. Mol. Ecol. 16 (21), 4523–4535.

Maclean, J., Hardy, B., Hettel, G., 2013. Rice Almanac: Source Book for the Most Important Economic Activities on Earth, fourth ed. IRRI, Los Baños, Philippines.

Mai, V., Chien, H.V., Van, V., Thi, V., Suong, T., Thiet, L.V., 2000. Rice seed contamination in Vietnam. In: Baki, B.B., Chin, D.V., Mortimer, M. (Eds.), Wild and Weedy Rice in Rice Ecosystems in Asia: A Review. Limited Proceedings No. 2. International Rice Research Institute, Los Baños (Philippines), pp. 17–19.

Maneechote, C., Jamjod, S., Rerkasem, B., 2004. Invasion of weedy rice in rice fields in Thailand: problems and management. IRRN 29, 20–22.

Mankin, S., Schofl, U., Hong, H.P., Wenck, A., Neuteboom, L., Whitt, S., et al., 2014. Herbicide-tolerant plants. US Patent 20140045686.

Massoni, P.F.S., Marchesan, E., Grohs, M., Roso, R., Coelho, L.L., Machado, S.L.O., et al., 2013. Influence of post-harvest management of rice crop on red rice seedbank. Planta Daninha 31, 89–98.

Mishra, S.K., Chandra, R., Chaturvedi, S.K., Singh, R., 2017. Weedy rice infestation (*Oryza sativa f. Spontanea L.*) in tribal areas of Madhya Pradesh. Int. J. Curr. Res. 9 (5), 51039–51042.

Mondal, S., Khan, M.I.R., Entila, F., Dixit, S., Sta, Cruz, P.C., et al., 2020. Responses of AG1 and AG2 QTL introgression lines and seed pre-treatment on growth and physiological processes during anaerobic germination of rice under flooding. Sci. Rep. 10, 10214.

Morishima, H., Sano, Y., Oka, H.I., 1984. Differentiation of perennial and annual types due to habitat conditions in the wild rice *Oryza* perennis. Plant. Syst. Evol. 144, 119–135.

Neik, T.X., Chai, J.Y., Tan, S.Y., Sudo, M., Cui, Y., Jayaraj, J., et al., 2019. When West meets East: the origins and spread of weedy rice between continental and island Southeast Asia. G3 9, 2941–2950.

Niruntrayakul, S., Rerkasem, B., Jamjod, S., 2009. Crossability between cultivated rice (*Oryza sativa*) and common wild rice (*O. rufipogon*) and characterization of F1 and F2 populations. Sci. Asia. 35, 161–169.

Noldin, J.A., 2000. Red Rice Status and Management in the Americas, in Wild and Weedy Rice in Rice Ecosystems in Asia-A Review. International Rice Research Institute, Los Baños, Philippines, pp. 21–24.

Noldin, J.A., Chandler, J.M., Mccauley, G.N., 1999. Red rice (*Oryza sativa*) biology. I. Characterization of red rice ecotypes. Weed Technol. 13, 12–18.

Noldin, J.A., Chandler, J.M., McCauley, G.N., 2006. Seed longevity of red rice ecotypes buried in soil. Planta Daninha 24, 611–620.

Norsworthy, J.K., Burgos, N.R., Scott, R.C., Smith, K.L., 2007. Weed management needs in Arkansas rice. In: Norman, R.J., Meullenet, J.F., Moldenhauer, K.A.K. (Eds.), B.R. Wells Rice Research Studies, 2006. University of Arkansas Division of Agriculture, Fayetteville.

Oerke, E.C., 2005. Crop losses to pests. J. Agric. Sci. 144, 31–43.

Oka, H.I., 1988. Origin of Cultivated Rice. Japan Sci. Soc. Press/Elsevier, Tokyo/Amsterdam.

Ottis, B.V., Smith, K.L., Scott, R.C., Talbert, R.E., 2005. Rice yield and quality as affected by cultivar and red rice (*Oryza sativa*) density. Weed Sci. 53, 499–504.

Pusadee, T., Schaal, B.A., Rerkasem, B., Jamjod, S., 2012. Population structure of the primary gene pool of *Oryza sativa* in Thailand. Genet. Resour. Crop. Evol. 60, 335–353.

Qiu, J., Zhu, J., Fu, F., Ye, C.Y., Wang, W., Mao, L., et al., 2014. Genome re-sequencing suggested a weedy rice origin from domesticated *indica-japonica* hybridization: a case study from southern China. Planta. 240 (6), 1353–1363.

Qiu, J., Zhou, Y., Mao, L., Ye, C., Wang, W., Zhang, J., et al., 2017. Genomic variation associated with local adaptation of weedy rice during de-domestication. Nat. Commun. Available from: https://doi.org/10.1038/ncomms15323.

Qiu, J., Jia, L., Wu, D., Weng, X., Chen, L., Sun, J., et al., 2020. Diverse genetic mechanisms underlie worldwide convergent rice feralization. Genome Biol. 21 (70). Available from: https://doi.org/10.1186/s13059-020-01980-x.

Rao, A.N., Brainard, D.C., Kumar, V., Ladha, K., Johnson, D.E., J.K., 2017. Preventive weed management in direct-seeded rice: targeting the weed seedbank. Adv. Agron. 144, 45–142.

Ratnasekera, D., 2015. Weedy rice: a threat to rice production in Sri Lanka. J. Uni. Ruhuna. 3 (1), 2–13.

Reagon, M., Thurber, C.S., Gross, B.L., Olsen, K.M., Jia, Y., Caicedo, A.L., 2010. Genomic patterns of nucleotide diversity in divergent populations of US weedy rice. BMC Evol. Biol. Available from: https://doi.org/10.1186/1471-2148-10-180.

RICE (CGIAR Research Program on Rice Agri-Food Systems). Rice Agri-Food Systems, CRP – Annexes, companion document, and partners' support letters. July 2016. International Rice Research Institute. http://ricecrp.org/wp-content/uploads/2017/03/RICE-phase-II-annexes.

Saha, S., Patra, B.C., Munda, S., Mohapatra, T., 2014. Weedy rice: problems and its management. Indian. J. Weed Sci. 46 (1), 14–22.

Scavino, A.F., Ji, Y., Pump, J., Klose, M., Claus, P., Conrad, R., 2013. Structure and function of the methanogenic microbial communities in Uruguayan soils shifted between pasture and irrigated rice fields. Environ. Microbiol. 15, 2588–2602.

Shivrain, V.K., Burgos, N.R., Gealy, D.R., Smith, K.L., Scott, R.C., Mauromoustakos, A., et al., 2009. Red rice (*Oryza sativa*) emergence characteristics and influence on rice yield at different planting dates. Weed Sci. 57, 94–102.

Shivrain, V.K., Burgos, N.R., Agrama, H.A., Lawton-Rauh, A., Lu, B., Sales, M.A., et al., 2010a. Genetic diversity of weedy red rice (*Oryza sativa*) in Arkansas, USA. Weed Res. 50, 289–302.

Shivrain, V.K., Burgos, N.R., Scott, R.C., Gbur, E.E., Estorninios, L.E., McClelland, M.R., 2010b. Diversity of weedy red rice (*Oryza sativa L.*) in Arkansas, USA in relation to weed management. Crop. Prot. 29 (7), 721–730.

Singh, M., Bhullar, M.S., Gill, G., 2018. Integrated weed management in dry-seeded rice using stale seedbeds and post sowing herbicides. Field Crop. Res. 224, 182–191.

Singh, K., Kumar, V., Saharawat, Y.S., Gathala, M., Ladha, J.K., Chauhan, B.S., 2013. Weedy rice: an emerging threat for direct-seeded rice production systems in India. J. Rice Res. Available from: https://doi.org/10.4172/jrr.1000106.

Song, X., Liu, L., Wang, Z., Qiang, S., 2009. Potential gene flow from transgenic rice (*Oryza sativa L.*) to different weedy rice (*Oryza sativa f. spontanea*) accessions based on reproductive compatibility. Pest. Manag. Sci. 65, 862–869.

Song, B.K., Chuah, T.S., Tam, S.M., Olsen, K.M., 2014. Malaysian weedy rice shows its true stripes: wild *Oryza* and elite rice cultivars shape agricultural weed evolution in Southeast Asia. Mol. Ecol. 23 (20), 5003–5017.

Sudianto, E., Beng-Kah, S., Ting-Xiang, N., Saldain, N.E., Scott, R.C., Burgos, N.R., 2013. Clearfield (R) rice: its development, success, and key challenges on a global perspective. Crop. Prot. 49, 40–51.

Suh, H.S., Sato, Y.I., Morishima, H., 1997. Genetic characterization of weedy rice (*Oryza sativa L.*) based on morpho-physiology, isozymes and RAPD markers. Theor. Appl. Genet. 94, 316–321.

Sun, J., Qian, Q., Ma, D., Xu, Z., Liu, D., Du, H., et al., 2013. Introgression and selection shaping the genome and adaptive loci of weedy rice in northern China. N. Phytol. 197 (1), 290–299.

Tang, L.H., Morishima, H., 1997. Genetic characterization of weedy rices and the inference on their origins. Jpn. J. Breed. 47 (2), 153–160.

Thurber, C.S., Reagon, M., Gross, B.L., Olsen, K.M., Jia, Y., Caicedo, A.L., 2010. Molecular evolution of shattering loci in US weedy rice. Mol. Ecol. 19, 3271–3284.

Tseng, T.M., Burgos, N.R., Shivrain, V.K., Alcober, E.A., Mauromoustakos, A., 2013. Inter- and intrapopulation variation in dormancy of *Oryza sativa* (weedy red rice) and allelic variation in dormancy-linked loci. Weed Res. 53 (6), 440–451.

Ulguim, AdaR., Carlos, F.S., Santos, R.AdaS., Zanon, A.J., Werle, I.S., Beck, M., 2018. Weed phytosociological in irrigated rice under different cultivation systems and crop rotation intensity. Ciência Rural. Available from: https://doi.org/10.1590/0103-8478cr20180230.

Vaughan, D.A., Watanabe, H., Abdullah Zain, M., Okuno, K., 1995. Evolution and genetic diversity of Malaysian weedy rice. In: Proceedings of the 15th Asian-Pacific Weed Science Society Conference, Tsukuba, Japan, pp. 461–466.

Vidotto, F., Ferrero, A., 2000. Germination behaviour of red rice (*Oryza sativa L.*) seeds in field and laboratory conditions. Agronomie. 20 (4), 375–382.

Vigueira, C.C., Li, W., Olsen, K.M., 2013. The role of Bh4 in parallel evolution of hull color in domesticated and weedy rice. J. Evolution. Biol. 26 (8), 1738–1749.

Vigueira, C.C., Qi, X., Song, B.K., Li, L.F., Caicedo, A.L., Jia, Y., et al., 2019. Call of the wild rice: *Oryza rufipogon* shapes weedy rice evolution in Southeast Asia. Evol. Appl. 12 (1), 93–104.

Watanabe, H., Vaughan, D.A., Tomooka, N., 1998. Weedy rice complexes: case study from Malaysia, Vietnam and Suriname. In: Proceedings of the International Symposium on Wild and Weedy Rices in Agroecosystems, Ho Chi Minh City, Vietnam.

Wedger, M.J., Pusadee, T., Wongtamee, A., Olsen, K.M., 2019. Discordant patterns of introgression suggest historical gene flow into Thai weedy rice from domesticated and wild relatives. J. Hered. 110, 601–609.

Wu, D.H., Gealy, D.R., Jia, M.H., Edwards, J.D., Lai, M.H., McClung, A.M., 2020. Phylogenetic origin and dispersal pattern of Taiwan weedy rice. Pest. Manag. Sci. 76 (5), 1639–1651.

Xia, H.B., Xia, H., Ellstrand, N.C., Yang, C., Lu, B.R., 2011. Rapid evolutionary divergence and ecotype diversification of germination behavior in weedy rice populations. New. Phytol. 191 (4), 1119–1127.

Zhang, Z., Gao, P., Dai, W., Song, X., Hu, F., Qiang, S., 2019. Effect of tillage and burial depth and density of seed on viability and seedling emergence of weedy rice. J. Integr. Agr. 18 (8), 1914–1923.

Zhao, C., Xu, W., Song, X., Dai, W., Dai, L., Zhang, Z., et al., 2018. Early flowering and rapid grain filling determine early maturity and escape from harvesting in weedy rice. Pest. Manag. Sci. 74 (2), 465–476.

Ziska, L.H., Gealy, D.R., Burgos, N., Caicedo, A.L., Gressel, J., Lawton-Rauh, A.L., et al., 2015. Weedy (red) rice: an emerging constraint to global rice production. Adv. Agron. 129, 181–228. Available from: https://doi.org/10.1016/bs.agron.2014.09.003.

Chapter 14

Parthenium hysterophorus

Amarpreet Kaur[1], Daizy Rani Batish[1], Bhagirath Singh Chauhan[2], Shalinder Kaur[1], Harminder Pal Singh[3] and Ravinder Kumar Kohli[1]

[1]*Department of Botany, Panjab University, Chandigarh, India,* [2]*The Centre for Crop Science, Queensland Alliance for Agriculture and Food Innovation (QAAFI) and School of Agriculture and Food Sciences (SAFS), The University of Queensland, Gatton, QLD, Australia,* [3]*Department of Environment Studies, Panjab University, Chandigarh, India*

Introduction

Parthenium hysterophorus L. (congress grass, carrot grass, ragweed parthenium; Asteraceae) is an annual herbaceous weed species known for its vast geographical distribution in nonnative biomes across five continents (Bajwa et al., 2016; Fig. 14.1). The obnoxious invasive weed species is indigenous to the Gulf of Mexico and Central and South America (Navie et al., 1996).

FIGURE 14.1 *Parthenium hysterophorus*. (A) seeds, (B) rosette stage, (C) trichomes, (D) flower, (E) flowering stage, and (F) whole plant.

Biology and Management of Problematic Crop Weed Species. DOI: https://doi.org/10.1016/B978-0-12-822917-0.00002-1

It has invaded almost every country of South Asia, Southeast Asia, Middle East Asia, Oceania, and Africa, mainly via accidental introductions (imported food grains or vehicular transportation), and in certain cases, through natural dispersal (Adkins and Shabbir, 2014). At present, it is a severe threat to the natural- and agroecosystems of nearly 50 tropical and subtropical countries of the world with India and Ethiopia being most terribly affected by its invasion (Belgeri and Adkins, 2015; Fig. 14.2).

The rampant spread of the weed is due to its innate characteristics, by virtue of which it can exploit resources of the invaded habitat much more efficiently compared with the indigenous plant species. The weed withholds a unique set of traits; for example, proficient reproductive ability, unusual seed biology, phytotoxicity, plasticity, and unpalatability, which are proven to be the fundamentals of its invasion success. Interactions of *P. hysterophorus* with above-ground and below-ground biota strongly influence the ecosystem processes and assemblages of local flora/fauna (Timsina et al., 2011; Ojija et al., 2019). The weed not only disturbs the biodiversity in the introduced region but also affects staple food crops, cash crops, vegetables, pulses, horticulture species, grassland species, etc., thus causing huge economic losses (Shi et al., 2015). Further, health issues induced by the weed among humans and livestock are another matter of massive concern (Kohli and Rani, 1994; Pablos et al., 2017).

Attempts have been made across the globe to control *P. hysterophorus* through integrated management approaches using mechanical, chemical, and biological methods, but due to a lack of constant efforts in most parts of its invaded range (particularly the developing countries), the weed has grown

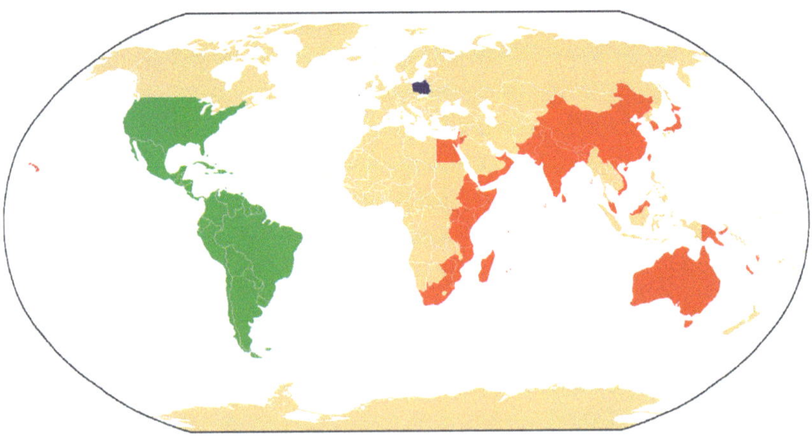

FIGURE 14.2 Global distribution of *Parthenium hysterophorus*. Areas shaded in green represent the native range of the weed, areas shaded in red represent the invasive range of the weed, and areas shaded in blue represent transient populations of the weed present in Europe (Bajwa et al., 2016).

unchecked (Campbell et al., 2018). With ongoing climatic changes, niche requirements and interference dynamics of the weed are rapidly shifting, resulting in its further expansion and dominance (Navie et al., 2005; Shrestha et al., 2015; Nguyen et al., 2017a; Bajwa et al., 2019). It is pertinent to put together information available on weed biology in the literature, particularly the reports assessing behavior of the weed in the climate change scenario. This will facilitate the appraisal of loopholes in current management practices and at the same time, breakthrough with new directives that can be used for betterment of these practices.

Global invasion

In the past six decades, *P. hysterophorus* has flourished in South Asian territories of India, Pakistan, Nepal, Bangladesh, Bhutan, and Sri Lanka (Bajwa et al., 2016; Fig. 14.2). The weed arrived in India before the 1950s along with an imported wheat (*Triticum aestivum* L.) grain stock from the United States (Kohli and Rani, 1994). It was first reported from the outskirts of Pune, India, in 1956 (Rao, 1956). However, its prolific spread has been credited to multiple reintroductions after 1956, along with the transportation of red wheat variety (*Milo*) from Mexico (Upadhyay et al., 2013). The possibility of multiple introductions has also been verified through genetic studies (Adkins and Shabbir, 2014). It has spread into neighboring countries via natural mechanisms of dispersal, vehicular movement, and export of contaminated seed lots (Shrestha et al., 2015). It was first reported from the Punjab province of Pakistan in 1980, but its major territorial expansion was witnessed in the 1990s (Bajwa et al., 2016). In Nepal, the presence of the weed was first confirmed in 1967 in Trishuli Valley, Central Nepal, and it has significantly expanded in the region in past two decades (Shrestha et al., 2015). In Sri Lanka, the weed entered in 1987 by being carried by a herd of animals with a Peace Keeping Force (Nishanthan et al., 2013), whereas in Bhutan and Bangladesh, it was reported lately in 2007 (Biswas and Das, 2007) and 2009 (Karim, 2009), respectively.

P. hysterophorus was reported from Vietnam in 1924, from where it has spread to the Yunnan Province of China (Li and Xie, 2002). Until 2004, its spread was limited to northern China (Hainan, Yunnan, Guangxi, Guangdong, Guizhou, and Fujian Provinces); but thereafter, it began to spread in the southern part of the country (Tang et al., 2009). The introduction of *P. hysterophorus* in other Eastern, Southeastern, and Middle-Eastern Asian countries, such as Yemen (Kilian et al., 2002), Malaysia (Karim, 2013), and the United Arab Emirates (Mahmoud et al., 2015) is fairly recent (Fig. 14.2).

In Australia, the introduction of *P. hysterophorus* dates back to 1950, and this population was established in Toogoolawah, Southeast Queensland, but has not spread beyond its introduced range. On a separate occasion, an introduction occurred in 1958, near Clermont in Central Queensland, from where the spread of *P. hysterophorus* began. The weed had supposedly arrived from Texas, USA, as a contaminant of pasture grass seed lots (Adkins and Shabbir, 2014). The different introductions of the weed categorize its population into two biotypes, of which the Clermont biotype is considered invasive (Bajwa et al., 2017). By 2003, the weed has been reported to cover an area of 520,000 km^2 in Queensland; however, consistent management efforts have reduced its infestation to 368,000 km^2 by 2013 (Dhileepan et al., 2018). The spread of the weed can also be noticed in Vanuatu, New Caledonia, Hawaii, Papua New Guinea, and Christmas Island (Adkins and Shabbir, 2014; Fig. 14.2).

In South Africa, the incident of infestation by *P. hysterophorus* is relatively recent. Although the species was present in this region since the 1880s, it was an uncommon weed, restricted to specific areas for almost a century. However, it turned problematic since the last two decades and has spread across four South African provinces and neighboring countries of Zimbabwe, Swaziland, Madagascar, and Mozambique (Terblanche et al., 2016; Rubaba et al., 2017; Fig. 14.2). In East Africa, the weed is predominantly present in Ethiopia, where it was introduced accidentally as a food grain contaminant transported under the USAID program in the 1980s (Rubaba et al., 2017; Fig. 14.2).

The spread of *P. hysterophorus* in the Euro-Mediterranean region is relatively sparse except in Israel (Dafni and Heller, 1982) and Egypt (Boulos and El-Hadidi, 1984). The weed has also been reported in Belgium and Poland but has not been found to naturalize yet, probably due to unfavorable climatic conditions (Brunel et al., 2014; Fig. 14.2).

Distribution patterns of *P. hysterophorus* across its invaded range

The weed is initially believed to have flourished in the disturbed arid/semiarid areas (wastelands, roadsides, railway tracks, deforested areas, etc.), later on spreading toward agricultural fields, rangelands, pastures, orchards, gardens, forests, riverbanks, and floodplains (Kohli and Rani, 1994; Navie et al., 1996; Tanveer et al., 2015; Fig. 14.3). Multiple and efficient dispersal mechanisms via wind, water, food grain stocks, vehicular movement, animals, etc., resulted in the wide distribution of *P. hysterophorus* (Mao et al., 2019). Its prevalence is more prominent in peri-urban regions that face more disturbances, anthropogenic pressure, and a lack of administrative attention (Kaur et al., 2019).

The weed has invaded both lowland and upland regions and is continuously paving its way towards the higher altitudes (Kohli et al., 2006; Shrestha et al., 2015). It has been commonly spotted in the lower parts of Kashmir, India (1600–1700 msl; Khuroo et al., 2007) and Nepal (1935 msl; Shabbir et al., 2019) in the past few decades. In India, *P. hysterophorus* has established itself in the lower and middle Himalayas, and lately, it has been

FIGURE 14.3 Infestation of *Parthenium hysterophorus* in various habitats: (A) agricultural field, (B) along the water source (C) wasteland (D) forest area (E) roadsides and (F) lower altitudes.

found to swiftly trespass the upper Himalayan zones where its monocultures can be seen trailing along the roadsides (Rathee et al., 2021).

Major economic and ecological sectors affected by *P. hysterophorus*

Agronomy

Being a weed of 42 cultivated species, including staple food crops, cash crops, oilseed crops, vegetables, and pulses, the weed severely affects agricultural productivity (Shi et al., 2015). To name a few, the weed has been found in the arable fields of cotton (*Gossypium hirsutum* L.), wheat, rice (*Oryza sativa* L.), sorghum (*Sorghum bicolor* (L.) Moench), tomato (*Solanum lycopersicum* L.), sugarcane (*Saccharum officinarum* L.), onion

(*Allium cepa* L.), groundnut (*Arachis hypogaea* L.), tobacco (*Nicotiana tabacum* L.), garlic (*Allium sativum* L.), maize (*Zea mays* L.), alfalfa (*Medicago sativa* L.), clover (*Trifolium repens* L.), cardamom (*Elettaria cardamomum* (L.) Maton), ginger (*Zingiber officinale* Roscoe), coconut (*Cocos nucifera* L.), areca nut (*Areca catechu* L.), mango (*Mangifera indica* L.), banana (*Musa* sp.), citrus (*Citrus* sp.), cucumber (*Cucumis sativus* L.), watermelon (*Citrullus lanatus* (Thunb.) Matsum. & Nakai), eggplant (*Solanum melongena* L.), common bean (*Phaseolus vulgaris* L.), capsicum (*Capsicum annuum* L.), etc. (Brunel et al., 2014; Shi et al., 2015).

However, direct competition of *P. hysterophorus* in the field conditions has been studied with only a few crop species, such as rice, sorghum, and maize. The weed, when allowed to compete with rice for 2, 4, 6, or 8 weeks, reduced the yield-related attributes of the crop by 5%–34% (Bajwa et al., 2020a). If the weed is allowed to grow uncontrolled for a complete season, different densities of *P. hysterophorus* (3–100 plants per m^2) may cause 69%–97% losses in grain yield of sorghum (Tamado et al., 2002a). A recent study agrees that the increasing density of the weed had a linear negative effect on the growth and yield of sorghum (Bajwa et al., 2020b). Yield losses in maize were reported to be between 21% and 53% due to competition with *P. hysterophorus* in a period ranging from 35 days after crop emergence to the end of the growing season (Safdar et al., 2016). Further, an increase in weed densities up to 20 plants per m^2 increased the dry weight of *P. hysterophorus* by 433% and the relative competition index by 145% (Safdar et al., 2015). Apart from the direct interference, the weed acts as an alternate host of several pests and pathogens, such as cotton mealybug and tobacco streak virus that lead to the spread of various diseases among crop species, thereby substantially affecting the crop productivity (Shabbir, 2014; Sharman et al., 2015; Muppudathi et al., 2020). Considering the negative impacts of the weed on agricultural production, it has been named "famine weed" in South Africa (Brunel et al., 2014).

Grasslands

Natural rangelands and pastures are under severe threat due to the invasion of *P. hysterophorus*. In Australia, the weed has been found to dominate the pasture species, reducing the grazing land and fodder production (McFadyen, 1992; Dhileepan, 2007). It has increased its spread in the South African rangelands by 671% within a decade, dramatically reducing the grass cover of the rangelands and their capacity to support livestock (O'Connor and van Wilgen, 2020). The seedbank of the weed established in the grasslands is an even larger problem for native grasses and forbs in the long-term (Nguyen et al., 2017b).

Biodiversity

The monospecific strands of *P. hysterophorus* disrupt the above-ground vegetation patterns, below-ground soil nutrient composition, and pollination networks (Nigatu et al., 2010; Timsina et al., 2011; Kaur et al., 2019; Ojija et al., 2019). The invasion of *P. hysterophorus* also interferes with ecosystem services. In some areas, its occurrence has attained a magnitude of epidemic, thereby suppressing various economically important endemic plant species (Shabbir and Bajwa, 2006). Its spread into protected areas, forest reserves, and national parks spoils the esthetics of the landscape and hinders conservation practices (Strathie et al., 2011).

Others

Continuous exposure to the weed can result in allergic reactions, respiratory issues, dermatitis, and hay fever in humans; whereas, if consumed by cattle, it can cause tainted meat, poor milk quality, and severe toxicity (Adkins and Shabbir, 2014). This is not only an extra load on health services but also a potential threat to the meat industry, particularly in Australia which is estimated to suffer a loss of AU$69 million in the future, only due to *P. hysterophorus* (Dhileepan, 2007). Apart from the economic burden imposed via disruption of agricultural and ecological services, the costs imposed by its management also incur huge monetary losses; however, empirical studies are relatively sparse to substantiate these losses.

Weed biology: what do we know about *P. hysterophorus* to date?

Studies have been carried out consistently to shed light on the innate characteristics that tend to make *P. hysterophorus* an invincible invasive weed species. Several unique characteristics such as proficient reproductive and dispersal strategies (Navie et al., 1996; Kohli et al., 2006), strong germination ability (Tamado et al., 2002b), persistent seedbanks (Navie et al., 2004), unique seed biology (Bajwa et al., 2016), tolerance to a wide range of biotic/abiotic stresses (Sharma et al., 2014; Bajwa et al., 2017, 2018), efficient resource capturing (Tamado et al., 2002a), plastic responses (Navie et al., 1996; Annapurna and Singh, 2003), advanced physiology (Bajwa et al., 2016), and phytotoxic potential (Kanchan and Jayachandra, 1980; Batish et al., 2002a,b, 2007; Singh et al., 2003; Belgeri and Adkins, 2015) provide a competitive superiority to *P. hysterophorus* in the invaded ecosystems.

Morphophysiological and reproductive attributes

The germination of the weed proceeds irrespective of the soil and temperature conditions (Tamado et al., 2002b; Annapurna and Singh, 2003; Kaur et al., 2017). Under an unsuitable environment, the weed suspends its growth at a juvenile rosette stage until the onset of favorable conditions, and therefore, it has a year-round prevalence (Kohli et al., 2006; Kaur et al., 2017; Khan et al., 2020; Fig. 14.4). It can adjust the duration of its life cycle

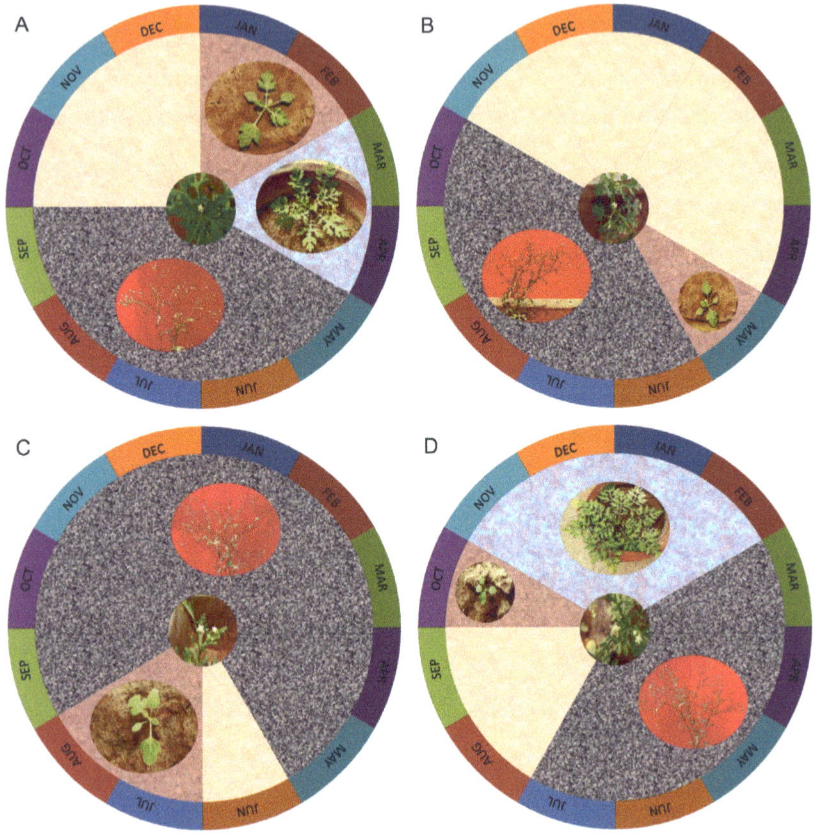

FIGURE 14.4 Life cycle of *Parthenium hysterophorus* in India in different seasons (S) varying in terms of month of sowing of the weed (M), and temperature (T) and relative humidity (RH) at the time of sowing (A) S1 (M December; T 12°C; RH 93.1%); (B) S2 (M May; T 30.3°C; RH 55%); (C) S3 (M July; T 30.2°C; RH 90.5%); (D) S4 (M October; T 25.1°C; RH 57%) (Kaur et al., 2017).

depending upon temperature, precipitation, and moisture variations, and therefore, it can persist in a much wider range of habitats (Tanveer et al., 2015; Kaur et al., 2017; Fig. 14.4). Massive flowering and seed production by the weed for longer durations and during most parts of the year lead to its extensive propagation (Adkins and Shabbir, 2014; Shrestha et al., 2015; Fig. 14.4). The weed is a prolific producer of small-sized, lightweight, innumerable seeds with a unique physiology that aids its dispersal and long-term persistence as soil seedbanks (Navie et al., 1996; Tamado et al., 2002b; Kohli et al., 2006).

Apart from that, trichomes are present on the leaves and stems of *P. hysterophorus* for secreting secondary metabolites/allelochemicals, angular woody stem provides mechanical strength to the weed, and the taproot system aids efficient resource capturing in the weed (Bajwa et al., 2016). It is not only the presence of superior morphofunctional traits that facilitate the spread of *P. hysterophorus*, but the weed is also flexible enough to modulate these attributes depending upon the ecological requirements (Annapurna and Singh, 2003; Kaur et al., 2017; Rathee et al., 2021).

Phytotoxic potential

P. hysterophorus is a phytotoxic weed species that tends to interfere with the growth and development of neighboring plants (crop plants, native weeds, grasses, and forbs) via the release of secondary metabolites (or allelochemicals) from its above- and below-ground parts (Batish et al., 2002a,b, 2007; Singh et al., 2003; Belgeri and Adkins, 2015; Shi et al., 2015). Different studies via laboratory trials (Singh and Sangeeta, 1991; Batish et al., 2002a; Singh et al., 2003; Hu et al., 2013), pot cultures (Batish et al., 2002a; Singh et al., 2005; Safdar et al., 2014), and field assays (Kanchan and Jayachandra, 1979; Tamado et al., 2002a) have established the strong allelopathic potential of *P. hysterophorus*. The phytotoxic ability of the weed has been proved against many agricultural crops, such as, wheat, rice, maize sorghum, tomato, *Brassica* sp., cowpea (*Vigna unguiculata* (L.) Walp.), sunflower (*Helianthus annuus* L.), pea (*Pisum sativum* L.), chickpea (*Cicer arientium* L.), spinach (*Spinacea oleracea* L.), radish (*Raphanus sativus* L.), teff (*Eragrostis tef* (Zucc.) Trotter), lettuce (*Lactuca sativa* L.), oats (*Avena sativa* L.), barley (*Hordeum vulgare* L.), and sorghum (*Glycine max* (L.) Merr.), as well as other economically important plants such as grasses (*Cenchrus ciliaris* L., *Urochloa panicoides* P.Beauv. etc.) and tree species (*Acacia leucophloea* (Roxb.) Willd.), *Casuarina equisetifolia* L., *Eucalyptus tereticornis* Sm., and *Leucaena leucocephala* (Lam.) de Wit etc. (Batish et al., 2002a; Singh et al., 2005; Shi et al., 2015; Hassan et al., 2018).

Most of these studies focus on growth parameters (germination, seedling length, seedling biomass, etc.), physiological processes (photosynthesis, respiration, etc.), and biochemical aspects (carbohydrate metabolism, protein metabolism, oxidative metabolism, etc.) of the test species. The allelopathic impact of *P. hysterophorus*, however, is not limited to these aspects, but its extracts have shown cytotoxic and genotoxic potential as well. The aqueous extracts lead to mitotic depression in meristematic cells of onion, resulting in chromosomal deformities and suppression of DNA transcription, which is a potential threat to the genomic balance of the test species (Sinha, 2009).

Parthenin, a sesquiterpene lactone, is identified as the major and most influential component imparting allelopathic properties to the weed (Khosla et al., 1980; Datta and Saxena, 2001; Batish et al., 2002b, 2007). Other than that, a wide variety of allelochemicals such as phenolic acids, flavonoids, tannins, pseudoguaianolides, and sesquiterpene lactones are produced by different parts of the weed, particularly the trichomes (Kanchan and

Jayachandra, 1980; Khaliq et al., 2015; Tanveer et al., 2015). Of late, volatile compounds such as β-myrcene, germacrene-D, *trans*-β-ocimene have also been extracted from the essential oils of *P. hysterophorus* (de Miranda et al., 2014). These allelochemicals act as novel weapons eliminating the resident flora (Timsina et al., 2011), modifying the soil chemical properties (Batish et al., 2002a; Timsina et al., 2011), and affecting the soil microbial communities (Jeyalakshmi et al., 2011). Allelopathy provides *P. hysterophorus* a strong competitive edge over the natives, which solidifies its dominion in the invaded habitat.

Miscellaneous

The ability of *P. hysterophorus* to perpetuate in diverse climatic regimes, soil types, and disturbed environments enhances its potential to occupy a wide array of habitat ranges (McFadyen, 1992; Annapurna and Singh, 2003). Another major reason for its unchecked spread is credited to the absence of natural pests, pathogens, and coadapted flora, which suppress its growth in native boundaries (Adkins and Shabbir, 2014). The weed also reflects variations in the morphological traits, biochemical attributes, reproductive biology, phenology, patterns of pollination, and genetic makeup within individuals (Hanif et al., 2012; Adkins and Shabbir, 2014; Bajwa et al., 2016; Kaur et al., 2019). These intraspecific variations are more prominently being noticed in Australian individuals, where these are being classified into two different biotypes, that is, Toogoolawah biotype and Clermont biotype (Bajwa et al., 2017). In India, such variations have been noticed in the population of *P. hysterophorus* in Chandigarh by Singla (1998) and Kaur et al. (2019). However, an explanation of how intraspecific variations can affect the invasiveness of the weed requires more exploration.

Response toward climate change: what can we expect from *P. hysterophorus* in future?

Studies projecting the response of *P. hysterophorus* under climate change scenarios have revealed that the growth, seed production, and overall performance of the weed will improve with an increase in temperature, carbon dioxide levels, and prolonged drought conditions (Navie et al., 2005; Shrestha et al., 2015; Nguyen et al., 2017a; Bajwa et al., 2019). Studies assessing the effect of drought on *P. hysterophorus* imply that the physiological and morphological responses in the juvenile and adult plants subjected to water-stressed conditions adopt drought avoidance strategies (Cowie et al., 2020a). Also, *P. hysterophorus* experienced enhanced growth and reproduction under elevated CO_2, whereas glyphosate efficacy decreased with increasing CO_2, suggesting that increasing CO_2 levels may hinder chemical control efforts used against the weed in the future (Cowie et al., 2020b).

It has also been expected that the weed will expand its invasion boundaries in the temperate northern hemisphere and high elevation equatorial regions (Kriticos et al., 2015). The recent reports of the occurrence of *P. hysterophorus* in previously uninvaded countries, such as Malaysia (Karim, 2013) and the United Arab Emirates (Mahmoud et al., 2015), indicate a strong probability of the weed expanding into parts of Southeast Asia, West Africa, and Eastern Europe (McConnachie et al., 2011). A recent study mapping current and potential distribution of *P. hysterophorus* in India has witnessed a considerable niche shifting in the weed in response to climatic changes (Ahmad et al., 2019).

Current management issues and potential solutions

Being one of the most aggressive and problematic weeds, *P. hysterophorus* has gained a lot of attention from agronomists, conservation managers, and invasion biologists all over the world. Not only the agricultural practitioners but various researchers, governmental/nongovernmental organizations, landscape managers, urban developmental authorities, media, and local people have been equally aware and enthusiastic about the management of the weed. Nevertheless, despite a lot of efforts, the menace of *P. hysterophorus* is still present; interfering with the agricultural practices, threatening the biodiversity, disrupting the ecosystem services, and incurring large monetary losses. This is an even larger problem in low-income countries such as India, Pakistan, Bangladesh, and Ethiopia, where economic and political factors affect management plans and policies. To understand the shortcomings of the management practices, it is crucial to assemble information on the success rate of the strategies which are commonly being employed to control the weed.

Mechanical control

Mechanical control methods involve uprooting, burning, slashing, cutting, or using larger mechanical equipment, depending on the intensity of spread of *P. hysterophorus*, size of the targeted area, and the life stage of the weed at which it is being removed. These methods are commonly practiced, despite a number of disadvantages associated with them. First, these methods raise health concerns if the labor involved does not follow proper safety measures (Kaur et al., 2014). Second, these methods do not offer a long-term solution, particularly if the plant has reached the reproductive stage and repetitive efforts are not being used (Batish et al., 2004). Third, the option is not feasible if the area involved is large, or the geographic conditions are rugged (Maharjan et al., 2020). Furthermore, mechanical removal is a labor-intensive and expensive method, and if decomposition of weed residues is not done properly, or the weed is burnt at the infestation site, it may contaminate the soil. The success of mechanical control methods is possible if an appropriate targeted area is selected and cleaned repetitively until the seedbank of the weed is completely exhausted. Care should also be taken so that management targets the plants at the vegetative stage, and the labor involved is using necessary precautions. Proper measures to decompose the weed should be taken, especially if the targeted area involves agricultural fields.

Chemical control

P. hysterophorus can be controlled by using a range of preemergence and postemergence herbicides. Postemergence herbicides applied at the rosette stage are found to be more effective than seedlings and flowering plants (Khan et al., 2012). The preemergence herbicides, norflurazon, clomazone, fluometuron, metribuzin, diuron, flumioxazin, chlorimuron, and quinclorac, and the postemergence herbicides, glyphosate, glufosinate, chlorimuron, trifloxysulfuron, halosulfuron, monosodium methanearsonate (MSMA), bromoxynil, 2,4-D, and flumioxazin, showed the tendency to control the weed by 60%–100% (Reddy et al., 2007). Prosulfuron at a reduced dose (20 g/ha) has proved to be an excellent control method for *P. hysterophorus* in sorghum (Rosales-Robles et al., 2005). The weed is susceptible to herbicides that inhibit photosynthesis, amino acid synthesis, and glutamine synthase rather than the herbicides with any other physiological action (Reddy et al., 2007).

Although chemical methods have a huge rate of success, particularly in agricultural systems, there is a strong need to reassess the dose of herbicides being employed, considering their negative effects on soil, environment, and human/animal health. Also, herbicide-resistant populations of *P. hysterophorus* have been identified in Mexico and the Dominican Republic, which showed resistance against glyphosate, picloram, dicamba, and 2,4-D (Palma-Bautista et al., 2019, 2020). This shows that the weed is rapidly evolving resistance to the applied chemicals, which could turn out to be a major problem for weed management in the future.

Certain modifications can improve herbicidal efficiency, thereby, reducing their dose by a large extent. Of late, antifoaming agents, buffering agents, organosilicon, and nitrogenous fertilizers are used as adjuvants in combination with herbicides to improve their performance (Adnan et al., 2020). Intercropping treatments, along with the use of herbicides, may also enhance the efficacy of the chemical control methods (Tadesse et al., 2010). Apart from that, the timing of the herbicidal application is of great importance as it ensures the success rate of the applied chemical. Several studies point out the critical period of weed control in a particular crop system. As suggested by Tamado et al. (2002a), the critical period for control of

P. hysterophorus in sorghum is suggested to be 19–57 days after the emergence of sorghum in upland regions, and 0–66 days in lowland regions. Similarly, Safdar et al. (2015, 2016) recommended the critical timing for removal of the weed from maize to be 8–23 days after crop emergence or when *P. hysterophorus* is present at density levels of 1.2 and 1.0 plants per m^2. Another study by Kaur et al. (2017) described the phenological calendar of *P. hysterophorus* in different seasons, which can be taken into consideration for targeting an appropriate and vulnerable growth stage for herbicidal application.

Biological control

As mentioned previously, *P. hysterophorus* has escaped its natural pests, pathogens, and herbivores, when introduced to an alien geographic range, which is attributed to be a major reason for its unchecked spread. Therefore, several countries have taken the extreme measures of transporting their natural enemies to the invaded range. A total of nine insect species and two rust fungi have been introduced into Australia since the mid-1970s for biological control of *P. hysterophorus* (Dhileepan et al., 2018; Fig. 14.5). Among these pests, the leaf-feeding beetle *Zygogramma bicolorata* Pallister (Coleoptera: Chrysomelidae) and the stem-galling moth *Epiblema strenuana* Walker (Lepidoptera: Tortricidae) were found to have a high impact on the weed population; whereas, the stem-boring weevil *Listronotus setosipennis* Hustache, the seed-feeding weevil *Smicronyx lutulentus* Dietz, the leaf-mining moth *Bucculatrix parthenica* Bradley, the sap-feeding bug *Stobaera concinna* (Stål), the stem-galling moth *Platphalonia mystica* (Razowski and Becker), the stem-galling weevil *Conotrachelus albocinereus* Fiedler, the root-feeding moth *Carmenta ithacae* (Beutenmüller), the winter rust *Puccinia abrupta* var. *partheniicola* (Jackson) Parmelee, and the summer rust *P. xanthii* var. *parthenii-hysterophorae* Seier, H.C. Evans & A. Romero, had limited effects (Dhileepan et al., 2018; Fig. 14.5). In India, biological control efforts were initiated in 1984 with the introduction of *Z. bicolorata* from Mexico (Dhileepan and Senaratne, 2009). Later, it was spread to the neighboring countries, Pakistan (Dhileepan and Senaratne, 2009), and Nepal (Shrestha et al., 2019). In South Africa, *Z. bicolorata* and *L. setosipennis* were released in 2013 after ensuring safety to nontarget plants (Mersie et al., 2019).

The success rate of biocontrol agents is moderate and, in some instances, a few nontarget species were found to be attacked by these pests. Yet, it is a useful, cost-effective, and eco-friendly approach. However, studies should be undertaken constantly to assess the behavioral changes in these pests. For instance, a recent study established that the body size and feeding ability of *Z. bicolorata* were reduced with an increase in temperature (Bhusal et al., 2020). Such challenges need to be timely addressed. Besides, new biocontrol

FIGURE 14.5 Biological control agents being used to target *Parthenium hysterophorus. Original source of image: Dhileepan, K., Callander, J., Shi, B., Osunkoya, O.O., 2018. Biological control of* Parthenium (Parthenium hysterophorus)*: the Australian experience. Biocontrol Sci. Technol. 28, 970–988. 10.1080/09583157.2018.1525486*

agents are being identified which might emerge as an effective pathogen against the weed. Among the fungal isolates collected from naturally infected plants, *Fusarium equiseti* can suppress the weed in its natural condition and it is also found to be host-specific in nature (Singh et al., 2020).

Another approach toward biological control of *P. hysterophorus* is the use of bioherbicides obtained from the plant species that have shown a strong allelopathic effect against the weed, such as *Imperata cylindrica* (L.) Raeusch., *Desmostachya bipinnata* (L.) Stapf, *Sorghum halepense* (L.) Pers., *Dichanthium annulatum* (Forssk.) Stapf, *Cenchrus pennisetiformis* Steud., and *Tagetes* sp. (Kaur et al., 2014). Tree species, *Ricinus communis* L. and *Ziziphus mauritiana* Lam., are also considered to be potential sources of bioherbicides against the weed (Safdar et al., 2013). Displacement of the weed by planting more competitive species such as *Bothriochloa insculpta*

(A.Rich.) A.Camus, *Cenchrus ciliaris* L., *Clitorea terneata* L., *Dichanthium aristatum* (Poir.) C.E. Hubb., and *Senna* sp. in its vicinity has also been suggested (Maharjan et al., 2020). However, these approaches are relatively new, and more studies will be required to ensure their successful application.

Integrated weed management

Integrated weed management is the best approach to monitor and regulate any weed, where mechanical, chemical, and biological methods are used in a suitable fashion and combination. For example, a recent study conducted by Shabbir et al. (2020) attempted to enhance the impact of a biocontrol agent, *E. strenuana* against *P. hysterophorus* using it in combination with a native grass, *Astrebla squarrosa* C.E. Hubb, and an introduced pasture plant, *Clitoria ternatea* L. Another study suggests that farmers can use approaches such as mulching, foliar application of table salt solution, hand pulling, slashing, and hand hoeing to control *P. hysterophorus* in fields along with chemical control methods (Natukunda et al., 2020). Preemergence treatment of atrazine (0.75 kg/ha) with wheat straw mulch (5 t/ha) significantly reduced the early growth of the weed in sorghum; whereas intercropping cowpea with preemergence treatment of pendimethalin (1 kg/ha) reduced *P. hysterophorus* growth 60–90 days after sowing (Tadesse et al., 2010). These practices are not being used on a very large scale, particularly due to a lack of awareness among agriculturists/conservationists. The approach of integrated weed management needs to be encouraged among researchers, farmers, and authorities involved in managing the weed.

Novel management approaches for the control of *P. hysterophorus*

A novel perspective suggests utilizing the invasive species for ecological/economic purposes, an alternative that can attract the involvement of diverse organizations and local people. In the case of *P. hysterophorus*, pharmacological properties are being extensively explored and it is considered to possess antibacterial, antineoplastic, and anticancerous activities (Das et al., 2007; Sivakumar et al., 2020), which could pave a way for its utilization in the future. The weed has also been studied for its phytoremediation ability against heavy metals and dyes (Ahmad and Al-Othman, 2014; Bapat and Jaspal, 2016). Activated carbon produced from *P. hysterophorus* can be utilized for the removal of nickel and chromium ions from wastewater (Lata et al., 2008; Bedada et al., 2020). A recent study suggests that thermophilic composting of the weed with cow dung may convert *P. hysterophorus* into a product with biomanure and biopesticide properties which can be utilized for sustainable agriculture production (Rai and Suthar, 2020). There could be more of such possibilities and research in this direction should be increased, as these options may emerge as better alternatives for weed management in the future.

Conclusions and future research directions

Despite the significant research that has been going on for the past half-century, *P. hysterophorus* is still a challenge for agronomists, ecologists, and conservation managers. The loopholes could be in the management practices, in their implementation, or in our understanding of invasive characteristics of the weed. It has been duly admitted that the weed possesses an efficient and unique invasion mechanism, yet only certain layers of this mechanism have been unveiled. Some unconventional and innovative studies are required to bridge the knowledge gap between our current interpretations of the weed and its factual potentialities. Constant monitoring of the weed is needed to improve our further understanding of the plants' invasiveness.

Apart from that, consistency in management efforts is the major issue, which needs to be addressed, particularly in countries where government participation in such environmental issues is minimal. Education and awareness should be spread, targeting the areas which have a low literacy rates and has farming as the main occupation. Integrated weed management is the best approach as suggested by several scientists, but it requires the participation of agronomists, researchers, local government bodies, nongovernmental organizations, academicians, conservationists, and landscape managers along with agriculturists, small scale farmers and local people, which are the major determinants of the success of any weed management program.

Acknowledgements

Amarpreet Kaur acknowledges the financial support provided by University Grants Commission (UGC), India.

References

Adkins, S., Shabbir, A., 2014. Biology, ecology and management of the invasive Parthenium weed (*Parthenium hysterophorus* L.). Pest Manag. Sci. 70, 1023–1029. Available from: https://doi.org/10.1002/ps.3708.

Adnan, M., Asif, M., Hussain, I., Hayyat, M.S., ul Haq, M.M., Hassan, M., et al., 2020. Chemical management of *Parthenium*: a review. Int. J. Bot. Stud. 5, 52–56.

Ahmad, A., Al-Othman, A.A.S., 2014. Remediation rates and translocation of heavy metals from contaminated soil through *Parthenium hysterophorus*. Chem. Ecol. 30, 317–327. Available from: https://doi.org/10.1080/02757540.2013.871269.

Ahmad, R., Khuroo, A.A., Hamid, M., Charles, B., Rashid, I., 2019. Predicting invasion potential and niche dynamics of *Parthenium hysterophorus* (Congress grass) in India under projected climate change. Biodivers. Conserv. 28, 2319–2344. Available from: https://doi.org/10.1007/s10531-019-01775-y.

Annapurna, C., Singh, J.S., 2003. Variation of *Parthenium hysterophorus* in response to soil quality: implications for invasiveness. Weed Res. 43, 190–198. Available from: https://doi.org/10.1046/j.1365-3180.2003.00332.x.

Bajwa, A.A., Chauhan, B.S., Farooq, M., Shabbir, A., Adkins, S.W., 2016. What do we really know about alien plant invasion? A review of the invasion mechanism of one of the world's worst weeds. Planta 244, 39–57. Available from: https://doi.org/10.1007/s00425-016-2510-x.

Bajwa, A.A., Chauhan, B.S., Adkins, S., 2017. Morphological, physiological and biochemical responses of two Australian biotypes of *Parthenium hysterophorus* to different soil moisture regimes. Environ. Sci. Pollut. Res. 24, 16186–16194. Available from: https://doi.org/10.1007/s11356-017-9176-1.

Bajwa, A.A., Chauhan, B.S., Adkins, S.W., 2018. Germination ecology of two Australian biotypes of ragweed *Parthenium* (*Parthenium hysterophorus*) relates to their invasiveness. Weed Sci. 66, 62–70. Available from: https://doi.org/10.1017/wsc.2017.61.

Bajwa, A.A., Wang, H., Chauhan, B.S., Adkins, S.W., 2019. Effect of elevated carbon dioxide concentration on growth, productivity and glyphosate response of *Parthenium* weed (*Parthenium hysterophorus* L.). Pest Manag. Sci. 75, 2934–2941. Available from: https://doi.org/10.1002/ps.5403.

Bajwa, A.A., Ullah, A., Farooq, M., Chauhan, B.S., Adkins, S., 2020a. Competition dynamics of *Parthenium hysterophorus* in direct-seeded aerobic rice fields. Exp. Agric. 56, 196–203. Available from: https://doi.org/10.1017/S0014479719000292.

Bajwa, A.A., Nawaz, A., Farooq, M., Chauhan, B.S., Adkins, S., 2020b. *Parthenium* weed (*Parthenium hysterophorus*) competition with grain sorghum under arid conditions. Exp. Agric. 56, 387–396. Available from: https://doi.org/10.1017/S0014479720000034.

Bapat, S.A., Jaspal, D.K., 2016. *Parthenium hysterophorus*: novel adsorbent for the removal of heavy metals and dyes. Glob. J. Environ. Sci. Manag. 2, 135–144. Available from: https://doi.org/10.7508/GJESM.2016.02.004.

Batish, D.R., Singh, H.P., Pandher, J.K., Arora, V., Kohli, R.K., 2002a. Phytotoxic effect of *Parthenium* residues on the selected soil properties and growth of chickpea and radish. Weed Biol. Manag. 2, 73–78. Available from: https://doi.org/10.1016/S0098-8472(01)00122-8.

Batish, D.R., Singh, H.P., Kohli, R.K., Saxena, D.B., Kaur, S., 2002b. Allelopathic effects of parthenin against two weedy species, *Avena fatua* and *Bidens pilosa*. Environ. Exp. Bot. 47, 149–155. Available from: https://doi.org/10.1016/S0098-8472(01)00122-8.

Batish, D.R., Singh, H.P., Kohli, R.K., Johar, V., Yadav, S., 2004. Management of invasive exotic weeds requires community participation. Weed Technol. 18, 1445–1448. Available from: https://doi.org/10.1614/0890-037X(2004)018[1445:MOIEWR]2.0.CO;2.

Batish, D.R., Singh, H.P., Kohli, R.K., Kaur, S., Saxena, D.B., Yadav, S., 2007. Assessment of phytotoxicity of parthenin. Z. Naturforsch. C. 62, 367–372. Available from: https://doi.org/10.1515/znc-2007-5-609.

Bedada, D., Angassa, K., Tiruneh, A., Kloos, H., Fito, J., 2020. Chromium removal from tannery wastewater through activated carbon produced from *Parthenium hysterophorus* weed. Energy Ecol. Environ. 5, 184–195. Available from: https://doi.org/10.1007/s40974-020-00160-8.

Belgeri, A., Adkins, S.W., 2015. Allelopathic potential of invasive *Parthenium* weed (*Parthenium hysterophorus* L.) seedlings on grassland species in Australia. Allelopath. J. 36, 1–14. Available from: https://doi.org/10.1046/j.1445-6664.2002.00050.x.

Bhusal, D.R., Ghimire, K.C., Patel, P., Bista, M., Upadhyay, R., Kumar, B., 2020. Temperature and altitude modulate feeding attributes of Mexican beetle, *Zygogramma bicolorata* Pallister on *Parthenium hysterophorus*. J. Therm. Biol. 89, 102540. Available from: https://doi.org/10.1016/j.jtherbio.2020.102540.

Biswas, R., Das, A.P., 2007. A note on the distribution of *Parthenium hysterophorus* L. in the Trongsa District of Bhutan. Pleione 1, 69.

Boulos, L., El-Hadidi, M.N., 1984. The Weed Flora of Egypt. American University of Cairo Press, Cairo.

Brunel, S., Panetta, D., Fried, G., Kriticos, D., Prasad, R., Lansink, A.O., et al., 2014. Preventing a new invasive alien plant from entering and spreading in the Euro-Mediterranean region: the case study of *Parthenium hysterophorus*. EPPO Bull. 44, 479–489. Available from: https://doi.org/10.1111/epp.12169.

Campbell, S.D., Vogler, W.D., Tana, T., 2018. Management: physical, cultural, chemical. In: Adkins, S.W., Shabbir, A., Dhileepan, K. (Eds.), Parthenium Weed: Biology, Ecology and Management. CABI, UK, pp. 157–176.

Cowie, B.W., Byrne, M.J., Witkowski, E.T.F., Strathie, L.W., Goodall, J.M., Venter, N., 2020a. Parthenium avoids drought: understanding the morphological and physiological responses of the invasive herb *Parthenium hysterophorus* to progressive water stress. Environ. Exp. Bot. 171, 103945. Available from: https://doi.org/10.1016/j.envexpbot.2019.103945.

Cowie, B.W., Venter, N., Witkowski, E.T.F., Byrne, M.J., 2020b. Implications of elevated carbon dioxide on the susceptibility of the globally invasive weed, *Parthenium hysterophorus*, to glyphosate herbicide. Pest Manag. Sci. 76, 2324–2332. Available from: https://doi.org/10.1002/ps.5767.

Dafni, A., Heller, D., 1982. Adventive flora of Israel—phytogeographical, ecological and agricultural aspects. Plant Syst. Evol. 140, 1–18. Available from: https://doi.org/10.1007/BF02409894.

Das, B., Reddy, V.S., Krishnaiah, M., Sharma, A.V.S., Kumar, K.R., Rao, J.V., et al., 2007. Acetylated pseudoguaianolides from *Parthenium hysterophorus* and their cytotoxic activity. Phytochemistry 68, 2029–2034. Available from: https://doi.org/10.1016/j.phytochem.2007.05.002.

Datta, S., Saxena, D.B., 2001. Pesticidal properties of parthenin (from *Parthenium hysterophorus*) and related compounds. Pest Manag. Sci. 57, 95–101.

de Miranda, C.A.S.F., Cardoso, M.G., de Carvalho, M.L.M., Figueiredo, A.C.S., Nelson, D.L., de Oliveira, C.M., et al., 2014. Chemical composition and allelopathic activity of *Parthenium hysterophorus* and *Ambrosia polystachya* weeds essential oils. Am. J. Plant Sci. 5, 1248–1257. Available from: https://doi.org/10.4236/ajps.2014.59137.

Dhileepan, K., 2007. Biological control of parthenium (*Parthenium hysterophorus*) in Australian rangeland translates to improved grass production. Weed Sci. 55, 497–501. Available from: https://doi.org/10.1614/WS-07-045.1.

Dhileepan, K., Senaratne, K.A.D.W., 2009. How widespread is *Parthenium hysterophorus* and its biological control agent *Zygogramma bicolorata* in South Asia? Weed Res. 49, 557–562. Available from: https://doi.org/10.1111/j.1365-3180.2009.00728.x.

Dhileepan, K., Callander, J., Shi, B., Osunkoya, O.O., 2018. Biological control of *Parthenium* (*Parthenium hysterophorus*): the Australian experience. Biocontrol Sci. Technol. 28, 970–988. Available from: https://doi.org/10.1080/09583157.2018.1525486.

Hanif, Z., Adkins, S.W., Prentis, P.J., Navie, S.C., O'Donnell, C., 2012. Characterization of the reproductive behavior and invasive potential of parthenium weed in Australia. Pak. J. Weed Sci. Res. 18, 767–774.

Hassan, G., Rashid, H.U., Amin, A., Khan, I.A., Shehzad, N., 2018. Allelopathic effect of *Parthenium hysterophorus* on germination and growth of some important crops and weeds of economic importance. Planta Daninha 36, e018176372. Available from: https://doi.org/10.1590/s0100-83582018360100132.

Hu, G., Zhang, Z.H., Hu, B.Q., 2013. Effect of aqueous leaf extract of *Parthenium hysterophorus* L. on the germination and shoot growth of two native species. Adv. Mater. Res. 726-731, 4348–4351. Available from: https://doi.org/10.4028/www.scientific.net/AMR.726-731.4348.

Jeyalakshmi, C., Doraisamy, S., Valluvaparidasan, V., 2011. Occurrence of soil microbes under *Parthenium* weed in Tamil Nadu. Indian J. Weed Sci. 43, 222–223.

Kanchan, S.D., Jayachandra, 1979. Allelopathic effects of *Parthenium hysterophorus* L. 1. Exudation of inhibitors through roots. Plant Soil 53, 27–35. Available from: https://doi.org/10.1007/BF02181876.

Kanchan, S.D., Jayachandra, 1980. Allelopathic effects of *Parthenium hysterophorus* L. Plant Soil 55, 67–75. Available from: https://doi.org/10.1007/BF02149710.

Karim, S.M.R., 2009. Parthenium weed: a new introduction to Bangladesh. In: Proceedings of the 21st Bangladesh Science Conference. Bangladesh Agricultural Research Institute, Gazipur, pp 43–44.

Karim, S.M.R., 2013. Obnoxious environmental pollutant, the Parthenium weed: A possibility of ecological disaster in Malaysia. In: Proceedings of the Joint Seminar on Natural Disasters: Experiences in Malaysia and Indonesia, Universiti Malaysia Kelantan, Jeli Campus, Malaysia.

Kaur, A., Batish, D.R., Kaur, S., Singh, H.P., Kohli, R.K., 2017. Phenological behaviour of *Parthenium hysterophorus* in response to climatic variations according to the extended BBCH scale. Ann. Appl. Biol. 171, 316–326. Available from: https://doi.org/10.1111/aab.12374.

Kaur, A., Kaur, S., Singh, H.P., Batish, D.R., Kohli, R.K., 2019. Phenotypic variations alter the ecological impact of invasive alien species: Lessons from *Parthenium hysterophorus*. J. Environ. Manag. 241, 187–197. Available from: https://doi.org/10.1016/j.jenvman.2019.03.129.

Kaur, M., Aggarwal, N.K., Kumar, V., Dhiman, R., 2014. Effects and management of *Parthenium hysterophorus*: a weed of global significance. Int. Sch. Res. Not. 2014, 368647. Available from: https://doi.org/10.1155/2014/368647.

Khaliq, A., Aslam, F., Matloob, A., Hussain, S., Tanveer, A., Alsaadawi, I., et al., 2015. Residual phytotoxicity of parthenium: impact on some winter crops, weeds and soil properties. Ecotoxicol. Environ. Saf. 122, 352–359. Available from: https://doi.org/10.1016/j.ecoenv.2015.08.019.

Khan, H., Marwat, K.B., Hassan, G., Khan, M.A., 2012. Chemical control of *Parthenium hysterophorus* L. at different growth stages in non-cropped area. Pak. J. Bot. 44, 1721–1726.

Khan, N., Jhariya, M.K., Yadav, D.K., Banerjee, A., 2020. Herbaceous dynamics and CO_2 mitigation in an urban setup—a case study from Chhattisgarh, India. Environ. Sci. Pollut. Res. 27, 2881–2897. Available from: https://doi.org/10.1007/s11356-019-07182-8.

Khosla, S.N., Singh, K., Sobti, S.N., 1980. Parthenin from *Parthenium hysterophorus* is phytotoxic too. Indian J. For. 3, 261–265.

Khuroo, A.A., Rashid, I., Reshi, Z., Dar, G.H., Wafai, B.A., 2007. The alien flora of Kashmir Himalaya. Biol. Invasions 9, 269–292. Available from: https://doi.org/10.1007/s10530-006-9032-6.

Kilian, N., Hein, P., Hubaishan, M.A., 2002. New and noteworthy records for the flora of Yemen, chiefly of Hadhramout and Al-Mahra. Willdenowia 32, 239–269. Available from: https://doi.org/10.3372/wi.32.32207.

Kohli, R.K., Batish, D.R., Singh, H.P., Dogra, K.S., 2006. Status, invasiveness and environmental threats of three tropical American invasive weeds (*Parthenium hysterophorus* L., *Ageratum conyzoides* L., *Lantana camara* L.) in India. Biol. Invasions 8, 1501–1510. Available from: https://doi.org/10.1007/s10530-005-5842-1.

Kohli, R.K., Rani, D., 1994. *Parthenium hysterophorus*- a review. Res. Bull. Panjab Univ. Sci. 44, 105–149.

Kriticos, D.J., Brunel, S., Ota, N., Fried, G., Lansink, A.G.J.M.O., Panetta, F.D., et al., 2015. Downscaling pest risk analyses: identifying current and future potentially suitable habitats for *Parthenium hysterophorus* with particular reference to Europe and North Africa. PLoS One 10, e0132807. Available from: https://doi.org/10.1371/journal.pone.0132807.

Lata, H., Garg, V.K., Gupta, R.K., 2008. Sequestration of nickel from aqueous solution onto activated carbon prepared from *Parthenium hysterophorus* L. J. Hazard. Mater. 157, 503–509. Available from: https://doi.org/10.1016/j.jhazmat.2008.01.011.

Li, Z., Xie, Y., 2002. Invasive Alien Species in China. Forestry Publishing Company of China, Beijing.

Maharjan, S., Shrestha, B.B., Devkota, A., Muniappan, R., Jha, P.K., 2020. Temporal and spatial patterns of research on a globally significant invasive weed *Parthenium hysterophorus* L.: A bibliographic review. Crop Prot. 135, 104832. Available from: https://doi.org/10.1016/j.cropro.2019.05.026.

Mahmoud, T., Gairola, S., El-Keblawy, A., 2015. *Parthenium hysterophorus* and *Bidens pilosa*, two new records to the invasive weed flora of the United Arab Emirates. J. N. Biol. Rep. 4, 26–32.

Mao, R., Nguyen, T.L.T., Osunkoya, O.O., Adkins, S.W., 2019. Spread pathways of the invasive weed *Parthenium hysterophorus* L.: the potential for water dispersal. Austral Ecol. 44, 1111–1122. Available from: https://doi.org/10.1111/aec.12774.

McConnachie, A.J., Strathie, L.W., Mersie, W., Gebrehiwot, L., Zewdie, K., Abdurehim, A., et al., 2011. Current and potential geographical distribution of the invasive plant *Parthenium hysterophorus* (Asteraceae) in eastern and southern Africa. Weed Res. 51, 71–84. Available from: https://doi.org/10.1111/j.1365-3180.2010.00820.x.

McFadyen, R.C., 1992. Biological control against parthenium weed in Australia. Crop Prot. 11, 400–407. Available from: https://doi.org/10.1016/0261-2194(92)90021-V.

Mersie, W., Alemayehu, L., Strathie, L., McConnachie, A., Terefe, S., Negeri, M., et al., 2019. Host range evaluation of the leaf-feeding beetle, *Zygogramma bicolorata* and the stem-boring weevil, *Listronotus setosipennis* demonstrates their suitability for biological control of the invasive weed, *Parthenium hysterophorus* in Ethiopia. Biocontrol Sci. Technol. 29, 239–251. Available from: https://doi.org/10.1080/09583157.2018.1545220.

Muppudathi, S.P., Natarajan, G., Ganesan, M.V., Sevugapperumal, N., Subbarayalu, M., Samuel, K.J., et al., 2020. Role of *Thrips palmi* and *Parthenium hysterophorus* pollen in active spread of tobacco streak virus in the cotton ecosystem. Virus Res. 284, 197979. Available from: https://doi.org/10.1016/j.virusres.2020.197979.

Natukunda, M.I., Natukunda, K., Kyeyune, G., Tusiime, S.M., Agbemafle, I., Bisikwa, J., 2020. Management strategies for the noxious invasive *Parthenium* weed (*Parthenium hysterophorus* L.) in Uganda. Afr. J. Agric. Res. 15, 1–9. Available from: https://doi.org/10.5897/AJAR2019.14569.

Navie, S.C., McFadyen, R.E., Panetta, F.D., Adkins, S.W., 1996. A comparison of the growth and phenology of two introduced biotypes of *Parthenium hysterophorus*. In: Shepherd, R.H. C. (Ed.), Eleventh Australian Weeds Conference Proceedings. Weed Science Society of Victoria, Australia, pp. 313–316.

Navie, S.C., Panetta, F.D., McFadyen, R.E., Adkins, S.W., 2004. Germinable soil seedbanks of central Queensland rangelands invaded by the exotic weed *Parthenium hysterophorus* L. Weed Biol. Manag. 4, 154–167. Available from: https://doi.org/10.1111/j.1445-6664.2004.00132.x.

Navie, S.C., McFadyen, R.E., Panetta, F.D., Adkins, S.W., 2005. The effect of CO_2 enrichment on the growth of a C_3 weed (*Parthenium hysterophorus* L.) and its competitive interaction with a C_4 grass (*Cenchrus ciliaris* L.). Plant Prot. Q. 20, 61–66. Available from: https://doi.org/10.2551/0815-2195.20.2.0506.

Nguyen, T., Bajwa, A.A., Navie, S., O'donnell, C., Adkins, S., 2017a. *Parthenium* weed (*Parthenium hysterophorus* L.) and climate change: the effect of CO_2 concentration, temperature, and water deficit on growth and reproduction of two biotypes. Environ. Sci. Pollut. Res. 24, 10727–10739. Available from: https://doi.org/10.1007/s11356-017-8737-7.

Nguyen, T., Bajwa, A.A., Belgeri, A., Navie, S., O'Donnell, C., Adkins, S., 2017b. Impact of an invasive weed, *Parthenium hysterophorus*, on a pasture community in south east Queensland, Australia. Environ. Sci. Pollut. Res. 24, 27188–27200. Available from: https://doi.org/10.1007/s11356-017-0327-1.

Nigatu, L., Hassen, A., Sharma, J., Adkins, S.W., 2010. Impact of *Parthenium hysterophorus* on grazing land communities in north-eastern Ethiopia. Weed Biol. Manag. 10, 143–152. Available from: https://doi.org/10.1111/j.1445-6664.2010.00378.x.

Nishanthan, K., Sivachandiran, S., Marambe, B., 2013. Control of *Parthenium hysterophorus* L. and its impact on yield performance of tomato (*Solanum lycopersicum* L.) in the Northern Province of Sri Lanka. Trop. Agric. Res. 25, 56–68. Available from: https://doi.org/10.4038/tar.v25i1.8030.

O'Connor, T.G., van Wilgen, B.W., 2020. The impact of invasive alien plants on rangelands in South Africa. In: van Wilgen, B., Measey, J., Richardson, D., Wilson, J., Zengeya, T. (Eds.), Biological Invasions in South Africa. Invading Nature - Springer Series in Invasion Ecology. Springer, Cham.

Ojija, F., Arnold, S.E., Treydte, A.C., 2019. Impacts of alien invasive *Parthenium hysterophorus* on flower visitation by insects to co-flowering plants. Arthropod-Plant Interact. 13, 719–734. Available from: https://doi.org/10.1007/s11829-019-09701-3.

Pablos, I., Eichhorn, S., Briza, P., Asam, C., Gartner, U., Wolf, M., et al., 2017. Proteomic profiling of the weed feverfew, a neglected pollen allergen source. Sci. Rep. 7, 6049. Available from: https://doi.org/10.1038/s41598-017-06213-z.

Palma-Bautista, C., Gherekhloo, J., Domínguez-Martínez, P.A., Domínguez-Valenzuela, J.A., Cruz-Hipolito, H.E., Alcántara-de la Cruz, R., et al., 2019. Characterization of three glyphosate resistant *Parthenium hysterophorus* populations collected in citrus groves from Mexico. Pestic. Biochem. Physiol. 155, 1–7. Available from: https://doi.org/10.1016/j.pestbp.2018.11.002.

Palma-Bautista, C., Hoyos, V., Plaza, G., Vázquez-García, J.G., Rosario, J., Rojano-Delgado, A. M., et al., 2020. Evolving multiple resistance to EPSPS, GS, ALS, PSI, PPO, and synthetic auxin herbicides in Dominican Republic *Parthenium hysterophorus* populations. A physiological and biochemical study. Agronomy 10, 554. Available from: https://doi.org/10.3390/agronomy10040554.

Rai, R., Suthar, S., 2020. Composting of toxic weed *Parthenium hysterophorus*: nutrient changes, the fate of faecal coliforms, and biopesticide property assessment. Bioresour. Technol. 311, 123523. Available from: https://doi.org/10.1016/j.biortech.2020.123523.

Rathee, S., Ahmad, M., Sharma, P., Singh, H.P., Batish, D.R., Kaur, S., Kaur, A., Yadav, S.S., Kohli, R.K. 2021. Biomass allocation and phenotypic plasticity are key elements of successful invasion of Parthenium hysterophorus at high elevation. Env. Exp. Bot. 184, 104392. Available from: https://doi.org/10.1016/j.envexpbot.2021.104392.

Rao, R.S., 1956. Parthenium: a new record for India. J. Bombay Nat. His. Soc. 54, 218–220.

Reddy, K.N., Bryson, C.T., Burke, I.C., 2007. Ragweed parthenium (*Parthenium hysterophorus* L.) control with Preemergence and Post-emergence herbicides. Weed Technol. 21, 982–986. Available from: https://doi.org/10.1614/WT-07-053.1.

Rosales-Robles, E., Sanchez-de-la-Cruz, R., Salinas-Garcia, J., Pecina-Quintero, V., 2005. Broadleaf weed management in grain sorghum with reduced rates of postemergence herbicides. Weed Technol. 19, 385–390. Available from: https://doi.org/10.1614/WT-04-170R1.

Rubaba, O., Chimbari, M., Mukaratirwa, S., 2017. Scope of research on *Parthenium hysterophorus* in Africa. S. Afr. J. Plant Soil 34, 323–332. Available from: https://doi.org/10.1080/02571862.2016.1266046.

Safdar, M.E., Tanveer, A., Khaliq, A., Naeem, M.S., Ahmad, S., 2013. Tree species as a potential source of bio-herbicides for controlling *Parthenium hysterophorus* L. Nat. Prod. Res. 27, 2154–2156. Available from: https://doi.org/10.1080/14786419.2013.790028.

Safdar, M.E., Tanveer, A., Khaliq, A., Naeem, M.S., 2014. Allelopathic action of parthenium and its rhizospheric soil on maize as influenced by growing conditions. Planta Daninha 32, 243–253. Available from: https://doi.org/10.1590/S0100-83582014000200001.

Safdar, M.E., Tanveer, A., Khaliq, A., Riaz, M.A., 2015. Yield losses in maize (*Zea mays*) infested with parthenium weed (*Parthenium hysterophorus* L.). Crop Prot. 70, 77–82. Available from: https://doi.org/10.1016/j.cropro.2015.01.010.

Safdar, M.E., Tanveer, A., Khaliq, A., Maqbool, R., 2016. Critical competition period of parthenium weed (*Parthenium hysterophorus* L.) in maize. Crop Prot. 80, 101–107. Available from: https://doi.org/10.1016/j.cropro.2015.11.002.

Shabbir, A., 2014. Ragweed emerging a major host for the cotton mealy bug in Pakistan. Indian J. Weed Sci. 46, 291–293.

Shabbir, A., Bajwa, R., 2006. Distribution of *Parthenium* weed (*Parthenium hysterophorus* L.), an alien invasive weed species threatening the biodiversity of Islamabad. Weed Biol. Manag. 6, 89–95. Available from: https://doi.org/10.1111/j.1445-6664.2006.00202.x.

Shabbir, A., Shrestha, B.B., Ali, M.H., Adkins, S.W., 2019. History and management–Southern Asia. In: Adkins, S., Shabbir, A., Dhileepan, K. (Eds.), Parthenium Weed: Biology, Ecology and Management. CABI, Wallingford, Oxfordshire, UK, pp. 231–252.

Shabbir, A., Dhileepan, K., Zalucki, M.P., Khan, N., Adkins, S.W., 2020. Reducing the fitness of an invasive weed, *Parthenium hysterophorus*: complementing biological control with plant competition. J. Environ. Manag. 254, 109790. Available from: https://doi.org/10.1016/j.jenvman.2019.109790.

Sharma, A.D., Bhullar, A., Rakhra, G., Mamik, S., 2014. Analysis of hydrophilic antioxidant enzymes in invasive alien species *Parthenium hysterophorus* under high temperature abiotic stress like conditions. J. Stress Physiol. Biochem. 10, 228–237.

Sharman, M., Thomas, J.E., Persley, D.M., 2015. Natural host range, thrips and seed transmission of distinct Tobacco streak virus strains in Queensland, Australia. Ann. Appl. Biol. 167, 197–207. Available from: https://doi.org/10.1111/aab.12218.

Shi, B., Aslam, Z., Adkins, S., 2015. The invasive potential of *Parthenium* weed: a role for allelopathy. In: Price, J.E. (Ed.), New Developments in Allelopathy Research. Nova Science Publishers, New York.

Shrestha, B.B., Shabbir, A., Adkins, S.W., 2015. *Parthenium hysterophorus* in Nepal: a review of its weed status and possibilities for management. Weed Res. 55, 132–144. Available from: https://doi.org/10.1111/wre.12133.

Shrestha, B.B., Pokhrel, K., Paudel, N., Poudel, S., Shabbir, A., Adkins, S.W., 2019. Distribution of *Parthenium hysterophorus* and one of its biological control agents (Coleoptera: *Zygogramma bicolorata*) in Nepal. Weed Res. 59, 467–478. Available from: https://doi.org/10.1111/wre.12384.

Singh, S.P., Sangeeta, 1991. Allelopathic potential of *Parthenium hysterophorus* L. J. Agron. Crop Sci. 167, 201–206. Available from: https://doi.org/10.1111/j.1439-037X.1991.tb00955.x.

Singh, H.P., Batish, D.R., Pandher, J.K., Kohli, R.K., 2003. Assessment of allelopathic properties of *Parthenium hysterophorus* residues. Agric. Ecosys. Env. 95, 537–541. Available from: https://doi.org/10.1016/S0167-8809(02)00202-5.

Singh, H.P., Batish, D.R., Pandher, J.K., Kohli, R.K., 2005. Phytotoxic effects of *Parthenium hysterophorus* residues on three *Brassica* species. Weed Biol. Manag. 5, 105–109. Available from: https://doi.org/10.1111/j.1445-6664.2005.00172.x.

Singh, S., Gill, P.K., Mishra, T., Dhaliwal, H.S., 2020. Biocontrol of *Parthenium hysterophorus* through different fungal isolates. Int. J. Microbiol. Res. 12, 1767–1770.

Singla, R.K., 1998. Biology and Control of *Parthenium hysterophorus* Linn. in Union Territory of Chandigarh. Dissertation. Panjab University, Chandigarh, India. Available from: http://shodhganga.inflibnet.ac.in/handle/10603/83480 (accessed 13 June 2020).

Sinha, V.S., 2009. *Parthenium hysterophorus* induced genotoxic hazards in *Allium cepa* L. Nat. Environ. Pollut. Technol. 8, 725–728.

Sivakumar, M., Surendar, S., Jayakumar, M., Seedevi, P., Sivasankar, P., Ravikumar, M., et al., 2020. *Parthenium hysterophorus* mediated synthesis of silver nanoparticles and its evaluation of antibacterial and antineoplastic activity to combat liver cancer cells. J. Clust. Sci. 32, 167–177. Available from: https://doi.org/10.1007/s10876-020-01775-x. In Press.

Strathie, L.W., McConnachie, A.J., Retief, E., 2011. Initiation of biological control against *Parthenium hysterophorus* L. (Asteraceae) in South Africa. Afr. Entomol. 19, 378–392. Available from: https://doi.org/10.4001/003.019.0224.

Tadesse, B., Das, T.K., Yaduraju, N.T., 2010. Effects of some integrated management options on *Parthenium* interference in sorghum. Weed Biol. Manag. 10, 160–169. Available from: https://doi.org/10.1111/j.1445-6664.2010.00380.x.

Tamado, T., Ohlander, L., Milberg, P., 2002a. Interference by the weed *Parthenium hysterophorus* L. with grain sorghum: influence of weed density and duration of competition. Int. J. Pest Manag. 48, 183–188. Available from: https://doi.org/10.1080/09670870110101739.

Tamado, T., Schutz, W., Milberg, P., 2002b. Germination ecology of the weed *Parthenium hysterophorus* in eastern Ethiopia. Ann. Appl. Biol. 140, 263–270. Available from: https://doi.org/10.1111/j.1744-7348.2002.tb00180.x.

Tang, S.Q., Wei, F., Zeng, L.Y., Li, X.K., Tang, S.C., Zhong, Y., et al., 2009. Multiple introductions are responsible for the disjunct distributions of invasive *Parthenium hysterophorus* in China: evidence from nuclear and chloroplast DNA. Weed Res. 49, 373–380. Available from: https://doi.org/10.1111/j.1365-3180.2009.00714.x.

Tanveer, A., Khaliq, A., Ali, H.H., Mahajan, G., Chauhan, B.S., 2015. Interference and management of parthenium: the world's most important invasive weed. Crop Prot. 68, 49–59. Available from: https://doi.org/10.1016/j.cropro.2014.11.005.

Terblanche, C., Nanni, I., Kaplan, H., Strathie, L.W., McConnachie, A.J., Goodall, J., et al., 2016. An approach to the development of a national strategy for controlling invasive alien plant species: the case of *Parthenium hysterophorus* in South Africa. Bothalia 46, a2053. Available from: https://doi.org/10.4102/abc.v46i1.2053.

Timsina, B., Shrestha, B.B., Rokaya, M.B., Münzbergova, Z., 2011. Impact of *Parthenium hysterophorus* L. invasion on plant species composition and soil properties of grassland communities in Nepal. Flora 206, 233–240. Available from: https://doi.org/10.1016/j.flora.2010.09.004.

Upadhyay, S.K., Ahmad, M., Singh, A., 2013. Ecological impacts of weed (*Parthenium hysterophorus* L.) invasion in saline soil. Int. J. Sci. Res. Publ. 3, 1–4.

Chapter 15

Phalaris minor and *Phalaris paradoxa*

Ankur Chaudhary[1], Simerjeet Kaur[2] and Bhagirath Singh Chauhan[3]
[1]*Chaudhary Charan Singh Haryana Agricultural University Regional Research Station, Uchani, India,* [2]*Punjab Agricultural University, Ludhiana, India,* [3]*The Centre for Crop Science, Queensland Alliance for Agriculture and Food Innovation (QAAFI) and School of Agriculture and Food Sciences (SAFS) The University of Queensland, Gatton, QLD, Australia*

Introduction

The genus *Phalaris* is found mostly in the Mediterranean climate, except *Phalaris arundinacea* L. which is observed in wet habitats (Baldini, 1995). *Phalaris* species grow in diverse habitats ranging from wild communities to disturbed areas in open grounds, agricultural fields, sandy soils, and waste beds (Baldini, 1995). Most of *Phalaris* species are serious weeds of winter crops especially in heavy soils (Viggiani and Angelini, 1993). Ecologically, *Phalaris* genus includes useful forage crop species (*P. canariensis* L. and *P. angusta* Nees ex Tris.), and some weedy and invasive species such as *P. arundinacea* (Voshell et al., 2011). In the *Phalaris* genus, *P. paradoxa* L. is the most polymorphic species and *P. minor* Retz. exhibit aneuploidy (Hanson and Hill, 1953).

Phalaris minor and *P. paradoxa* are important winter season (annual) agricultural weeds. Globally, *P. minor* has been reported in more than 60 countries of the world (Anderson, 1961; Baldini, 1995) and is the most severe weed in wheat. *Phalaris paradoxa* has been found in places with variable rainfall (Thurley and Chancellor, 1985; Taylor et al., 2005) and soil types (Michael et al., 2010); Finot and Pedreros, 2012). *Phalaris minor* and *P. paradoxa* are well distributed in Europe, Asia, South and North America, East and South Africa, and Australia (Table 15.1).

Botanical description and yield losses

Phalaris minor, a self-pollinated ($2n = 28$) annual grass weed, is also known as small canarygrass, Mediterranean canarygrass, and littleseed canarygrass. *Phalaris paradoxa*, also known as hood canarygrass, awned canarygrass,

Biology and Management of Problematic Crop Weed Species. DOI: https://doi.org/10.1016/B978-0-12-822917-0.00018-5

TABLE 15.1 Global distribution of *Phalaris minor* and *P. paradoxa*.

Continent/ region	Countries		References
	Phalaris minor	*Phalaris paradoxa*	
Asia	India, Pakistan, China, Israel, Iran, Sri Lanka, Afghanistan, Nepal, Myanmar, Bhutan, and Bangladesh	Iran, Korea, Nepal, China, and Israel	Anderson (1961), Baldini (1995), Ryu et al. (2018)
Africa	Egypt, South Africa, and Zimbabwe	Ethiopia	Anderson (1961), Smit and Cairns (2000), Negewo et al. (2011)
North America	Mexico and USA	Canada, Mexico, and USA	Hitchcock (1950), Baldini (1995)
South America	Argentina, Brazil, Colombia, and Uruguay	Morroco, Bolivia, Argentina, and Uruguay	Hajjaj and El Oualkadi (2019)
Europe	Greece, Italy, Spain, France, and Portugal	Italy, Spain, Norway, Chile, and Britain	Anderson (1961), Thurley and Chancellor (1985), Baldini (1995), Finot and Pedreros (2012)
Oceania	Australia, and New Zealand	Australia	Anderson (1961), Martin et al. (1988), (Alemseged et al., 2001), Taylor et al. (2005)

paradoxa grass, bristle-spiked canarygrass, paradoxical canarygrass, and variable canarygrass, is mainly self-pollinated ($2n = 14$) with up to 8% outcrossing rate (Baldini, 1995). These two annual grass species reproduce through seed production. *Phalaris minor* dispersed their caryopses by shattering from their glumes, and *P. paradoxa* disarticulated spikelets from inflorescences at maturity (Matus-Cádiz and Hucl, 2002).

Both *P. minor* and *P. paradoxa* have long panicles, short glumes, short fertile florets, and small caryopses. *Phalaris paradoxa* has two sterile florets per spikelet, whereas *P. minor* has only one sterile floret per spikelet. *Phalaris minor* has pubescent short sterile florets, whereas *P. paradoxa* has obsolete sterile florets with sparse pubescence. *Phalaris minor* lacks sterile spikelets in their inflorescences, whereas various combinations of sterile spikelet types make up four types of inflorescence in *P. paradoxa* (Anderson, 1961; Matus-Cádiz and Hucl, 2002). *Phalaris paradoxa* possessed the greatest variability for panicle length/width and glume length/width relative to *P. minor* (Matus-Cádiz and Hucl, 2002).

Crop–weed competition is defined by several factors such as crop cultivation practices, nature of crops, and varieties and climatic factors as well. The initial period of 4–6 weeks after sowing is most critical for weed competition in wheat (*Triticum aestivum* L.). Similarly, crop yield loss depends on type of weed species present, weed density and dry matter accumulation, emergence time, nature of crop cultivar, planting density, and other management practices (Singh et al., 1999; Malik and Singh, 1995). Further, types of cropping and tillage systems along with soil moisture at the time of crop sowing also decide the level of weed infestation and subsequent yield losses. Weed risk assessment studies categorized *P. paradoxa* under the "high risk and invasive" category for the United States (USDA, 2016) and India (Singh and Priyadarshi, 2014). Yield losses in wheat due to *P. minor* alone may be up to 50%, and heavy infestations may lead to total crop failure (Malik and Singh, 1995; Singh, 2007). The infestation of *P. paradoxa* caused a 40% wheat yield reduction in Australia (Dellow and Milne, 1986). Wheat crop biomass was significantly reduced with an infestation of 200 *P. minor* plants/m^2, whereas 100 plants of *P. paradoxa* were sufficient to significantly reduce wheat yield (Bhan et al., 2006). The economic injury level of *P. paradoxa* in winter wheat varied from 56–112 plants/m^2 under two different field experiments in Spain (Portugal et al., 2009). The economic threshold levels of *P. minor* and *P. paradoxa* were reported to be 3.1 and 2.6 plants/m^2, respectively, for manual weeding in China (Xu et al., 2010).

Collateral damage is also reported due to the presence of this weed in pastures as there are reports of the death of horses which were feeding on a fallow-wheat paddock with 80% population of *P. paradoxa* in Australia (Bourke et al., 2003). As contaminants of wheat flour, carcinogenicity was associated with *P. paradoxa* in Iran (Sangster et al., 1983). Also, *P. paradoxa* is reported to be an alternate host for barley yellow dwarf virus and its associated aphid vectors in Ethiopia (Bekele, 2011), which can affect wheat and oat (*Avena sativa* L.) crops also. *Phalaris paradoxa* may impact trade and it is listed as a weed of quarantine importance in many countries.

Biology and ecology

Seed dormancy, germination, and persistence are affected by management factors such as tillage systems, water regimes, stubble/straw management, and environmental factors. The emergence of *P. paradoxa* was nearly 95% within 2 months of seedbed preparation (Gasparetto et al., 2003). In India, *P. minor* emerges in multiple flushes and its emergence starts with wheat emergence in November month and later flushes emerge with subsequent irrigation in the winter season (Singh, 2007). The rate of biomass accumulation of *P. minor* was slow during the initial 60 days, and the rate was greater between 60 and 90 days after sowing (Malik and Singh, 1995). The thousand seed weight of *P. minor* varied from 1.5 to 2.1 g (Bhan and Choudary, 1976), and the plant invested 27% of aboveground biomass in seed production, irrespective of plant size (Franke et al., 2007). The aboveground biomass failed to influence the thousand seed weight (1.8 g) of *P. minor* and even small plants also produced seeds of equal individual seed weights; but, the numbers of seeds were reduced. *Phalaris paradoxa* produced 5105 seeds/m^2 in wheat fields and only 50 seeds/m^2 in barley (*Hordeum vulgare* L.) fields in Australia (Walker et al., 2001). A variation in seed germination was recorded between geographically disjunctive separate seed collections (north and south of Aljouf region in the north of Saudi) of *P. minor* which might be associated with environmental conditions experienced by the mother plants before seed dispersal (Alshallash, 2018).

Tillage system and distribution of *Phalaris* species

Tillage practices alter soil water content, temperature aeration, degree of mixing of crop residues, and composition of weed flora (Buhler, 1995; Cardina et al., 2002). In the rice (*Oryza sativa* L.)−wheat system, it has been observed that weed seed distribution in the upper 10 cm of the soil was found similar between zero-till and conventional-till wheat crops (Franke et al., 2007). This might be due to intensive tillage and puddling performed during the preceding rice crop in both tillage systems. However, zero tillage in wheat considerably reduced the first flush of *P. minor* seedling emergence to about 50%, irrespective of the soil texture and aeration (Singh et al., 1999; Franke et al., 2007). The reduced rate of *P. minor* seedling emergence under zero-till could be associated with variation in relative moisture distribution within the upper soil profile along with the formation of a crust on the soil surface that mechanically impeded weed seedling emergence from lower layers as compared to conventional tillage. *Phalaris minor* seedling emergence was greater from 1 cm soil depth, and thereafter, a drastic reduction in emergence was recorded for the seeds placed below 10 cm (Hari et al., 2003). Overall, *P. minor* seedling emergence decreased with the increase in soil depth (Ahmadi, 2016). In conventional or intensive tillage systems, rapid alteration in soil physicochemical regimes, soil temperature, mechanical strength, and red/far-red ratio resulted in breaking seed dormancy and greater weed seedling emergence (Franke et al., 2007).

A series of varying burial experiments demonstrated that only 1%−4% of *P. paradoxa* seeds survived after 2 years, and after 3 years, no seeds survived (Walker et al., 2006). Cultivation not only increased emergence by 40% of naturally shed seeds; but, it also changed the periodicity of emergence. Tillage in March (autumn) promoted seedling emergence of *P. paradoxa*, and the majority of seedlings emerged in May (at the beginning of the winter crop season) under subtropical conditions of the north-east grain region of Australia (Taylor et al., 2005). In uncultivated plots the majority of seedlings emerged in July which is the middle of the winter cropping season. Seed burial experiments showed that seeds of *P. paradoxa* is relatively short-lived, with <10% of seeds germinated after 2 years. Approximately 90% of seeds were either lost due to predation or fungal infection, and emergence accounted for only 5% of this loss. Seed germination was least from seeds on the soil surface, which was due to the fact that high-intensity white light on the soil surface suppressed seed germination of *P. paradoxa* (Taylor et al., 2004). It was suggested that seeds shed onto the soil surface persisted for a longer time and strategic cultivation may be used as one of many weed management tools to increase seed germination and alter the periodicity of emergence.

Seed dormancy and viability

Seed dormancy and viability in *P. minor* are profoundly affected by environment, edaphic, and management factors (Om et al., 2005; Singh, 2007). Dormancy in *P. minor* seeds under field conditions was maintained for less than 60 days (Om et al., 2005), whereas earlier studies showed that freshly harvested mature seeds of *P. minor* remained dormant for 3–6 months (Singh, 1998; Chhokar, 1998). The collected seeds from soil samples after wheat harvesting showed 80%–96% under field conditions, whereas dormancy ranged from 3 to 6 months under controlled conditions. The viability of *P. minor* seeds was only for the forthcoming season or maximum of 10 months under field conditions in rice–wheat crop rotation (Hari et al., 2003). Seed predation, germination, and decay are the most important factors determining seed losses in the soil under natural conditions. Further, the germination of dark brown *P. minor* seeds was more rapid as compared to light yellow and green-colored seeds (Mehra and Gill, 1988).

The loss in germinability and/or seed longevity under field conditions is affected by soil type, its physical conditions (such as aeration, moisture, temperature, and texture), and seeding depth. For example, buried seeds in one location with a sandy loam aerobic soil (Hisar, Haryana, India) under fallow-wheat rotation completely lost germinability in less than 1 year, whereas buried seeds of *P. minor* at this time still germinated in another location (Karnal, Haryana, India) with the soil characterized with an anaerobic environment under rice–wheat rotation (Franke et al., 2007). The exhumed seeds of *P. minor* completely lost the ability to germinate within 2 years of their burial at 30 cm soil depth in the rice–wheat rotation, whereas it could be viable even up to 8 years under laboratory conditions (Franke et al., 2007). A 72% seed germination was observed from 5 years old seeds stored under laboratory conditions and germination reduced to 0% after 8 years (Chhokar, 1998). The enormous variation is likely due to several environmental factors (soil temperature, moisture, light, microbial populations, etc.) under natural conditions, which are reflected poorly under a controlled environment.

Germination of *P. paradoxa* was stimulated by a range of smoked water concentrations of 5%–100% (Adkins et al., 2000). The persistence of *P. paradoxa* seeds was very low; less than 1% of shed seeds remained viable after 2 years (Taylor et al., 2005). Seeds of *P. paradoxa* were found to be very short-lived and 75%–98% of seeds were lost in the first year and 97%–99% seeds after the second year. The restricted viability and short longevity of *Phalaris* species provide opportunities for efficient crop diversification.

Effect of environmental factors on germination and growth

To optimize a weed-control strategy, knowledge of the mechanisms by which nonchemical control methods affect the weed population dynamics in the field is a prerequisite. The distribution of both *Phalaris* species across different continents reflects that both species are well adapted to the varied environmental conditions.

Temperature

Temperature significantly affects germination percentage, rate, and time to reach 50% of maximum germination. The germination rate of wheat (associated crop) was 89% at 5°C; however, both *Phalaris* species (*P. minor* and *P. paradoxa*) could not germinate in the same conditions (Xu et al., 2011). *Phalaris minor* seeds germinated over a range of 10°C–25°C; however, the maximum germination was recorded at 15°C (Bhan and Choudary, 1976; Derakhshan et al., 2014). The optimum temperature range was higher for seeds which were on the soil surface and this value decreased with an increase in burial depth of seeds (Singh and Ghosh, 1982). *Phalaris minor* germination showed a decreasing trend with a rise in temperature, and maximum reduction in germination was observed at 46°C (Hari et al., 2003). Further, the temperature in the upper 2-cm soil profile of a field that was plowed and kept dry after wheat harvesting was nearly 10°C higher than the ambient air temperature (42°C) and decreased subsequently through the soil profile. The duration and magnitude of the heating effect for reducing seed bank of *P. minor* in shallow depths can be amplified with the use of plastic mulch.

The germination of *P. minor* seeds increased from 53% at 5°C to 97% at 20°C under laboratory conditions (Sethi and Kaur, 2016). Seed germination was faster at 20°C than at 10°C during the initial period; but after 3 weeks, the germination rate was similar at both temperatures. The germination of *P. minor* was significantly lower at constant temperatures under dark conditions than at diurnal temperature variations (Alshallash, 2018). Seed germination of *P. minor* was highest (96%) at 15/25°C (night/day), and the lowest germination (0%) was observed at 20/35°C (Ahmadi, 2016). Thus it is worth understanding how *P. minor* and *P. paradoxa* seeds respond to different temperatures in different continents, so that better weed management strategies can be developed.

Light

White and red light enhanced the germination of *P. minor* seeds than the seeds wrapped in aluminum foil (Mishra et al., 2005). Germination of *P. minor* seeds was much lower in the dark (13%) as compared to 76% germination in the light (Singh et al., 1999). Germination of *P. minor* increased from 28% to 57% with an increase in light intensity from 0 (full darkness) to 1000 $\mu mol/m^2/s^1$ (Ohadi et al., 2010). Even a short time exposure (0.1 s) showed a stimulating effect on germination (45%), and it was 1.6 times higher than germination of seeds in darkness. This has practical implications for the germination of *P. minor* under field conditions as frequent tillage operations which may enhance the exposure of seeds to light that subsequently, could trigger germination of the seeds. So, night time tillage may result in the emergence of fewer seedlings.

In another study, Australian and England biotypes of *P. paradoxa* were collected and these biotypes showed different levels of innate dormancy (Taylor et al., 2004). Seed germination of one Australian biotype was stimulated by red and white light, whereas red light did not stimulate seed germination in other deeply dormant Australian and English biotypes. Seed sensitivity to light can be increased by a period of burial, and increased light sensitivity of buried seeds may stimulate further seed germination after cultivation. Dormancy in *P. paradoxa* biotypes was overcome by placing imbibed seeds in darkness at 16°C for either 7 or 14 days and followed by exposure to red or white light for a 15-min period (Taylor et al., 2005). Narrowing the crop row width, increasing the crop density, sowing of competitive cultivars can also result in a dense canopy soon after emergence and may considerably reduce the quantity of light reaching the soil surface (Norsworthy and Oliver, 2002). This is particularly more important for conservation cropping systems and sowing of wheat in rice residue (as mulch, e.g., Happy seeder-sown wheat) may help in cutting light exposure to *P. minor* seeds and thus inhibiting its germination.

Soil moisture stress

Both *Phalaris* species exhibit some tolerance to drought stress and over 60% germination of *P. minor* and *P. paradoxa* seeds was observed when soil water content exceeded 12.5% (Xu et al., 2011). Germination of *P. minor* seeds was initiated by the availability of soil moisture at various depths; however, the greatest proportion (60%) of seedlings emerged from the upper 2-cm layer after irrigation (Kumar and Kataria, 1977). Heavy textured soil with a high moisture level (e.g., the microclimate of the rice–wheat cropping system) favored vigorous growth and infestation of *P. minor* (Malik and Singh, 1995). Water deficit significantly decreased the growth of *P. minor* and net photosynthesis was decreased due to a decrease in stomatal conductance and alteration in the concentration of substomatal CO_2 (Iqbal and Wright, 1997). Further, it was observed that net photosynthesis was recovered following re-watering in wheat but failed in *P. minor*. This has a practical implication as temporal water deficit may increase the relative competitive ability of wheat.

Soil pH

Both *Phalaris* species exhibit some tolerance to pH and over 60% germination of seeds was observed within a pH range of 5–10 (Xu et al., 2011). Maximum germination (96%) of *P. minor* was observed at 7 pH and minimum (30%) at 5 pH (Ahmadi, 2016). Similarly, in another study, it was observed that increasing pH from 5 to 7 increased the germination of *P. minor* seeds from 33% to 73%, with a decrease in germination following a further increase in pH (Sethi and Kaur, 2016). At pH 5, the speed of germination was slow and the mean germination time was higher than at pH 6–8. The pH of either <6 or >7 significantly increased the membrane leakage and the total soluble sugar content of seedling shoots to tolerate unfavorable conditions. This information may help in devising ecological weed management strategies.

Osmotic stress

The germination rate of wheat (associated crop) was 53% at sodium chloride (NaCl) concentrations of 0.25 mol/L; however, seed germination and seedling growth of *P. minor* and *P. paradoxa* was more at NaCl concentrations under 0.05 mol/L (Xu et al., 2011). The lowest germination (23%) of *P. minor* was observed at an osmotic potential of −0.4 MPa, and it was completely inhibited at −0.6 or lower values, whereas 85% germination was observed at −0.1 MPa and 68% at −0.2 MPa (Sethi and Kaur, 2016). This information may help in predicting the potential distribution of *Phalaris* species to new areas and assist in formulating ecological weed-control measures.

Global scenario of herbicide resistance

The evolution of herbicide resistance in weeds depends on biology, ecology, and genetics of weeds, herbicidal application and other biological components (Beckie, 2006). The first case of herbicide resistance in *P. minor* was confirmed in India during 1991 against isoproturon(Malik and Singh, 1995).

To date, globally 11 and 8 unique cases of herbicide-resistant *P. minor* and *P. paradoxa*, respectively, have been confirmed (Table 15.2; Heap, 2020).

Phalaris minor evolved resistance against isoproturon after its successive applications for more than 15 years in India, covering an area of 0.8–1.0 million ha of wheat-growing regions of north-western India (Malik and Singh, 1995; Bhullar and Walia, 2004a; Franke et al., 2007). Wheat yield decreased substantially by 25%–50% due to resistance evolution in *P. minor* and under heavy infestations, farmers were forced to plow their fields in the absence of effective herbicides in north-west India. Wheat productivity was restored and a 35%–50% increase in wheat yield was reported in initial years of application of fenoxaprop, sulfosulfuron, and clodinafop at farmers' fields (Singh, 2007). But with sole reliance on these herbicides, the increase in yield remained 23%–27%, 12%–23%, and 17%–35%, respectively, within 7–8 years compared with the weedy plots. The GR_{50} (herbicide rate inhibiting plant growth by 50% with respect to the untreated control) values also showed a 10, 8, and 3–4 fold increase for sulfosulfuron, fenoxaprop, and clodinafop, respectively, compared to the initial values indicating the evolution of cross-resistance to these herbicides in *P. minor* (Dhawan et al., 2009). Further, a bioassay test also showed decreased susceptibility of *P. minor* to these herbicides as *P. minor* mortality declined from 85%–100% to 10%, 30%–40% and 75% with fenoxaprop, sulfosulfuron and clodinafop, respectively.

Phalaris paradoxa, the third most difficult grass weed of cropping systems with wheat and winter pulse crops in subtropical Australia, has evolved multiple herbicide resistance across two sites of actions (Walker et al., 2001). There are reports of poor control of *P. paradoxa* with ACCase inhibiting herbicides in durum wheat fields in central and southern Italy (Gasparetto et al., 2003). Both *P. minor* and *P. paradoxa* have evolved resistance to ACCase inhibitors in wheat fields in Guanajuato, Mexico (Table 15.2). In Iran, both *Phalaris* species have evolved resistance against ACCase inhibitors in the winter wheat crop. In Pakistan, ACCase inhibitor herbicides have been successfully applied to control *P. minor* in wheat for more than 20 years (Yasin et al., 2011), but, only 51%–71% mortality was found even at the double dose of fenoxaprop for *P. minor* biotypes, with a resistance index of 2.1–6.0 (Abbas et al., 2016, 2017).

Mechanism of herbicide resistance

The basis of resistance in isoproturon-resistant biotypes of *P. minor* in India was reported as metabolic (Singh et al., 1998; Kumar and Guru, 2016). However, no difference was found in the absorption, translocation, and metabolism of the ^{14}C-diclofop-methyl between the resistant and susceptible biotypes of *P. minor* in Mexico. A point mutation (gly-2096-ser) was detected in the CT domain of the plastid localized ACCase gene of resistant *P. minor* biotypes in Mexico (Cruz-Hipolito et al., 2012). In India, two

TABLE 15.2 Confirmation of unique cases of *Phalaris minor* and *P. paradoxa* resistant to different herbicide modes of actions.

Year	Country	Situation	Sites of action
Phalaris minor			
1991	India	Wheat	PSII inhibitors
1993	Israel		ACCase inhibitors
1994	India		
1996	Mexico		
1999	South Africa	Pastures and wheat	Multiple resistance: two sites of action ACCase inhibitors ALS inhibitors
2001	United States	Onions	ACCase inhibitors
2004	Iran	Winter wheat	
2006	India	Wheat	Multiple resistance: three sites of action ACCase inhibitors ALS inhibitors PSII inhibitors
2012	Australia	Wheat	ACCase inhibitors
2013	India	Wheat	ALS inhibitors
2015	Pakistan	Wheat	ACCase inhibitors
Phalaris paradoxa			
1979	Israel	Railways and roadsides	PSII inhibitors
1996	Mexico	Wheat	ACCase inhibitors
1997	Australia	Wheat and winter pulses	
1998	Italy	Wheat	
2004	Israel	Peas and wheat	
2007	Iran	Winter wheat	
2012	Australia	Spring barley and wheat	Multiple resistance: two sites of action ACCase inhibitors ALS inhibitors
2015	Syria	Faba beans, potatoes and wheat	ACCase inhibitors

Note: PSII inhibitors: herbicides belonging to the photosystem II inhibitors group; ACCase inhibitors: herbicides belonging to the acetyl-CoA carboxylase inhibitors group; ALS inhibitors: herbicides belonging to the acetolactate synthase inhibitors group.
Source: Adapted from Heap, I., 2020. The International Herbicide-Resistant Weed Database. Online. Sunday, October 4, Available www.weedscience.org.

mutations, Trp-2027-Cys and Ile-2041-Asn were observed in the resistant *P. minor* biotypes to ACCase inhibitors (Raghav et al., 2016). The target site-based and molecular studies revealed the substitution of Trp-2027-Cys and Asp-2078-Gly amino acids endowed resistance to ACCase inhibitors in Iranian *P. minor* biotypes (Gherekhloo et al., 2012). Further, resistance factors for these biotypes were 5.67–5.95, 8.72–9.4, and 7.65–8.13 for diclofop-methyl, fenoxaprop and clodinafop, respectively. Similarly, Ile-1781-Leu and Asp-2078-Gly point mutations were identified as the prime cause of resistance to aryloxyphenoxypropionates and cyclohexanediones herbicides, as well as for phenylpyrazoline herbicides (Cruz-Hipolito et al., 2015). In another study, it was observed that different resistant *P. minor* biotypes showed amino acid substitutions at different positions (e.g., Ile-2041-Asn, Asp-2078-Gly, Ile-1781-Leu, and Trp-2027-Cys). Further, these resistant biotypes of *P. minor* exhibited earlier and faster seedling growth as compared to susceptible biotypes and there was up to a 20% reduction in biomass allocation to roots (Torres-García et al., 2018).

In Israel, the substitution of a mixture of Ile-2041-Asn or Asp-2078-Gly in the CT domain of the chloroplastic ACCase gene was responsible for resistance in *P. paradoxa* biotypes (Hochberg et al., 2009). In another study, it was observed that substitutions of Ile -1781-Val and Asp-2078-Gly in the resistant biotype of *P. paradoxa* collected from Greece endowed resistance to different ACCase inhibitors with varying resistance levels (Collavo et al., 2011). In Norway, Ile-2041-Asn mutation conferred resistance to aryloxyphenoxypropionates, cyclohexanediones, and phenylpyrazoline in *P. paradoxa*, (Hochberg et al., 2007). This is unlike the previous reports where point mutation in position 2041 confers resistance to aryloxyphenoxypropionates but not to cyclohexanediones.

Biological fitness of resistant biotypes

Reduced vigor is regarded as an intrinsic feature of the triazine resistance phenomenon. Contrary to this, a triazine-resistant biotype of *P. paradoxa* was superior in seed germination and seedling emergence than triazine-susceptible biotypes (Schönfeld et al., 1987b). A point mutation in PsbA gene resulted in triazine resistance in *P. paradoxa* in Israel; but, this was not accompanied by low photosynthetic rates or reduced growth (Schönfeld et al., 1987a). Differences in plant fitness associated with herbicide resistance were also observed in Mexican *P. minor* biotypes. The embryo size of resistant biotypes was reduced up to 50% (on a dry weight basis), which resulted in a decrease in seed vigor and establishment of seedlings (Torres-García et al., 2015). Contrary to this, Indian isoproturon-resistant biotypes of *P. minor* germinated faster and earlier as compared to the susceptible biotypes, which was associated with higher amylase activity in seeds of the resistant biotypes (Dhawan et al., 2003). The susceptible *P. minor* biotypes were associated with a lower endogenous content of gibberellic acid (GA_3), and exogenous application of GA_3 stimulated germination in these biotypes. Such studies based on quantitative genetics could give a better understanding of the relationship between herbicide resistance and the ecological cost.

Cross and/or multiple resistance

A fenoxaprop-resistant biotype of *P. minor* from Israel showed cross-resistance to other ACCase inhibitors but was susceptible to propanil, isoproturon, and methabenzthiazuron which do not inhibit this enzyme (Tal et al., 1996). The dose-response assay conducted in Iran showed that among the collected 34 *P. minor* biotypes, only three biotypes were resistant to fenoxaprop, diclofop-methyl, and clodinafop, whereas 14 populations were resistant to fenoxaprop only and seven biotypes showed resistance against both fenoxaprop and diclofop-methyl (Gherekhloo et al., 2011). In Greece, 60% of collected biotypes of *P. minor*, were resistant to diclofop-methyl with some biotypes having high resistance indices (up to 12.4) along with cross-resistance to other herbicides (Travlos, 2012). Multiple resistances (against two to three herbicide sites of action) in *P. minor* against ACCase, ALS, and PSII inhibitors have been confirmed in India (Kaur et al., 2016) Chhokar and Sharma, 2008) and South Africa (Pieterse, 2010).

Integrated management

Herbicide inputs need to be reduced to minimize environmental contamination. To achieve this, the integration of chemical and nonchemical control methods is the best practice. Integrated weed management (IWM) options using competitive crops and good agronomic practices may be used to control resistant weeds. The inclusion of stale seedbed techniques and more competitive crops in the rotation will help in suppressing weed growth which will have possible synergistic effects on weed-control efficiency of chemical weed management. The prediction of emergence patterns of weed species in relation to soil environmental conditions decides the inclusion of cultural weed-control strategies such as tillage, sowing time, seeding density, etc., in an IWM program.

Preventive weed-control measures

Preventive measures are very important to avoid the entering of weeds in fields. Various preventive weed-control strategies can be integrated to prevent weed infestation to new fields, for example, use of weed-free crop seeds, prevention of seed dispersal and dissemination, use of well-decomposed compost, cleaning of agricultural implements, controlling weeds on irrigation ditches and bunds, prevention of weed seed rain, and subsequent buildup of the weed seed reserve in soil by mechanically cutting inflorescence or seed crushing (harvest weed-control systems).

Cultural weed-control measures or good agronomic practices

Alternate crops reduced soil weed seed bank through its influence on seed germination and mortality (Maxwell et al., 1990). A survey of farmers' fields conducted by Malik and Singh (1995) revealed that incidence of isoproturon resistance in *P. minor* was lower in cropping patterns such as rice-berseem (*Trifolium alexandrinum* L.)–sunflower (*Helianthus annuus* L.)–wheat (8%), sugarcane (*Saccharum officinarum* L.)–vegetables–wheat (9%), and cotton (*Gossypium arboreum* L.)–pigeonpea (*Cajanus cajan* L.)–wheat (16%) as compared with rice–wheat crop rotation (67%). Barley has a competitive edge for suppressing seed production of *P. paradoxa* compared with wheat in the subtropical grain region of Australia (Walker et al., 2001).

The rice residue retention of 5–7 t/ha reduced *P. minor* density by 24%–54% in a zero-till wheat crop (Brar and Walia, 2008). Under a zero-till system, most weed seeds (seed bank) are concentrated within the upper soil layer and thus, are more vulnerable to depletion/damage by germination, predation, and decomposition as compared to the conventional tillage system. Seedling emergence and the seed distribution pattern of weeds are profoundly influenced by tillage practices. Further, 50%–60% of seeds of *P. minor* were subjected to postdispersal seed predation under zero tillage with rice residue retention compared to conventional-till wheat (Kumar et al., 2013).

The crop can be made competitive against weeds by adopting good agronomic practices. Soil solarization is the best nonchemical method to control weeds in conventional and organic agriculture systems (Arora and Yaduraju, 1998). The temperature under polythene sheets exceeded 60°C at a 5 cm depth that decreased to 45°C at 15 cm soil depth. False seedbed preparation is the principal component of IWM programs to control *P. paradoxa* in Italy (Gasparetto et al., 2003). A gradual reduction in the germination of *P. minor* was observed with the increased duration of submergence of seeds (Chhokar, 1998). It was observed that continuous submergence conditions for at least 20 days were required to maintain in order to reduce the weed seed bank (Om et al., 2005; Prabhakar et al., 2012).

Wheat crop density affected the level of competition from *P. paradoxa*, and a density of 80–150 plants/m^2 resulted in the maximum crop yield and reduction in weed seed production (Walker et al., 2002). An increase in seed rate by 50% (150 kg/ha) improved the competitiveness of wheat and reduced biomass of *P. minor* by 35%, compared with a seed rate of 100 kg/ha (Bhullar and Walia, 2004b). Similarly, *P. minor* dry biomass in a wheat crop was significantly reduced by 10%, 25%, and 28% at crop seeding rates of 140, 160, and 180 kg/ha, respectively, than a seed rate of 100 kg/ha

(Mansoori, 2019). The shoot dry weight of *P. minor* was 89% lower in the 25th October sown crop than the 10th November sown wheat crop; however, it was 22% higher than the 20th December sown wheat crop, emphasizing the greater competitiveness of *P. minor* in late sowing of wheat (Yadav and Singh, 2005). *Phalaris minor* tillers were reduced by 39%–52% and dry weight by 40%–46% at 180 kg N/ha as compared to 100 kg N/ha (Raj et al., 2020). This reduction in weed density and biomass was due to vigorous wheat crop growth as a result of more availability of nitrogen to wheat at the higher dose of nitrogen, which boosted up crop competitiveness.

Allelopathy

Use of allelopathic crops in rotation, as cover crops or water extract application, offer eco-friendly and cost-effective weed control. Green manuring of sesbania (*Sesbania aculeate* L.) and sunflower reduced *P. minor* density by 100% under laboratory conditions (Om et al., 2002). Further, sesbania extracts in the 1:6 (wt./vol.) ratio on a dry weight basis reduced growth of resistant biotypes of *P. minor* by 16%–34% in a pot study. Green manuring with sunflower resulted in a 35%–48% reduction in *P. minor* population and improved wheat grain yield by 19% as compared to summer fallow. Further, a survey of farmers' fields showed that the sunflower effect was so pronounced that even 20% of farmers did not even apply any recommended herbicide for the control of *P. minor* in the wheat crop after its rotation. Some broad-leaved weeds such as *Chenopodium album* L., *Medicago denticulata* L., *Melilotus indica* L. and *Convolvulus arvensis* L. showed considerable allopathic potential against *P. minor* by inhibiting its germination by 100% (Om et al., 2002). The emergence of *P. minor* under laboratory conditions decreased and completely ceased using a very low concentration (0.0714%, vol./vol.) of eucalyptus (*Eucalyptus globules* L.) oil (Batish et al., 2007). The application of eucalyptus oil on 21 days old *P. minor* resulted in chlorosis, necrosis with rapid electrolyte leakage from leaf tissue and plants failed to recover at concentrations higher than 5% (vol./vol.).

Allelopathic crop mulches enhanced weed seed decay by 23%–52% and reduced viability of buried *P. minor* seeds by 41%–69% (Abbas et al., 2018). Wheat cultivars had a differential allelopathic inhibition activity against *P. minor* through the production of phenolic compounds (Kashif et al., 2015). Rice cultivars showed strong inhibitory allelopathic potential against *P. minor*. The water extract of scented cultivars (*basmati* types) had a pronounced inhibitory effect on *P. minor* as compared to nonbasmati (nonscented), and the effect was more conspicuous at higher concentrations (Om et al., 2002). Rice cv. PR 115 exhibited the maximum inhibitory effect and reduced the germination and seedling vigor index of *P. minor* by 33% and 78%, respectively, to control, whereas *Pusa Basmati*-1509 reduced the germination and seedling vigor index by 38% and 69%, respectively, as compared to control (Kaur et al., 2017). Allelopathic crop and weed species may be exploited for devising weed-control strategies under organic agriculture systems.

Biological weed control

The metabolites of fungi such as *Drechslera hawaiiensis, D. biseptata, D. holmii, and D. australiensis* exhibited herbicidal activity and significantly reduced seed germination, shoot, and root growth of *P. minor* seedlings (Akbar and Javaid, 2012). *D. australiensis* resulted in a 94% reduction of growth parameters, whereas *D. hawaiiensis* reduced length and dry biomass of shoot by 60%. The consortia of *Pseudomonas* strains reduced root length and increased *P. minor* seedling mortality by >50% (Dar et al., 2020). Some pathogenesis effects (leaf blotch symptoms) of *Septoria phalaridis*, a fungus, were observed on *P. paradoxa* plants (Seifbarghi et al., 2010). The use of these fungi may be used for development of mycoherbicides for control of *Phalaris* species.

Chemical weed management

The repeated use of the same mode of action based herbicides is the key factor involved in the fast evolution of herbicide resistance. Controlling weeds in the field using herbicide mixtures with two or more sites of action in rotations and/or mixtures can avoid and delay the development of herbicide resistance (Diggle et al., 2003; Beckie and Reboud, 2009). The timing of postemergence herbicides for the management of *P. minor* in wheat is of prime importance. The application of sulfosulfuron at Z12 (<2 leaves) and Z13 (2−3 leaves) wheat stages (before first irrigation) provided >80% control of *P. minor* (Rasool et al., 2017). While pinoxaden gave >90% control of *P. minor* at Z23 (7−8 leaf stages) wheat stage. Pinoxaden, fenoxaprop plus metribuzin and mesosulfuron plus iodosulfuron were effective when applied at Z21 (3−4 leaf) wheat stage. Flufenacet 250−300 g/ha applied at Z00 and Z13 wheat stages resulted in more than 85%−90% mortality of *P. minor* (Rasool et al., 2019). The alternate herbicide for control of herbicide-resistant weeds was also identified, for example, iodosulfuron controlled resistant biotypes of *P. minor* biotypes which evolved resistance against ACCase inhibitors since 1996 in South Africa (Smit and Cairns, 2000). Pyroxasulfone 127.5 g/ha as preemergence and metribuzin 200 g/ha as postemergence provided effective control of *P. minor* (Pandey, 2002). In another study, it was observed that the preemergence application of flumioxazin 125−150 g/ha effectively controlled *P. minor* and broad-leaved weeds (Chhokar et al., 2019). To prolong the effectiveness of these herbicides, their rotational use at an optimum dose and time with a proper application technology should be integrated with other weed-control tactics. The list of herbicides being currently used to control these weeds is given in Table 15.3.

TABLE 15.3 Commonly used and recommended herbicides against *Phalaris minor* and *P. paradoxa*.

Herbicide site of action	Chemical class/ family	Herbicides (Examples)	Time of application
Acetolactate synthase inhibitor	Sulfonylurea	Sulfosulfuron, iodosulfuron	Postemergence
Acetyl-CoA carboxylase inhibitor	Aryloxyphenoxy propionate	Clodinafop, diclofop, fenoxaprop	Postemergence
	Cyclohexanedinone	Tralkoxydim	
	Phenylpyrazoline	Pinoxaden	
Photosystem II inhibitor	Ureas	Chlorotoluron, isoproturon, metoxuron	Preemergence and early postemergence
	Triazine	Metribuzin, terbutryn	
Microtubule-assembly inhibitor	Dinitroaniline	Pendimethalin	Preemergence
Very long-chain fatty acid elongase inhibitor	Chloroacetamide	Pyroxasulfone	Preemergence
Protoporphyrinogen oxidase inhibitors	N-Phenylphthalimides	Flumioxazin	Preemergence

Conclusions and way forward

The periodic variability in dormancy and viability in seeds of *Phalaris* species is affected by various environmental factors including moisture, temperature, light, water submergence, residue management, harvesting method, soil texture, soil structure, depth, and period of seed burial in the soil. The seeding of wheat in rice residues results in an immediate reduction in weed plant densities in the field, besides savings in labor and soil cultivation costs. The development of weed emergence and density models could serve as the basis for making decisions on the use of weed management strategies (Singh et al., 2020). To provide a sustainable and environmentally friendly weed management strategy, the cropping system should be designed in a way that regulates the temporal and spatial distribution of invasive weed populations. The potential role of seed and seedling microclimate conditions in regulating seed dormancy and/or germination needs to be studied in a systematized way. Crop rotation and consequently different weed management methods would be the best way to control resistant weeds. As it may not be possible to shift to another crop, rotation of herbicides and applying herbicides with different modes of action (from different chemical families) would be useful to control resistant *Phalaris* in the short term. Harvest weed seed control is a novel way to reduce the reservoir of the soil weed seed bank. An understanding of soil weed seed bank dynamics is crucial as this will dictates the length of time that management strategies are required to reduce the reservoir of weed seeds to the minimum. Hence, an integrated management strategy is required to prevent further escalation in the magnitude of the problem of the evolution of herbicide resistance in *Phalaris* species.

References

Abbas, T., Nadeem, M.A., Tanveer, A., Ahmad, R., 2016. Evaluation of fenoxaprop-p-ethyl resistant littleseed canarygrass (*Phalaris minor*) in Pakistan. Planta Daninha 34, 833–838.

Abbas, T., Nadeem, M.A., Tanveer, A., Ali, H.H., Matloob, A., 2017. Evaluation and management of acetyl-CoA carboxylase inhibitor resistant little seed canary grass (*Phalaris minor*) in Pakistan. Arch. Agron. Soil. Sci. 63, 1613–1622.

Abbas, T., Nadeem, M.A., Tanveer, A., Ali, H.H., Farooq, N., 2018. Role of allelopathic crop mulches and reduced doses of tank-mixed herbicides in managing herbicide-resistant *Phalaris minor* in wheat. Crop. Prot. 110, 245–250.

Adkins, S.W., Davidson, P.J., Matthew, L., Navie, S.C., Wills, D.A., Taylor, I.N., et al., 2000. Smoke and germination of arable and rangeland weeds. In: Proc Sixth International Workshop on Seeds, 'Seed biology: advances and applications' at Merida, Mexico, pp. 347–359.

Ahmadi, A., 2016. The effect of temperature, salinity and soil burial on canary grass (*Phalaris minor*) germination. Jordan. J. Agric. Sci. 405, 1–10.

Akbar, M., Javaid, A., 2012. Evaluation of herbicidal potential of fungal metabolites against *Phalaris minor*. Afr. J. Microbiol. Res. 6, 4053–4057.

Alshallash, K.S., 2018. Germination of weed species (*Avena fatua*, *Bromus catharticus*, *Chenopodium album* and *Phalaris minor*) with implications for their dispersal and control. Ann. Agric. Sci. 63, 91–97.

Anderson, D.E., 1961. Taxonomy and distribution of the genus *Phalaris*. Iowa State J. Sci. 36, 1–96.

Baldini, R.M., 1995. Revision of the genus *Phalaris* L. (Gramineae). Webbia 49, 265–329.

Batish, D.R., Singh, H.P., Setia, N., Kohli, R.K., Kaur, S., Yadav, S.S., 2007. Alternative control of little seed canary grass using eucalypt oil. Agron. Sustain. Dev. 27, 171–177.

Beckie, H.J., 2006. Herbicide-resistant weeds: management tactics and practices. Weed Tech. 20, 793–814.

Beckie, H.J., Reboud, X., 2009. Selecting for weed resistance: herbicide rotation and mixture. Weed Tech. 23, 363–370.

Bekele, B., 2011. Achievements of research on barley and cereal yellow dwarf viruses infecting barley in Ethiopia. Barley Research and Development in Ethiopia. 2nd National Barley Research and Development Review Workshop, Holetta, Ethiopia.

Bhan, V.M., Choudary, D.B., 1976. Germination, growth and reproductive behaviour of *Phalaris minor* Retz. as affected by date of planting. Indian J. Weed Sci. 8, 126–130.

Bhan, R., Froud-Williams, R., Preston, C., Watts, J., Crossman, N., 2006. *Phalaris* spp. competition with wheat using an additive design series. 15th Australian Weeds Conference, Adelaide, South Australia.

Bhullar, M.S., Walia, U.S., 2004a. Classification of isoproturon resistant biotypes of *Phalaris minor* Retz. Indian J. Weed Sci. 36, 112–114.

Bhullar, M.S., Walia, U.S., 2004b. Effect of seed rate and row spacing on the efficacy of clodinafop for combating isoproturon resistant *Phalaris minor* Retz. in wheat. Plant. Prot. Q. 19, 143–146.

Bourke, C.A., Colegate, S.M., Slattery, S., Oram, R.N., 2003. Suspected *Phalaris paradoxa* (paradoxa grass) poisoning in horses. Aust. Vet. J. 81, 635–637.

Brar, A.S., Walia, U.S., 2008. Effect of rice residue management techniques and herbicides on nutrient uptake by *Phalaris minor* Retz. and wheat (*Triticum aestivum* L.). Indian J. Weed Sci. 40, 121–127.

Buhler, D.D., 1995. Influence of tillage systems on weed population dynamics and management in corn and soybean in the central USA. Crop. Sci. 35, 1247–1258.

Cardina, J., Herms, C.P., Doohan, D.J., 2002. Crop rotation and tillage system effects on weed seed banks. Weed Sci. 50, 448–460.

Chhokar, R.S., 1998. Biology and control of isoproturon resistant littleseed canarygrass (*Phalaris minor*) (PhD dissertation). CCS Haryana Agricultural University, Hisar, India, pp. 1–122.

Chhokar, R.S., Sharma, R.K., 2008. Multiple herbicide resistance in littleseed canarygrass (*Phalaris minor*): a threat to wheat production in India. Weed Biol. Manage 8, 112–123.

Chhokar, R.S., Sharma, R.K., Gill, S.C., Singh, G.P., 2019. Flumioxazin and flufenacet as possible options for the control of multiple herbicide-resistant littleseed canarygrass (*Phalaris minor* Retz.) in wheat. Weeds J. Asian-Pacific Weed Sci. Soc. 1, 45–60.

Collavo, A., Panozzo, S., Lucchesi, G., Scarabel, L., Sattin, M., 2011. Characterisation and management of *Phalaris paradoxa* resistant to ACCase-inhibitors. Crop. Prot. 30, 293–299.

Cruz-Hipolito, H., Domínguez-Valenzuela, J.A., Osuna, M.D., De Prado, R., 2012. Resistance mechanism to acetyl coenzyme A carboxylase inhibiting herbicides in *Phalaris paradoxa* collected in Mexican wheat fields. Plant. Soil. 355, 121–130.

Cruz-Hipolito, H., Fernandez, P., Alcantara, R., Gherekhloo, J., Osuna, M.D., De Prado, R., 2015. Ile-1781-Leu and Asp-2078-Gly mutations in ACCase gene, endow cross-resistance to APP, CHD, and PPZ in *Phalaris minor* from Mexico. Int. J. Mol. Sci. 16, 21363–21377.

Dar, A., Zahir, Z.A., Asghar, H.N., Ahmad, R., 2020. Preliminary screening of rhizobacteria for biocontrol of little seed canary grass (*Phalaris minor* Retz.) and wild oat (*Avena fatua* L.) in wheat. Can. J. Microbiol. 66, 368–376.

Dellow, J.J., Milne, B.R., 1986. Control of *Phalaris paradoxa* in wheat. Aust. Weeds 31, 22–23.

Derakhshan, A., Gherekhloo, J., Vidal, R.A., De Prado, R., 2014. Quantitative description of the germination of little seed canary grass (*Phalaris minor*) in response to temperature. Weed Sci. 62, 250–257.

Dhawan, R.S., Dhawan, A.K., Kajla, S., Moudgil, R., 2003. Amylase induction during seed germination in isoproturon susceptible and resistant biotypes of *Phalaris minor* Retz. Indian J. Exp. Biol. 41, 373–375.

Dhawan, R.S., Punia, S.S., Singh, S., Yadav, D., Malik, R.K., 2009. Productivity of wheat (*Triticum aestivum*) as affected by continuous use of new low dose herbicides for management of littleseed canarygrass (*Phalaris minor*). Indian J. Agron. 54, 58–62.

Diggle, A.J., Neve, P.B., Smith, F.P., 2003. Herbicides used in combination can reduce the probability of herbicide resistance in finite weed populations. Weed Res. 43, 371–382.

Finot, V.L., Pedreros, J.A., 2012. *Phalaris paradoxa* L. (Poaceae: Phalaridinae), a new introduced weed species in Central Chile {(*Phalaris paradoxa* L. (Poaceae: Phalaridinae), *nueva maleza introducida en Chile Central*)}. Gayana Bot. 69, 193–196.

Franke, A.C., Singh, S., McRoberts, N., Nehra, A.S., Godara, S., Malik, R.K., et al., 2007. *Phalaris minor* seed bank studies: longevity, seedling emergence and seed production as affected by tillage regime. Weed Res. 47, 73–83.

Gasparetto, M.A., Sattin, M., Campagna, C., Filì, V., Innocenti, M., Alberati, D., 2003. Dinamica di emergenza delle infestanti del frumento duro. L'informatore agrario 12, 77–80.

Gherekhloo, J., Mohassel, R., Mahalati, M.H., Zand, M.N., Ghanbari, E., Osuna, M.D., et al., 2011. Confirmed resistance to aryloxyphenoxypropionate herbicides in *Phalaris minor* populations in Iran. Weed Biol. Manag. 11, 29–37.

Gherekhloo, J., Osuna, M.D., De Prado, R., 2012. Biochemical and molecular basis of resistance to ACCase-inhibiting herbicides in Iranian *Phalaris minor* populations. Weed Res. 52, 367–372.

Hajjaj, B., El Oualkadi, A., 2019. Effect of clodinafop-propargyl on *Phalaris paradoxa* L. (awned canary-grass) in wheat crop. Int. J. Environ. Agric. Biotech. 4, 1820–1822.

Hanson, A.A., Hill, H.D., 1953. The occurrence of aneuploidy in *Phalaris* spp. Bull. Torrey Bot. Club 80, 16–20.

Hari, O.M., Dhiman, S.D., Hemant, K., Sajjan, K., 2003. Biology and management of *Phalaris minor* in wheat under a rice-wheat system. Weed Res. 43, 59–67.

Heap, I., 2020. The International Herbicide-Resistant Weed Database. Online. Sunday, October 4, Available www.weedscience.org.

Hitchcock, A.S. (Ed.), 1950. Manual of Grasses of the United States. United States Government Printing Office, Washington, USA.

Hochberg, O., Sibony, M., Tal, A., Rubin, B., Fløistad, E., 2007. Molecular bases for the resistance to ACCase inhibiting herbicides in *Phalaris paradoxa*. 14th EWRS Symposium, European Weed Research Society, Hamar, Norway, p. 150.

Hochberg, O., Sibony, M., Rubin, B., 2009. The response of ACCase-resistant *Phalaris paradoxa* populations involves two different target site mutations. Weed Res. 49, 37–46.

Iqbal, J., Wright, D., 1997. Effects of nitrogen supply on competition between wheat and three annual weed species. Weed Res. 37, 391–400.

Kashif, M.S., Cheema, Z.A., Muhammad, F., 2015. Allelopathic interaction of wheat (*Triticum aestivum*) and little seed canary grass (*Phalaris minor*). Int. J. Agri. Biol. 17, 363–368.

Kaur, H., Kaur, N., Sethi, R., 2017. Allelopathic potential of rice varieties against major weeds of rice and wheat. Indian J. Weed Sci. 49, 179–181.

Kumar, N., Guru, S.K., 2016. Herbicide resistance mechanism of *Phalaris minor* in Uttarakhand. J. Crop. Weed 12, 129–133.

Kumar, V., Kataria, O.P., 1977. Germination pattern and growth of *Phalaris minor* L. seeded at different depths. 6th Asian-Pacific Weed Science Society Conf., Indonesia, vol. 1, pp. 75–79.

Kumar, V., Singh, S., Chhokar, R.S., Malik, R.K., Brainard, D.C., Ladha, J.K., 2013. Weed management strategies to reduce herbicide use in zero-till rice-wheat cropping systems of the Indo-Gangetic Plains. Weed Tech. 27, 241–254.

Malik, R.K., Singh, S., 1995. Little seed canary grass (*Phalaris minor*) resistance to isoproturon in India. Weed Tech. 9, 419.

Mansoori, I., 2019. The effect of *Phalaris minor* (canary grass) densities and seed rates of wheat on yield and yield components of wheat (*Triticum aestivum*). Biol. Forum: An Int. J. 11, 176–182.

Martin, R.J., McMillan, M.G., Cook, J.B., 1988. Survey of farm management practices of the northern wheat belt of New South Wales. Aust. J. Exp. Agric. 28, 499–509.

Matus-Cádiz, M., Hucl, P., 2002. Morphological variation within and among five annual *Phalaris* species. Can. J. Plant. Sci. 82, 85–88.

Maxwell, B.D., Roush, M.L., Radosevich, S.R., 1990. Predicting the evolution and dynamics of herbicide resistance in weed population. Weed Technol. 4, 2–13.

Mehra, S.P., Gill, H.S., 1988. Effect of temperature on germination of *Phalaris minor* Retz. and its competition in wheat. J. Res. Punjab Agric. Univ. 25, 529–533.

Michael, D.R., Lindenmayer, D.B., Cunningham, R.B., 2010. Managing rock outcrops to improve biodiversity conservation in Australian agricultural landscapes. Ecol. Manag. Restor. 11, 43–50.

Mishra, J.S., Mathew, R., Yaduraju, N.T., 2005. Effect of puddling on distribution of rabi weed seeds and their emergence. Indian J. Weed Sci. 37, 36–39.

Negewo, T., Feyissa, A., Liben, M., Zemichael, B., 2011. Achievements of research on weeds and their management in barley in Ethiopia. Barley Research and Development in Ethiopia. 2nd National Barley Research and Development Review Workshop, Holetta, Ethiopia.

Ohadi, S., Rahimian Mashhadi, H., Tavakkol-Afshari, R., Beheshtian Mesgaran, M., 2010. Modelling the effect of light intensity and duration of exposure on seed germination of *Phalaris minor* and *Poa annua*. Weed Res. 50, 209–217.

Om, H., Dhiman, S.D., Kumar, S., Kumar, H., 2002. Allelopathic response of *Phalaris minor* Retz. to crop and weed plants in rice-wheat system. Crop. Prot. 21, 699–705.

Om, H., Kumar, S., Dhiman, S.D., 2005. Dormancy and viability of *Phalaris minor* seed in a rice-wheat cropping system. Weed Res. 45, 140–148.

Pandey, J., 2002. Effect of dose and mode of metribuzin application on *Phalaris minor* and yield of wheat (*Triticum aestivum*). Indian J. Agric. Sci. 72, 11–13.

Pieterse, P.J., 2010. Herbicide resistance in weeds—a threat to effective chemical weed control in South Africa. South. Afr. J. Plant. Soil. 27, 66–73.

Portugal, J., Vasconcelos, T., Forte, P., de Sousa, E., Calha, I., Moreira, I., et al., 2009. Economic injury level and tolerance threshold of *Phalaris paradoxa* in wheat. XII Congressa da Sociedad Española de Malherbologia, pp. 611–613.

Prabhakar, S.V., Yaduraju, N.T., Dadlani, M., Gopal, M., 2012. Effect of different cropping conditions, depth of burial, age of seeds, and allelopathic effect of different crops on the germination behavior of little seed canary grass (*Phalaris minor* Retz.) seeds. Plant. Breed. Seed Sci. 65, 99–112.

Raghav, N., Singh, R., Chhokar, R.S., Sharma, D., Kumar, R., 2016. Mutations in the plastidic ACCase gene endowing resistance to ACCase-inhibiting herbicide in *Phalaris minor* populations from India. Biotech 6, 12.

Raj, R., Das, T.K., Kaur, R., Shekhawat, K., Singh, R., Singh, V.K., 2020. Effects of nitrogen and densities on interference and economic threshold of *Phalaris minor* in wheat. Crop. Prot. 135, 105215.

Rasool, R., Bhullar, M.S., Gill, G.S., 2017. Growth stage of *Phalaris minor* Retz. and wheat determines weed control and crop tolerance of four post-emergence herbicides. Span. J. Agric. Res. 15, 1001.

Rasool, R., Bhullar, M.S., Singh, M., Gill, G.S., 2019. Flufenacet controls multiple herbicide resistant Phalaris minor Retz. in wheat. Crop. Prot. 121, 127–131.

Ryu, T., Lee, S., Kim, D., Choi, D., Kim, N., 2018. An unrecorded Alien species in Korea: *Phalaris paradoxa* L. (Poaceae). Korean J. Plant. Res. 31 (1), 52–55.

Sangster, A.G., Hodson, M.J., Parry, D.W., 1983. Silicon deposition and anatomical studies in the inflorescence bracts of four *Phalaris* species with their possible relevance to carcinogenesis. N. Phytol. 93, 105–122.

Schönfeld, M., Yaacoby, T., Ben-Yehuda, A., Rubin, B., Hirschberg, J., 1987a. Triazine resistance in *Phalaris paradoxa:* physiological and molecular analyses. Z. Naturforsch 42c, 779–782.

Schönfeld, M., Yaacoby, T., Michael, O., Rubin, B., 1987b. Triazine resistance without reduced vigor in *Phalaris paradoxa*. Plant. Physiol. 83, 329–333.

Seifbarghi, S., Razavi, M., Abbasi, M., Zare, R., 2010. First report of leaf blotch caused by *Septoria phalaridis* on *Phalaris paradoxa*. Plant. Dis. 94, 375.

Sethi, R., Kaur, N., 2016. Germination ecology of herbicide-resistant population of little seed canary grass from north-western India. J. Crop. Imp. 30, 274–286.

Singh, S., 1998. Studies on the mechanism of isoproturon resistance in Phalaris minor (littleseed canarygrass) (PhD. University of Strathclyde, Glasgow, UK.

Singh, S., 2007. Role of management practices on control of isoproturon-resistant littleseed canarygrass (*Phalaris minor*) in India. Weed Technol. 21, 339–346.

Singh, R.D., Ghosh, A.K., 1982. Soil profile distribution and effect of temperature and soil depth on germination of *Phalaris minor* Retz. Indian Society of Weed Science Annual Conf., Hisar, Abstract, pp. 41–42.

Singh, S., Kirkwood, R.C., Marshall, G., 1998. Effect of monooxygenase inhibitor piperonylbutoxide on herbicidal activity and metabolism of isoproturon in resistant biotypes of *Phalaris minor*. Pestic. Biochem. Physiol. 59, 143–153.

Singh, S., Kirkwood, R.C., Marshall, G., 1999. Biology and control of *Phalaris minor* Retz. (littleseed canarygrass) in wheat. Crop. Prot. 18, 1–16.

Singh, M., Kaur, S., Chauhan, B.S., 2020. Weed interference models. In: Chantre Guillermo, R., González-Andújar José, L. (Eds.), Decision Support Systems for Weed Management. Springer Nature, Switzerland AG, pp. 117–142.

Singh, M.C., Priyadarshi, M.B., 2014. Predicting invasive plants using weed risk assessment. Indian J. Weed Sci. 46, 91–95.

Smit, J.J., Cairns, A.L.P., 2000. Resistance of little seeded canary grass (*Phalaris minor* Retz.) to ACCase inhibitors. S. Afr. J. Plant. Soil. 17, 124–127.

Tal, A., Zarka, S., Rubin, B., 1996. Fenoxaprop-P resistance in *Phalaris minor* conferred by an insensitive acetyl-coenzyme A carboxylase. Pesticide Biochem. Physiol. 56, 134–140.

Taylor, I.N., Peters, N.C.B., Adkins, S.W., Walker, S.R., 2004. Germination response of *Phalaris paradoxa* L. seed to different light qualities. Weed Res. 44, 254–264.

Taylor, I.N., Walker, S.R., Adkins, S.W., 2005. Burial depth and cultivation influence emergence and persistence of *Phalaris paradoxa* seed in an Australian sub-tropical environment. Weed Res. 45, 33–40.

Thurley, B., Chancellor, R.J., 1985. Observations on the occurrence of *Phalaris paradoxa* in England and Wales. Ann. Appl. Biol. 107, 79–86.

Torres-García, J.R., Núñez-Farfán, J., Uscanga-Mortera, E., Trejo, C., Conde-Martínez, V., Kohashi-Shibata, J., et al., 2015. Competition for canopy cover between accessions of *Phalaris minor* that are susceptible and resistant to ACCase inhibiting herbicides. Nordic J. Bot. 33, 615–623.

Torres-García, J.R., Segura-León, O., Uscanga-Mortera, E., Trejo, C., Conde-Martínez, V., Kohashi-Shibata, J., et al., 2018. Evolution, growth and phenology of *Phalaris minor* biotypes resistant to ACCase-inhibiting herbicides in Mexico. Bot. Sci. 96, 95–102.

Travlos, I., 2012. Evaluation of herbicide-resistance status on populations of little seed canary grass (*Phalaris minor* Retz.) from southern Greece and suggestions for their effective control. J. Plant. Prot. Res. 52, 308–313.

USDA, 2016. Weed Risk Assessment for *Phalaris paradoxa* L. (Poaceae)-Awned canary-grass. United States Department of Agriculture (USDA), Animal and Plant Health Inspection Service (APHIS), Plant Protection and Quarantine (PPQ). Version 1, p. 22.

Viggiani, P., Angelini, R., 1993. Erbe spontanee ed infestanti: tecniche di riconoscimento-Graminacee. Ed agricole, Bologna.

Voshell, S.M., Baldini, R.M., Kumar, R., Tatalovich, N., Hilu, K.W., 2011. Canary grasses (*Phalaris*, Poaceae): molecular phylogenetics, polyploidy and floret evolution. Taxon 60, 1306–1316.

Walker, S.R., Robinson, G.R., Medd, R.W., 2001. Management of *Avena ludoviciana* and *Phalaris paradoxa* with barley and less herbicide in subtropical Australia. Aust. J. Exp. Agric. 41, 1179–1185.

Walker, S.R., Medd, R.W., Robinson, G.R., Cullis, B.R., 2002. Improved management of *Avena ludoviciana* and *Phalaris paradoxa* with more densely sown wheat and less herbicide. Weed Res. 42, 257–270.

Walker, S.R., Wilson, B., Wu, H., Widderick, M., Taylor, I., 2006. Weed seed persistence with changing farming practices in southern Queensland. 15th Australian Weeds Conference, Adelaide, South Australia.

Xu, G.F., Zhang, F.D., Li, T.L., Shan, L., Zhang, Y., Wu, D., 2010. Biological characteristics, influence on growth of wheat and its economical threshold of *Phalaris paradoxa* L. and *Phalaris minor* Retz. Sci. Agric. Sin. 43, 4409–4417.

Xu, G.F., Zhang, F.D., Li, T.L., Shen, S.C., Zhang, Y.H., 2011. Influence of environmental factors on seed germination of *Phalaris paradoxa* and *Phalaris minor*. Acta Bot. Bor. Occidentalia Sin. 31, 1458–1465.

Yadav, S.K., Singh, G., 2005. Studies on the depth and periodicity of *Phalaris minor* emergence in wheat under different crop establishment methods. Indian J. Weed Sci. 37, 29–32.

Yasin, M., Iqbal, Z, Safdar, M.H., Rehman, A., Ali, A., Asif, M., Aziz, M., Tanveer, A., Pervez, M.A., et al., 2011. *Phalaris minor* control, resistance development and strategies for integrated management of resistance to fenoxaprop-ethyl. Afr. J. Biotechnol. 10, 11802–11807.

Further reading

Alemseged, Y., Jones, R.E., Medd, R.W., 2001. A farmer survey of weed management and herbicide resistance problems of winter crops in Australia. Plant. Prot. Q. 16, 21–25.

Arora, A., Yaduraju, N.T., 1998. High-temperature effects on germination and viability of weed seeds in soil. J. Agron. Crop. Sci. 181, 35–43.

Kaur, N., Kaur, T., Kaur, S., Bhullar, M.S., 2016. Development of cross-resistance in isoproturon resistant *Phalaris minor* Retz. in Punjab. Agri. Res. J. 53, 69–72.

Norsworthy, J.K., Oliver, L.R., 2002. Effect of irrigation, soybean (*Glycine max*) density, and glyphosate on hemp sesbania (*Sesbania exaltata*) and pitted morning glory (*Ipomoea lacunosa*) interference in soybean. Weed Technol. 16, 7–17.

Chapter 16

Solanum nigrum and *Solanum physalifolium*

Alireza Taab
Department of Agronomy and Plant Breeding, Faculty of Agriculture, Ilam University, Ilam, Iran

Introduction

There are several weed species from the genus *Solanum* (Family: Solanaceae) including *Solanum nigrum* L. (black nightshades), *S. physalifolium* Rusby (hairy nightshade or hoe nightshade), *S. sarrachoides* Sendtner, *S. villosum* L., *S. americanum* Mill., *S. luteum* Mill., *S. ptycanthum* Dun., *S. retroflexum* Dunal., and *S. triflorum* Nutt. They form a species group centering around *S. nigrum* known as section *Solanum* (Maurella) (Edmonds, 1977) or Morelloid clade of *Solanum* (Särkinen et al., 2018) because of a shared ecological niche. Within the *Solanum* species group, some characteristics are phenotypically plastic in some species but genetically controlled in others (Jagatheeswari et al., 2013). Therefore there is taxonomic confusion around *S. nigrum* and its component species (Edmonds and Chweya, 1997), due to the historical factors, phenotypic plasticity, genetic variation, presence of polyploidy series, and possibilities of the interspecific hybridization (Edmonds, 1977). Thus the information regarding to characteristics of these species need to be interpreted with caution.

Solanum nigrum appears as a more important weed than that of *S. physalifolium*. Both are dicotyledonous plants (Fig. 16.1) with a C_3 photosynthetic pathway. *Solanum nigrum* is one of the worst world's weeds that is found in over 37 crops in 61 countries around the world (Holm et al., 1991). *Solanum nigrum* causes both reduction in crop quantity through competition for recourses and in crop quality by contaminating harvested crops (Edmonds and Chweya, 1997; Defelice, 2003). For instance, the economic threshold for *S. nigrum* in soybean [*Glycine max* (L.) Merr.] was reported to be more than 1 plant/m^2 (Sartorato et al., 1996). Full season *S. nigrum* competition with cotton may cause 60%–100% yield reduction (Keeley and Thullen, 1989). To prevent yield reduction in cotton, the crop must be kept *S. nigrum* free for a 3-week period at planting (Keeley and Thullen, 1989).

Biology and Management of Problematic Crop Weed Species. DOI: https://doi.org/10.1016/B978-0-12-822917-0.00005-7

FIGURE 16.1 From top to bottom: seed, mature berry, flower, and plant of *Solanum nigrum* (right) and *S. physalifolium* (left).

Members of Solanaceae family contain solanine, which is toxic to humans (Defelice, 2003) and animals (Jagatheeswari et al., 2013). *Solanum nigrum* and *S. physalifolium* both are problematic when pea crops are processed (Bithell, 2004). Immature fruits are similar in size, color, and gravity to the peas, making it difficult to distinguish and separate during harvesting and processing, which subsequently leads to loss in value (Elliot, 1972; Knott, 1986; Geoffrion, 2000; Coleman et al., 2020). Also, the separation results in crop downgrading and product loss (Elliot, 1972; Knott, 1986; Coleman et al., 2020). Moreover, species belonging to *Solanum* may contaminate crops, such as various beans or even the wool of grazing sheep by staining from the berry juice, which greatly reduces both the quality and economic return to farmers (Watt and Breyer-Brandwijk, 1962; Boik, 1996; Jagatheeswari et al., 2013).

Solanum nigrum acts as a host for insects, nematodes, and disease pests that attack crops (Defelice, 2003). It also hosts plant diseases like powdery mildew (Sudha and Lakshmanan, 2007). *Solanum physalifolium* also was found to be a host for *Phytophthora infestans* (Andersson et al., 2003) that causes potato and tomato disease. Moreover, the allelopathic effect of *S. nigrum* has been reported on garden pea (*Pisum sativum* L.), finger millet (*Eleusine coracana* (L.) Gaertn), fenugreek (*Trigonella foenumgraecum* L.) (Girija and Gowri, 2008). and chickpea (*Cicer arieti*num L.) (Kadioglu et al., 2005). Leaf extract of *S. nigrum* showed inhibitory effect on seed germination of these species (Girija and Gowri, 2008; Kadioglu et al., 2005).

Solanum nigrum is used as leafy vegetables in some countries (Kimaru et al., 2014). Various parts of the plant species belong to section *Solanum* specially shoots and berries are widely used as food crops or for various medicinal and local uses in many countries (Jagatheeswari et al., 2013), whereas it is known as a toxic plant in some other countries (Defelice, 2003).

Data on *S. nigrum* seedlings emergence timing relative to the crop in the field is of important values as it has implications for the control and survival of the species. Some *S. nigrum* plants may escape or survive weed control measures, whereas some plants could successfully produce seeds prior to the crop harvest (Bithell, 2004). Therefore information on biology, life cycle, and seedling emergence characteristics of the weeds is necessary to optimize weed management strategies (Bithell, 2004; Taab, 2009).

Biology and life cycle

Species belonging to genus *Solanum* are distributed from temperate to tropical regions of the world and from sea level to altitudes over 3500 m (Holm et al., 1991; Edmonds and Chweya, 1997). *Solanum nigrum* is a summer annual weed that sometimes appears as a perennial plant but it reproduces by seeds only (Defelice, 2003; Coleman et al., 2020). *Solanum nigrum* plants are susceptible to frost (Edmonds and Chweya, 1997) and usually die in winter. They are, however, capable of surviving in relatively mild winter conditions, and because of this, they may appear as biennial or perennial plants for 5 years under such conditions (Coleman et al., 2020). Its life cycle takes place from spring to late summer and early autumn (Edmonds and Chweya, 1997; Coleman et al., 2020). Therefore it interferes in summer crops. It has two subspecies including subspecies *nigrum* and *schultesii* (Defelice, 2003). Särkinen et al. (2018) described *S. nigrum* as sympatric with *S. villosum*. They suggested that the simplest distinguishing character between the two species is mature berry color. *Solanum villosum* has red, orange, or yellow berries, whereas *S. nigrum* has black or green berries. These two species quite often are found in the same field (Taab, personal observations).

Solanum nigrum may grow up to 1 m in height (Defelice, 2003; Coleman et al., 2020) and produce 600–168,000 seeds per plant (Keeley and Thullen, 1983; Holm et al., 1991; Defelice, 2003). However, it is estimated that it can produce between 5000 and 800,000 seeds per plant, depending on the growing conditions (Coleman et al., 2020). Seed survival of *S. nigrum* in the soil has been reported to be 8 years (Coleman et al., 2020) or 18–39 years, depending on soil disturbances (Ogg and Rogers, 1989; Defelice, 2003). Its seeds also remain viable after being eaten by birds and livestock (Roberts and Lockett, 1978; Defelice, 2003).

Solanum nigrum seeds may be dispersed through contaminated agricultural products, by water, birds, and animals who feed on its succulent berries (Roberts and Lockett, 1978; Edmonds and Chweya, 1997; Defelice, 2003; Coleman et al., 2020). Stankiewicz et al. (2001) suggested that the seeds of *S. nigrum*, which are resistant to triazines, might have spread in Europe through migratory birds. Therefore there is a risk of its seed dispersion by different means.

Solanum nigrum presents herbaceous stems with spreading to erect forms of growth habit, fibrous roots, and variable leaf and stem shapes, depending on environmental growing conditions and genotypes (Ogg et al., 1981). It presents a high degree of phenotypic plasticity in response to habitat conditions (Krstić et al., 2002; Kieloch and Gołębiowska, 2018). Rogers and Ogg (1981) studied morphological and anatomical characteristics in vegetative organs of some species belonging to the *S. nigrum* complex (*S. nigrum*, *S. sarrachoides*, *S. americanum*, and *S. ptycanthum*). They noticed that these species varied in relation to environmental conditions. *Solanum nigrum* plants produce larger leaves when growing in the shade than the plant growing in full sunlight (Fortuin and Omta, 1980). In addition, plants growing in low light have increased stem weight, internode length, and plant height (Crotser et al., 2003). Thus the morphology of the plant is affected by the growing condition.

Solanum nigrum is a short day plant (Bassett and Munro, 1985; Crotser et al., 2003) that flowers after mid-summer through early autumn. Its plants may flower 40–65 days after emergence; however, flowering may be delayed in spring emerged seedlings when compared to summer emerged individuals (Keeley and Thullen, 1983; Roberts and Lockett, 1978). The flowering may occur at 600 growing degree days (GDD) during a 6-week period, and its seed maturity may take place at approximately 1000 GDD with a base temperature of 6°C (McGiffen and Masiunas, 1992). With the same base temperature, Bithell et al. (2014) reported that its flower bud appeared at 434 GDD and flowered at 633 GDD.

Solanum physalifolium is an annual herbaceous plant with a prostrate growth habit and many branches (Edmonds, 1986; Sobrino and Del Monte, 1994; Edmonds and Chweya, 1997). It has two varieties including *Solanum physalifolium* var. nitidibaccatum (Bitter) Edmonds, syn. *S. nitidibaccatum* Bitter, and *Solanum physalifolium var. physalifolium*. The plant with teeth in the edge of the leaf blade is considered as the former variety and the plant with entire leaf as the latter variety. Fruits take the form of berries and are green to green-brown or dark green in color when they mature (Edmonds and Chweya, 1997).

Solanum physalifolium seedlings emerge in spring (Taab, 2009), and it flowers and fruits during the summer and autumn after which the plant may die with freezing temperatures in northern hemisphere (Edmonds, 1986). Bithell et al. (2014) reported that its flower bud appearance to take place at 352 GDD and its flowering at 509°C with the base temperature of 6°C.

Seed dormancy

Primary seed dormancy was shown in both *S. nigrum* and *S. physalifolium*. Seeds of *S. physalifolium* were found to be more dormant than that of *S. nigrum* (Taab, 2009). Difference in dormancy levels between the two species could be associated with a trade-off between dormancy level and seed number or size. *Solanum nigrum* produces a larger number of seeds than *S. physalifolium* but the latter species has seeds with a thicker and darker seed coat and are comparatively bigger (Taab, 2013). Bithell et al. (2002) also reported the presence of primary dormancy in *S. physalifolium*. The level of dormancy in *S. nigrum* was found to be weaker than *S. physalifolium* (Andersson and Yahya, 2003; Taab and Andersson, 2009a). Freshly harvested seeds of *S. nigrum* have primary dormancy (Roberts and Lockett, 1978; Bithell et al., 2002; Andersson and Yahya, 2003; Taab, 2009). The level of primary dormancy may vary depending on populations (Taab and Andersson, 2009c) and mother plant growing conditions (Kieloch and Gołębiowska, 2018). Kieloch and Gołębiowska (2018) reported that *S. nigrum* seeds collected from plants grown in conditions with access to light above the crop canopy are more germinable. However, there are reports on lack of primary dormancy in freshly harvested seeds of *S. nigrum* (Keeley and Thullen, 1983; Givelberg et al., 1984; Ogg and Rogers, 1989; Agong, 1993; Defelice, 2003). This could be due to differences among populations, seed handling duration, and conditions, or the germination conditions adapted to test dormancy. Seeds of this species readily germinates in light and at alternating temperatures (Taab, 2009). The level of dormancy may be reduced during storage at room temperature (Taab, personal observation).

According to Baskin and Baskin (1998), the type of seed dormancy in *S. nigrum* and *S. sarrachoides* is nondeep physiological and the seeds are conditionally dormant depending on conditions required to release dormancy. *Solanum sarrachoides* is closely related to *S. physalifolium* var. *nitidibaccatum*, which was separated from *S. sarrachoides* (Edmonds, 1986). Temperature was known as the most important factor in regulating the changes in dormancy (Bouwmeester, 1990; Bouwmeester and Karssen, 1992, 1993; Baskin and Baskin, 1998; Benech-Arnold et al., 2000). A seasonal change in dormancy, with the reduction during autumn, winter, and spring and induction during summer, has been reported in buried seeds of *S. nigrum* (Roberts and Lockett, 1978; Taab and Andersson, 2009a) and *S. physalifolium* (Taab and Andersson, 2009a). This confirms the presence of dormancy cycle in seeds that gradually changes from conditional dormancy to nondormancy and vice versa (Baskin and Baskin, 1998) depending on the conditions that seeds are exposed to.

The degree of primary seed dormancy may vary among seeds produced on the same individual plant, position on the plant, time of maturation, and environmental condition during development as it varies greatly between and within weed species (Håkansson, 2003). The level of abscisic acid during the primary dormancy development (Hilhorst and Karssen, 1992) and gene expression in the maternal parent (Munir et al., 2001) were shown to affect dormancy. During seed maturation, temperature (Fenner, 1991; Roach and Wulff, 1987; Steadman et al., 2004) and photoperiod (Munir et al., 2001) were found to influence seed dormancy status substantially. For instance, the higher level of dormancy in seeds was associated with the lower temperature during seed development on mother plant (Roach and Wulff, 1987). Therefore seeds of both *S. nigrum* and *S. physalifolium* have primary dormancy that develops during seed maturation process on mother plant depending on the environmental conditions and the genetic makeup of the plant. Seeds may enter the state of secondary dormancy after dormancy reduction if the required conditions for germination are not met. Thus thereafter seeds may experience yearly cycles in dormancy between conditionally dormant and nondormant statuses depending on the environmental conditions.

Seed germination

There are several reports confirming successful germination of *S. nigrum* seeds at constant temperatures (Givelberg et al., 1984; Agong, 1993; Benvenuti and Macchia, 1993; del Monte and Tarquis, 1997). In contrast, others reported poor or no germination at constant temperatures (Roberts and Lockett, 1978; Wagenvoort and Van Opstal, 1979; Teketay, 1998; Kremer and Lotz, 1998a; Bithell et al., 2002; Taab and Andersson, 2009b).

Alternating temperatures was found to increase germination of *S. nigrum* (Teketay, 1998; Bithell et al., 2002; Taab and Andersson, 2009b; Dong et al., 2020) and alternating temperatures with amplitudes ranging from 5°C to 15°C was also shown best to promote its germination substantially (Taab and Andersson, 2009b). For example, fluctuating temperatures in the range of 9/25°C for 16/8 h (Wagenvoort and Van Opstal, 1979), 20/12°C, and 30/12°C for 12/12 h (Teketay, 1998), and at 21/6°C, 23.5/8.5°C, 26/11°C, 28.5/13.5°C, 31/16°C, 33.5/18.5°C, 36/21°C, and 38.5/23.5°C in 16/8 h light/darkness (Taab and Andersson, 2009b) and at of 25/15°C and 30/20°C in light/darkness (Dong et al., 2020) were found to give almost complete germination in *S. nigrum* seeds. However, the reduction in germination was observed at temperature regimes of 36/21°C, 38.5/23.5°C (Taab and Andersson, 2009b), 35/25°C and 40/30°C (Dong et al., 2020). The reduction in seed germination at higher alternating temperatures was associated with the inhibitory effect of high temperatures composing the alternating temperature with wide amplitude on seed germination (Taab and Andersson, 2009b).

Temperature requirements for seed germination depend on the level of dormancy that can be reduced by stratification (Taab and Andersson, 2009b) or after ripening (dry storage) (Taab, personal observations). It has been shown that stratification at 5°C for 6 weeks (Wagenvoort and Van Opstal, 1979) and at 5°C and 15°C for 5.4 weeks (Taab and Andersson, 2009b) increased the rate and total germination of *S. nigrum* seeds.

The temperature range over which the germination can occur is affected by seed dormancy changes (Vleeshouwers and Bouwmeester, 2001) and the induction of dormancy results in a narrowing of this range (Vleeshouwers and Bouwmeester, 2001), whereas loss of dormancy results in a widening of the range or lower the base temperature for germination (Batlla and Benech-Arnold, 2003; Steadman and Pritchard, 2004). Seeds of *S. nigrum* have been reported to germinate at base temperatures ranging from 7.5°C to 15°C with an optimum temperature range of 20°C–30°C and a maximum temperature range from 30°C to 37°C as the maximum temperature at which germination may occur (Givelberg et al., 1984; Benvenuti and Macchia, 1993; Del Monte and Tarquis, 1997; Kremer and Lotz, 1998a; Bithell, 2004; Larina, 2008; Taab and Andersson, 2009b; Dong et al., 2020). Thus variations observed in the minimum range of temperatures for germination could be due to the differences in the dormancy status of the studied populations.

Solanum nigrum seeds are photoblastic (Roberts and Lockett, 1978; Givelberg et al., 1984; Kazinczi and Hunyadi, 1990; Teketay, 1998; Bithell et al., 2002; Taab, 2009) and light stimulates its germination substantially (Taab and Andersson, 2009a,b,c). Roberts and Lockett (1978) reported higher germination (73%) in light than in darkness (11%) at alternating temperatures in *S. nigrum* seeds. Taab and Andersson (2009b) observed increased germination in light than in darkness in both constant and alternating temperatures in *S. nigrum* seeds. However, according to Kieloch and Gołębiowska

(2018), seeds of *S. nigrum* germinated better in darkness than in light regardless of their origin. Dong et al. (2020) also observed that at the alternating temperatures of 15/5°C and 20/10°C, germination was greater in the darkness than in the light/dark treatment in *S. nigrum* seeds. However, germination did not differ between light and dark treatments at temperatures of 25/15°C and 30/20°C. Darkness may inhibit germination in *S. nigrum* seeds exhibiting a deep level of dormancy (Taab and Andersson, 2009a,b). Moreover, low R:FR ratios, due to plant canopy shading, may inhibit *S. nigrum* seed germination (Teketay, 1998). Therefore seed response to light and darkness is associated with dormancy degree and might also be linked to population characteristics.

Gibberellin (GA) (Roberts and Lockett, 1978; Givelberg et al., 1984) and potassium nitrate (KNO_3) (Roberts and Lockett, 1978) may stimulate germination of *S. nigrum* seeds in darkness. Seeds of *S. nigrum* germinate at water potential (Ψ) above −0.8 MPa (Givelberg et al., 1984; Dong et al., 2020). Guillemin et al. (2013) also reported an estimated base water potential of −0.89 MPa for *S. nigrum* seed germination. A soil moisture content of 35%–40% has been reported for optimum seed germination in *S. nigrum* (Wakhloo, 1964). Dong et al. (2020) studied the effect of salinity and alkalinity stress on *S. nigrum* seeds showed that seed germination was inhibited at concentrations greater than 200 mM sodium chloride and sodium hydrogen carbonate and germination was high at concentrations less than 50 mM. They also observed that its seeds germinate more than 95% at the pH range from 5 to 10.

Del Monte and Tarquis (1997) reported a temperature regime of 30/15°C over 8/16 h with a 12/12 h light/dark as the optimal condition for seed germination of *S. physalifolium*. They also reported a base temperature of 21°C for its germination that was affected by seed dormancy. Some authors reported that germination in *S. physalifolium* is not affected by GA (Edmonds, 1986; Bithell et al., 2002), whereas Sobrino and Del Monte (1994) reported GA as a germination enhancing factor in *S. physalifolium*. However, KNO_3 (Bithell et al., 2002) and light (Taab and Andersson, 2009c) were shown to enhance seed germination in *S. physalifolium*.

Seedling emergence

Solanum nigrum is known as a late emerging weed species (Buhler et al., 1997), and its seedlings usually emerge over an extended period (Taab, 2009). Its seedling emergence has been reported over 10 months in New Zealand (Popay et al., 1995), 5 months from early May to September in the United Kingdom (Roberts and Lockett, 1978), 8 months from March through October in California (Keeley and Thullen, 1983), 6 months from April through September in Washington in the United States (Ogg and Dawson, 1984), over 3 months from May through July in Sweden (Taab and

Andersson, 2009c), and almost the same period in the Netherlands (Kremer and Lotz, 1998a). A soil temperature ranging from 13°C to 20°C preceding the start of emergence has been reported as the requirement for seedlings recruitment of *S. nigrum* (Roberts and Lockett, 1978; Keeley and Thullen, 1983; Ogg and Dawson, 1984, Popay et al., 1995; Kremer and Lotz, 1998a; Taab and Andersson, 2009c). A fluctuating temperature above this range with an amplitude between 10°C and 15°C increases its seedling emergence (Roberts and Lockett, 1978; Edmonds and Chweya, 1997; Taab and Andersson, 2009c). Bi- or three-modal recruitment pattern with sporadic seedling emergence over an extended period in *S. nigrum* was associated with variation among seed batches in dormancy state with subsequent variation in germination requirements (Taab and Andersson, 2009c). Seed dormancy induction due to high temperatures during summer months in *S. nigrum* (Taab and Andersson, 2009a) prevents late emergence in summer otherwise the plants would die because of low temperatures in autumn before completing their life cycle. Therefore proportions of the seeds remain in the seedbank for the future that ensure the survival of the species (Taab, 2009).

Solanum nigrum seedlings may emerge well from soil depths up to 4 cm (Kremer and Lotz, 1998b), 5 cm (Grundy and Mead, 1998), and 6 cm (Dong et al., 2020). Dong et al. (2020) observed low emergence for *S. nigrum* seeds placed at the soil surface and it was highest for seeds at 1 cm depth followed by a decline at depths down to 6 cm. (Lati et al., 2012) also reported an optimal seedling emergences in seeds of *S. nigrum* buried at depths of 1–2 cm, which declined at approximately 2.5 cm and it was minimal at 4–5-cm soil depth. Therefore seedling recruitment of the species mainly takes place from seeds buried at shallower soil depths and deep burial by tillage may inhibit seedling emergence. However, it should be considered that deep burial of seeds may contribute in the buildup of a long-term seed pool.

Soil disturbance can increase the seedling emergences of *S. nigrum* (Ogg and Dawson, 1984; Roberts and Lockett, 1978). However, the seedling emergence of *S. nigrum* can be reduced by night-time cultivation (Scopel et al., 1994). In addition, according to Ogg and Dawson (1984), shallow tillage during the first part of April increased the overall emergence of *S. nigrum*, but later tillage had no effect. Thus late tillage has no effect on seedling emergence if timed to a period of dormancy induction in buried seeds (Taab and Andersson, 2009a).

Management

Several weed control measures including chemical and nonchemical options have been used to control *S. nigrum* and related weed species in many crops. Many herbicides have been registered to control *S. nigrum*, for example, atrazine, bentazone, chloridazon, chlorthal-dimethyl, clomazone, cyanazine, dicamba, dimethenamid-p, diuron, eptam (EPTC), fluroxypyr, linuron, metham, metolachlor, oxyfluorfen, pendimethalin, phenmedipham, prometryn, s-metolachlor, and terbutryn (Coleman et al., 2020). In addition, *S. nigrum* is reported to be controlled with acetochlor, aciflourfen, ametryn, bromacil, chlorimuron, cycloate, 2,4-D, ethofumesate, imazaquin, ioxynil, lactofen, MCPA, metobromuron, metribuzin, paraquat, simazine, terbacil, terbuthylazine, and trifluralin (CABI, 2020). The use of metham sodium as a soil fumigant is found very effective to control *S. nigrum* in tomatoes. However, it has to be applied to the soil 20 days before planting to avoid possible injury to the tomato (Zaragoza, 2003). Due to differences in regulations to use herbicides by countries and also to look for possibility of using an herbicide, information provided by herbicide producers and local advisors must be noted.

Herbicides have been used extensively targeting *S. nigrum* as well as many other broadleaved species in the cultivations. Therefore there are reports of *S. nigrum* biotypes resistant to atrazine, simazine, cyanazine, prometryn, terbuthylazine, and paraquat in several countries (Heap, 2020). Although herbicides are an efficient option for weed control, they have to be used with caution. To reduce environmental hazards and herbicide resistance development, they need to be used in integrated weed management (IWM) programs that combine nonchemical measures. Moreover, herbicide rotation that considers herbicides with different modes of action also need to be taken into account in managing *S. nigrum* and related species.

Preventive measures should be adapted where *S. nigrum* and related weed species have not been introduced and established. Therefore preventive measures such as sowing certified or weed free crop seeds, cleaning farm equipment before transporting between the farms, screening irrigation water system to prevent weed seed transportation, using weed seed free manure, and controlling weeds before setting seeds must be taken into account. Care should be taken during transplanting of crop seedlings like tomatoes and vegetables, which can also be sources of weed contamination. Farm hygiene and weed control operations in noncrop areas should also be considered to prevent further distribution of the weed species.

Soil disturbance can be applied to reduce the presence of *S. nigrum* in the seedbank (Ogg and Rogers, 1989; Defelice, 2003). However, large numbers of viable seeds may survive in the soil for 5 (Roberts and Lockett, 1978) to 39 years (Weller and Phipps, 1978/79). Thus a long-term strategy is needed to manage such a weed where it has established its seedbank. Cultivation in dark/night has been reported to reduce the field germination of S. *nigrum* and *S. sarrachoides* (Scopel et al., 1994). However, in other studies, cultivation in dark or soil stirring did not seems to promote seedling emergence in *S. nigrum* and *S. physalifolium* (Bithell, 2004; Taab and Andersson, 2009c). Seeds of both species were shown to germinate up to 100% both in light and in darkness when dormancy was fully reduced and seed germinability was lower in darkness during which dormancy was partly induced. Buried seeds also did not germinate in light or darkness when dormancy was fully induced (Taab and Andersson, 2009a). Therefore the cultivation in dark might be sufficient to reduce weed seedling emergence during the transition period from dormancy to nondormancy states and vice versa. However, cultivation during daytime can be applied to stimulate more seedling emergences, during which dormancy has been partly reduced to manage the weed seedbank (Taab, 2009). Changes in seed dormancy is under control of environmental conditions (e.g., mainly seasonal temperature), and this could vary from region to region depending on the climate. Therefore information on how the seasonal temperature effects dormancy changes in seeds of the species has to be taken into account.

Sowing date is a measure that can be optimized to control weeds. *Solanum nigrum* (Buhler et al., 1997) and *S. physalifolium* (Taab, 2009) are comparatively late emerging weed species. The late seasonal emergence of *S. nigrum* has been noted as a factor, which may reduce its contamination in late sown versus early sown peas (Elliot, 1972; Knott, 1986). Thus sowing date can be considered to reduce the risk of *Solanum* weed berries contamination in crops (Elliot, 1972; Knott, 1986). Moreover coinciding of *S. nigrum* fruiting with pea maturation can be avoided by early sowing of pea in the spring because *S. nigrum* usually emerges late in spring and the pea can be harvested before *S. nigrum* fruit maturity. Therefore as seasonal temperature is an important environmental factor regulating weed emergence timing, the crops should be sown early in the season, when the environmental conditions allow.

Hand weeding is a labor-intensive operation but it can be used in small scale farming and as a site-specific weed control option to control *Solanum* weed species (e.g., *S. nigrum* and *S. physalifolium*) especially before flowering. *Solanum* weed species usually produce large numbers of long-lived seeds that may increase the weed problems in the future if they are not controlled. In addition, site-specific weed control options including hand weeding, herbicides, and flamers might be considered to control weed patches and weeds that may escape earlier weed control operations. This also highlights the importance of weed survey techniques and programs during the crop growing season.

Plant density can be adjusted through an increased crop seed sowing rate or reduced crop planting row distance to promote rapid plant canopy closure and reduce the opportunity space to be used by the weed. However, Coleman et al. (2020) recommended balancing the weed control benefits of increased planting density against potential adverse effects of intraspecific competition and the risk of soil and plant diseases.

Cover crops can be used to suppress weeds and affect the weed community compositions (Bàrberi and Mazzoncini, 2001; Mennan et al., 2020), which can be grown in crop or between crop seasons. In addition, the cover crop residues can be used as mulch to control weeds. Campiglia et al. (2010) reported 69% reduction in weed above-ground biomass in tomato when using hairy vetch as dead organic mulch. Karkanis et al. (2012) showed that between 72% and 85% control of *S. nigrum* can be achieved using mulching with barley straw in leek. Therefore cover crop, or cover crop residues, can be considered to control *S. nigrum* as well as other weed species. In addition, it has to be mentioned that although plastic mulches are also effective means of weed control, the living and biodegradable mulches are preferred over plastic mulches due to practical problems after crop harvest to remove the mulch from the field and the subsequent environmental issues.

Mechanical weed control measures including tillage, mowing, and hoeing are frequently used to control weeds. However, timing of the weed control operation is vital. Ogg and Dawson (1984) reported that tillage a few days after seed germination reduced *S. nigrum* seedling emergence, whereas shallow tillage prior to germination may have increased the number of emerging seedlings. Kurstjens et al. (2002) showed that intra-row mechanical weed measures are more effective to control *S. nigrum* early in the stage of growth than the later stages. However, as *S. nigrum* and *S. physalifolium* usually emerge late in the season, interrow cultivation should be considered to control the weeds in row crops.

Nonchemical weed control methods, such as cultural and physical measures, are preferred over herbicides due to the development of herbicide-resistant biotypes and related environmental hazardous. Crop rotation provides the opportunity to use long-term, ample weed control techniques which should be considered in cropping systems. *Solanum nigrum* is reported to be more dominant in transplanted tomatoes than in direct-seeded (Zaragoza, 2003). Therefore planting type could also be taken into account to reduce the species interference in crops. To effectively control *S. nigrum* and *S. physalifolium*, as well as other weeds, formulating IWM is of important value in the cultivations. For instance, soil incorporation of prometryn herbicide into moist soil in combination with a powered rotary tiller was found to result in excellent control of *S. nigrum* in cotton (Keeley and Thullen, 1991). Tei et al. (1999) recommended a row application of soil-acting herbicides such as pendimethalin and oxifluorfen integrated with interrow hoeing and/or by split low-dose of metribuzin and rimsulfuron to control *S. nigrum* at early growth stage in tomato. Integration of mulching with barley straw and oxyfluorfen herbicide was reported to provide satisfactory control of weeds in leek (Karkanis et al., 2012). Finally, it is recommended that as weed populations and efficacy of weed control measures are affected by local conditions of a field, the feasibility of adapting sustainable weed control strategies should be tested under local conditions.

References

Agong, S.G., 1993. Seed extraction procedures for long term conservation of *Solanum nigrum*. Seed Sci. Technol 21, 447–451.

Andersson, L., Yahya, A., 2003. Primary dormancy in *Solanum nigrum* and *Solanum physalifolium*. Asp. Appl. Biol. 69, 229–235.

Andersson, B., Johansson, M., Jönsson, B., 2003. First report of *Solanum physalifolium* as a host plant for *Phytophthora infestans* in Sweden. Plant. Dis. 87, 1538. Available from: https://doi.org/10.1094/PDIS.2003.87.12.1538B.

Bàrberi, P., Mazzoncini, M., 2001. Changes in weed community composition as influenced by cover crop and management system in continuous corn. Weed Sci. 49, 491–499. Available from: https://doi.org/10.1614/0043-1745(2001)049[0491:CIWCCA]2.0.CO;2.

Baskin, C.C., Baskin, J.M., 1998. Seeds: Ecology, Biogeography, and Evolution of Dormancy and Germination. Academic Press, San Diego, USA.

Bassett, I., Munro, D., 1985. The biology of Canadian Weeds.: 67. *Solarium ptycanthum* Dun., *S. nigrum* L. and *S. sarrachoides* Sendt. Can. J. Plant. Sci. 65, 401–414.

Batlla, D, Benech-Arnold, R.L., 2003. A quantitative analysis of dormancy loss dynamics in *Polygonum aviculare* L. seeds. Development of a thermal time model based on changes in seed population thermal parameters. Seed Sci. Res. 13, 55–68.

Benech-Arnold, R.L., Sanchez, R.A., Forcella, F., Kruk, B.C., Ghersa, C.M., 2000. Environmental control of dormancy in weed seed banks in soil. Field Crop. Res. 67, 105–122.

Benvenuti, S., Macchia, M., 1993. Calculation of threshold temperature for the development of various weeds. Agric. Mediterr. 123, 252–256.

Bithell, S.L., 2004. An evaluation of *Solanum nigrum* and *S. physalifolium* biology and management strategies to reduce nightshade fruit contamination of process pea crops (Master thesis). Lincoln University, New Zealand.

Bithell, S.L., McKenzie, B.A., Bourdot, G.W., Hill, G.D., Wratten, S.D., 2002. Germination requirements of laboratory stored seeds of black nightshade and hairy nightshade. In: Proceedings of the 55th Conference of the New Zealand Plant Protection. Rotorua, New Zealand New Zealand Plant Protection Society, Inc. pp. 222–227.

Bithell, S.L., Hill, G.D., McKenzie, B.A., Wratten, S.D., 2014. Influence of black nightshade (*Solanum nigrum*) and hairy nightshade (*Solanum physalifolium*) phenology on processed pea contamination. N. Z. J. Crop. Hortic. Sci. 42, 38–49. Available from: https://doi.org/10.1080/01140671.2013.844719.

Boik, J., 1996. Cancer and Natural Medicine. Oregon Medical Press, pp. 220–222.

Bouwmeester, H.J., 1990. The effect of environmental conditions on the seasonal dormancy pattern and germination of weed seeds (PhD thesis). Wageningen Agricultural University, The Netherlands.

Bouwmeester, H.J., Karssen, C.M., 1992. The dual role of temperature in the regulation of the seasonal changes in dormancy and germination of seeds of *Polygonum persicaria* L. Oecologia 90, 88–94.

Bouwmeester, H.J., Karssen, C.M., 1993. Seasonal periodicity in germination of seeds of *Chenopodium album* L. Ann. Bot. 72, 463–473.

Buhler, D.D., Hartzler, R.G., Forcella, F., Gunsolus, J., 1997. Relative Emergence Sequence for Weeds of Corn and Soybeans. Iowa State University Publication on Pest Management, Ames, IA, File 9 (SA- 11).

CABI, 2020. Solanum nigrum (black nightshade). In: CABI Invasive Species Compendium. Available at https://www.cabi.org/isc/datasheet/50540 (Verified 22 August 2020).

Campiglia, E., Caporali, F., Radicetti, E., Mancinelli, R., 2010. Hairy vetch (*Vicia villosa* Roth.) cover crop residue management for improving weed control and yield in no-tillage tomato (*Lycopersicon esculentum* Mill.) production. Eur. J. Agron. 33, 94–102.

Coleman, M., Kristiansen, P., Sindel, B., Fyfe, C., 2020. Blackberry Nightshade (*Solanum nigrum*): Weed Management Guide for Australian Vegetable Production. School of Environmental and Rural Science, University of New England, Armidale.

Crotser, M.P., Witt, W.W., Spomer, L.A., 2003. Neutral density shading and far redradiation influence black nightshade (*Solanum nigrum*) and eastern blacknightshade (*Solanum ptycanthum*) growth. Weed Sci. 51, 208–213. Available from: https://doi.org/10.1614/0043-1745(2003)051[0208:NDSAFR]2.0.CO;2.

Defelice, M.S., 2003. The black nightshade, *Solanum nigrum* L. et al.-Poison, poultice, and Pie. Weed Technol. 17, 421–427.

Del Monte, J.P., Tarquis, A.M., 1997. The role of temperature in the seed germination of two species of the *Solanum nigrum* complex. J. Exp. Bot. 48, 2087–2093.

Dong, H., Ma, Y., Wu, H., Jiang, W., Ma, X., 2020. Germination of *Solanum nigrum* L. (Black Nightshade) in response to different abiotic factors. Planta Daninha 38, e020219463Epub June 15, 2020. Available from: https://doi.org/10.1590/s0100-83582020380100049.

Edmonds, J.M., 1977. Taxonomic studies on *Solanum* L. Section *Solanum (Maurella)*. Bot. J. Linn. Soc. 75, 141–178.

Edmonds, J.M., 1986. Biosystematics of *Solanum sarrachoides* Sendtner and *S. physalifolium* Rusby (*S. nitidibaccatum* Bitter). Bot. J. Linn. Soc. 92, 1–38.

Edmonds, J.M., Chweya, J.A., 1997. Black nightshades (*Solanum nigrum* L.) and related species, Promoting the Conservation and Use of Underutilized and Neglected Crops, 15. Institute of Plant Genetics and Crop Plant Research, Gatersleben/International Plant Genetic Resources Institute, Rome, Italy.

Elliot, A.H.A., 1972. The effects of weeds on vegetable crops grown for processing. 11th British Weed Control Conference 3, pp. 932–938.

Fenner, M., 1991. The effects of the parent environment on seed germinability. Seed Sci. Res. 1, 75–84.

Fortuin, F.T.J.M., Omta, S.W.P., 1980. Growth analysis and shade experiment with *Solanum nigrum* L, The Black Nightshade, a leaf and fruit vegetable in West-Java. Neth. J. Agri. Sci. 28, 199–210.

Geoffrion, R., 2000. Toxic plants in crops. Keep a watchful eye in plots in which weed control is poor or zero. Phytoma 532, 8–11.

Girija, G., Gowri, S., 2008. Allelopathic effect of *Solanum nigrum* on *Pisum Sativum, Eleusine Coracana* and *Trigonella Foenum Graecum*. Biomed. Pharmacol. J. Available from: http://biomedpharmajournal.org/?p = 303.

Givelberg, A., Horowitz, M., Poljakoff-Mayber, A., 1984. Germination behavior of *Solanum nigrum* seeds. J. Exp. Bot. 35, 588–598.

Grundy, A.C., Mead, A., 1998. Modelling the effects of seed depth on weed seedling emergence. Asp. Appl. Biol. 51, 75–82.

Guillemin, J.P., Gardarin, A., Granger, S., Reibel, C., Munier-Jolain, N., Colbach, N., 2013. Assessing potential germination period of weeds with base temperatures and base water potentials. Weed Res. 53, 76–87.

Håkansson, S., 2003. Weeds and Weed Management on Arable Land: An Ecological Approach. CABI Publishing, Wallingford, Oxon, UK, p. 288.

Heap, I., 2020. The International Herbicide-Resistant Weed Database. Online. Tuesday, August 18, 2020. <www.weedscience.org>.

Hilhorst, H.W.M., Karssen, C.M., 1992. Seed dormancy and germination: the role of abscisic acid and gibberellins and the importance of hormone mutants. Plant. Growth Regul. 11, 225–238.

Holm, L.G., Pluckneet, D.L., Pancho, J.V., Herberger, J.P., 1991. The World's Worst Weeds: Distribution and Biology. Krienger Publishing (The University Press of Hawaii, Honolulu), Malabar, FL, p. 609.

Jagatheeswari, D., Bharathi, T., Ali, H.S.J., 2013. Black night shade (*Solanum nigrum* L.)—an updated overview. Int. J. Pharm. Biol. Arch. 4, 288–295.

Kadioglu, I., Yanar, Y., Asav, U., 2005. Allelopathic effects of weeds extracts against seed germination of some plants. J. Environ. Biol. 26, 169–173.

Karkanis, A., Bilalis, D., Efthimiadou, A., Katsenios, N., 2012. Comparison between conventional and organic weed management: growth and yield of leek (*Allium porrum* L.). Hortic. Sci. (Prague) 39, 81–88.

Kazinczi, G., Hunyadi, K., 1990. Germination of seeds of black nightshade (*Solanum nigrum* L.) from different coloured berries. Zeitschrift fur Pflanzenkrankheiten und Pflanzenschutz, Sonderheft 12, 83–88.

Keeley, P.E., Thullen, R.J., 1983. Influence of planting date on the growth of black nightshade (*Solanum nigrum*). Weed Sci. 31, 180–184.

Keeley, P., Thullen, R., 1989. Growth and Competition of Black Nightshade (*Solanum nigrum*) and Palmer Amaranth (*Amaranthus palmeri*) with Cotton (*Gossypium hirsutum*). Weed Sci. 37, 326–334. Available from: https://doi.org/10.1017/S0043174500072003.

Keeley, P., Thullen, R., 1991. Biology and control of black nightshade (*Solanum nigrum*) in cotton (*Gossypium hirsutum*). Weed Technol. 5 (4), 713–722. Available from: https://doi.org/10.1017/S0890037X00033741.

Kieloch, R., Gołębiowska, H., 2018. Influence of environmental conditions and crop competition on morphological and biological diversity of *Avena fatua* L. and *Solanum nigrum* L. Acta Agrobot. 71, 1740. Available from: https://doi.org/10.5586/aa.1740.

Kimaru, S.L., Kimenju, J.W., Kilalo, D.C., Onyango, C.M., 2014. Growth and yield response of selected species of African leafy vegetables infested with root knot nematodes (*Meloidogyne incognita*). Glob. J. Biol. Agri. Health Sci. 3, 1–6.

Knott, C.M., 1986. Weed control in peas—current practices and future prospects. Asp. Appl. Biol. 12, 159–169.

Kremer, E., Lotz, L.A.P., 1998a. Germination and emergence characteristics of triazine-susceptible and triazine-resistant biotypes of *Solanum nigrum*. J. Appl. Ecol. 35, 302–310.

Kremer, E., Lotz, L.A.P., 1998b. Emergence depth of triazine susceptible and resistant *Solanum nigrum* seeds. Ann. Appl. Biol. 132, 277–288.

Krstić, L., Merkulov, L.S., Boža, P.P., 2002. The variability of leaf anatomical characteristics of *Solanum nigrum* L. (Solanales, Solanaceae) from different habitats. Proc. Nat. Sci., Matica Srpska 102, 59–70. Available from: https://doi.org/10.2298/ZMSPN0201059K.

Kurstjens, D.A.G., Vermeulen, G.D., Bleeker, P.O., 2002. Effect of plant dry mass on uprooting by intra-row weeders. In Proceedings of 5th Workshop EWRS Physical and Cultural Weed Control. Pisa, Italy. pp. 246–266.

Larina, S.Yu., (All Russian Institute of Plant Protection, Russia) 2008. Solanum nigrum. In Afonin, A.N., Greene, S.L., Dzyubenko, N.I. & Frolov A.N. (eds.). Interactive Agricultural Ecological Atlas of Russia and Neighboring Countries. Economic Plants and their Diseases, Pests and Weeds [Online]. <http://www.agroatlas.spb.ru/en/content/weeds/Solanum_nigrum/>.

Lati, R.N., Filin, S., Eizenberg, H., 2012. Black nightshade (*Solanum nigrum*) emergence and development is affected by seed density and burial depth. Phytoparasitica 195–203. Available from: https://doi.org/10.1007/s12600-011-0210-4.

McGiffen, M.E., Masiunas, J.B., 1992. Prediction of black and eastern black nightshade (*Solanum nigrum* and *S. ptycanthum*) growth using degree-days. Weed Sci. 40, 86–89.

Mennan, H., Jabran, K., Zandstra, B., Pala, F., 2020. Non-chemical weed management in vegetables by using cover crops: a review. Agron 10, 257. Available from: https://doi.org/10.3390/agronomy10020257.

Munir, J., Dorn, L.A., Donohue, K., Schmitt, J., 2001. The effect of maternal photoperiod on seasonal dormancy in *Arabidopsis thaliana* (Brassicaceae). Am. J. Bot. 88, 1240–1249.

Ogg Jr., A.G., Dawson, J.H., 1984. Time of emergence of eight weed species. Weed Sci. 32, 327–335.

Ogg Jr., A.G., Rogers, B.S., 1989. Taxonomy, distribution, biology, and control of black nightshade (*Solanum nigrum*) and related species in the United States and Canada. Rev. Weed Sci. 4, 25–58.

Ogg Jr., A.G., Rogers, B.S., Schilling, E.E., 1981. Characterization of black nightshade (*Solanum nigrum*) and related species in the United States. Weed Sci. 29, 27–32.

Popay, A.I., Cox, T.I., Ingle, A., Kerr, R., 1995. Seasonal emergence of weeds in cultivated soil in New Zealand. Weed Res. 35, 429–436.

Roach, D.A., Wulff, R.D., 1987. Maternal effects in plants. Annu. Rev. Ecol. Syst. 18, 209–235.

Roberts, H.A., Lockett, P.M., 1978. Seed dormancy and field emergence in *Solanum nigrum* L. Weed Res. 18, 231–241.

Rogers, B.S., Ogg, A.G., 1981. Biology of weeds of the *Solanum nigrum* complex (*Solanum* section *Solanum*) in North America. Agricultural Reviews and Manuals. US Department of Agriculture, Science and Educational Administration, pp. 1–30, Western Series No 23.

Särkinen, T., Poczai, P., Barboza, G.E., Van Der Weerden, G.M., Baden, M., Knapp, S., 2018. A revision of the Old World Black Nightshades (Morelloid clade of *Solanum* L., Solanaceae). PhytoKeys 106, 1–223. Available from: https://doi.org/10.3897/phytokeys.106.21991.

Sartorato, I., Berti, A., Zanin, G., 1996. Estimation of economic thresholds for weed control in soybean [*Glycine max* (L.) Merr]. Crop. Prot. 15, 63–68. Available from: https://doi.org/10.1016/0261-2194(95)00114-x.

Scopel, A.L., Ballare, C.L., Radosevich, S.R., 1994. Photostimulation of seed germination during soil tillage. N. Phytol. 126, 145–152.

Sobrino, E., Del Monte, J.P., 1994. Two alien *Solanum* species new to the Spanish flora, and their characterization within the *Solanum nigrum* complex (Solanaceae). Flora Mediterr. 4, 101–109.

Stankiewicz, M., Gadamski, G., Gawronski, S.W., 2001. Genetic variation and phylogenetic relationships of triazine-resistant and triazine-susceptible biotypes of *Solanum nigrum*-analysis using RAPD markers. Weed Res. 41, 287–300.

Steadman, K.J., Ellery, A.J., Chapman, R., Moore, A., Turner, N.C., 2004. Maturation temperature and rainfall influence seed dormancy characteristics of annual ryegrass (*Lolium rigidum*). Aust. J. Agric. Res. 55, 1047–1057.

Steadman, K.J., Pritchard, H.W., 2004. Germination of *Aesculus hippocastanum* seeds following cold-induced dormancy loss can be described in relation to a temperature-dependent reduction in base temperature (T_b) and thermal time. New Phytol. 161, 415–425.

Sudha, A., Lakshmanan, P., 2007. *Solanum nigrum*, a new host for powdery mildew disease of *Capsicum annuum* in the Madurai district of Tamil Nadu, India. Aust. Plant. Dis. Notes 2, 97–98. Available from: https://doi.org/10.1071/DN07040.

Taab, A., 2009. Seed dormancy and germination characteristics of *Solanum nigrum* and *S. physalifolium* as affected by temperature (PhD thesis). The Swedish University of Agricultural Sciences, Uppsala, Sweden.

Taab, A., 2013. Weed seed dormancy, germination and seedling emergence characteristics. In: Taab, A. (Ed.), Weeds and Their Ecological Functions. Nova Science Pub Inc, USA, pp. 1–54.

Taab, A., Andersson, L., 2009a. Seasonal changes in seed dormancy of *Solanum nigrum* and *Solanum physalifolium*. Weed Res. 49, 90–97.

Taab, A., Andersson, L., 2009b. Seed dormancy dynamic and germination characteristics of *Solanum nigrum*. Weed Res. 49, 490–498.

Taab, A., Andersson, L., 2009c. Primary dormancy and seedling emergence of black nightshade (*Solanum nigrum*) and hairy nightshade (*S. physalifolium*). Weed Sci. 57, 526–532.

Tei, F., Baumann, D.T., Dobrzanski, A., Giovinazzo, R., Kleifeld, Y., Roche, F., et al., 1999. Weeds and Weed Management in Tomato. A review. 11th European Weed Research Society Symposium, Basel, Switzerland. p. 132.

Teketay, D., 1998. Environmental factors that control the germination of five Solanum species from Ethiopia. Trop. Ecol. 39, 79–87.

Vleeshouwers, L.M., Bouwmeester, H.J., 2001. A simulation model for seasonal changes in dormancy and germination of weed seeds. Seed Sci. Res. 11, 77–92.

Wagenvoort, W.A., Van Opstal, N.A., 1979. The effect of constant and alternating temperatures, rinsing, stratification and fertilizer on germination of some weed species. Sci. Hortic. 10, 15–20.

Wakhloo, J.L., 1964. Ecological and physiological studies on two species of *Solanum*: I. Germination and development of *S. xanthocarpum* Schrad and Wendl and *S. nigrum* L. Flora 155, 237–249.

Watt, J.M., Breyer-Brandwijk, M.G., 1962. *Solanum nigrum* L. The Medicinal and Poisonous Plants of Southern and Eastern Africa. E. and S. Livingstone Ltd., Edinburgh and London, pp. 996–1000.

Weller, R.F., Phipps, R.H., 1978. A review of the black nightshade (*Solanum nigrum* L.). Prot. Ecol. 1, 121–139.

Zaragoza, C., 2003. Weed management in vegetables. In: Labrada, R. (Ed.), Weed Management for Developing Countries, 2003. Food and Agriculture Organization of the United Nations, Rome. Available at: http://www.fao.org/3/y5031e/y5031e0b.htm#bm11 (19-08-2020).

Further reading

Albouchi, F, Attia, M, Hanana, M, Hamrouni, L, 2018. Ethnobotanical Notes and Phytopharmacologiques on *Solanum nigrum* Linn. (Family: Solanaceae). Am J Phytomed Clin Ther 6. Available from: https://doi.org/10.21767/2321-2748.100341.

Henderson, R.J.F., 1974. *Solanum nigrum* L. (Solanaceae) and related species in Australia. Contributions from the Queensland Herbarium. Queensland Herbarium, pp. 1–78, No. 16.

Hutchinson, P.J.S., 2014. Hairy nightshade critical interference period in potatoes. Weed Technology 28 (3), 543–551. Available from: https://doi.org/10.1614/WT-D-13-00160.1.

Schilling, E.E., 1981. Systematics of Solanum Sect. Solanum (Solanaceae) in North America. Systematic Botany 6 (2), 172–185. Available from: https://www.jstor.org/stable/2418547; <https://doi.org/10.2307/2418547>.

Chapter 17

Sonchus oleraceus

Gurpreet Singh Khalsa[1], Sahil Dahiya[1] and Bhagirath Singh Chauhan[2]
[1]Queensland Alliance for Agriculture and Food Innovation (QAAFI), The University of Queensland, Gatton, QLD, Australia, [2]The Centre for Crop Science, Queensland Alliance for Agriculture and Food Innovation (QAAFI) and School of Agriculture and Food Sciences (SAFS), The University of Queensland, Gatton, QLD, Australia

Introduction

Unwanted plants within a crop field interfere with the crop, creating competition for the available resources (nutrients, illumination, moisture, etc.) thereby hampering crop growth and development and ultimately reducing yield. In Australian agricultural production, weed management is a costly constraint (Medd, 1997). Due to recent developments in herbicide resistance in Australia, particularly in the north-eastern grain region of Queensland and New South Wales, weed management costs have increased (Adkins et al., 1997; Matthews, 1994; Preston, 2000). The most common weed in this region is *Sonchus oleraceus* L., which grows well in the summer as well as winter, and is one of the most common weeds in crops like wheat (*Triticum aestivum* L.), chickpea (*Cicer arietinum* L.), and also in fallows of summer and winter seasons (Osten et al., 2007). *Sonchus oleraceus* is considered a noxious invasive weed due to its genetic diversity and morphological features in association with herbicide resistance (Adkins et al., 1997).

Sonchus oleraceus is native to Europe and North Africa (Garcke, 1972; Matthei et al., 1993), but is now widely distributed across North America, South America, Africa, Asia, Australia, and New Zealand (Hyatt, 2006). It can complete its lifecycle over a diverse range of habitats (Holm and Centre, 1977). It is an extremely common weed all over England and Wales and is also well-established up to altitudes of 365 m in the Yorkshire region of England (Lewin, 1948). Its occurrence has been observed in 13 out of 272 arable fields examined in north-east Switzerland (Buchli, 1936). In Egypt, *S. oleraceus* is one of the most common weeds of old and reclaimed land (Abd El-Gawad, 2014; Abd El-Ghani et al., 2014; Gomaa et al., 2012). In Australia, it is a common weed which has increased its geographic range over a period of time (Manalil et al., 2017). Subtropical to tropical weather conditions with plenty of rainfall give way to the

Chauhen-BMPCWS-1632754. DOI: https://doi.org/10.1016/B978-0-12-822917-0.00001-X

rapid emergence of this weed and increased seed dispersal by exploiting favorable environmental conditions, which has allowed for its wide dispersal throughout Australia (Werth et al., 2011; Widderick et al., 2010). *Sonchus oleraceus* inhabits agricultural lands like crop fields, orchards, and vineyards. It can be found in plowed fertile fields and alongside paths, but not in unrunied green lands or in forests (with an exception of being found in exposed soil due to animal activities or in areas of the forests cleared by fire) (Lewin, 1948). This weed can also be found in urban gardens, houses, walls, and along roadsides, particularly where the soil has been made bare by dung or burning (Shenstone, 1912). It grows on a wide variety of substrates in North-West Europe (Dadd, 1962). In desert areas, it is usually only found near waterways (Halvorson and Guertin, 2003).

Several physiological traits in plants help to understand the mechanisms involved in the growth and development under different environmental conditions. Changes in physiological parameters at different growth stages enhance the capability of plants to withstand environmental stresses. With the development of agricultural operating systems, various concepts have arisen from time to time to minimize input costs for the benefit of producers. Changing the soil condition is likely to affect the germination and establishment of weed flora (Holzner, 1978). Reduced tillage systems are one such concept which has gained popularity within the last few decades in Australia and worldwide. Implementation of this concept helps to control soil erosion, improve the physical and chemical characteristics of the soil, reduce moisture losses, and cut fuel costs. Besides the soil condition and other agronomic advantages, variations in tillage practices can also have a significant effect on the relative prevalence of weeds in the field. Seed recovery, either through seedlings or viable seeds, is greatly influenced by tillage (Widderick et al., 2002). The transition to conservation tillage has placed a heavy reliance on herbicidal weed control, which has resulted in the development of herbicide-resistant weed biotypes (Adkins et al., 1997).

Other than employing a reduced tillage system, various other techniques, like cover crop, intercropping, and mulches, have also been adopted to suppress the weed growth. Cover crops are also used to control weed populations by maintaining seasonal vegetative cover (Moore et al., 1994). Cover crops have so many merits, as they help growers to meet year-round requirements of conversation tillage, prevent soil erosion, improve organic matter content, as well as soil structure, and suppress weed emergence and growth (Akemo et al., 2000; Krutz et al., 2009; Norsworthy et al., 2011). Weed growth can also be reduced by adopting intercropping techniques. In comparison to one crop, intercropping combinations (growing two or more crops simultaneously) also reduce weed growth by using more of the available resources (Ballare and Casal, 2000; Liebman and Dyck, 1993; Melander et al., 2005). Furthermore, mulching is also an option to control weed populations. Mulches inhibit weed growth by inhibiting light from reaching the soil surface (Bangarwa et al., 2009, 2011). Other than these options, modified planting dates, seeding rates, and row spacings also have a great impact on the suppression of weed growth.

Modifications in weed flora composition due to improvements in soil fertility resulted in increasing weed–crop competition (Jones, 1966; Landi, 1975; Thurston, 1976). A range of biological and ecological traits of a plant species determine its potential to be invasive, but the seed characteristics of a plant, such as number of seeds, persistence of seed, and its ability to disperse seeds, has the largest influence on invasion potential (Gordon et al., 2008).

Sonchus oleraceus is one such weed species, which has small and lightweight seeds and dispersed through wind to spread over several kilometers (Jakobsson and Eriksson, 2003). *Sonchus oleraceus* poses serious threats to different pulses, grains, and other valuable crops. It is highly competitive, difficult to control, and has a tendency to adapt very well to varied cropping regions due to its germination and growth acceptance to wide temperature regimes and its tolerance to osmotic and saline conditions (Chauhan et al., 2006; Manalil et al., 2018). There are various favorable traits like profuse seed production, germination throughout the year, the absence of innate dormancy, allelopathic effects, and a high efficiency of using resources in competition with crops (Chauhan et al., 2006; Gomaa et al., 2014; Song et al., 2017; Widderick et al., 2010). Due to such traits, this weed poses a serious threat to crops and has become 8th, 10th, and 11th most significant in terms of infested area (596,000 ha), yield losses (1.81×10^7 kg), and economic losses (AU$4.9 million), in the Australian grain cropping regions (Balick et al., 2000; Manalil et al., 2017; Weber, 2017).

Currently, one of the major threats to weed control faced by farmers worldwide is the development of resistance to various herbicidal groups, which in turn becomes the biggest challenge for scientists. The question arises of how the best management practice can be integrated into decision-making and how complex economic and regulatory criteria can be tackled. Canopy architecture modifications, crop competition enhancement through cultivation of high weed-competitive cultivars, increased crop seed rates, and narrow row spacing are some of the common weed management strategies in conservation agricultural systems (Bajwa et al., 2017; Brennan et al., 2001; van der Meulen and Chauhan, 2017).

Taxonomy and description

Sonchus oleraceus (Compositae) is an annual herb (Holm and Centre, 1977). The achenes in *S. oleraceus* are winged less at margins and narrow (Hutchinson et al., 1984). *Sonchus oleraceus* is an amphidiploid, with a reported chromosome number of $2n = 32$; which originated from a hybridization between *Sonchus asper* L. ($2n = 22$) and *Sonchus tenerrimus* L. ($2n = 14$) (Boulos, 1976; Hutchinson et al., 1984). It is also believed that the increased limitations of environmental conditions and adaptations of

S. oleraceus is the result of the combination of *S. asper* and *S. tenerrimus* genomes (Cho et al., 2019).

The genus *Sonchus* has shown wide variations in morphological characters (Elkamali et al., 2011). In this genus, *S. oleraceus* and *S. asper* are difficult to distinguish using only morphological characteristics. However, marginal spiky leaf margins with downward pointing auricles in the leaves and bracts are the main distinguishing features of *S. oleraceus*. A basal rosette is formed by old leaves near the ground and later leaves appear on the flowering stem, which terminates in an inflorescence. The initial orbicular leaves have minute serrated margins, whereas fully grown leaves are pinnatifid and have irregularly toothed margins. These fully grown leaves become lobed thereby attain 10–25 cm length. Rapid growth of internodes raise the plant height up to 1 m or more after the appearance of first floral bud (Holm and Centre, 1977).

Seed biology and seed bank dynamics

Sonchus oleraceus is an annual plant which reproduces only by seed. Its flowers are hermaphrodite and self-fertile (Harper, 1977). Pollination occurs through insects, mainly by bees and flies, and both selfing and outcrossing are involved in seed production (Hutchinson et al., 1984). The average number of seeds per capitulum is around 140, and the mean number of capitula per plant is 4.4 (Cici and Van Acker, 2009). *Sonchus oleraceus* seeds germinate approximately 2 weeks after sowing and require light for germination (Hutchinson et al., 1984). Their germination is not greatly influenced by temperature, and the maximum germination has been documented to occur when seeds are sown on the soil surface. *Sonchus oleraceus* has a potential to emerge all year round from the top 2 cm layer of soil, but most of the emergence takes place from the top 1 cm layer (Widderick et al., 2004). Under a no-tillage system, seedling recruitment of *S. oleraceus* was found to be significantly higher in comparison to minimum tillage systems, which is attributed to its small size and the light weight of the seeds (0.323 g/1000 seeds), making it difficult to emerge from greater depths (Chauhan et al., 2006).

Due to long fruiting periods in *S. oleraceus*, seed collections made at different times are likely to give differing results in terms of emergence (Roberts and Neilson, 1981). Soon after harvest, 65%–100% of the seeds of *S. oleraceus* germinated, suggesting that this weed species has very little innate dormancy (Widderick et al., 2004). Open flowers appeared to be capable of producing viable seeds (Gill, 1938). Seed persistence was found to be greater in the top 2 cm layer of soil under zero-tillage conditions than systems with tillage and the burial of seeds below 2 cm during tillage operations also increased seed persistence but reduced emergence (Widderick et al., 2004).

In England and Canada, seedling emergence of *S. oleraceus* takes place mainly during the late spring [April (12/4°C) to June (19/10°C)], but the appearance of seedlings is seen throughout the period of March to November (Hutchinson et al., 1984; Roberts and Neilson, 1981). The highest germination of *S. oleraceus* was observed at 25/10°C, whereas a very poor germination response was observed at constant temperatures of 7°C−35°C (Andersen, 1968). In Italy, above 90%, germination has been recorded under wide temperature conditions ranging between 10°C and 35°C; but above 40°C, germination was nil (Gresta et al., 2010). In subtropical grain regions of Australia, germination of *S. oleraceus* takes place readily across wide temperature regimes (5°C−35°C) (Widderick et al., 2004). In another study in Queensland, Australia, germination exceeded to 84% under different day/night temperature regimes of 20/10°C, 25/15°C, and 30/20°C (Manalil et al., 2018). Just 2% of the seeds were viable on the soil surface after 6 months, and 12% remained intact at a depth of 10 cm after 30 months, in the eastern part of Australia (Widderick et al., 2010).

Light plays an important role in the germination and growth of *S. oleraceus*, whereas those that germinated in shade grew poorly (Lewin, 1948). In Sheffield (United Kingdom), studies revealed that the germination of *S. oleraceus* was much lower (12%) in a tall turf grass sward in comparison to short-turfed grass (36%) due to the light interception response(Fenner, 1978). Complete darkness significantly reduced germination under different temperature conditions (Manalil et al., 2018).

Osmotic adjustment, defined as lowering the osmotic potential in plants due to the accumulation of net solutes, enables plants to endure moderate water deficits and grow in regions of limited water availability (Ackerson, 1981; Girma and Krieg, 1992). Compatible solutes, such as proline, soluble sugar, and soluble protein, are important physiological indicators to determine the ability of a plant's osmotic changes (Kadkhodaie et al., 2014). It has been observed that *S. oleraceus* possesses a higher tolerance to prolonged drought in comparison to some of its related species due to its potential to withstand oxidative damage to cellular components, osmoregulation, and photosynthetic ability in different growth stages (Jia et al., 2018). However, *S. oleraceus* prefers moist conditions, as it stimulates germination and growth. It can germinate under varying pH levels and is not completely suppressed by saline soil conditions (Manalil et al., 2018). Germination of *S. oleraceus* decreased from 95% to 11% under increased osmotic stress from 0 to −0.6 MPa and was inhibited completely at −0.8 MPa (Chauhan et al., 2006; Widderick et al., 2010). It favors slightly alkaline to neutral soil conditions rich in organic matter. Seed germination was highest (>90%) when the soil pH ranged from 5 to 9, but fell to 77% at a pH of 10, thereby indicating that a wide pH range in most soils is not a limiting factor for successful germination of *S. oleraceus* seeds (Chauhan et al., 2006).

Interference with agronomic production and invasiveness

Sonchus oleraceus has emerged as a major problematic weed in various agroecosystems around the world (Abdullah et al., 2017; Widderick et al., 2010). Widespread dispersal of the achenes of *S. oleraceus* is attributed to its features like tall scapes (generally 1 m tall) and light-weight achenes with dense filamentous pappi (Hutchinson et al., 1984). Under congenial conditions and mismanagement, *S. oleraceus* emerges, set seeds, and generates an abundant seed bank, which will continue to be a problem in the northern grain region of Australia (Manalil et al., 2018). This weed can proliferate in a fallow phase due to noncompetition with crops where it can grow abundantly, resulting in tremendous seed production thereby making it difficult to control in the following crop. *Sonchus oleraceus* utilizes ample stored moisture, resulting in valuable yield losses in succeeding crops (Chauhan et al., 2006; Widderick et al., 2010). In soybean (*Glycine max* L.), 50% yield losses have been reported at *S. oleraceus* density of 18–20 plants/m^2 (Song et al., 2017), whereas similar yield losses have been recorded in wheat at *S. oleraceus* density of 50 plants/m^2 (Mobli et al., 2019). It has also been reported as an important weed in vegetable crop production, particularly in loa-my and clay soils in regions of Queensland and New South Wales, Australia (Henderson, 1998).

Invasive weed species dominate agroecosystems and therefore not only limit crop productivity but may also kill existing native species, alter soil properties, change the population and behavior of microorganisms, and affect the profile of nutrients in the soil (Lorenzo et al., 2013). The invasion of any weed species also depends on the ability and the response to natural selection to extend the range of physiological tolerance. Mechanisms for invasion include competence, physiological dominance, morphological plasticity, and allelopathy (Lorenzo et al., 2013). The establishment of *S. oleraceus* in crops can lead to severe threats to cereals, grains, cotton, and vegetables where light, nutrients, and water availability are limited at different phases of crop production. In annual or perennial cultivation of crops, *S. oleraceus* can occur as a weed, especially in those crops with wider spacing or a longer vegetation duration, such as sunflower (*Helianthus annus* L.), tobacco (*Nicotiana* spp.), groundnut (*Arachis hypogea* L.), frenchbean (*Phaseolus vulgaris* L.), and citrus (*Citrullus lanatus* L). It competes with crop plants during the fructification stage and has a damaging impact on yield (Holm and Centre, 1977). It grows well in maize (*Zea mays* L.) crops having optimum interspace (Lorenzoni, 1963). In Australia the emerging dominance of *S. oleraceus* in weed–crop competition is thought to be the result of a gradual shift from traditional farming systems to conservation systems (Widderick et al., 2010). The mechanisms of seed biology, herbicide resistance, and microclimate alternations created by conservation systems have made it a difficult weed to manage in Australia (Adkins et al., 1997; Widderick et al., 1999, 2010). In Australia and New Zealand, introduced populations of *S. oleraceus* have outperformed the native populations in terms of various growth parameters like higher biomass, leaf number, height, large number of seeds, and higher germination rates, which have led to successful invasion (Ollivier et al., 2020). In the north-eastern grain region of Australia, *S. oleraceus* is listed among the main weeds in agricultural land of 4,300,000 ha.

Allelopathy

In agro-systems, weeds have a significant role in the introduction of toxins into the soil which in turn influences germination, growth, and establishment by creating competition for resources, namely, minerals, nutrients, and water (Patterson, 1986). Leaves of *S. oleraceus* possess various bioactive compounds including alkaloids, tannins, steroids, phenols, flavonoids, and tarpenes (Singh, 2010; Yin et al., 2007). As an example, *S. oleraceus* affects the legume (*Trifolium alexandrinum* L.) and its related weed species *Melilotus indicus* L. and *Trifolium resupinatum* L.) through the possible release of phytotoxins from root exudates and decaying residues (Gomaa et al., 2015). Studies conducted in the North Western part of India on the allelopathic effect of *S. oleraceus* and other weeds observed significant suppression and a negative response on seed germination, seedling growth, and seedling dry weight of wheat (Gupta and Mittal, 2012).

Management

Cultural control

Vertical distribution of weed seeds is affected by tillage systems employed within the field (Pareja et al., 1985), and such differential distribution can result in altering dynamics of the weed population in the soil profile. Tillage practices influence the exposure of weed seeds to light and manipulate the physical as well as chemical environments (Buhler and Mester, 1991; Harper, 1957) of the weed seeds which results in inhibiting their emergence. Thus employing minimum tillage, in addition to summer and autumn plowing, is effective in controlling *S. oleraceus* by distributing its seed bank in the upper soil surface (Widderick et al., 2002, 2010).

A zero-tillage system is an effective method for controlling *S. oleraceus*, as the majority of its seeds are present in the uppermost 2 cm soil layer and the seed bank can be rapidly depleted. However, the effectiveness of this system depends upon proper destruction of the seed bank, which can be achieved by completely controlling all seedlings of this biotype for approximately 2 years so that there is no replenishment of the seed bank.

Seedling emergence can be reduced by effectively burying the weed seeds below 2 cm using strategic inversion tillage. But the practice should not be repeated for at least 2 years. This is necessary for avoiding reoccurrence of the persistent weed seeds back to the upper soil layer (Chauhan et al., 2006; Widderick et al., 2010). In regions, where the weed is not a major threat, strategic tillage techniques can help to minimize the threat of weed infestation in the future (Widderick et al., 2002). Repeated shallow tillage practices before planting can also help in reducing the emergence and survival of the weed (De Cauwer et al., 2008; Widderick et al., 2002).

To enhance crop competition against weeds, it is important to consider cover crops, plant density, and sowing times to achieve maximum crop yields (Chauhan and Johnson, 2010; Williams, 2009). Competitive crops, like barley and wheat when grown at narrow spacings (e.g. 25 cm) with a high plant density ($\geq$100 plants/m^2), significantly reduced the density, biomass, and seed production of *S. oleraceus* (Widderick et al., 2004). A recent study reported that an increased density of mungbean to 328 plants/m^2 produced 1185 *S. oleraceus* seeds per plant, which were much lower than produce at standard plant density (Mobli et al., 2019). Thus the choice to grow competitive crops in the field is one of the most beneficial options in terms of economic returns, where *S. oleraceus* is suspected to be a major problematic weed.

Chemical control

Various weed management practices are available to control *S. oleraceus*, but herbicides are mostly used for the effective control of this weed. Various preemergence and postemergence herbicides are used for its control. Glyphosate, either alone or within a mixture, is commonly used for this weed in both fallows and glyphosate-tolerant crops.

The herbicides carfentrazone, tribenuron methyl, bromoxynil octanate, sulfentrazone, etc. are effective postemergent herbicides (El-Kholy et al., 2013), and flumioxazin and isoxaflutole are effective preemergence herbicides for controlling this weed (Chauhan and Jha, 2020). Farmers use different herbicides (e.g., glyphosate, paraquat, saflufenacil, etc.) as desiccants at crop maturity to control *S. oleraceus*. Glyphosate proved to be an effective herbicide in controlling *S. oleraceus*, either alone or in combination with metsulfuron-methyl, 2,4-D, and/or triclopyr plus picloram, by providing 99%–100% control (Widderick et al., 2004). In the case of glyphosate-resistant (GR) biotypes, glyphosate on its own may not provide effective control. Thus to improve the control of GR biotypes, paraquat is applied in sequence as a double-knock. To get the best results, an interval of 7–10 days is required between the knocks. Thus GR *S. oleraceus* biotypes are effectively controlled (95%–100%) with the double-knock application of fluroxypyr, picloram plus 2,4-D, and glyphosate followed by paraquat and diquat for the second knock application (Churchett et al., 2014).

Glyphosate is a commonly used herbicide in the fallow phase and glyphosate-tolerant crops due to the development of resistance against ALS inhibitor herbicides (Neve et al., 2003). Due to its nonselective nature, it is used as a preplant herbicide. It is also used as a preharvest herbicide in mature crops or as postharvest herbicide in wheat stubble (Steinmann et al., 2012). In a reduced tillage system, glyphosate is commonly used as a preemergence herbicide to control the weeds which survived the previously harvested crops (Baylis, 2000).

Herbicidal application at the rosette stage is recommended by the majority of weed researchers for the successful control of *S. oleraceus*, as the effectiveness decreases at later life stages. Intensive use of glyphosate for weed control increases the threat of resistance development in *S. oleraceus* biotypes, creating the need for alternative herbicide options to control GR weed biotypes (Churchett et al., 2014). For example, a GR *Amaranthus rudis* L. biotype was not controlled by glyphosate, but some alternate postemergence herbicides provided up to 70% control of this biotype (Heap, 2002).

The limited availability of different herbicide groups can result in unwanted problems in the future regarding the management of weeds in both cropping and noncropping zones. In summer and winter crops, different herbicide management options are available. Herbicides with different modes of action should be used in rotation by farmers for the successful control of this weed. However, there is a lack of information regarding the formulation, the evaluation of new herbicide groups, successful herbicide combinations, and doses that reduce weed infestation not only in crop zones, but noncropped zones as well.

Biological control

Contrary to the strategies of chemical and cultural regulation, a few attempts have been made to biologically control *S. oleraceus*. The use of biological agents for integrated weed management (IWM) is probably the most important development for the future of weed management planning. In conjunction with herbicidal and cultural practices, exotic and native natural enemies can be used to reduce the incidence of weed. However, biological control introductions are often irreversible and permanent and can reduce target populations as well as nontargets. Therefore before any biological control project can be implemented, it is highly advisable to determine the range of organisms already present. Throughout Australia, two organisms have been found to be detrimental to *Sonchus* spp.: an undescribed species of eriophyid mite (Aceria sp.) and a rust fungus (*Miyagia pseudosphaeria*), which causes extensive curling and leaf-rolling damage which can reduce *S. oleraceus* growth up to 89%. Both appear to be highly specific to *S. oleraceus* and could be considered a conservative approach to biological control (McCarren and Scott, 2008). The leaf-gall fly (*Cystiphora sonchi*) produces viable galls and has been demonstrated to be a promising biocontrol agent for not only for *S. oleraceus* but also for *S. arvensis* L. and *S. asper* in Europe and was also released in Canada in 1981 to control various *Sonchus* species (Peschken, 1982). Further, *Ensina sonchi* L., an introduced fly in the Hawaii region, attacks the flowerheads of *S. oleraceus* L., thereby resulting in a 20%–30% reduction in the seed yield of *S. oleraceus* (Duan and Messing, 1998).

Conclusions

The dynamic nature of weed management results in changing requirements for weeds and local cultivation systems. To identify the priorities and opportunities, there is an immense need to explore the multifaceted impact of weeds on crop production and management. Flowers of *S. oleraceus* produce seeds through both selfing and outcrossing, and these seeds have the capability to disperse through wind and can travel long distances due to their lightweight (Hutchinson et al., 1984). Presently, there is no information available regarding the herbicide-resistant gene flow. Thus future research of *S. oleraceus*, biology and ecology, with focus on using ALS mutations as a biotechnological tool, can play a pivotal role for the development of strategic management of *S. oleraceus*. In addition to this, literature of *S. oleraceus* lacks information regarding the effect of different crop densities on growth, development, and seed production of resistant and susceptible biotypes which could aid in formulating IWM techniques.

As there are reports of development of resistance against glyphosate in *S. oleraceus*, particularly where zero-tillage practices are prevalent. In future, if farmers adopt an IWM approach to control this weed, the risk of resistance development can be greatly minimized. Furthermore, alternative approaches like glyphosate mix or nonchemical options must be explored to control seed setting of *S. oleraceus*. Also, in mixed cropping systems, resource competition of weeds with enhanced cropping systems can limit the growth of weeds. Therefore further studies to elucidate competitive traits (like allelopathy) involved in weed suppression must be carried. In light of biological control, various factors like temperature specificity of insect-pests also play a very crucial role in the establishment of the biological control agents. For instance, *Miyagia pseudosphaeria* is already being widespread in Australia but is absent in Canada due to extreme cold temperatures. Thus strategic molecular approaches, along with inclusive inoculation and temperature studies, are required to find more suitable pathotypes of the biocontrol agent species.

References

Abd El-Gawad, A.M., 2014. Ecology and allelopathic control of Brassica tournefortii in reclaimed areas of the Nile Delta. Egypt. Turk. J. Botany 38, 347–357.

Abd El-Ghani, M., Soliman, A., Fattah, R.A., 2014. Spatial distribution and soil characteristics of the vegetation associated with common succulent plants in Egypt. Turk. J. Botany 38, 550–565. Available from: https://doi.org/10.3906/bot-1309-49.

Abdullah, M., Rafay, M., Hussain, T., Ahmad, H., Tahir, U., Rasheed, F., et al., 2017. Nutritive potential and palatability preference of browse foliage by livestock in arid rangelands of Cholistan desert (Pakistan). JAPS, J. Anim. Plant. Sci. 27, 1656–1664.

Ackerson, R.C., 1981. Osmoregulation in cotton in response to water stress: ii. Leaf carbohydrate status in relation to osmotic adjustment. Plant. Physiol. 67, 489–493. Available from: https://doi.org/10.1104/pp.67.3.489.

Adkins, S.W., Wills, D., Boersma, M., Walker, S.R., Robinson, G., Mcleod, R.J., et al., 1997. Weeds resistant to chlorsulfuron and atrazine from the north-east grain region of Australia. Weed Res. 37, 343–349. Available from: https://doi.org/10.1046/j.1365-3180.1997.d01-56.x.

Akemo, M.C., Regnier, E.E., Bennett, M.A., 2000. Weed suppression in spring-sown rye (secale cereale): pea (pisum sativum) cover crop mixes. Weed Technol. 14, 545–549.

Andersen, R., 1968. Germination and Establishment of Weeds for Experimental Purposes. Weed Science Society of America, Urbana, IL.

Bajwa, A., Walsh, M., Chauhan, B., 2017. Weed management using crop competition in Australia. Crop. Prot. 95, 8–13.

Balick, M., Nee, M., Atha, D., 2000. Checklist of the Vascular Plants of Belize: With Common Names and Uses.

Ballare, C.L., Casal, J.J., 2000. Light signals perceived by crop and weed plants. F. Crop. Res. 67, 149–160. Available from: https://doi.org/10.1016/s0378-4290(00)00090-3.

Bangarwa, S.K., Norsworthy, J.K., Gbur, E.E., 2009. Integration of a Brassicaceae cover crop with herbicides in plasticulture tomato. Weed Technol. 23, 280–286. Available from: https://doi.org/10.1614/WT-08-165.1.

Bangarwa, S.K., Norsworthy, J.K., Gbur, E.E., Zhang, J., Habtom, T., 2011. Allyl isothiocyanate: a methyl bromide replacement in polyethylene-mulched bell pepper. Weed Technol. 25, 90–96. Available from: https://doi.org/10.1614/WT-D-10-00076.1.

Baylis, A.D., 2000. Why glyphosate is a global herbicide: strengths, weaknesses and prospects. Pest. Manag. Sci. 56, 299–308. Available from: https://doi.org/10.1002/(SICI)1526-4998(200004)56:4(299::AID-PS144)3.0.CO;2-K.

Boulos, L., 1976. Sonchus L. Flora Europaea, Volume 4 Plantaginaceae to Compositae (and Rubiaceae). pp. 327–328.

Brennan, J.P., Lemerle, D., Martin, P., 2001. Economics of increasing wheat competitiveness as a weed weapon, in: Proceedings of the 45th Annual Conference of the Australian Agricultural and Resource Economics Society. Adelaide, pp. 1–15.

Buchli, M., 1936. Oekologie der Ackerunkrauter der Nordost-Schweiz. (Beitr. Geobot. Landesaufn. No. 19). Bern.

Buhler, D.D., Mester, T.C., 1991. Effect of tillage systems on the emergence depth of giant (Setaria faberi) and green foxtail (Setaria viridis). Weed Sci. 39, 200–203. Available from: https://doi.org/10.1017/S0043174500071472.

Chauhan, B.S., Jha, P., 2020. Glyphosate resistance in *Sonchus oleraceus* and alternative herbicide options for its control in Southeast Australia. Sustainability 12, 8311.

Chauhan, B.S., Johnson, D.E., 2010. Responses of rice flatsedge (Cyperus iria) and Barnyardgrass (Echinochloa crus-galli) to rice interference. Weed Sci. 58, 204–208. Available from: https://doi.org/10.1614/WS-D-09-00050.1.

Chauhan, B.S., Gill, G., Preston, C., 2006. Factors affecting seed germination of annual sowthistle (Sonchus oleraceus) in southern Australia. Weed Sci. 54, 854–860. Available from: https://doi.org/10.1614/WS-06-047R.1.

Cho, M.-S., Kim, J.H., Kim, C.-S., Mejías, J.A., Kim, S.-C., 2019. Sow thistle chloroplast genomes: insights into the plastome evolution and relationship of two weedy species, Sonchus asper and Sonchus oleraceus (Asteraceae). Genes 10. Available from: https://doi.org/10.3390/genes10110881.

Churchett, J.D., McLean, A.R., Widderick, M.J., 2014. Exploring herbicide alternatives to delay glyphosate resistance developing in common sowthistle (Sonchus oleraceus) in the northern grain region. In: Baker M. (ed) Science, Community and Food Security: the Weed Challenge., in: Proceedings of 19th Australasian Weeds Conference. Hobart, Tasmania, p. 413.

Cici, S.Z.H., Van Acker, R.C., 2009. A review of the recruitment biology of winter annual weeds in Canada. Can. J. Plant. Sci. 89, 575–589. Available from: https://doi.org/10.4141/CJPS08131.

Dadd, C.V., 1962. Weed control in cereals: a review, in: Proc. 6th Br. pp. 123–140.

De Cauwer, B., Reheul, D., Nijs, I., Milbau, A., 2008. Management of newly established field margins on nutrient-rich soil to reduce weed spread and seed rain into adjacent crops. Weed Res. 48, 102–112. Available from: https://doi.org/10.1111/j.1365-3180.2007.00607.x.

Duan, J., Messing, R.H., 1998. Effect of Tetrastichus giffardianus (Hymenoptera: Eulophidae) on nontarget flowerhead-feeding tephritidis (Diptera: Tephritidae). Environ. Entromol. 1022–1028.

Elkamali, H.H., El-Kheir, M.S.M., Habeballa, R.S., Hamaza, N.B., Abdalla, I.E., Ahmedani, E. I., et al., 2011. Genetic relationships of two Sonchus Species collected from two locations in Khartoum state using RAPD markers. Curr. Res. J. Biol. Sci. 3, 95–99.

El-Kholy, R., Abouamer, W., Ayoub, M., 2013. Efficacy of some herbicides for controlling broad-leaved weeds in wheat fields. J. Appl. Sci. Res. 9, 945–951ISSN 1819-544X. Available from: http//www.aensiweb.com/jasr/jasr/2013/945-951.pdf.

Fenner, M., 1978. Susceptibility to shade in seedlings of colonizing and closed turf species. N. Phytol. 81, 739–744. Available from: https://doi.org/10.1111/j.1469-8137.1978.tb01648.x.

Garcke, A., 1972. Illustrierte Flora. Verlag Paul Parey, Hamburg, Germany.

Gill, N., 1938. The viability of weed seeds at various stages of maturity. Ann. Appl. Biol. 25, 447–456.

Girma, F.S., Krieg, D.R., 1992. Osmotic adjustment in Sorghum: II. Relationship to gas exchange rates. Plant. Physiol. 99, 583–588. Available from: https://doi.org/10.1104/pp.99.2.583.

Gomaa, N., Sherif, E., Hegazy, A., Hassan, M., 2012. Floristic diversity and vegetation analysis of Brassica nigra (L.) Koch communities.

Gomaa, N.H., Hassan, M.O., Fahmy, G.M., González, L., Hammouda, O., Atteya, A.M., 2014. Allelopathic effects of Sonchus oleraceus L. on the germination and seedling growth of crop and weed species. Acta Bot. Brasilica 28, 408–416.

Gomaa, N.H., Hassan, M.O., Fahmy, G.M., González, L., Hammouda, O., Atteya, A.M., 2015. Flavonoid profiling and nodulation of some legumes in response to the allelopathic stress of Sonchus oleraceus L. Acta Bot. Brasilica 29, 553–560.

Gordon, D.R., Onderdonk, D.A., Fox, A.M., Stocker, R.K., 2008. Consistent accuracy of the Australian weed risk assessment system across varied geographies. Divers. Distrib. 14, 234–242. Available from: https://doi.org/10.1111/j.1472-4642.2007.00460.x.

Gresta, F., Cristaudo, A., Onofri, A., Restuccia, A., Avola, G., 2010. Germination response of four pasture species to temperature, light, and post-harvest period. Plant. Biosyst.—An. Int. J. Deal. all Asp. Plant Biol. 144, 849–856. Available from: https://doi.org/10.1080/11263504.2010.523549.

Gupta, A., Mittal, C., 2012. Effect of allelopathic leaf extract of some selected weed flora of Ajmer district on seed germination of Triticum aestivum L. Sci. Res. Rep. 2, 311–315.

Halvorson, W., Guertin, P., 2003. "Factsheet for: Oxalis stricta L." USGS Weeds in the West project: Status of Introduced Plants in Southern Arizona Parks. Tucson, Arizona. Tucson, Arizona.

Harper J.L., 1957. The ecological significance of dormancy and its importance in weed control, in: Proceedings of the 4th International Congress of Crop Protection. pp. 415–420.

Harper, J.L., 1977. Population Biology of Plants. John L. Harper. Academic Press, London; New York.

Heap, I., 2002. International survey of herbicide resistant weeds. URL http://www.weedscience.com (accessed 6.19.02).

Henderson, C., 1998. Integrated weed management in vegetable production/Craig Henderson, Final report (Horticultural Research & Development Corporation); VG406. Horticultural Research & Development Corporation, Gordon, N.S.W.

Holm and Centre, L.G., 1977. The World's Worst Weeds: Distribution and Biology. East-West centre by the University Press of Hawaii.

Holzner, W., 1978. In: van der Maarel, E., Werger, M.J.A. (Eds.), Weed Species and Weed Communities BT—Plant Species and Plant Communities. Springer, Netherlands, Dordrecht, pp. 119–126.

Hutchinson, I.A., Colosi, J., Lewin, R.A., 1984. The biology of Canadian weeds: 63. Sonchus asper (L.) Hill and S. oleraceus L. Can. J. Plant. Sci. 64, 731–744. Available from: https://doi.org/10.4141/cjps84-100.

Hyatt, P., 2006. Sonchus oleraceus L. Pp. 275. In: Flora of North America North of Mexico (Flora of North America Editorial Committee, Eds. 1993 +) 19, 275.

Jakobsson, A., Eriksson, O., 2003. Trade-offs between dispersal and competitive ability: a comparative study of wind-dispersed Asteraceae forbs. Evol. Ecol. 17, 233–246. Available from: https://doi.org/10.1023/A:1025526903281.

Jia, P.Y., Zhang, L.X., Huang, Z., Tian, F.P., Hu, Y., Wu, G.L., 2018. Physiological characteristics of three wild sonchus species to prolonged drought tolerance in arid regions. Pak. J. Bot. 50, 9–17.

Jones, R., 1966. Effect of seed bed preparation on the weedflora of spring barley, in: Proc. 8th British Weed Control Conference. pp. 227–228.

Kadkhodaie, A., Zahedi, M., Razmjoo, J., Pessarakli, M., 2014. Changes in some anti-oxidative enzymes and physiological indices among sesame genotypes (Sesamum indicum L.) in response to soil water deficits under field conditions. Acta Physiol. Plant. 36, 641–650.

Krutz, L.J., Locke, M.A., Steinriede Jr., R.W., 2009. Interactions of tillage and cover crop on water, sediment, and pre-emergence herbicide loss in glyphosate-resistant cotton: implications for the control of glyphosate-resistant weed biotypes. J. Environ. Qual. 38, 1240–1247. Available from: https://doi.org/10.2134/jeq2008.0342.

Landi, R., 1975. Eflects of some nutritional factors on wheatweed competition. Rivta Agron. 9, 146–152.

Lewin, R.A., 1948. Biological flora of the British Isles Sonchus L. (Sonchus oleraceus L. and S. asper (L.) Hill). J. Ecol. 36, 203–223. Available from: https://doi.org/10.2307/2256663.

Liebman, M., Dyck, E., 1993. Weed management: a need to develop ecological approaches. Ecol. Appl. 3, 39–41. Available from: https://doi.org/10.2307/1941790.

Lorenzo, P., Hussain, M.I., González, L., 2013. In: Cheema, Z.A., Farooq, M., Wahid, A. (Eds.), Role of Allelopathy During Invasion Process by Alien Invasive Plants in Terrestrial Ecosystems BT—Allelopathy: Current Trends and Future Applications. Springer Berlin Heidelberg, Berlin, Heidelberg, pp. 3–21. Available from: https://doi.org/10.1007/978-3-642-30595-5_1.

Lorenzoni, G.G., 1963. La Vegetazione Infestante Del Mais in Friuli, Veneto E Lombardia. The Weed Vegetation of Maize in Frili, Veneto, and Lombardy 8, 1–54.

Manalil, S., Werth, J., Jackson, R., Chauhan, B.S., Preston, C., 2017. An assessment of weed flora 14 years after the introduction of glyphosate-tolerant cotton in Australia. Crop. Pasture Sci. 68, 773–780. Available from: https://doi.org/10.1071/CP17116.

Manalil, S., Haider Ali, H., Chauhan, B.S., 2018. Germination ecology of turnip weed (Rapistrum rugosum (L.) All.) in the northern regions of Australia. PLoS One 13, e0201023.

Matthei, O., Stuessy, T., Marticorena, C., 1993. La flora adventicia del Archipielago de Juan Fernandez. Gayana Bot. Chile Univ. Concepc. 2, 69–102.

Matthews J.M., 1994. Management of herbicide resistant weed populations. In: Powels, S.B., Holtum, J.A.M. (Eds.), 'Herbicide resistance in plants – biology and biochemistry'. Boca Raton, FLorida.

McCarren, K.L., Scott, J., 2008. Two biological control options for Sonchus oleraceus in Australia. In: van Klinken, R.D., Osten, V.A., Panetta, F.D., Scanlan, J.C. (Eds.), Hot Topics in the Tropics. Queensland Weeds Society, Cairns, Queensland, pp. 259–261. in: Proceedings of the 16th Australian Weeds Conference.

Medd, R.W., 1997. Biological change: weeds. In: 'Sustainable crop production in the sub-tropics—an Australian perspective.' Brisbane, Australia.

Melander, B., Rasmussen, I.A., Bàrberi, P., 2005. Integrating physical and cultural methods of weed control—examples from European research. Weed Sci. 53, 369–381. Available from: https://doi.org/10.1614/WS-04-136R.

Mobli, A., Matloob, A., Chauhan, B.S., 2019. The response of glyphosate-resistant and glyphosate-susceptible biotypes of annual sowthistle (*Sonchus oleraceus*) to mungbean density. Weed Sci. 67, 642–648. Available from: https://doi.org/10.1017/wsc.2019.47.

Moore, M.J., Gillespie, T.J., Swanton, C.J., 1994. Effect of cover crop mulches on weed emergence, weed biomass, and soybean (Glycine max) development. Weed Technol. 8, 512–518. Available from: https://doi.org/10.1017/S0890037X00039609.

Neve, P., Diggle, A.J., Smith, F.P., Powles, S.B., 2003. Simulating evolution of glyphosate resistance in Lolium rigidum II: past, present and future glyphosate use in Australian cropping. Weed Res. 43, 418–427. Available from: https://doi.org/10.1046/j.0043-1737.2003.00356.x.

Norsworthy, J.K., McClelland, M., Griffith, G., Bangarwa, S.K., Still, J., 2011. Evaluation of cereal and brassicaceae cover crops in conservation-tillage, enhanced, glyphosate-resistant cotton. Weed Technol. 25, 6–13. Available from: https://doi.org/10.1614/WT-D-10-00040.1.

Ollivier, M., Kazakou, E., Corbin, M., Sartori, K., Gooden, B., Lesieur, V., et al., 2020. Trait differentiation between native and introduced populations of the invasive plant Sonchus oleraceus L. (Asteraceae). NeoBiota 55, 85–115.

Osten, V.A., Walker, S.R., Storrie, A., Widderick, M., Moylan, P., Robinson, G.R., et al., 2007. Survey of weed flora and management relative to cropping practices in the north-eastern grain region of Australia. Aust. J. Exp. Agric. 47, 57–70.

Pareja, M.R., Staniforth, D.W., Pareja, G.P., 1985. Distribution of weed seed among soil structural units. Weed Sci. 33, 182–189. Available from: https://doi.org/10.1017/S0043174500082060.

Patterson, D.T., 1986. Research Methods in Weed Science. Southern Weed Science Society of America, Illinois.

Peschken, D., 1982. Host specificity and biology of cystiphora sonchi [Dip.: Cecidomyiidae], a candidate for the biological control of Sonchus species. BioControl 27, 405–415. Available from: https://doi.org/10.1007/BF02372063.

Preston, C., 2000. Herbicide mode of action and herbicide resistance. Aust. Weed Manag. Syst. 209–226.

Roberts, H.A., Neilson, J.E., 1981. Seed survival and periodicity of seedling emergence in twelve weedy species of compositae. Ann. Appl. Biol. 97, 325–334. Available from: https://doi.org/10.1111/j.1744-7348.1981.tb05119.x.

Shenstone, J.C., 1912. The flora of London building-sites. J. Bot. 50, 117–124.

Singh, S., 2010. Phytochemical investigation of sonchus oleraceus leaves and citrullus colocynth root.

Song, J.-S., Kim, J.W., Im, J.H., Lee, K.J., Lee, B.W., Kim, D.S., 2017. Effects of single- and multiple-weed interference on soybean yield in the far-eastern region of Russia. Weed Sci. 65.

Steinmann, H., Dickeduisberg, M., Theuvsen, L., 2012. Uses and benefits of glyphosate in German arable farming. Crop. Prot. 42, 164–169. Available from: https://doi.org/10.1016/j.cropro.2012.06.015.

Thurston, J., 1976. Weeds in cereals in relation to agricul-tural practices. Ann. Appt. Biot. 83, 338–341.

van der Meulen, A., Chauhan, B.S., 2017. A review of weed management in wheat using crop competition. Crop. Prot. 95, 38–44. Available from: https://doi.org/10.1016/j.cropro.2016.08.004.

Weber, E., 2017. Invasive Plant Species of the World: A Reference Guide to Environmental Weeds, 2nd ed. University of Postdam, Germany.

Werth, J., Thornby, D., Walker, S., 2011. Assessing weeds at risk of evolving glyphosate resistance in Australian sub-tropical glyphosate-resistant cotton systems. Crop. Pasture Sci. 62, 1002–1009.

Widderick, M., Sindel, B., Walker, S., 1999. Distribution, importance and management of Sonchus oleraceus (common sowthistle) in the northern cropping region of Australia., in: A. C. Bishop, M. Boersma and C. D. Barns (Eds.), Twelfth Australian Weeds Conference. Tasmania, Australia.

Widderick, M., Sindel, B., Walker, S., Jacob, H., Dodd, J., Moore, J.H., 2002. Emergence of Sonchus oleraceus (common sowthistle) is favoured under zero tillage farming systems.

Widderick, M., Walker, S., Sindel, B.M., 2004. Better management of "Sonchus oleraceus" L. (common sowthistle) based on the weed's ecology, in: 14th Australian Weeds Conference Papers and Proceedings: Weed Management—Balancing People, Planet, Profit. Weed Society of New South Wales, pp. 535–537 LB-E1 Refereed Scholarly Conference Pub.

Widderick, Mi.J., Walker, S.R., Sindel, B.M., Bell, K.L., 2010. Germination, emergence, and persistence of Sonchus oleraceus, a major crop weed in subtropical Australia. Weed Biol. Manag. 10, 102–112. Available from: https://doi.org/10.1111/j.1445-6664.2010.00370.x.

Williams, M.M., 2009. Within-season changes in the residual weed community and crop tolerance to interference over the long planting season of sweet corn. Weed Sci. 57, 319–325. Available from: https://doi.org/10.1614/WS-08-164.1.

Yin, J., Kwon, G.J., Wang, M.H., 2007. The antioxidant and cytotoxic activities of Sonchus oleraceus L. extracts. Nutr. Res. Pract. 1, 189–194. Available from: https://doi.org/10.4162/nrp.2007.1.3.189.

Chapter 18

Sorghum halepense

Caleb C. Squires and Michael J. Walsh
University of Sydney, School of Life and Environmental Science, Sydney Institute of Agriculture, Camden, NSW, Australia

Introduction

Sorghum halepense (L.) Pers, commonly known as Johnsongrass, is a warm-season C_4 grass (Downton, 1975) which provides a role as an animal forage in some areas. Unfortunately, it is also an environmental weed and a major weed in crops, being labeled as one of the worst weeds worldwide (Holm et al., 1977). *Sorghum halepense* is a problem in a diverse range of crops and is characterized by proliferous seed production and extensive rhizomatous growth. When such characteristics are combined with herbicide resistance, control of *S. halepense* becomes a serious challenge.

Over the last 40 years, there have been at least five reviews on the biology and management of *S. halepense* (Ceseski et al., 2017; Travlos et al., 2019; Warwick and Black, 1983; Monaghan, 1979; Peerzada et al., 2017). Three of those reviews date from the last 3 years, showing an increasing interest in this weed as it becomes more problematic. Such reviews contain a wealth of information, and little value is to be gained by repetition in full here. However, it is noteworthy that the majority of chemical management references and suggestions in these reviews, date between the 1960's and the late 1980's. More recent herbicide-centric publications deal primarily with reporting and characterizing cases of herbicide resistance. The chemical control suggestions have remained mostly constant for over 30 years, despite increasing cases of herbicide resistance (Ceseski et al., 2017; Heap, 2020). Few new management techniques for *S. halepense* have been reported, and nonchemical recommendations typically revert to suggestions first publicized in the early 1900's (Cates and Spillman, 1907)

The objective of this chapter is to determine key biological attributes of *S. halepense* that are potential targets for current and future weed control practices. These will be identified through a review of relevant literature and the development of a clear understanding of the growth and development of

Biology and Management of Problematic Crop Weed Species. DOI: https://doi.org/10.1016/B978-0-12-822917-0.00012-4

this weed as it occurs in agriculture and the environment. Detailed management suggestions for the plethora of crops infested by *S. halepense* would be beyond the scope appropriate here, especially given the worldwide distribution of *S. halepense*, with different production systems, environments, regulations, herbicide product availabilities, and product labels. Consequently, the chemical management section of this chapter will instead highlight a few of the dominant herbicide groups that are currently useful and list some factors for consideration with herbicide use. Nonchemical management alternatives with proven value will be reviewed, as well as potential new approaches for additional investigation.

A weed of natural and disturbed systems

Sorghum halepense thrives in warm conditions, with optimal germination and growth at 30°C (Hull, 1970). It prefers moist but well-drained growing conditions, and excels in regrowth, or colonizing of disturbed areas. *Sorghum halepense* outcompetes native species with its size and rapid growth (Schwinning et al., 2017), as even a single seedling is capable of spreading during warm months at a rate of 1.3 m^2/month (Horowitz, 1973). Although initially desired as a forage grass, once its weedy characteristics were understood, it became unwelcome in pastures (Mcwhorter, 1971). *Sorghum halepense* also infests noncrop areas such as field margins, ditches, and irrigation canals. While infestations in all the previously mentioned situations are troublesome, problems are most severe in cropping systems.

The biggest problems posed by *S. halepense* are in cropping systems, due to competition and interference with crops. Crop yield and quality are reduced by uncontrolled or poorly controlled weeds, causing direct losses in-crop value. Management of *S. halepense* also incurs costs associated with methods of control and from forced management decisions regarding crop and rotation options. Small-scale cropping practices, such as organic vegetable production, may have some advantages however with inherent flexibility to use alternative, potentially labor intensive nonchemical weed control techniques.

In large-scale cropping systems, including grain and cotton (*Gossypium hirsutum L.*) production, *S. halepense* is of particular concern for a few primary reasons. Broad-acre agriculture is dependent on mechanization and savings associated with economies of scale. Consequently, labor intensive but effective weed control methods such as manual pulling, repeated cultivation, and solarization become unfeasible due to cost. The result is increased dependence on herbicides for both fallow and in-crop weed control. Heavy reliance on the most inexpensive herbicides invariably results in evolution of herbicide resistance, which in turn, directly threatens production and sustainability of long-term large-scale cropping. With few alternative weed control methods in use in large-scale cropping systems, there is potential for *S. halepense* to be even more problematic in the future as herbicide resistance proliferates.

Biology

Description

As described by Warwick and Black (1983) and illustrated in Fig. 18.1, *S. halepense* is a perennial grass which grows 0.5–2.5 m tall, with rhizomes up to 1 cm in diameter and 2 m in length. Leaves are cauline, and often sheaths have a waxy white secretion at the base. Leaf veins are prominent, whereas the leaf surface is hairless and ribbed. Margins are open, ligules membranous, auricles are absent; blades are rolled in bud, flat, bright green, or slightly purple, 20–60 cm long, and 1.0–3.3 cm wide. The Inflorescence is a large, erect panicle, at first compact, then later spreading and open, up to 50 cm long and 20 cm wide. At maturity, the grain or caryopsis remains enclosed by reddish brown to shiny black glumes 4.0–6.6 mm long, 2.0–2.6 mm wide. Glumes are also marked with fine lines on the surface. Plants regenerate from partial rhizomes, preferentially from the apical bud, although germination can occur from any node (Hull, 1970). Initial growth

FIGURE 18.1 A) Rhizome, B) seedling, and C) inflorescence of *S. halepense*.

from rhizomes is more vigorous than growth of young shoots which originated from seeds (McWhorter, 1961).

The morphology of *S. halepense* contributes to its weediness and impact on crops. At early growth stages, young seedlings appear similar to corn (*Zea mays* L.) or sorghum (*Sorghum bicolor* (L.) Moench), but with narrower leaves. Positive seedling identification in such cases depends on uprooting and examination of the seed still attached to the roots, which complicates identification and timely management. The rapid growth, spreading nature, and tall stature ensures a competitive advantage over low stature and slow growing crops. Vigorous growth from even small rhizome fragments also complicates management, as improper and infrequent tillage can spread and increase *S. halepense* biotypes which can grow faster than crop seedlings.

Distribution and expansion

Sorghum halepense distribution is worldwide, in all warm-season agricultural areas, and other than water availability, is restricted in range only by freezing temperatures in the −3°C to −9°C range (Hull, 1970; McWhorter, 1972; Stoller, 1977). Aided by climate change, and the resulting warmer weather, models suggest that *S. halepense* distribution will continue to advance poleward in range, by a further 200–600 km by the year 2100 (McDonald et al., 2009).

Range expansion is also aided by apparent high genetic diversity (Karimi and Saeidi, 2016), with high gene flow maintaining variation and ability to adapt. Individual biotypes have been noted to vary in water use efficiency, carbon assimilation rate, and other photosynthetic traits which provide advantages in differing crop and noncrop climates (Kelly et al., 2020). Another example of biotype variation is rhizome depth, which has been identified as key to winter survival of *S. halepense* biotypes in cold temperatures (Warwick et al., 1986), allowing expansion into otherwise hostile climates. Gene flow is made possible through key reproductive characteristics. *Sorghum halepense* is able to self-fertilize, but outcrossing via wind dispersion was found to be the dominant mode of pollination, aided also by a wide variety of insects (Kaur and Soodan, 2017). Pollen can be viable up to 10 h after anther dehiscence (Kaur and Soodan, 2017), providing ample time for long distance gene flow.

Origin and hybridization

Extensive molecular analysis has confirmed that *S. halepense* ($2n = 40$ or $2n = 20$) is the result of a hybridization between cultivated S*orghum bicolor* of African origin and a perennial wild grass *Sorghum propinquum* of Southeast Asian origin (Paterson et al., 1995, 2020), inheriting both high seed production and extensive rhizome growth. The lineage and compatibilities create possibilities for direct gene flow from sorghum (Arriola and Ellstrand, 1996), made worse by feral sorghum hybrids which naturally segregate for male sterility (Ohadi et al., 2018), providing opportunities for hybridization with *S. halepense*. The problem is twofold in that herbicide resistance, and potentially local adaptations, are passed between the two species, compounding weed management problems (Paterson et al., 2009; Ohadi et al., 2017).

Seed dormancy, longevity, and production

Seed production from one plant has been reportedly as high as 28,000 (Horowitz, 1973), and seeds are distributed by a variety of methods. In an undisturbed corn crop, 98% of *S. halepense* seeds were found to naturally fall within 5 m of the mother plant, but mechanized harvesting equipment moved seeds up to 22 m (Barroso et al., 2012), illustrating again the additional problems *S. halepense* poses in large-scale cropping which are dependent on such equipment. Movement via water is also common, as *S. halepense* seeds float, and have been found to be viable after spending 12 months in irrigation water (Erbaş and Doğan, 2017). Even a single-irrigation event has been shown to move seed 300 m down a vegetated irrigation channel (Rudi et al., 2018).

In Stoneville, Mississippi, 48% of *S. halepense* seeds were found to be viable, based on germination and tetrazolium tests, after field burial for 5.5 years (Egley and Chandler, 1983). In Argentina, seeds at depth (22.5 cm) survived more than 6 years, whereas only surviving less than 2.5 years at the surface or at 7.5 cm (Leguizamon, 1986). The primary cause of dormancy is the physical restriction of the seed coat, although changes in temperature also stimulate germination (Taylorson and McWhorter, 1969). In laboratory tests, seeds physically nicked, or treated with sodium hypochlorite (NaOCl), sulfuric acid (H_2SO_4), or potassium nitrate (KNO_3) solutions broke dormancy; however, the total germination percentage ($<35\%$) was relatively low (Balicevic et al., 2016). Due to long seed dormancy, unexpected outbreaks can occur, years after *S. halepense* populations have been brought under control.

Impact on crop yield, pests, and livestock

Sorghum halepense primarily affects crop growth through competition for resources, but there are also additional risks from pests and risks for livestock. The largest threat in crops is due to infestations which can cause massive reductions in-crop yield. For example, corn yield reductions of up to 86%, along with substantial reductions in grain protein and starch content due to *S. halepense* infestations have been reported (Karkanis et al., 2020).

Similarly, full season infestations in soybean (*Glycine max* L. Merrill) crops reduced yields by up to 80% in Tennessee (Williams and Hayes, 1984), whereas competition with cotton in California for 25 weeks reduced cotton yield by 90% (Keeley and Thullen, 1989). Sugarcane (*Saccharum spp.*) fields in Louisiana which were infested with *S. halepense* had cane yield reduced by 36% (Ali et al., 1986).

Yield can also be affected by cereal pests or pathogens which are capable of using *S. halepense* as an alternate host. Therefore even if *S. halepense* only infests field margins, it acts as a green bridge to spread diseases and pests to crops. Some examples are Goss's bacterial wilt and leaf blight of corn (Ikley et al., 2015) which can cause up to 44% yield loss in the United States (Carson and Wicks, 1991), sorghum ergot in Mexico (Montes-belmont et al., 2003), corn rootworm in Argentina (Cabrera Walsh, 2007), spider-mite in Texas (Smith and Otto, 1973), and fall armyworm in Brazil (Moraes et al., 2020).

While the focus here is primarily the impact of *S. halepense* on crop production, grazing can play an integral part in management, but not without risks to livestock. Animals grazing on *S. halepense* can be poisoned in two different ways. First, *S. halepense* is a plant which accumulates high nitrate levels in its foliage (Zoltani, 2018), which is converted to nitrite during animal digestion, leading to methemoglobin formation and subsequent suffocation from lack of oxygen in the blood. Second, when *S. halepense* plants undergo severe stress such as drought or cold, high levels of cyanogenic glucosides accumulate (Arnold et al., 2014). These compounds are then broken down into cyanide and also inhibit the oxygen carrying capacity of the blood (AL-Sultan, 2003; Gensa, 2019). Grazing during adverse conditions can be avoided, but complicates grazing management.

Residues of *S. halepense* break down and leach biological compounds into the soil and can suppress or impede other plants from growing. Effects have been documented in wheat (*Triticum aestivum* L.), where uncharacterized extracts of *S. halepense* were found to supress seedling growth in laboratory conditions (Nouri et al., 2012). The allelopathic effect of volatile oils distilled from *S. halepense* was also investigated in laboratory conditions on wheat, soybean, *Echinochloa crus-galli (L.) P. Beauv.*, and *Amaranthus retroflexus* L, and the responsible chemicals were identified as various sesquiterpenes that also increased bacterial load in surrounding soil (QingHui et al., 2009).

Management

Prevention

Potential sources of *S. halepense* infestations include contaminated crop seed, importation in plant or soil debris on dirty farming equipment, movement by animals, and travel through flood or irrigation water (Ball, 1902). Risks associated with these sources are easily mitigated in comparison with efforts to eradicate established populations, but require attention to seed quality, equipment and machinery hygiene, installation and maintenance of irrigation water screens, and fence-line weed control.

Chemical management

Chemicals have long provided a valuable tool for management of *S. halepense* in cropping systems, and currently are heavily relied on, particularly in large-scale agriculture. Unfortunately, under strong selection pressure, herbicide resistance has become increasingly problematic, with potential to become worse. A comprehensive review of all herbicides and modes of action available for *S. halepense* is beyond the scope appropriate here. There are a few key herbicide groups which have been, and continue to be especially important for *S. halepense* control. These include inhibitors of 5-enolpyruvylshikimate-3-phosphate synthase, acetyl-CoA carboxylase, and acetolactate synthase, all of which have known *S. halepense* instances of herbicide resistance (Heap, 2020). An important management principle to prolong the usefulness of these and other herbicides for *S. halepense* control, is to minimize the risk of a population evolving resistance by rotation and mixing of compatible herbicides (Beckie and Reboud, 2009).

Nonchemical management

With the increasing challenges brought by herbicide resistance, nonchemical weed management tactics are increasing in value. When each tactic is looked at in isolation, the effect on weeds may seem unimpressive compared to typically high herbicide efficacy. However, as part of an integrated weed management program, the combined effect can greatly reduce weed pressure and even prolong the usefulness of remaining herbicides.

Soil solarization

The power of the sun can be harnessed via soil solarization, a weed management tactic which can be valuable in some high value or organic cropping situations. During summer, the soil is covered by a black, or preferably clear plastic material, and left for extended periods of time. Incoming shortwave solar radiation penetrates the soil covering and is re-emitted as longwave radiation which cannot pass back through the covering (Mahrer, 1979). The solar energy is then trapped and effectively raises the temperature of the soil. Under optimal moist conditions, soil temperatures can rise over 10°C more than ambient soil temperature and can effectively kill pathogens, weeds and even *S. halepense* rhizomes and seeds. *Sorghum halepense seeds* were 100% nonviable after solarization for 4 weeks in pot experiments (Khan et al., 2012; Kapoor, 2020). In heavily infested okra (*Abelmoschus esculentus* L.) in the field, 8 weeks of solarization reduced *S. halepense* cover from 42% to 13% (Law et al., 2008). Solarization in cauliflower (*Brassica oleracea* L. var. Botrytis) fields for 8 weeks reduced *S. halepense* density from 10 plants/m^2 to 1.6 plants/m^2 (Khan et al., 2012). Negative aspects of this method is the long time period needed for treatment, cost and disposal of the copious amounts of plastic used, and loss of crop or productivity while land is undergoing solarization unless crops are planted strategically into holes cut into the film. A more recent development in solarization is the introduction and use of biodegradable films to reduce weeds (Martín-Closas et al., 2017), which have been found to last long enough to have an effect on some weed species (Martín-Closas et al., 2017; Candido et al., 2012; Bonanomi et al., 2008). Some tests have also been conducted on biodegradeable films which can be applied as a spray which then hardens to form a temporary solid layer (Sartore et al., 2013). Detailed investigations on the use of biodegradable solarization on *S. halepense* rhizomes and seed is scarce and should be further pursued in various cropping scenarios to determine efficacy and cost.

Crop competition

Manipulating crop density or row spacing is a traditional method of reducing weed biomass. Sorghum crops planted at high densities were found to out compete *S. halepense* (Hoffman et al., 2002) and in mango (*Mangifera indica* L.) orchards, planting bushy legumes under trees reduced *S. halepense* biomass by up to 90% (Agreda et al., 2018). Soybeans in narrow 0.25 m rows with high *S. halepense* density of 32 plants/m^2 had approximately 100 kg/ha higher yield than soybeans in 1 m row spacings (Williams and Hayes, 1984)., narrow row spacings in sugarcane of 60 and 90 cm reduced *S. halepense* biomass and density by 8% compared to 120 cm rows (Munsif et al., 2015); however, cane yield also decreased by 13%.

Allelopathy

Plant residues of some brassica species are known to leach allelopathic residues, which suppress weed growth. In laboratory tests, dry shoot extract of rapeseed (*Brassica napus L. oleifera DC.*) reduced germination of *S. halepense* seeds by half, whereas in field conditions, rapeseed residues suppressed *S. halepense* germination and growth, with a 63% or higher level of control across several years, even at up to 60 days after incorporation (Uremis et al., 2009).

Fallow, grazing, and mowing

One of the earliest texts on management of *S. halepense* (Cates and Spillman, 1907) dates from before herbicides were an option for control. No specific weed densities were given, but a heavily infested paddock was split in half, with each half undergoing separate management practices. One half was kept mown short for 3 years, whereas the other half was planted with cotton for 2 years, then corn. Then both were planted with cotton. *Sorghum halepense* which had been repeatedly mowed developed very shallow root systems and was easily cultivated and eradicated. However, weed control failed in the other half of the field, due to the deep root systems that *S. halepense* had developed, likely from the occasional cultivation during the intervening years. Subsequent research has reaffirmed that validity of frequent mowing, grazing, or clipping to deplete rhizomes (Horowitz, 1972; Burt, 1954), which can then be coupled with strategic tillage to bring rhizome fragments to the surface where they die from exposure (Hull, 1970). Additional benefits of such a strategy are reduced winter survival, which depends on rhizome depth (Warwick et al., 1986) and a reduction of inputs to the soil seed bank from mowing or grazing before flowering and subsequent seed production. An additional consideration is that *S. halepense* is known to divert extra resources to root and root length growth in response to low water availability (Acciaresi and Guiamet, 2010); so this tactic may prove less effective in short-term drought conditions.

Biological control

Biological control methods involve using naturally occurring pests or diseases in sufficient numbers to reduce weed fitness and survival. Successful biological control stories are few, but a prime example in Australia is the use of the moth *Cactoblastis cactorum* to target prickly pear *(Opuntia stricta* Haw.) which had overrun most of north east Australia (Dodd, 1929). Introduction of foreign species is not without risks, and the majority of biological control efforts with *S. halepense* have investigated pests or diseases native to the area, but deployed in increased numbers. An example is *Schizaphis graminum*, also known as greenbug. Specific *S. graminum* biotypes are reportedly not virulent on cultivated sorghum but are virulent on *S. halepense* and reduced both growth and plant water content in greenhouse experiments in Texas (Gorena, 2004). Although *S. halepense* was not controlled by greenbug, it was theorized that reduced fitness could help increase other control measures. While initial studies using *Bipolaris sorghicola* as a fungal mycoherbicide were promising (Winder and Van Dyke, 1990), with one isolate (BS1) showing high virulence on *S. halepense* seedlings which resulted in 100% death in greenhouse experiments. However, challenges were experienced when trying to obtain enough material for inundative inoculation in field studies. Another potential fungal biocontrol *Sporisorium cruentum reduced S. halepense* rhizome and aerial parts by 43% and 67%, respectively, when individual plants were inoculated in a greenhouse setting. While only briefly overviewed here, biological control for *S. halepense* is an area where further investigation is warranted.

In summary, existing available weed control options all have severe limitations. Herbicides, while inexpensive, are challenged by increasing herbicide resistance. Frequent tillage is expensive and destructive to soil structure and soil conservation. Grazing can have negative effects from toxicity and mowing consistently over time lowers the productivity by precluding the growth of crops. Allelopathy and biological control options show potential, but are currently unable to effectively control *S. halepense*. In short, additional development is needed for new weed control technologies.

Site-specific weed control

A new area of research is that of site-specific weed control. Various technological advances in image analysis, computer software, and other detection technologies provide opportunities to recognize and spatially target individual weeds instead of treating the whole field. Such methods may rely on optical reflectance, color, or more complex image analysis to specifically target individual weeds.

Recognition and detection of *S. halepense* has been demonstrated in corn using ultrasonic sensors (Andújar et al., 2011), light detection and ranging techniques (Andújar et al., 2013), and basic image analysis based on row detection (Burgos-Artizzu et al., 2011). One of the most difficult scenarios is differentiation at early growth stages from corn or sorghum due to morphological similarities. However, as cameras and computer processing speed increases, analysis and identification of individual plants should be possible, given a few unique features. For example, as part of domestication, most corn varieties have been bred with high apical dominance and thus reduced tillering (Zhang et al., 2019; Kebrom and Brutnell, 2015). The domestication of sorghum likewise resulted in few tillers in comparison with the extensive tillering and different plant architecture *S. halepense* displays even quite early in development. Algorithms trained on images which capture those differences should in theory be able to differentiate and offer selectivity for a multitude of weed control options. Detection and recognition of *S. halepense* is an area of research priority given the potential for worldwide impact and the value of additional opportunities to manage herbicide resistance.

Harvest weed seed control

Harvest weed seed control (HWSC) is the collection and/or destruction of weed seeds at crop harvest to reduce inputs to the soil seed bank

(Walsh et al., 2013). Such methods include chaff collection, chaff lining, chaff tramlining, and windrow burning. In particular, mechanical destruction of seeds via an integrated impact mill [e.g., integrated Harrington seed destructor (iHSD)], is effective in killing 99% of seeds. As an example of potential efficacy, *Lolium rigidum* (Gaud.) densities in Australian wheat crops were reduced in by an average of 60% due to HWSC practices in one season (Walsh et al., 2017). Over time, due to gradual reduction of the soil seed bank, HWSC practices continually reduce weed numbers. Initial studies show strong potential for targeting *S. halepense* via HWSC, with 96% of seeds retained at or above canopy level at harvest time in common crops such as corn (Walsh et al., 2018). Narrow windrow burning was found to cause lethal temperatures for *S. halepense* seeds (Norsworthy et al., 2020), and when seeds of *S. halepense* were processed through a stationary iHSD, 99.9% of seeds were destroyed (Schwartz-lazaro and Norsworthy, 2018). Further research on HWSC methods on *S. halepense* will provide more information on the suitability and effectiveness in various crops.

Conclusion

Sorghum halepense poses management challenges worldwide, due to its unique biology and weedy characteristics. Some effective management methods such as grazing or mowing have not changed over the last century, but may have limited applicability. Chemical control methods have dominated recent management practices, and despite increasing cases of herbicide resistance, still remain effective in the majority of cases. Nonchemical management methods are undergoing a renaissance of sorts, with new methods or developments made in areas such as biological control, soil solarization, HWSC, and site-specific control technologies. As *S. halepense* adapts and expands its range, additional creativity in tactics and methods will be required to ensure long-term sustainability of cropping systems.

References

Acciaresi, H.A., Guiamet, J.J., 2010. Below- and above-ground growth and biomass allocation in maize and *Sorghum halepense* in response to soil water competition. Weed Res. 50 (5), 481–492.

Agreda, F.M., De Coss, A.L., Pohlan, J., 2018. Bio-control of *Rottboellia cochinchinensis* (Lour.) and *Sorghum halepense* (L.) Pers. with multipurpose bushy legumes in "Ataulfo" mango (Mangifera indica L.). Indian J. Horticult. 8 (1).

AL-Sultan, S.I., 2003. *Sorghum halepenses* and its cyanide content. Pak. J. Nutr. 2 (3), 123–124.

Ali, A.D., et al., 1986. Effects of Johnsongrass (*Sorghum halepense*) density on sugarcane (Saccharum officinarum) yield. Weed Sci. 34 (3), 381–383.

Andújar, D., et al., 2011. Weed discrimination using ultrasonic sensors. Weed Res. 51 (6), 543–547.

Andújar, D., et al., 2013. Potential of a terrestrial LiDAR-based system to characterise weed vegetation in maize crops. Comput. Electron. Agric. 92, 11–15.

Arnold, M., et al., 2014. Cyanide poisoning in ruminants. Agric. Nat. Resour. Publ. 168, 1–2.

Arriola, P.E., Ellstrand, N.C., 1996. Crop-To-weed gene flow in the genus *Sorghum* (Poaceae): spontaneous interspecific hybridization between Johnsongrass, *Sorghum halepense*, and crop Sorghum, *S. bicolor*. Am. J. Botany 83 (9), 1153–1159.

Balicevic, R., Ravlic, M., Balic, A., 2016. Dormancy and germination of Johnson grass seed (*Sorghum halepense* (L.) Pers.). J. Cent. Eur. Agric. 17 (3), 725–733. Available from: http://jcea.agr.hr/volumes.php?search = Article%3A1769.

Ball, C.R., 1902. Johnson Grass: Report of Investigations Made During the Season of 1901, vol. 11. US Government Printing Office.

Barroso, J., et al., 2012. Johnsongrass (*Sorghum halepense*) seed dispersal in corn crops under mediterranean conditions. Weed Sci. 60 (1), 34–41.

Beckie, H.J., Reboud, X., 2009. Selecting for weed resistance: herbicide rotation and mixture. Weed Technol. 23 (3), 363–370. Available from: https://www.cambridge.org/core/product/identifier/S0890037X00000932/type/journal_article.

Bonanomi, G., et al., 2008. Soil solarization with biodegradable materials and its impact on soil microbial communities. Soil. Biol. Biochem. 40 (8), 1989–1998.

Burgos-Artizzu, X.P., et al., 2011. Real-time image processing for crop/weed discrimination in maize fields. Comput. Electron. Agric. 75 (2), 337–346. Available from: http://doi.org/10.1016/j.compag.2010.12.011.

Burt, E.O., 1954. The Control of Johnsongrass (Sorghum halepense L. Pers.) by Herbicides and cultural Practices. The Ohio State University.

Cabrera Walsh, G., 2007. *Sorghum halepense* (L.) Persoon (Poaceae), a new larval host for the South American corn rootworm Diabrotica speciosa (Germar) (Coleoptera: Chyrsomelidae). Coleopterists Bull. 61 (1), 83–84.

Candido, V., et al., 2012. Effect of different solarizing materials on weed suppression and lettuce response. Phytoparasitica 40 (2), 185–194.

Carson, M.L., Wicks, Z.W., 1991. Relationship between leaf freckles and wilt severity and yield losses in closely related maize hybrids. Phytopathology 81 (1), 95.

Cates, J.S., Spillman, W.J., 1907. A Method of Eradicating Johnson Grass. U.S. Department of Agriculture, no. 279.

Ceseski, A., Al-Khatib, K., Dahlberg, J.A., 2017. Biology and management of Johnsongrass (*Sorghum halepense*). Biology and Management of Johnsongrass (Sorghum halepense). UC Agriculture & Natural Resources.

Dodd, A.P., 1929. The Progress Of Biological Control Of Pricklypear In Australia. Government Printer, Brisbane, Australia.

Downton, W.J.S., 1975. The occurrence of C4 photosynthesis among plants. Photosynthetica 9, 96–105.

Egley, G.H., Chandler, J.M., 1983. Longevity of weed seeds after 5.5 years in the stoneville 50-year buried-seed study. Weed Sci. 31 (2), 264–270.

Erbaş, F., Doğan, M.N., 2017. Viability and buoyancy of seeds of some weed species in water. Turkish J. Weed Sci. 20 (1), 48–60.

Gensa, U., 2019. Review on cyanide poisoning in ruminants. J. Biol., Agric. Healthc. 9 (6). Available from: www.iiste.org.

Gorena, R.L., 2004. Characterization of Schizaphis graminum (Rondani) (Homoptera: Aphididae) Biotype Evolution via Virulence and Fitness on Sorghum bicolor (L.) Moench and Sorghum halepense (L.) Persoon. Texas A&M University. Available from: http://www.ncbi.nlm.nih.gov/pubmed/18944109.

Heap, I.M., 2020. *The International Survey of Herbicide Resistance Weeds, Online*, viewed 24 July 2020, <www.weedscience.com>.

Hoffman, M.L., Buhler, D.D., Regnier, E.E., 2002. Utilizing Sorghum as a functional model of crop–weed competition. II. Effects of manipulating emergence time or rate. Weed Sci. 50 (4), 473–478.

Holm, L.R.G., et al., 1977. The World's Worst Weeds. Distribution and Biology. University Press of Hawaii, Honolulu, Hawaii.

Horowitz, B.Y.M., 1972. Effects of frequent clipping on three perennial weeds, *Cynodon dactylon* (L.) Pers., *Sorghum halepense* (L.) Pers. and *Cyperus rotundus* L. Exp. Agric. 8 (3), 225–234.

Horowitz, M., 1973. Spatial growth of *Sorghum Halepense* (L.) Pers. Weed Res. 13 (2), 200–208.

Hull, R.J., 1970. Germination control of Johnsongrass rhizome buds. Weed Sci. 18 (1), 118–121.

Ikley, J.T., Wise, K.A., Johnson, W.G., 2015. Annual Ryegrass (*Lolium multiflorum*), Johnsongrass (*Sorghum halepense*), and large Crabgrass (*Digitaria sanguinalis*) are alternative hosts for *Clavibacter michiganensis* subsp. nebraskensis, causal agent of Goss's Wilt of corn. Weed Sci. 63 (4), 901–909.

Kapoor, R.T., 2020. Effect of soil solarization for weed management in *Abelmoschus esculentus* L. Moench. Plant. Arch. 20, 1641–1647.

Karimi, A., Saeidi, H., 2016. Genetic diversity of *Sorghum halepense* (L.) Pers. in Iran as revealed by IRAP markers. Plant. Genet. Resources: Character. Utilis. 14 (2), 132–141.

Karkanis, A., et al., 2020. Johnsongrass (*Sorghum halepense* (L.) pers.) interference, control and recovery under different management practices and its effects on the grain yield and quality of maize crop. Agronomy 10 (2).

Kaur, R., Soodan, A.S., 2017. Reproductive biology of *Sorghum halepense* (L.) Pers. (Poaceae; Panicoideae; Andropogoneae) in relation to invasibility. Flora: Morphology, Distribution, Funct. Ecol. Plants 229, 32–49. Available from: http://doi.org/10.1016/j.flora.2017.02.009.

Kebrom, T.H., Brutnell, T.P., 2015. Tillering in the sugary sweet corn is maintained by overriding the teosinte branched1 repressive signal. Plant. Signal. Behav. 10 (12).

Keeley, P.E., Thullen, R.J., 1989. Growth and Interaction of Johnsongrass (*Sorghum halepense*) with cotton (*Gossypium hirsutum*). Weed Sci. 37 (3), 339–344.

Kelly, S., Fletcher, R.A., Barney, J.N., 2020. Intraspecific, ecotypic and home climate variation in photosynthetic traits of the widespread invasive grass Johnsongrass. AoB Plants 12 (3), 1–9.

Khan, M.A., et al., 2012. Soil solarization: an organic weed-management approach in cauliflower. Commun. Soil. Sci. Plant. Anal. 43 (13), 1847–1860.

Law, D.M., et al., 2008. Evaluating solarization and cultivated fallow for johnsongrass (*Sorghum halepense*) control and nitrogen cycling on an organic farm. Biol. Agric. Horticult. 26 (2), 175–191.

Leguizamon, E.S., 1986. Seed survival and patterns of seedling emergence in *Sorghum halepense* (L.) Pers. Weed Res. 26 (6), 397–404.

Mahrer, Y., 1979. Prediction of soil temperatures of a soil mulched with transparent polyethylene. J. Appl. Meteorol. 18 (10), 1263–1267.

Martín-Closas, L., Costa, J., Pelacho, A.M., 2017. Agronomic effects of biodegradable films on crop and field environment. In: Malinconico, M. (Ed.), Soil Degradable Bioplastics for a Sustainable Modern Agriculture. pp. 169–185. Available from: http://link.springer.com/10.1007/978-3-662-54130-2.

McDonald, A., et al., 2009. Climate change and the geography of weed damage: analysis of U.S. maize systems suggests the potential for significant range transformations. Agricult., Ecosyst. Environ. 130 (3–4), 131–140.

McWhorter, C.G., 1961. Morphology and development of Johnsongrass plants from seeds and rhizomes. Weeds 9 (4), 558.

Mcwhorter, C.G., 1971. Introduction and spread of Johnsongrass in the United States. Weed Sci. 19 (5), 496–500.

McWhorter, C., 1972. Factors affecting Johnsongrass rhizome production and germination. Weed Sci. 20 (1), 41–45.

Monaghan, N., 1979. The biology of Johnson grass (*Sorghum halepense*). Weed Res. 19 (4), 261–267.

Montes-belmont, R., et al., 2003. Alternate hosts of claviceps africana frederickson, mantle and de millano, causal agent of sorghum ergot in the State of Morelos, Mexico. Rev. Mexicana de. Fitopatología 21 (1), 63–66.

Moraes, A., et al., 2020. Survival and development of fall Armyworm (Lepidoptera: Noctuidae) in weeds during the off-season. Fla. Entomologist 103 (2), 288–292.

Munsif, F., et al., 2015. Influence of row spacing on weed density, biomass and yield of chip bud settling of sugarcane. Pak. J. Weed Sci. Res. 53 (9), 1689–1699.

Norsworthy, J.K., et al., 2020. Seed destruction of weeds in southern US crops using heat and narrow-windrow burning. Weed Technol. 1–8.

Nouri, H., Talab, Z.A., Tavassoli, A., 2012. Effect of weed allelopathic of sorghum (*Sorghum halepense*) on germination and seedling growth of wheat, Alvand cultivar. Ann. Biol. Res. 3 (3), 1283–1293.

Ohadi, S., et al., 2017. Gene flow and its consequences in *Sorghum* spp. Crit. Rev. Plant. Sci. 36 (5–6), 367–385.

Ohadi, S., et al., 2018. Surveying the spatial distribution of feral sorghum (*Sorghum bicolor* L.) and its sympatry with johnsongrass (*S. halepense*) in South Texas. PLoS One 13 (7), 1–14.

Paterson, A.H., et al., 1995. The weediness of wild plants: molecular analysis of genes influencing dispersal and persistence of johnsongrass, *Sorghum halepense* (L.) Pers. Proc. Natl Acad. Sci. U. S. A. 92 (13), 6127–6131.

Paterson, A.H., et al., 2009. The *Sorghum bicolor* genome and the diversification of grasses. Nature 457 (7229), 551–556.

Paterson, A.H., et al., 2020. The evolution of an invasive plant, *Sorghum halepense* L. ("Johnsongrass"). Front. Genet. 11 (May), 1–10.

Peerzada, A.M., et al., 2017. Eco-biology, impact, and management of *Sorghum halepense* (L.) Pers. Biol. Invasions 1–19.

QingHui, M., et al., 2009. Chemical compositions and allelopathic potential of volatile oil from *Sorghum halepense* (L.) Pers. Acta Phytophylacica Sin. 36 (3), 277–282.

Rudi, G., et al., 2018. Characterization of the long-distance dispersal of Johnsongrass (*Sorghum halepense*) in a vegetated irrigation channel. River Res. Appl. 34 (9), 1219–1228.

Sartore, L., Vox, G., Schettini, E., 2013. Preparation and performance of novel biodegradable polymeric materials based on hydrolyzed proteins for agricultural application. J. Polym. Environ. 21 (3), 718–725.

Schwartz-lazaro, L.M., Norsworthy, J.K., 2018. Utilization of the integrated harrington seed destructor on weeds commonly. Ark. Soybean Res. Stud. 6, 80–81. Available from: https://scholarworks.uark.edu/aaesser/17.

Schwinning, S., et al., 2017. Accelerated development in Johnsongrass seedlings (*Sorghum halepense*) suppresses the growth of native grasses through sizeasymmetric competition. PLoS One 12 (5), 1–18.

Smith, K., Otto, W., 1973. A review of the spider-mite problem on grain sorghum and corn in West Texas. J. Educ. Res. 1149.

Stoller, E.W., 1977. Differential cold tolerance of quackgrass and Johnsongrass rhizomes. Weed Sci. 25 (4), 348–351.

Taylorson, R.B., McWhorter, C.G., 1969. Seed dormancy and germination in ecotypes of Johnsongrass. Weed Sci. 17 (3), 359–361.

Travlos, I.S., et al., 2019. Key aspects on the biology, ecology and impacts of Johnsongrass [*Sorghum halepense* (L.) Pers] and the role of glyphosate and nonchemical alternative practices for the management of this weed in Europe. Agronomy 9 (11), 1–12.

Uremis, I., et al., 2009. Allelopathic potentials of residues of 6 brassica species on johnsongrass. Afr. J. Biotechnol. 8 (15), 3497–3501. Available from: http://www.ajol.info/index.php/ajb/article/view/61834.

Walsh, M., Newman, P., Powles, S., 2013. Targeting weed seeds in-crop: a new weed control paradigm for global agriculture. Weed Technol. 27 (3), 431–436.

Walsh, M.J., Aves, C., Powles, S.B., 2017. Harvest weed seed control systems are similarly effective on rigid ryegrass. Weed Technol. 31 (2), 178–183.

Walsh, M.J., et al., 2018. Opportunities and challenges for harvest weed seed control in global cropping systems. Pest. Manag. Sci. 74, 2235–2245.

Warwick, S.I., Black, L.D., 1983. The biology of Canadian weeds: 61. *Sorghum halepense* (L.) Pers. Can. J. Plant. Sci. 63 (4), 997–1014.

Warwick, S.I., Phillips, D., Andrews, C., 1986. Rhizome depth: the critical factor in winter survival of *Sorghum halepense* (L.) Pers. (Johnson grass). Weed Res. 26 (6), 381–388.

Williams, C.S., Hayes, R.M., 1984. Johnsongrass (*Sorghum halepense*) competition in soybeans (*Glycine max*). Weed Sci. 32 (4), 498–501.

Winder, R.S., Van Dyke, C.G., 1990. The pathogenicity, virulence, and biocontrol potential of two bipolaris species on Johnsongrass (*Sorghum halepense*). Weed Sci. 38 (1), 89–94.

Zhang, X., et al., 2019. The tin1 gene retains the function of promoting tillering in maize. Nat. Commun. 10 (1), 1–13. Available from: http://doi.org/10.1038/s41467-019-13425-6.

Zoltani, C.K., 2018. Cardiovascular toxicity. Veterinary Toxicology: Basic and Clinical Principles: Third Edition. Elsevier Inc., pp. 227–238. Available from: http://doi.org/10.1016/B978-0-12-811410-0.00014-3.

Index

Note: Page numbers followed by "*f*" and "*t*" refer to figures and tables, respectively.

B

C

D

E

F

Q

R

S

Kolibri 360 GmbH
Ferdinand-Jühlke-Straße 7
99095 Erfurt, Deutschland
produktsicherheit@kolibri360.de